Serotonin: Molecular Biology, Receptors and Functional Effects

Edited by
John R. Fozard
Pramod R. Saxena

Birkhäuser Verlag
Basel · Boston · Berlin

Volume Editors' Addresses:

John R. Fozard
Sandoz Pharma Ltd.
Preclinical Research
P.O. Box 6
CH-4002 Basel

Pramod R. Saxena
Erasmus University Rotterdam
Institute of Pharmacology
P.O. Box 1738
NL-3000 DR Rotterdam

Library of Congress Cataloging-in-Publication Data

Serotonin—molecular biology, receptors, and functional effects/
edited by John R. Fozard, Pramod R. Saxena.
  p. cm.
  Includes bibliographical references and index.
  ISBN 3-7643-2607-7—ISBN 0-8176-2607-7
  1. Serotonin—Mechanism of action.   2. Serotonin receptors. 3. Serotoninergic mechanisms.   I. Fozard, John R.   II. Sazema, Pramod R.
QP364.7.S47   1991
615'.71—dc20

Deutsche Bibliothek Cataloging-in-Publication Data

Serotonin: molecular biology, receptors and functional effects
/ed. by John R. Fozard; Pramod R. Saxena. – Basel; Boston;
Berlin: Birkhäuser, 1991
  ISBN 3-7643-2607-7 (Basel. . .)
  ISBN 0-8176-2607-7 (Boston)
NE: Fozard, John R. [Hrsg.]

Product Liability: The publisher can give no guarantee for information about drug dosage and
application thereof contained in this book. In every individual case the respective user must
check its accuracy by consulting other pharmaceutical literature.

The use of registered names, trademarks, etc. in this publication does not imply, even in the
absence of a specific statement, that such names are exempt from the relevant protective laws
and regulations and therefore free for general use.

© 1991 Birkhäuser Verlag
P.O. Box 133
CH-4010 Basel/Switzerland

Printed in Germany on acid-free paper

ISBN 3-7643-2607-7
ISBN 0-8176-2607-7

Preface

The Second IUPHAR Satellite Meeting on Serotonin was held under the auspices of the Serotonin Club in Basel, Switzerland in July 1990. The scope was wide, ranging from molecular biology through *in vitro* and *in vivo* pharmacology to new drug tools and their clinical significance. There were three invited review lectures, by J. M. Palacios, D. I. Wallis and A. Kaumann, and S. Peroutka gave the first Serotonin Club Irvine H. Page Lecture. The rest of the oral programme was put together by the Scientific Organizing Committee based on volunteered research contributions. The invited review lecturers, the platform speakers and selected poster contributors were invited to write up their contributions for inclusion in this volume. Most complied and this book is the result of their efforts.

When instructing the authors prior to the meeting, we emphasized that selected new data should be put in the context of the literature findings. In this way we hoped to achieve topicality yet preserve the review perspective which facilitates its appreciation by the non-specialist. It was truly a pleasure to read the interesting papers which resulted and to prepare them for publication. We believe they convey to a remarkable degree the spirit of what was generally felt to be a highly stimulating exchange of information on matters serotonergic which took place in Basel last July.

April 1991

J. R. Fozard

P. R. Saxena

Contents

Part I: Molecular Pharmacology, Cell Biology, Binding Sites and Receptors

S. J. Peroutka
A tribute to Dr. Irvine H. Page............................. 3

J. M. Palacios, C. Waeber, G. Mengod and M. Pompeiano
Molecular neuroanatomy of 5-HT receptors 5

T. Branchek, T. Zgombick, M. Macchi, P. Hartig and
  R. Weinshank
Cloning and expression of a human $5\text{-}HT_{1D}$ receptor 21

B. L. Roth, M. W. Hamblin, R. Desai and R. D. Ciaranello
Developmental and synaptic regulation of $5\text{-}HT_2$ and $5\text{-}HT_{1C}$
  serotonin receptors...................................... 33

P. M. Whitaker-Azmitia and E. C. Azmitia
Serotonin trophic factors in development, plasticity and aging ... 43

F. C. Zhou and E. C. Azmitia
A neurotrophic factor – SNTF – for serotonergic neurons....... 50

G. Reiser
Molecular mechanisms of action induced by $5\text{-}HT_3$ receptors in a
  neuronal cell line and by $5\text{-}HT_2$ receptors in a glial cell line ... 69

J. A. Peters, H. M. Malone and J. J. Lambert
Characterization of $5\text{-}HT_3$ receptor mediated electrical responses
  in nodose ganglion neurones and clonal neuroblastoma cells
  maintained in culture.................................... 84

M. Hamon, S. El Mestikawy, M. Riad, H. Gozlan, G. Daval and
  D. Vergé
Specific antibodies as new tools for studies of central $5\text{-}HT_{1A}$
  receptors... 95

C. Waeber and J. M. Palacios
$5\text{-}HT_{1C}$, $5\text{-}HT_{1D}$ and $5\text{-}HT_2$ receptors in mammalian brain:
  Multiple affinity states with a different regional distribution ... 107

D. Hoyer, H. Boddeke and P. Schoeffter
Second messengers in the definition of 5-HT receptors 117

viii

M. D. Gershon, P. R. Wade, E. Fiorica-Howells, A. L. Kirchgessner and H. Tamir
5-HT$_{1P}$ receptors in the bowel: G protein coupling, localization and function 133

M. Baez, L. Yu and M. L. Cohen
Is contraction to serotonin mediated via 5-HT$_{1C}$ receptor activation in rat stomach fundus? 144

H. O. Kalkman and J. R. Fozard
Further definition of the 5-HT receptor mediating contraction of rat stomach fundus: relation to 5-HT$_{1D}$ recognition sites 153

D. J. Prentice, V. J. Barrett, S. J. MacLennan and G. R. Martin
Temperature dependence of agonist and antagonist affinity constants at 5-HT$_1$-like and 5-HT$_2$ receptors 161

R. Y. Wang, C. R. Ashby, Jr. and E. Edwards
Characterization of 5-HT$_3$-like receptors in the rat cortex: Electrophysiological and biochemical studies 174

R. A. Glennon, S. J. Peroutka and M. Dukat
Binding characteristics of a quarternary amine analog of serotonin: 5-HTQ 186

D. van Heuven-Nolsen, C. M. Villalón, M. O. den Boer and P. R. Saxena
5-HT$_1$-like receptors unrelated to the known binding sites? 192

Part II: Peripheral and Central Pharmacology

D. I. Wallis and P. Elliott
The electrophysiology of 5-HT 203

J. Bockaert, L. Fagni, M. Sebben and A. Dumuis
Pharmacological characterization of brain 5-HT$_4$ receptors: Relationship between the effects of indole, benzamide and azabicycloalkylbenzimidazolone derivatives 220

D. E. Clarke, G. S. Baxter, H. Young and D. A. Craig
Pharmacological properties of the putative 5-HT$_4$ receptor in guinea-pig ileum and rat oesophagus: Role in peristalsis 232

K. H. Buchheit and T. Buhl
5-HT$_{1D}$ and 5-HT$_4$ receptor agonists stimulate the peristaltic reflex in the isolated guinea pig ileum 243

C. M. Villalón, M. O. den Boer, J. P. C. Heiligers and P. R. Saxena
The 5-HT$_4$ receptor mediating tachycardia in the pig 253

E. Hamel and D. Bouchard
Contractile 5-HT$_{1D}$ receptors in human brain vessels............ 264

N. Toda
Human, monkey and dog coronary artery responses to serotonin
and 5-carboxamidotryptamine 275

D. Cambridge, M. V. Whiting and L. J. Butterfield
5-Carboxamidotryptamine induced renal vasoconstriction
in the dog... 282

A. K. Mandal, K. T. Kellar and R. A. Gillis
The subretrofacial nucleus: A major site of action for the
cardiovascular effects of 5-HT$_{1A}$ and 5-HT$_2$ agonist drugs 289

G. H. Dreteler, W. Wouters and P. R. Saxena
Cardiovascular effects of injection of 5-HT, 8-OH-DPAT and
flesinoxan into the hypothalamus of the rat 300

P. M. Larkman and J. S. Kelly
Pharmacological characterization of the receptor mediating 5-HT
evoked motoneuronal depolarization *in vitro*................ 310

P. Blandina, J. Goldfarb and J. P. Green
Stimulation of 5-HT$_3$ receptors inhibits release of endogenous
noradrenaline from hypothalamus........................ 322

R. W. Fuller
Antagonism of serotonin agonist-elicited increases in serum
corticosterone concentration in rats...................... 330

P. Chaouloff, V. Baudrie and D. Laude
Influence of 5-HT$_{1A}$ and 5-HT$_2$ receptor agonists on blood glucose
and insulin levels.................................... 339

*M. J. Millan, K. Bervoets, S. Le Maroville-Girardon, C. Grevoz
and F. C. Colpaert*
Novel *in vivo* models of 5-HT$_{1A}$ receptor-mediated activity:
8-OH-DPAT-induced spontaneous tail-flicks and inhibition of
morphine-evoked antinociception........................ 346

G. A. Higgins, C. C. Jordan and M. Skingle
Evidence that the unilateral activation of 5-HT$_{1D}$ receptors in the
substantia nigra of the guinea-pig elicits contralateral rotation . 355

Part III: Pathophysiological Roles and Prospects for New Therapies

A. J. Kaumann
5-Hydroxytryptamine and the human heart.................. 365

B. Johansen, J. M. Lyngsø and K. Bech
The effects of BRL 24924 (renzapride) on secretion of gastric acid
   and pepsin in dogs 374

G. J. Sanger, K. A. Wardle, S. Shapcott and K. F. Yee
Constipation evoked by 5-HT$_3$ receptor antagonists 381

L. M. Petrovic, S. A. Lorens, M. G. T. Cabrera, B. H. Gordon,
   R. J. Handa, D. B. Campbell and J. Clancy, Jr.
Subchronic D-fenfluramine treatment enhances the immunological
   competence of old female Fischer 344 rats................... 389

T. A. Kent, A. Jazayeri and J. M. Simard
Serotonin as a vascular smooth muscle cell mitogen 398

F. C. Tanner, V. Richard, M. Tschudi, Z. Yang and T. F. Lüscher
Serotonin, the endothelium and the coronary circulation 406

J. L. Amezcua, E. Hong and R. A. Bobadilla
Effects of selective 5-HT$_2$ receptor antagonists on some
   haemodynamic changes produced by experimental pulmonary
   embolism in rabbits 414

P. P. A. Humphrey, W. Feniuk, M. Motevalian, A. A. Parsons and
   E. T. Whalley
The vasoconstrictor action of sumatriptan on human isolated
   dura mater 421

B. Costall and R. J. Naylor
Influence of 5-HT$_3$ receptor antagonists on limbic-cortical
   circuitry 430

L. M. Pinkus and J. C. Gordon
Utilization of zacopride and its R- and S-enantiomers in studies
   of 5-HT$_3$ receptor "subtypes"............................ 439

J. E. Macor, C. A. Burkhart, J. H. Heym, J. L. Ives, L. A. Lebel,
   M. E. Newman, J. A. Nielsen, K. Ryan, D. W. Schulz,
   L. K. Torgersen and B. K. Koe
CP-93,129: A potent and selective agonist for the serotonin
   (5-HT$_{1B}$) receptor and rotationally restricted analog of
   RU-24,969 449

T. Glaser, J. M. Greuel, E. Horvàth, R. Schreiber and J. De Vry
Differentiation of 8-OH-DPAT and ipsapirone in rat models of
   5-HT$_{1A}$ receptor function.............................. 460

M. J. Boyce, C. Hinze, K. D. Haegele, D. Green and P. J. Cowen
Initial studies in man to characterise MDL 73,000EF, a novel
   5-HT$_{1A}$ receptor ligand and putative anxiolytic......... 471

K. F. Martin and D. J. Heal
8-OH-DPAT-induced hypothermia in rodents. A specific model of
   5-HT$_{1A}$ autoreceptor function? 483

C. W. Callaway, D. E. Nichols, M. P. Paulus and M. A. Geyer
Serotonin release is responsible for the locomotor hyperactivity in
   rats induced by derivatives of amphetamine related to MDMA . . 491

Index   507

Part I

Molecular Pharmacology, Cell Biology, Binding Sites and Receptors

Serotonin: Molecular Biology, Receptors and Functional Effects
ed. by J. R. Fozard/P. R. Saxena
© 1991 Birkhäuser Verlag Basel/Switzerland

A Tribute to Dr. Irvine H. Page

by

S. J. Peroutka

*Department of Neurology, Stanford University, Stanford, CA 94305, USA, and
Neurology Service (127), Palo Alto Veteran's Administration Hospital, 3801 Miranda Avenue,
Palo Alto, CA 94304, USA*

SCIENCE IS A SUPERB EDIFICE THAT RISES,
LEVEL UPON LEVEL, BUILT BY THE UNTIRING,
CREATIVE BUT CRITICAL INVESTIGATORS OF
EACH GENERATION.

Irvine H. Page, 1978

On the occasion of the First Irvine H. Page Lectureship at The Second
IUPHAR Satellite Meeting on Serotonin, it is fitting that the contributions of Dr. Page be reviewed by current investigators. As Dr. Page
points out in the above quotation, the significant and unparalleled
advances made in the field of 5-hydroxytryptamine (5-HT) research
during the past decade have been a direct result of more than 40 years
of 5-HT research. Although 5-HT has intrigued scientists for more than
a century (see Table 1), the modern scientific "birth" of 5-HT can be
dated to Dr. Page's *Science* paper of 1948. He and his co-authors stated:
"We would like provisionally to name it *serotonin*, which indicates that
its source is serum and its activity is one of causing constriction"
(Rapport et al. 1948).

From this initial paper, 5-HT research reports have grown to the extent
that more than 1500 appeared in 1989. The body of work extends from
molecular biology to clinical pharmacology and has outpaced similar
studies in the dopaminergic, adrenergic histammergic and cholinergic
fields. Unlike in the era of Dr. Page, it is impossible to attribute this
progress to any single individual or to any group of investigators.

Table 1. Early history of serotonin

1868	Difibrinated blood increased vascular resistance
1932	"Vasotonin" (Bayliss and Ogden)
1933	"Enteramine" (Vialle and Erspamer)
1948	"Serotonin" (Rapport, Green and Page)

4

Most importantly, the future looks extremely bright for continued advances in the field of 5-HT research due to the constant introduction of both new techniques and investigators. As summarized by Dr. Page himself:

SEROTONIN, IN SHORT, HAS TAKEN ON A LIFE OF ITS OWN AND WE ARE NO LONGER LIVING TOGETHER. I HAVE NO REGRETS, BECAUSE A LONG LIFE HAS TAUGHT ME THAT THE NATURAL HISTORY OF ONE'S ACTIVE PARTICIPATION IN A DISCOVERY IS ABOUT 5 TO 10 YEARS. THEN THE SUBJECT GROWS COMPLICATED, NEW VERY BRIGHT YOUNG FACES APPEAR WITH THEIR BETTER METHODS AND THEY TAKE OVER. IF THEY ARE AWARE THAT ANYTHING PRECEDED THEIR OWN WORK, THEY GIVE NO INDICATION OF IT − WHICH IS NATURE'S WAY OF PREVENTING CONSTIPATION OF THE MIND.

Irvine H. Page, 1985

References

Rapport, M. M., Green, A. A., and Page, I. H. (1948). Crystalline serotonin. Science 108: 329.
Page, I. H. (1978). Foreward to: Serotonin in Health and Disease Volume III: The Central Nervous System. Essman, W. B. (ed.) New York: Spectrum Publications.
Page, I. H. (1985). The neonatology of serotonin. In: Vanhoutte, P. M. (ed.), Serotonin and the Cardiovascular System. New York: Raven Press, pp. xiii−xv.

Serotonin: Molecular Biology, Receptors and Functional Effects
ed. by J. R. Fozard/P. R. Saxena

Molecular Neuroanatomy of 5-HT Receptors

J. M. Palacios, C. Waeber, G. Mengod, and M. Pompeiano

Preclinical Research, SANDOZ Pharma Ltd., CH-4002, Basel, Switzerland

Summary. Serotoninergic terminals are found throughout the brain, suggesting an involvement of 5-HT in many brain functions. The effects of 5-HT on a given neuron are dependent on the different types of receptors expressed by this neuron, their location on the neuron and the signal transduction mechanism used. In order to provide information on the localization of 5-HT receptors both at the regional and cellular levels, we have combined ligand binding and *in situ* hybridization autoradiography to identify neurons expressing the different 5-HT receptor mRNAs and to localize 5-HT binding sites. The information gathered is discussed with reference to functional data in some well characterized anatomical systems: the striato-nigral pathway, the hippocampal formation and the brainstem, with special emphasis on the cranial nerve nuclei.

Introduction

Most of the neurons synthesizing 5-HT in the CNS are located in the raphé nuclei of the brainstem, but serotoninergic nerve terminals can be found in virtually every brain region (Steinbush 1981). This wide serotoninergic innervation contrasts with the small number of 5-HT producing cells (11,500 cells in the rat dorsal raphé; Descarries et al. 1982), although it can account for the multifarious brain functions where this amine has been reported to be involved (Osborne and Hamon 1988). The apparent discrepancy between the scarcity of 5-HT neurons and the extended distribution of their terminals raises questions about the mechanisms of discrimination between several potential actions triggered by the release of 5-HT. One first and important device is the very connectivity with neural elements located upwards and downwards from the raphé nuclei. The spatial distribution of receptors on a single cell is another factor that can affect the response to 5-HT. Mechanisms based on topological factors certainly achieve a great deal of selectivity, but the fact that evolution has retained and conserved several subtypes of receptors for 5-HT indicates that this multiplicity is beneficial, probably because it provides the organisms with enhanced fine-tuning capabilities. Distinct receptor subtypes can exhibit different affinities and thus respond to lower or higher concentrations of the transmitter. Additionally, a receptor which is part of an ion channel (i.e. 5-HT_3 receptor; Derkach et al. 1989) obviously acts faster than one which is linked to a channel via an intermediate molecule (i.e. 5-HT_{1A}

receptor; Andrade et al. 1986), while receptors coupled to the synthesis of second messenger molecules (i.e. 5-HT_{1B}, 5-HT_{1C}, 5-HT_2, 5-HT_4; Peroutka 1988, Dumuis et al. 1989) influence ionic conductances only indirectly (Sternweis and Pang 1990) and can eventually produce long-term changes in the target cell.

Although receptor subtypes can display the same affinity for their endogenous ligand, differences in their primary structures affect the properties of their binding to other compounds. This allows the pharmacologist to study independently (ideally) distinct receptors and helps the anatomist to investigate their precise regional and cellular distributions using autoradiographic or other techniques.

The understanding of the actions of 5-HT in different brain regions and neuronal populations requires the definition of the receptor types expressed by these cells and their localization both in the neuronal circuits and on the different parts of the neurons. This paper describes the distribution of the different subtypes of 5-HT receptors in choosen regions of the rat brain, when possible at the cellular level (using selective lesions or *in situ* hybridization histochemistry with adequate probes, see Material and Methods) and attempts to relate these findings to known functional data.

Materials and Methods

The technique of *in vitro* autoradiography is basically the same as that of ligand binding on homogenates, but it can provide valuable anatomical data, due to the relative preservation of the tissue samples (Palacios et al. 1988). Brains from laboratory animals such as rats are placed on dry ice immediately after sacrifice. Frozen tissues are then cut using a cryostat-microtome (at 10 or 20 μm thickness) and mounted on gelatin-coated glass slides. Sections are thawed out immediately prior to incubation in the presence of radiolabelled ligand. This incubation step is normally preceded by a preincubation in buffer, to eliminate endogenous ligand. Eventually, loosely or non specifically bound radioligand is washed away by successive immersions in fresh buffer. Buffer salts, which can generate autoradiographic artifacts, are eliminated by a final dipping in distilled water. After drying, slides bearing labelled sections along with radioactive standards are placed in an X-ray cassette under a radiation-sensitive film. After development, the optical densities of the autoradiograms can be determined using computer-aided microdensitometry, converted to the concentration of the radioligand and hence to that of the binding sites in particular brain areas.

In situ hybridization histochemistry is a related procedure which uses [32]P- or ^{35}S-labelled oligonucleotide probes complimentary to mRNA coding for a receptor protein (Vilaró et al. 1991). The use of this method is obviously conditioned by the knowledge of the sequence of the gene coding for the receptor. The oligonucleotides we use are normally 48 nucleotides long and are choosen for their lack of homology with published sequences of other G-protein-coupled receptor genes; they are frequently derived from the region that codes for the amino- or carboxy-terminal portion or the third intracytoplasmic loop of the receptor. Prior to the hybridization step, tissue sections are fixed in paraformaldehyde and the mRNA molecules they contain are made more accessible to the probe by proteolytic treatment. The specificity of the hybridization signal is related to the stringency conditions during the hybridization and the washing procedures. Autoradiograms can be generated as with radioligand-labelled sections. Alternatively, one can take advantage of the stability of the hybrids at 42°C to dip the sections in melted photographic emulsion. After drying, exposure and development, the autoradiographic grains are located directly on the tissue, which can be stained and observed under the microscope, providing a valuable level of anatomical resolution.

When information about the sequence of a receptor gene is not available, the cellular localization of receptor can be inferred by performing selective lesions (Palacios and Dietl 1988). For instance, the injection of 6-hydroxydopamine in the substantia nigra causes the selective death of dopaminergic neurons: decreased densities of receptors in the substantia nigra or its projection areas one month after surgery can indicate that these receptors are located on dopaminergic neurons. Conversely, an increased density of 5-HT receptors after lesion of serotoninergic raphé neurons with 5,7-dihydroxytryptamine can be taken as a sign of postsynaptic receptor up-regulation.

Results and Discussion

Analysis of autoradiograms generated with [^{3}H]5-HT reveals concentration of binding sites in two forebrain regions: the basal ganglia and the hippocampus. Conversely, the first available 5-HT$_3$ radioligands labelled significant densities of receptors only in some nuclei of the brainstem. Given the enrichment of 5-HT receptors in these regions, their distribution in these selected areas will be discussed in detail.

Striato-Nigral System

The basal ganglia are a component of what was formerly called extrapyramidal pathway, which, together with the pyramidal pathway, controls motor activity. The connectivity of the basal ganglia can be described as several interconnected loops comprising, among other regions, the caudate-putamen, the globus pallidus and the substantia nigra. Most of the neurotransmitters used by the neurons involved in this circuitry are now well characterized (Graybiel 1990).

Caudate-Putamen and Globus Pallidus

The majority of $5\text{-}HT_1$ receptors in the basal ganglia (striatum and globus pallidus) is accounted for by $5\text{-}HT_{1B}$ receptors (Figure 1G). It should be noticed that these receptors have only been detected in the rat, mouse, hamster and opossum brain (unpublished data for the latter 2 species), while $5\text{-}HT_{1D}$ are present in the other vertebrates investigated until now (Waeber et al. 1989a). The highest density of $5\text{-}HT_{1B}$ sites is observed in the globus pallidus (Figure 1G). This abundance of 5-HT receptors matches that of serotoninergic terminals (Lavoie and Parent 1990). The density of [^{3}H]citalopram binding sites is lower in the globus pallidus than in the caudate-putamen (Figure 1I). In the latter region however the decreasing medio-lateral gradient of labelling intensity seems to correlate with the pattern of 5-HT innervation described by Lavoie and Parent (1990).

There is no direct evidence to establish unequivocally the nature of the cells expressing $5\text{-}HT_{1B}/5\text{-}HT_{1D}$ receptors in the basal ganglia. Neurotoxin injections in the caudate-putamen have been reported to reduce the number of $5\text{-}HT_{1B}/5\text{-}HT_{1D}$ receptors in this region, indicating their localization on the cell body or dendrites of striatal neurons (Quirion and Richard 1987; Waeber et al. 1989a; b).

In contrast with the enrichment in $5\text{-}HT_{1B}$ sites, $5\text{-}HT_{1A}$ sites in the striatum are present only at low densities (Figure 1D), in agreement with the virtual absence of mRNA coding for this receptor (Figure 1A). $5\text{-}HT_2$ binding sites present an homogeneous distribution in the caudate-putamen, but are not detected in the globus pallidus (Figure 1F). The presence of mRNA coding for this receptor in the caudate-putamen (Figure 1C) would point to a localization of these receptors on the cell body or dendrites of striatal neurons. While the labelling pattern of $5\text{-}HT_{1B}$ and $5\text{-}HT_2$ binding sites in the striatum is homogeneous, patches of $5\text{-}HT_{1C}$ binding sites are found in this region, particularly concentrated in its medial aspects (Figure 1E); the neurochemical identity of these cells still has to be established. A thin layer of dense labelling is present on the most lateral caudate-putamen (subcallosal streak), while an intermediate level of $5\text{-}HT_{1C}$ sites is observed in the surrounding striatal tissue, as well as in the globus pallidus. *In situ* hybridization reveals clusters of autoradiographic grains (Figure 1B) which seem to correspond to $5\text{-}HT_{1C}$ binding sites. However, this technique demonstrates a decreasing medio-lateral gradient of $5\text{-}HT_{1C}$ mRNA concentration in the striatum, at variance with the apparently homogeneous distribution of [³H]mesulergine binding sites. A thin layer of hybridization signal is also observed in the lateral caudate-putamen. In the rat brain, the 5-HT₃ receptor selective ligand [³H](S)zacopride does not reveal a significant density of sites (Figure 1H). However, species differences probably exist, as this ligand reveals the presence of intermediate densities of these sites in the human and rabbit striatum (not shown).

Substantia Nigra

The substantia nigra presents one of the densest 5-HT innervation of the rat brain (Steinbush 1981). This observation is in agreement with the high density of [³H]citalopram binding sites found in this region (Figure 2E, 3I). Immunohistochemical studies performed using antibodies to 5-HT conjugates in the rat and the primate brain demonstrate that 5-HT terminals are denser in the pars reticulata than in the pars compacta (Steinbush 1981; Lavoie and Parent 1990), a pattern which is similar to that shown in Figures 2E and 3I. The high density of 5-HT$_{1B}$ receptors found in the substantia nigra pars reticulata (Figure 2F) coincides with the concentration of [³H]citalopram binding sites in this region (Figure 2E). This is reminiscent of the fact that 5-HT$_{1B}$ receptors have been proposed to correspond to presynaptic autoreceptors in the rat cortex (Engel et al. 1986). Destruction of 5-HT innervation using 5,7-dihydroxytryptamine has led to contradictory results. Some groups (Blackburn et al. 1984 as well as Weissman et al. 1986) have reported an up-regulation of nigral 5-HT$_{1B}$ sites, whereas a decrease of [³H]5-HT binding sites after such lesions has also been described (Vergé et al. 1986). It should be noted that these groups did not find any alteration of 5-HT$_1$ receptor concentration in the cortex and in other target areas of the raphé. On the other hand, a selective lesion of nigral dopaminergic neurons with 6-hydroxydopamine did not alter the density of 5-HT$_{1B}$/5-HT$_{1D}$ sites either in the substantia nigra or in the caudate-putamen (Palacios and Dietl 1988; Waeber et al. 1989a; b), indicating that these receptors are probably not located on dopaminergic neurons. On the contrary, excitotoxic lesions of caudate-putamen intrinsic neurons resulted in a loss of 5-HT$_{1B}$/5-HT$_{1D}$ sites both close to the injection site and in the ipsilateral substantia nigra, suggesting a presynaptic localization of these receptors on terminals of striatal neurons. The use of *in situ* hybridization using adequate probes can provide information on the cellular localization of 5-HT$_{1C}$ and 5-HT$_2$ receptors. 5-HT$_{1C}$ receptor mRNA (Figure 2A) is abundant in the pars compacta and pars lateralis and is observed in some cells in the pars reticulata, as seen from nuclear emulsion-dipped sections. Moreover there seem to be a rostro-caudal gradient in the hybridization signal, the mRNA being more abundant at more caudal levels. 5-HT$_{1C}$ receptors binding sites (Figure 2B) are distributed throughout the SN, although they are more abundant in its dorsal aspects. 5-HT$_2$ receptor mRNA (Figure 2C) and binding sites (Figure 2D) have an overlapping distribution, as they are present at intermediate densities in the pars compacta. Neither 5-HT$_{1A}$ receptor mRNA nor binding sites have been detected in the substantia nigra (Figure 3A, D). Finally, a low level of [^{3}H](S)zacopride labelling in the substantia nigra pars reticulata (Figure 3H) can provide an anatomical substrate for the effects of 5-HT$_3$ receptor antagonists, when injected in this region (Costall et al. 1988).

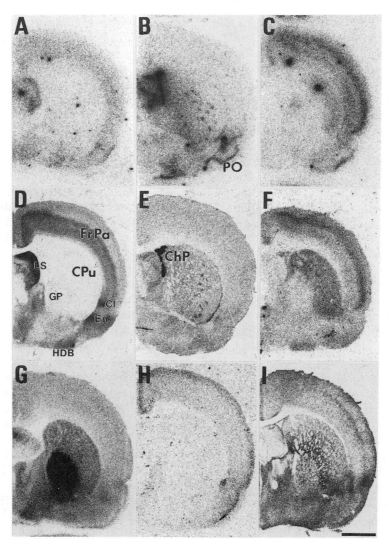

Taking these autoradiographic data in the basal ganglia and substantia nigra together, it is clear that 5-HT can exert its effects at different points of this loop: locally in the caudate via 5-HT$_{1B}$, 5-HT$_{1C}$, 5-HT$_2$ receptors, modulating the output to the globus pallidus and substantia nigra; in the descending pathway at the level of terminals in the substantia nigra and locally in the substantial nigra through the control of dopaminergic cell bodies or of nigral interneurons.

Hippocampal Region

The entire hippocampal formation receives a dense serotoninergic innervation originating mainly in the dorsal raphé and median raphé nuclei (Moore and Halaris 1975). In good agreement with the described distribution of serotoninergic terminals (Moore and Halaris 1975; Swanson et al. 1987; Freund et al. 1990), 5-HT uptake sites (Figure 3I) are very abundant in strata oriens and lacunosum moleculare of CA1 and CA2 and in strata oriens and radiatum of CA3 and CA4, while the other layers of the Ammon's horn and the dentate gyrus contained a much lower density of sites. The ventral hippocampus in general contains a higher density of sites than dorsal regions, while the pyramidal

---

Figures 1 to 4 are pictures from autoradiograms generated as described in Dietl and Palacios (1988) (for most of the subtypes of 5-HT binding sites), Waeber et al. (1989a and 1990) (for 5-HT$_{1D}$ and 5-HT$_3$ binding sites), D'Amato et al (1987) (for 5-HT uptake sites) and Vilaró et al (1991) (for in situ hybridization studies). Abbreviations are: 7, facial nucleus; 10, dorsal motor nucleus of vagus; CA1, field CA1 of Ammon's horn; CA2, field CA2 of Ammon's horn; CA3, field CA3 of Ammon's horn; Cb, cerebellum; ChP, choroid plexus; Cl, claustrum; CPu, caudate-putamen; DČo, dorsal cochlear nucleus; DG, dentate gyrus; En, endopiriform nucleus; Ent, entorhinal cortex; FrPa, frontoparietal cortex; GP, globus pallidus; HDB, nucleus of the horizontal limb of the diagonal band of Broca; Int, interpositus cerebellar nucleus; IO, inferior olive; IP, interpeduncular nucleus; LatC, lateral cerebellar nucleus; LS, lateral septum; LVe, lateral vestibular nucleus; MG, medial geniculate nucleus; MVe, medial vestibular nucleus; PCRt, parvocellular reticular nucleus; PH, posterior hypothalamic nucleus; Pn, pontine nuclei; PO, primary olfactory cortex; PrH, prepositus hypoglossal nucleus; ROb, raphé obscurus nucleus; RPa, raphé pallidus nucleus; S, subiculum; SNc, substantia nigra pars compacta; SNr, substantia nigra pars reticulata; Sol, nucleus of the solitary tract; Sp5, nucleus of the spinal tract of the trigeminus; Str, striate cortex; SuG, superficial grey layer of the superior colliculus; SuVe, superior vestibular nucleus; VCo, ventral cochlear nucleus.

Figure 1. Autoradiograms generated by incubating rat brain sections at the level of the caudate-putamen either with radioligands or with ^{32}P-labelled oligonucleotide probes. Sections in A, B and C have been respectively hybridized with oligonucleotides complementary to 5-HT$_{1A}$, 5-HT$_{1C}$ and 5-HT$_2$ receptor mRNA. D, E and F are sections consecutive to those used for in situ hybridization; they have been incubated with [^{3}H]8-OH-DPAT (to label 5-HT$_{1A}$ sites; D), [^{3}H]mesulergine in the presence of 100 nm spiperone (to label 5-HT$_{1C}$ sites; E) and [^{3}H]LSD in the presence of 1 μM sulpiride and 300 nM 5-HT (to label 5-HT$_2$ sites; F). Section in G has been incubated with [^{125}I]I-CYP in the presence of 3 μM isoproterenol to label 5-HT$_{1B}$ receptors, while 5-HT$_3$ receptors (in H) have been labelled using [^{3}H](S)zacopride. 5-HT uptake sites are selectively visualized using [^{3}H]citalopram (I). Scale bar is 2 mm.

12

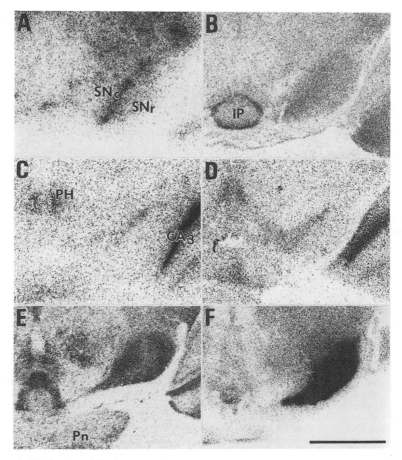

Figure 2. Higher magnification autoradiograms generated by incubating rat brain sections at the level of the substantia nigra either with radioligands or with ^{32}P-labelled oligonucleotide probes. Sections in A and C have been respectively hybridized with oligonucleotides complementary to 5-HT$_{1C}$ and 5-HT$_2$ receptor mRNA. B and D are sections consecutive to those used for *in situ* hybridization; they have been incubated with [^{3}H]mesulergine in the presence of 100 nM spiperone (to label 5-HT$_{1C}$ sites; B) and [^{3}H]ketanserin (to label 5-HT$_2$ sites; D). Section in F has been incubated with [^{125}I]I-CYP in the presence of 3 μM isoproterenol to label 5-HT$_{1B}$ receptors, while 5-HT uptake sites are selectively visualized using [^{3}H]citalopram (E). Scale bar is 2 mm.

cell layer of Ammon's horn and the granule cell layer of the dentate gyrus show the lowest density of [^{3}H]citalopram binding sites.

In situ hybridization studies show that the density of 5-HT$_{1A}$ receptor mRNA (Figure 3A) is very high in the granule cell layer of the dentate gyrus and the pyramidal cell layer of CA1 and CA4 fields and of CA3 (the latter only in its ventral aspects). In contrast, 5-HT$_{1A}$ receptor binding sites (Figure 3D) are very concentrated in the molecular layer

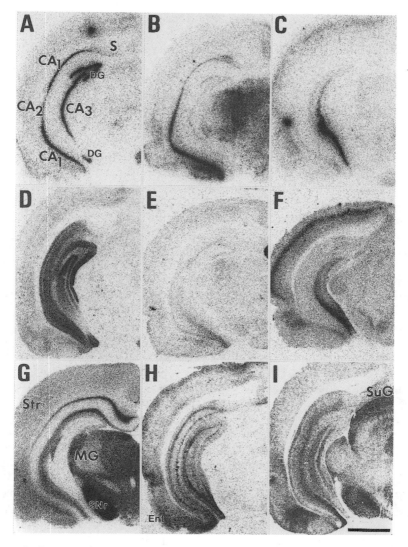

Figure 3. Autoradiograms generated by incubating rat brain sections at the level of the posterior hippocampus either with radioligands or with ³²P-labelled oligonucleotide probes. Conditions are the same as in Figure 1 (section in F has been incubated with [³H]ketanserin). Scale bar is 2 mm.

of the dentate gyrus and in strata oriens and radiatum of Ammon's horn. The regional distribution of [³H]8-OH-DPAT binding sites in the different CA fields parallels that of the mRNA. Much lower densities of receptors are seen in the hilus and the granule cell layer of the dentate gyrus as well as the pyramidal cell layer of the Ammon's horn, indicating that 5-HT$_{1A}$ receptors are rather present in the dendritic fields of pyramidal and granule cells.

5-HT$_{1C}$ receptor mRNA (Figure 3B) is very abundant in the pyramidal cell layer of the ventral and posterior part of CA1 and CA2 and of the anterior part of the CA3, while it is very low in the pyramidal cell layer of the dorsal and anterior region of CA1 and CA2, posterior part of CA3 and in the granule cell layer of the dentate gyrus. 5-HT$_{1C}$ binding sites (Figure 3E) show a regional distribution in agreement with that of the hybridization signal, being concentrated on the pyramidal cell layer of CA1 and CA2 (at ventral levels) and the granule cell layer of the dentate gyrus. However, receptors are also seen in the stratum lacunosum moleculare of the CA1 and CA3 fields at anterior and dorsal level.

5-HT$_2$ receptor mRNA (Figure 3C) is observed in the pyramidal cell layer of the Ammon's horn: it is abundant in the CA3 field, particularly in its ventral tip, and less abundant in CA2 and CA4. Very low levels of mRNA are also detected in the granule cell layer of the dentate gyrus. 5-HT$_2$ binding sites (Figure 3F) are present at very high concentrations in strata radiatum and oriens of the ventral tip of CA3, while the CA4 field exhibits a lower density of receptors in the same layers. Only a very low density of receptors is detected in strata radiatum and oriens of the CA2 field.

5-HT$_{1B}$ receptors are very abundant in the pyramidal cell layer and stratum lacunosum moleculare of the dorsal CA1 and CA2 fields and of the ventral CA3 field (Figure 3G). A lower density of receptors is seen in the molecular layer of the dentate gyrus.

5-HT$_3$ receptors are abundant in the pyramidal cell layer and in the stratum lacunosum moleculare of CA1, CA2 and CA3 as well as in the granule cell layer of the dentate gyrus (Figure 3H). Kainic acid injection in the dorsal hippocampus leads to a decrease of 5-HT$_3$ binding sites (Waeber and Palacios, unpublished), suggesting a location of these receptors on cells intrinsic to the Ammon's horn.

Recently, a novel 5-HT receptor subtype, termed 5-HT$_4$, originally described in cultured neurons of mouse colliculi (Dumuis et al. 1989), has been described in rat hippocampal membranes (De Vivo and Mayaani 1990). This subtype has also been identified electrophysiologically on CA1 pyramidal cells (Chaput et al. 1990).

The anatomy of the serotoninergic innervation of the hippocampus cannot be easily described. Differences have been reported in the distribution pattern of fibers coming from different raphé nuclei (Freund et al. 1990): fibers originating from the dorsal raphé are rather diffuse compared to those arising from the median raphé. Moreover, the dorsal raphé is responsible for the innervation of the principal cells of the Ammon's horn, as shown in electrophysiological experiments (see Hirose et al. 1990 and references therein) while the median raphé innervates a particular subset of GABAergic interneurons of the hippocampus

(Freund et al. 1990). Another type of hippocampal GABAergic interneuron, the basket cell, does not show contacts with serotoninergic fibers (Freund et al. 1990).

Paralleling these intricate anatomical relationships, the electrophysiological effects of 5-HT in the hippocampus are also quite complex. Microiontophoretically applied 5-HT produces three distinct actions on CA1 pyramidal cells (Andrade and Nicoll 1987; Colino and Halliwell 1987): 1) a hyperpolarization; 2) a rebound depolarization with spike generation which follows the above described hyperpolarizing phase in a subset of pyramidal cells normally quiescent (Hirose et al. 1990); 3) a reduction of the long-lasting afterhyperpolarization. While the hyperpolarizing response is mediated by 5-HT_{1A} receptors (Colino and Halliwell 1987) which seem to be linked via G-protein to a K^+-channel (Andrade et al. 1986), 5-HT_{1B} and 5-HT_3 blocking agents prevent the depolarizing action of 5-HT. 5-HT_{1A}, 5-HT_{1B}, 5-HT_2 or 5-HT_3 compounds do not interfere with the afterhyperpolarization (Colino and Halliwell 1987). Recently, a 5-HT_4 receptor selective compound has been shown to antagonize potently the 5-HT-induced depolarization and reduction of the afterhyperpolarization (Chaput et al. 1990).

In agreement with these electrophysiological findings, our autoradiographic studies indicate that: 5-HT_{1A} receptors are present on the dendrites of CA1 pyramidal cells; 5-HT_2 receptors are expressed on dendrites of pyramidal cells mainly of CA3 and only in the caudoventral part but not of CA1; both 5-HT_{1B} and 5-HT_3 receptors are present in the CA1 pyramidal cell layer. However 5-HT_{1C} and 5-HT_2 receptors, which are also expressed by pyramidal cells and are present mainly on their cell bodies, have not yet been associated with any of the functional actions of 5-HT described.

In conclusion, the action of 5-HT in the hippocampus is far from simple, for heterogeneities are found both on the presynaptic (5-HT innervation) and the postsynaptic (target cells, 5-HT receptor subtypes) components. 5-HT fibers do not only innervate pyramidal cells but also certain interneurons. Pyramidal cells themselves are not physiologically homogeneous: quiescent and spontaneously firing populations have been described (Colino and Halliwell 1987). Interestingly, all the different 5-HT receptor subtypes described to date have been detected in the hippocampus, though with distinct distributions. The physiological actions of 5-HT in the hippocampus may depend on the firing of particular subsets of raphé neurons, on the type of target cell, but also on the subtype of receptor activated and its cellular localization.

Brainstem

The brainstem represents the cranial continuation of the spinal cord, not only anatomically but also functionally. It contains the cranial nerve nuclei, some of which are the origin of viscero- or somatomotor fibers, while others are the target of viscero- or somatosensitive fibers. In addition, neurons located in certain brainstem nuclei are involved in the functions of the inner ear.

5-HT$_{1A}$ receptor mRNA and binding sites in the brainstem (Figure 4A, D) are particularly enriched in all raphé nuclei. In addition, they are also very abundant in the nucleus of the spinal tract of the trigeminus and nucleus of the solitary tract. The highest densities of 5-HT$_{1C}$ receptor mRNA and sites observed in Figure 4B, E are in fact located over the choroid plexus. These receptors and their mRNAs are diffusely distributed in the reticular formation at all levels of the brainstem. They are also found in the vestibular nuclei, facial, hypoglossal and sensory trigeminal nuclei. 5-HT$_2$ receptor mRNA and binding sites (Figure 4C, F) have a more restricted distribution, being found at high concentrations in almost all the motor nuclei innervating muscles both of myotomic (oculomotor nuclei and hypoglossal nucleus) and branchial developmental origin (motor trigeminal nucleus and facial nucleus). Low concentrations of 5-HT$_2$ receptor mRNA and binding sites are seen in the cochlear nuclei. Intermediate densities of 5-HT$_{1B}$ receptors (Figure 4G) are present diffusely in the reticular formation; higher concentrations can be observed in the vestibular and solitary tract nuclei. Of the different 5-HT receptor subtypes, 5-HT$_3$ sites (Figure 4H) are the ones exhibiting the most limited distribution in the brainstem, but also the highest density values: they are extremely enriched in the most lateral aspects of the nucleus of the spinal tract of the trigeminus (a part which seems to correspond to that displaying a high density of 5-HT terminals (Steinbush 1981)), in the nucleus of the solitary tract and in the dorsal motor nucleus of the vagus. Unilateral nodose ganglionectomy has been reported to reduce the density of 5-HT$_3$ receptors in the ipsilateral nucleus tractus solitarius, suggesting a presynaptic location of these sites on vagal terminal afferents (Pratt and Bowery 1989). While these lesions do not discriminate between vagal afferents from different sources, bilateral subdiaphragmatic vagotomy (which only removes abdominal afferents) abolishes 5-HT$_3$ receptor binding in the dorsal vagal complex of the ferret (Reynolds et al. 1989); these receptors might consequently be located on vagal afferents of gastric origin. It is worth mentioning that at the level of the spinal cord, administration of capsaicin to neonatal rats (a treatment inducing a degeneration of unmyelinated primary sensory fibres) also results in a loss of 5-HT$_3$ receptors in the substantia gelatinosa (Hamon et al. 1989). Interestingly, 5-HT$_{1A}$ receptors, which are similarly enriched in the substantia gelatinosa of the spinal cord, also disappear after the same treatment (Daval et al. 1987). 5-HT$_{1A}$ as well as 5-HT$_3$ receptors thus appear to be presynaptically located on primary afferent fibres in the spinal cord.

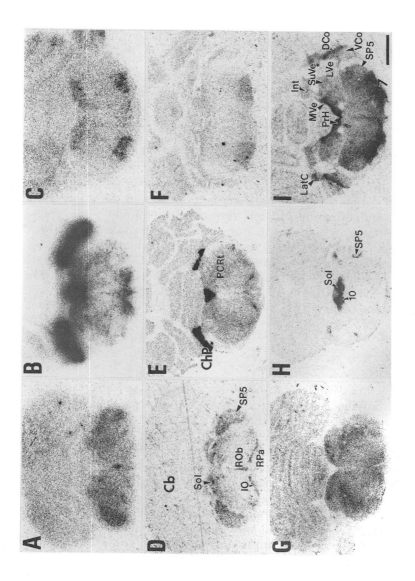

Figure 4. Autoradiograms generated by incubating rat brain sections at two levels of the brainstem either with radioligands or with ³²P-labelled oligonucleotide probes. Conditions are the same as in Figure 3. Scale bar is 2 mm.

18

In agreement with the distribution of 5-HT positive fibers (Steinbush 1981) and 5-HT content (Palkowits et al. 1974), the highest density of [^{3}H]citalopram sites in the brainstem is found in the motor nuclei, medial vestibular nuclei and dorsal cochlear nuclei (Figure 4I) and solitary tract nucleus. Intermediate densities of 5-HT uptake sites are observed in the reticular formation and nucleus of the spinal tract of the trigeminus while low densities are seen in the other vestibular nuclei and in the ventral cochlear nucleus.

At variance with the different parts of the hippocampal formation, particular roles can be ascribed to the various nuclei of the brainstem. Taking into account these roles and the distribution of the different subtypes of 5-HT receptors in the brainstem, gross functional correlates can be tentatively attributed to each subtype. Thus, it might be inferred that somatic sensory functions are modulated by 5-HT principally acting via 5-HT$_{1A}$ and 5-HT$_3$ receptors and somatic motor functions via 5-HT$_2$ receptors. A similar design could be present in the spinal cord, where 5-HT$_{1A}$ and 5-HT$_3$ binding sites are present in the dorsal horn (probably modulating inputs from sensory afferents) and 5-HT$_2$ receptors predominate in the ventral horn. Concerning visceral functions, 5-HT$_3$ receptors are involved in the motor aspect while different receptors are implicated in sensory functions (5-HT$_{1A}$, 5-HT$_{1B}$ and 5-HT$_3$ are found in the solitary tract nucleus). Different 5-HT receptors seem to be involved in the control of vestibular functions (5-HT$_2$, 5-HT$_{1C}$ and 5-HT$_{1B}$), whereas an action of 5-HT on cochlear nuclei is mediated mainly via 5-HT$_2$ receptors.

Conclusions

The introduction of molecular tools such as ligands and oligonucleotide probes to study the characteristics of the serotoninergic system reveals a complex organization both at the gross anatomical and cellular levels, which is at the present time difficult to integrate into a comprehensive functional diagram. We have presented three examples of regions where subtypes of 5-HT receptors coexist, though with different fine anatomical distribution, cellular localization and physiological functions. Progress towards an understanding of 5-HT functions in the brain requires still further development of tools such as drugs acting selectively at particular receptor subtype and specific antibodies against the different receptor proteins and, more importantly, requires precise definition of the complete catalogue of 5-HT receptor subtypes.

Acknowledgements

M. P. was supported by a fellowship from the Consiglio Nazionale delle Ricerche, Rome, Italy.

References

Andrade, R., Malenka, R. C., and Nicoll, R. A. (1986). A G protein couples serotonin and GABA$_B$ receptors to the same channels in the hippocampus. Science 234: 1261–1265.

Andrade, R., and Nicoll, R. A. (1987). Novel anxiolytics discriminate between postsynaptic serotonin receptors mediating different physiological responses on single neurons of the rat hippocampus. Naunyn-Schmiedeberg's Arch. Pharmacol. 336: 5–10.

Blackburn, T. P., Kemp, J. D., Martin, D. A., and Cox, B. (1984). Evidence that 5-HT agonist-induced rotational behaviour in the rat is mediated via 5-HT$_1$ receptors. Psychopharmacol. 83: 163–165.

Chaput, Y., Araneda, R. C., and Andrade, R. (1990). Pharmacological and functional analysis of a novel serotonin receptor in the rat hippocampus. Eur. J. Pharmacol. 182: 441–456.

Colino, A., and Halliwell, J. V. (1987). Differential modulation of three separate K-conductances in hippocampal CA1 neurons by serotonin. Nature 328: 73–77.

Costall, B., Kelly, M. E., Naylor, R. J., and Tyers, M. B. (1988). Modification of locomotor activity following drug action on 5-HT$_3$ receptors in the substantia nigra of the mouse. Neurosci. Lett. Suppl. 29: S70.

D'Amato, R. J., Largent, B. L., Snowman, A. M., and Snyder, S. H. (1987). Selective labelling of serotonin uptake sites in rat brain by [^{3}H]citalopram contrasted to labelling of multiple sites by [^{3}H]imipramine. J. Pharmacol. Exp. Ther. 242: 364–371.

Daval, G., Vergé, D., Basbaum, A. I., Bourgoin, S., and Hamon, M. (1987). Autoradiographic evidence of serotonin$_1$ binding sites on primary afferent fibres in the dorsal horn of the rat spinal cord. Neurosci. Lett. 83: 71–76.

Descarries, L., Watkins, K. C., Garcia, S., and Beaudet, A. (1982). The serotonin neurons in nucleus raphé dorsalis of adult rat: a light and electron microscope radioautography study. J. Comp. Neurol. 207: 239–254.

Derkach, V., Surprenant, A., and North, R. A. (1989). 5-HT$_3$ receptors are ion channels. Nature 339: 706–709.

De Vivo, M. and Maayani, S. (1990). Stimulation and inhibition of adenylyl cyclase by distinct 5-hydroxytryptamine receptors. Biochem. Pharmacol. 40: 1551–1558.

Dumuis, A., Bouhelal, R., Sebben, M., Cory, R., and Bockaert, J. (1989). A nonclassical 5-hydroxytryptamine receptor positively coupled with adenylate cyclase in the central nervous system. Mol. Pharmacol. 34: 880–887.

Engel, G., Göthert, M., Hoyer, D., Schlicker, E., and Hillebrand, K. (1986). Identity of inhibitory presynaptic 5-hydroxytryptamine (5-HT) autoreceptors in the rat brain cortex with 5-HT$_{1B}$ binding sites. Naunyn-Schmiedeberg's Arch. Pharmacol. 332: 1–7.

Freund, T. F., Gulyás, A. I., Acsádi, L., Görcs, T., and Tóth, K. (1990). Serotoninergic control of the hippocampus via local inhibitory interneurons. Proc. Natl. Acad. Sci. USA 87: 8501–8505.

Graybiel, A. M. (1990). Neurotransmitters and neuromodulators in the basal ganglia. Trends Neurosci. 13: 244–254.

Hamon, M., Gallisot, M. C., Menard, F., Gozlan, H., Bourgoin, S., and Vergé, D. (1989). 5-HT$_3$ receptor binding sites are on capsaicin-sensitive fibres in the rat spinal cord. Eur. J. Pharmacol. 164: 315–322.

Hirose, A., Sasa, M., Akaike, A., and Takaori, S. (1990). Inhibition of hippocampal CA1 neurons by 5-hydroxytryptamine, derived from the dorsal raphé nucleus and 5-hydroxytryptamine$_{1A}$ agonist SM-3997. Neuropharmacol. 29: 93–101.

Lavoie, B., and Parent, A. (1990). Immunohistochemical study of the serotoninergic innervation of the basal ganglia in the squirrel monkey. J. Comp. Neurol. 299: 1–16.

Moore, R. Y., and Halaris, A. E. (1975). Hippocampal innervation by serotonin neurons of the midbrain raphé in the rat. J. Comp. Neurol. 164: 171–184.

Osborne, N. N., and Hamon, M. (eds) (1988). Neuronal Serotonin. Chichester: John Wiley & Sons.

Palacios, J. M., and Dietl, M. M. (1988). Autoradiographic studies of serotonin receptors. In Sanders-Bush, E. (ed.). The serotonin receptors. Clifton, NJ: The Humana Press, pp. 89–138.

Palacios, J. M., Cortés, R., and Dietl, M. M. (1988). A laboratory guide for the in vitro labelling of receptors in tissue sections for autoradiography. In Van Leeuwen, F. W., Buijs, R. M., Pool, C. W., and Pach, O. (eds), Molecular Neuroanatomy. Amsterdam: Elsevier, pp. 95–106.

20

Palkovits, M., Brownstein, M., and Saavedra, J. M. (1974). Serotonin content of the brain stem nuclei in the rat. Brain Res. 80: 237–249.

Peroutka, S. J. (1988). 5-hydroxytryptamine receptors subtypes: molecular, biochemical and physiological characterization. Trends Neurosci. 11: 496–500.

Pratt, G. D., and Bowery, N. G. (1989). The 5-HT$_3$ receptor ligand, [^{3}H]BRL 43694, binds to presynaptic sites in the nucleus tractus solitarius of the rat. Neuropharmacol. 28: 1367–1376.

Quirion, R., and Richard, J. (1987). Differential effects of selective lesions of cholinergic and dopaminergic neurons on serotonin-type 1 receptors in rat brain. Synapse 1: 124–130.

Reynolds, D. J. M., Andrews, P. L. R., Leslie, R. A., Harvey, J. M., Grasby, P. M. and Grahame-Smith, D. G. (1989). Bilateral abdominal vagotomy abolishes binding of [^{3}H]BRL 43694 in ferret dorsovagal complex. Br. J. Pharmacol. 98: 692P.

Steinbush, H. W. M. (1981). Distribution of serotonin-immunoreactivity in the central nervous system of the rat – Cell bodies and terminals. Neurosci. 6: 557–618.

Sternweis, P. C., and Pang, I. H. (1990). The G protein-channel connection. Trends Neurosci. 13: 122–126.

Swanson, L. W., Köler, C., and Björklund, A. (1987). The limbic region. I: The septo-hippocampal system. In Björklund, A., Hökfelt, T., and Swanson, L. W. (eds), Handbook of chemical neuroanatomy, Vol. 5, Integrated systems of the CNS, part I. Amsterdam: Elsevier, pp. 125–277.

Vergé, D., Daval, G., Marcinkiewicz, M., Patey, A., El Mestikawy, S., Gozlan, H., and Hamon, M. (1986). Quantitative autoradiography of multiple 5-HT$_1$ receptor subtypes in the brain of control or 5,7-dihydroxytryptamine-treated rats. J. Neurosci. 6: 3474–3482.

Vilaró, M. T., Martinez-Mir, M. I., Sarasa, M., Pompeiano, M., Palacios, J. M., and Mengod, G. (1991). Molecular neuroanatomy of neurotransmitter receptors: the use of in situ hybridization histochemistry for the study of their anatomical and cellular localization. In Osborne, N. N. (ed.), Current aspects of the Neurosciences, Vol. 3, Basingstoke: Macmillan Press, pp. 1–36.

Waeber, C., Dietl, M. M., Hoyer, D., and Palacios, J. M. (1989a). 5-HT$_1$ receptors in the vertebrate brain: regional distribution examined by autoradiography. Naunyn-Schmiedeberg's Arch. Pharmacol. 340: 486–494.

Waeber, C., Zhang, L., and Palacios, J. M. (1989b). 5-HT$_{1D}$ receptors in the guinea-pig brain: pre- and postsynaptic localizations in the striato-nigral pathway. Brain Res. 528: 197–206.

Waeber, C., Pinkus, L. M., and Palacios, J. (1990). The (S)-isomer of [^{3}H]zacopride labels 5-HT$_3$ receptors with high affinity in rat brain. Eur. J. Pharmacol. 181: 283–287.

Weissman, D., Mach, E., Oberlander, C., Demassey, Y., and Pujol, J. F. (1986). Evidence for a hyperdensity of 5-HT$_{1B}$ binding sites in the substantia nigra of the rat after 5,7-dihydroxytryptamine intraventricular injection. Neurochem. Int. 9: 191–200.

Serotonin: Molecular Biology, Receptors and Functional Effects
ed. by J. R. Fozard/P. R. Saxena
© 1991 Birkhäuser Verlag Basel/Switzerland

Cloning and Expression of a Human 5-HT$_{1D}$ Receptor

T. Branchek, J. Zgombick, M. Macchi, P. Hartig, and R. Weinshank

Neurogenetic Corporation, Paramus, NJ 07652, USA

Summary. A gene encoding a serotonin 5-HT$_{1D}$ receptor was cloned from a human brain cDNA library and functionally expressed in a heterologous expression system. Radioligand binding experiments were used to evaluate the pharmacological properties of membrane fragments expressing this cloned human receptor. [^{3}H]5-HT bound with high affinity ($K_d = 4.3$ nM); the site density was 5.2 pmoles/mg protein. Competition studies revealed the following affinities (K_i in nM): 5-CT $= 0.67$; 5-HT $= 4.2$; 5-methoxytryptamine $= 4.7$; yohimbine $= 25$; spiperone > 1000; pindolol > 1000; zacopride $> 10,000$. High affinity binding was sensitive to the addition of guanine nucleotides, indicating functional coupling to a G-protein. 5-HT was found to inhibit forskolin-stimulated adenylate cyclase activity in L-M(TK-) cells expressing the cloned receptor.

These data are consistent with the assignment of the newly cloned gene as a 5-HT$_{1D}$ receptor. Since the deduced amino acid sequence of this gene exhibits only moderate homology to the 5-HT$_{1A}$ receptor and low homology to 5-HT$_2$ receptor subfamily (5-HT$_2$ and 5-HT$_{1C}$ receptors), the human 5-HT$_{1D}$ receptor apparently represents the first member of a new subfamily of serotonin receptor genes. Since subtypes of the 5-HT$_{1D}$ receptor have been hypothesized, this human 5-HT$_{1D}$ clone will facilitate the isolation of any additional 5-HT$_{1D}$ receptor subtypes in the human brain.

Introduction

Receptors for 5-hydroxytryptamine can be divided into 4 groups (5-HT$_1$; 5-HT$_2$/5-HT$_{1C}$; 5-HT$_3$; 5-HT$_4$) based on their deduced amino acid sequences (where known), pharmacological properties, second messenger coupling, and electrophysiological properties. The 5-HT$_1$ group appears to be the largest and most diverse. The three well-established members of this group are the 5-HT$_{1A}$, 5-HT$_{1B}$, and 5-HT$_{1D}$ receptors (in molecular terms, the 5-HT$_{1C}$ receptor is a member of the 5-HT$_2$ group). These receptors all display high affinity for [^{3}H]5-HT in radioligand binding assays, couple to the inhibition of adenylate cyclase, and function as autoreceptors in certain brain regions. The 5-HT$_{1B}$ and 5-HT$_{1D}$ receptors are currently presumed to be species homologs of the same receptor subtype, since they share a similar anatomical localization in the brain and are both coupled to adenylate cyclase inhibition (Hoyer and Middlemiss 1989). Other putative 5-HT$_1$ receptors include the 5-HT$_{1E}$ (Leonhardt et al. 1989), 5-HT$_{1R}$ (Xiong and Nelson 1989), 5-HT$_{1DSV}$ (Sumner and Humphrey 1989; Humphrey et al. 1990), and 5-HT$_{1P}$ (Mawe et al. 1986; Branchek et al. 1988) receptors.

The 5-HT$_{1D}$ receptor subtype was first reported in the bovine caudate by Heuring and Peroutka (1987). It has been defined by its high affinity binding of [^{3}H]5-HT in the presence of ligands to mask the 5-HT$_{1A}$, 5-HT$_{1B}$, and 5-HT$_{1C}$ receptors. In functional assays it has been found to couple to the inhibition of adenylate cyclase activity (Hoyer and Schoeffter 1988) with a rank order of potencies of 5-CT > 5-HT > sumatriptan > yohimbine > 8-OH-DPAT (Schoeffter et al. 1988; Schoeffter and Hoyer 1989). This receptor has also been described in the human, monkey, pig, dog, cat, rabbit, pigeon, and guinea pig (Herrick-Davis and Titeler 1988; Waeber et al. 1988a; b, 1989a; b; c; Xiong and Nelson 1989; Middlemiss et al. 1990). In the brain, it is most abundant in the caudate, cortex, and substantia nigra (Heuring and Peroutka 1987; Herrick-Davis and Titeler 1988; Waeber et al. 1988a, 1989a; c). In addition to the brain, 5-HT$_{1D}$-like receptors have been described in coronary arteries (Molderings et al. 1989; Schoeffter and Hoyer 1990; Toda and Okamura 1990; Kaumann et al. 1990), dog saphenous vein (Humphrey et al. 1988; Sumner and Humphrey 1990), and several other tissues (Buchheit and Bühl 1990; Schoeffter and Sahin-Erdemli 1990).

5-HT$_{1D}$ receptors are located both pre- and postsynaptically (Fenuik et al. 1985; Schlicker et al. 1989; Herrick-Davis et al. 1989). Some investigators have proposed that the terminal autoreceptor is a distinct entity from the postsynaptic 5-HT$_{1D}$ receptor. Others believe that they are the same molecular species and that other properties of the cells on which they reside, such as receptor reserve, confer the differences which have been observed in their pharmacological properties (Humphrey et al. 1988). Still other researchers have grouped a large collection of these similar receptors as members of the 5-HT$_1$-like family and believe that the species differences in pharmacology between the different tissue preparations reveals a highly heterogeneous collection of receptors (Bradley et al. 1986). In the complex tissue preparations from different species which are used to define this receptor family, and with the different methods of detection used by different investigators, it seems difficult if not impossible to resolve these issues. Molecular cloning, expression, and characterization of all potential members of this important receptor family provides a powerful method to address these questions and to design truly site-selective drugs for the 5-HT$_{1D}$ and other 5-HT$_1$ receptor subfamilies.

Three subtypes of receptor for 5-HT have been cloned and characterized to date. These are the 5-HT$_{1A}$ (Kobilka et al. 1987; Fargin et al. 1988), 5-HT$_{1C}$ (Lübbert et al. 1987; Julius et al. 1988), and 5-HT$_2$ receptors (Pritchett et al. 1988; Kao et al. 1989; Julius et al. 1990). All three receptors have been shown to belong to the large family of G-protein coupled receptors. Characteristic of this family is a seven transmembrane spanning motif with an extracellular amino terminus and an intracellular carboxy terminus. Since 5-HT$_{1D}$ receptor is also known to be a G protein-coupled receptor, it was postulated to have the same structural motif. This information was employed in the design of a cloning strategy for the human 5-HT$_{1D}$ receptor. A preliminary account of these data has been presented (Branchek et al. 1990a).

Material and Methods

A new gene belonging to the G protein coupled receptor family was isolated by PCR and homology screening techniques utilizing selected oligonucleotides based on serotonin receptor sequences. A full-length clone for this new gene was then obtained from a human brain cDNA library. Nucleotide sequence analysis was performed using the Sanger dideoxy nucleotide chain termination method on double stranded plasmid templates using Sequenase (US Biochemical Corp.). The entire coding region was cloned into a eucaryotic expression vector (pSVL) and transiently transfected into Cos 7 cells using the DEAE dextran method (Cullen 1987). Following transfection, cells were allowed to grow for 48 h, were harvested by scraping, and a membrane fraction was prepared. Radioligand binding assays were performed using ^{3}H-5-HT to determine if this DNA encoded a receptor with 5-HT$_{1D}$ binding properties (i.e. high affinity ^{3}H-5-HT binding in the presence of pindolol and SCH 23390). High affinity binding was detected in these transiently expressing cells.

To facilitate additional characterization, a cell line which continuously expressed the 5-HT$_{1D}$ receptor (stable transfectant) was constructed. A plasmid containing the nucleotide sequence for the human 5-HT$_{1D}$ receptor was transfected into murine fibroblasts (L-M[tk-]) using the calcium phosphate method. A selectable marker (aminoglycoside transferase) was co-transfected and stable transfectants were isolated by selection in the antibiotic G418. Each resistant clonal cell line was then screened for its ability to bind ^{3}H-5-HT (28.4 Ci/mmole; DuPont-NEN) specifically. Sham-transfected cells were studied in paralled to rule out the possibility of endogenous 5-HT$_{1D}$ receptors or their induction by the transfection protocol. No background was obtained in these cells and thus any specific binding was attributed to the exogenous gene. Since this procedure yields cells with a broad range of expression densities, the clone with the highest site density was chosen for further characterization. Radioligand binding assays were performed according to the method of Herrick-Davis and Titeler (1988). All samples were run in triplicate with a minimum of three repetitions. Data were analyzed by computer-assisted nonlinear analysis (Accufit and Accucomp; Lundon software, Chagrin Falls, OH). Adenylate cyclase assays were performed according to the method of Salomon (1979). Protein was determined by the method of Bradford (1976).

Results

A new G-protein-coupled receptor gene has been isolated from a human brain cDNA library. Two clones were isolated and were characterized by restriction endonuclease mapping and DNA sequence analysis. Complementary portions of the clones were then ligated to obtain a full-length clone. Hydropathy analysis of the protein sequence suggests the presence of seven membrane-spanning domains. A comparison of this sequence to the amino acid sequences of other human biogenic amine receptors is shown in Figure 1. Within the transmembrane domains, the newly cloned receptor displayed the highest transmembrane homology to the 5-HT_{1A} receptor (56%) with lesser homology to the 5-HT_2 receptor (43%) or to the dopamine D_2 receptor (46%).

Analysis of the ligand binding profile of the protein product encoded by the newly isolated human gene was required for assignment of receptor subtype. Therefore, the gene was cloned into the pSVL expression vector and transfected into Cos-7 (monkey kidney) cells as transient transfectants to verify that the gene could be expressed and to determine its pharmacological profile. Transient transfectants of the human 5-HT_{1D} gene were found to bind $[^3H]5\text{-HT}$ with high affinity and this binding was resistant to competition by specific 5-HT_{1A}, 5-HT_{1B}, 5-HT_{1C}, and 5-HT_{1P} ligands (data not shown). In addition, it was found to have high affinity for 5-carboxamidotryptamine. On this basis, the gene was tentatively assigned as a 5-HT_{1D} receptor candidate and further analysis was done on stable transfectant cell lines.

Within a Receptor Family:			*% Homology*
alpha$_1$	vs	alpha$_2$	44
	vs	beta$_1$	45
	vs	beta$_2$	42
Within a Subfamily:			
alpha$_{2A}$	vs	alpha$_{2B}$	75
	vs	alpha$_{2C}$	75
5-HT_2	vs	5-HT_{1C}	78
5-HT_{1D} Clone:			
5-HT_{1D}	vs	5-HT_{1A}	56
	vs	5-HT_2	43
	vs	alpha$_1$	44
	vs	alpha$_{2A}$	45
	vs	D_2	46

Figure 1. Relationship between pharmacological subtypes and amino acid sequence of G protein-coupled receptors. Percent (%) homology indicates identical amino acids in the transmembrane spanning region of the receptor pair.

Stable expression of the receptor in clonally-derived cell lines was accomplished by co-transfection of the 5-HT$_{1D}$ gene and a selectable marker into murine L-M[tk-] fibroblasts as a host cell. The clone with the highest expression density was selected for further study. A membrane fraction was prepared from cells expressing the cloned gene and radioligand binding assays were performed.

Membranes derived from transfected cells bound [^{3}H]5-HT saturably, reversibly, and with high affinity (Figure 2A) The equilibrium dissociation constant was 4.3 ± 0.6 nM and the site density was 5.02 ± 0.86 pmol/mg protein (n = 5). At the midpoint concentration of the radioligand, the specific binding of [^{3}H]5-HT represented approximately 95% of the total binding (Figure 2A). In order to examine the coupling of the cloned human receptor to endogenous G-proteins in murine fibroblasts, the effect of non-hydrolyzable guanine nucleotide analog Gpp(NH)p was measured. Gpp(NH)p was found to produce a dose-dependent decrease in the specific high affinity binding of [^{3}H]5-HT, with a midpoint effective concentration (IC$_{50}$) of 4 μM (data not shown).

In order to characterize the pharmacological properties of the high affinity binding site for [^{3}H]5-HT, radioligand binding competition studies were performed. As seen in Figure 2B, monophasic competition

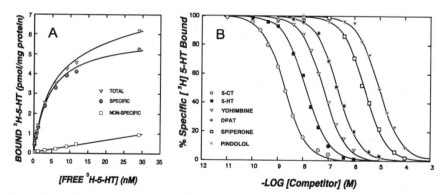

Figure 2. A. Measurement of the equilibrium dissociation constant of [^{3}H]5-HT binding to membrane fragments harvested from murine cells (L-M[tk-]) stably transfected with the human 5-HT$_{1D}$ receptor gene. Bound radioligand is plotted as a function of free radioligand. The symbols are indicated on the graph. Paramenters were determined by computer-assisted nonlinear curve fitting (Lundon Software, Chagrin Falls, OH). $K_d = 4.3 \pm 0.6$ nM; $B_{max} = 5.02 \pm 0.86$ pmol/mg protein. Specific binding = 98% of total binding at the K_d concentration of radioligand.

B. Measurement of affinities of various 5-HT receptor ligands for the cloned human 5-HT$_{1D}$ receptor. [^{3}H]5-HT (5 nM) was incubated in the presence of the competitors and the specific binding was plotted as a function of competitor concentration. Monophasic competition curves ($n_H = 1.0$) were observed. Values of K_i were determined by computer-assisted nonlinear curve fitting (Lundon Software, Chagrin Falls, OH) and are presented in Table 1.

Table 1. Binding affinities of various 5-HT receptor ligands for the cloned human 5-HT_{1D} receptor are compared to inhibition constants reported in native human brain cortical membrane preparations. The ligands were chosen to exclude the possibility that $^3\text{H-5-HT}$ binding was to a number of other known serotonin receptor subtypes.

Compound	K_1(nM) Cloned Receptor	K_1(nM) Human Cortex[a,b]	Rules Out
5-CT	0.67 ± 0.07	1.1^a	5-H5_{1C}, 5-HT_{1E}, 5-HT_{1P}, 5-HT_2, 5-HT_4
5-HT	4.2 ± 0.1	2.4^a	5-HT_4
5-MT	4.7 ± 0.1	4.3^a	5-HT_{1B}, 5-HT_{1P}, 5-HT_2
Yohimbine	25 ± 2	42^a	
DOI/DOB	720 ± 140	670^b	5-HT_2, 5-HT_{1C}
Spiperone	1014 ± 135	—	5-HT_{1A}, 5-HT_2, D_2
Pindolol	4000 ± 1100	$> 10,000^b$	5-HT_{1A}, 5-HT_{1B},
Zacopride	$> 10,000$	—	5-HT_3
ICS 205–930	—	$> 10,000^b$	

[a]Heuring and Peroutka (1987); [b]Herrick-Davis and Titeler (1988); Abbreviations = 5-MT = 5-methoxytryptamine; DOB = 4-bromo-2,5-dimethoxyphenylisopropylamine; DOI = 4-iodo-2,5-dimethoxyphenylisopropylamine.

curves for [^{3}H]5-HT binding ($n_H = 1.0$) were obtained for a series of compounds. 5-carboxyamidotryptamine was the most potent compound tested, followed by 5-HT itself. As is typical for the 5-HT_{1D} receptor, yohimbine exhibited relatively high affinity (Table 1). The rank order of potencies (Figure 2B; Table 1) is consistent with designation of this [^{3}H]5-HT binding as a 5-HT_{1D} binding site. In addition, the inhibition constants for the cloned 5-HT_{1D} receptor were compared to those reported for a series of 5-HT_1 receptors (Hoyer 1989). The best correlation was obtained with the 5-HT_{1D} binding site reported in the native calf caudate membranes ($r = 0.91$), further supporting the assignment of this gene as encoding a human 5-HT_{1D} receptor.

The 5-HT_{1D} receptor has been shown to couple to the inhibition of adenylate cyclase activity Hoyer and Schoeffter (1988). To verify further that this DNA sequence encoded a human 5-HT_{1D} receptor, the functional coupling of the cloned transfected receptor was determined by the effect of 5-HT on forskolin stimulated adenylate cyclase activity (FSAC). 5-HT (1 μM) was found to produce a significant inhibition of FSAC. This effect was reversed by the non-selective 5-HT_1 antagonist methiothepin.

Discussion

The serotonin 5-HT_{1D} receptor has recently been implicated in a variety of important clinical problems including migraine, anxiety, depression, anorexia, cardiac function, and movement control, in addition to its potential role as an autoreceptor. It has been difficult to evaluate the importance of this receptor and to develop truly site-selective drugs due to its low abundance and lack of selective pharmacological agents and clean model systems. Pharmacological investigations have indicated that there may be multiple subtypes of 5-HT_{1D} receptor. In order to acquire an improved understanding of this receptor subfamily and to develop selective reagents which may be of therapeutic value we designed a strategy which was successful in the isolation of the first human 5-HT_{1D} receptor gene.

The deduced amino acid sequence of the newly isolated gene was analyzed to uncover relationships between it and other known monoamine receptor sequences. The membrane spanning domains of the 5-HT_{1D} receptor were most homologous to the sequence of the 5-HT_{1A} receptor and distinctly less closely related to the other two known serotonin receptor sequences (the 5-HT_2 and 5-HT_{1C} receptors). In fact, the 5-HT_{1D} receptor was more closely related to the sequence of the dopamine D2 receptor, and several other adrenergic receptors, than it was to the sequences of the serotonin 5-HT_2 receptor. These relationships between G protein-coupled receptor sequences reveal interesting information about their interrelationships and functions. As shown in Figure 1, receptor subtypes that share the same neurotransmitter but exhibit different second messenger coupling and pharmacolocical properties generally exhibit 40–45% amino acid homology in their transmembrane regions. Receptors of the same subfamily, which share common second messenger pathways and pharmacologic properties, exhibit a much closer relationship between their primary amino acid sequences, with transmembrane homologies of 70–80%. The transmembrane homology between the 5-HT_{1D} receptor and the 5-HT_{1A} receptor (56% homology) was much much lower than the degree of homology between receptors of the same subfamily. Thus, it appears that the 5-HT_{1D} receptor represents the first member of a new serotonin receptor subfamily, distinct from both the 5-HT_{1A} and 5-HT_2 receptor subfamilies.

In order to identify and evaluate the properties of the newly cloned receptor, pharmacological characterization was performed on the human gene transfected into a murine host cell line (LM-(tk-)). We have previously determined that expression of a human receptor gene in a murine environment does not alter the binding properties of the 5-HT_2 receptor, which displays significant species differences in pharmacology (Kae et al. 1989; Hartig et al. 1990; Branchek et al. 1990b). The ligand binding properties provide compelling evidence that the newly cloned gene encoded a human 5-HT_{1D} receptor. The expression levels obtained in this heterologous expression system (5 pmol/mg protein) provide a convenient and homogenous serotonin receptor preparation without the contamination by other serotonin receptor subtypes typically encountered in 5-HT_{1D} receptor studies. In the human brain, 5-HT_{1D} receptors exhibit a site density twenty fold lower than that reported here, as well as shallow competition curves and poor specific to nonspecific binding ratios.

The non-hydrolyzable analog of GTP, Gpp(NH)P, was able to reduce the high affinity binding of [^{3}H]5-HT to the transfected human 5-HT$_{1D}$ receptor, indicating that this receptor is capable of existing in both high and low affinity agonist binding states. Thus, in the murine host, the human receptor appears to couple to endogenous G-proteins. However, the specific midpoint of the inhibition curve ($4\,\mu$M) and degree of maximum inhibition (25%) are somewhat different from values obtained in studies using native human brain membranes. It is likely that these apparent discrepancies are the result of the high receptor expression levels obtained in the transfection host cells relative to the supply of endogenous G-proteins. Lower expression levels may maximize the proportion of receptors in the G-protein coupled state and thus display larger and more sensitive responses to guanine nucleotides. Such effects have been reported for the muscarinic (Mei et al. 1989) and for the 5-HT$_2$ receptors (Branchek et al. 1990b; Adham et al. 1990). Further evidence for functional coupling of the 5-HT$_{1D}$ receptor was obtained by a preliminary investigation of adenylate cyclase activity in which 5-hydroxytryptamine was found to inhibit forskolin-stimulated adenylate cyclase activity. This effect was blocked by addition of methiothepin, consistent with expectation for the coupling of a 5-HT$_{1D}$ receptor.

The "5-HT$_{1D}$ receptor", as defined by pharmacological criteria, appears to be a set of related receptors rather than a unique entity (Figure 3). The original description of this receptor in the bovine caudate (Heuring and Peroutka 1987) has been followed by a rapid proliferation of information on receptors with similar but not identical properties. A 5-HT$_{1D}$-like receptor has been studied in the pig cortex, where it is thought to function as a terminal autoreceptor (Schlicker et al. 1989). The rank order of agonist potencies for this autoreceptor, as measured by the method of electrically evoked [^{3}H]5-HT overflow from pre-loaded tissue displays a higher efficacy of 5-HT relative to 5-CT than that detected in adenylate cyclase assays on the caudate neclaus. In addition, the two preparations display other subtle differences. The pre-junctional 5-HT$_{1D}$-like receptor in the isolated perfused rat kidney (Charlton et al. 1986; Bond et al. 1989) may be another member of this family. The dog saphenous vein preparation also has pre- and postjunctional receptor components (Feniuk et al. 1985). However, they differ only quantitatively in agonist potencies, not in rank order of potencies or these agonists (Humphrey et al. 1988). The rabbit brain has been shown to have population of 5-HT$_{1D}$ binding sites designated 5-HT$_{1R}$ (Xiong and Nelson 1989). These receptors exhibit some pharmacologic differences when compared with the bovine or human 5-HT$_{1D}$ receptor, which may represent species variations among related genes, or may indicate the presence of yet another novel 5-HT$_1$ receptor subtype.

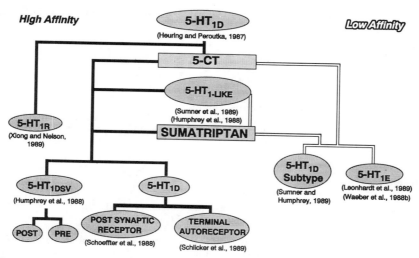

Figure 3. Relationships between receptors of the 5-HT$_{1D}$ subfamily, organized according to their affinities for the ligands 5-carboxamidotryptamine (5-CT) and sumatriptan. The pharmacologically-defined 5-HT$_{1D}$ receptor subtype appears to be a collection of closely related receptor proteins, whose identities and properties remain to be fully characterized.

A related group of receptors with certain properties in common with the 5-HT$_{1D}$ subtype have also been reported. They have been designated the 5-HT$_{1E}$ receptor (Leonhardt et al. 1989), a "5-HT$_{1D}$ subtype" (Sumner and Humphrey 1989) and a "subpopulation of 5-HT$_{1D}$ recognition sites" (Middlemiss et al. 1990). The presence of a similar binding site was illustrated in the work of Waeber and colleagues (1988b). Each is characterized in radioligand binding studies by a high affinity for [^{3}H]5-HT under 5-HT$_{1D}$ binding conditions and displays low affinity for 5-CT, and/or ergotamine, sumatriptan, and derivatives of spiperone. The 5-HT$_{1E}$ receptor is modulated by guanine nucleotides (Leonhardt et al. 1989), as would be predicted for a G protein-coupled receptor.

Several other receptors for 5-HT have been described either by radioligand binding assays or by functional response models (Berry-Kravis and Dawson 1983; Bradley et al. 1986; Mawe et al. 1986; Gershon et al. 1989; Branchek et al. 1988). Although the pharmacological tools have not been available to fully define the relationships amongst these receptors and the 5-HT$_{1D}$ subfamily, information derived from molecular cloning may help to clarify these relationships. In addition, this cloned human 5-HT$_{1D}$ receptor can be used to evaluate the molecular relationship between the 5-HT$_{1D}$ and 5-HT$_{1B}$ receptors. At present, it is not possible to evaluate which member of the 5-HT$_{1D}$ family is encoded by the newly cloned gene. Further analysis of data from functional assays, mRNA localization, and studies of similar genes from other species will be required to answer the many remaining questions.

Acknowledgements

The authors wish to thank Ms. Barbara Dowling, Mr. Harvey Lichtblau, Ms. Anastasia Kokkinakis, Ms. Lisa Gonzalez and Ms. Jenny Xanthos for their excellent technical assistance.

References

Adham, N., Macchi, M., Kao, H-T., Hartig, P., and Branchek, T. (1990). Density-dependent regulation of affinity states of the cloned human 5-HT$_2$ receptor. Soc. Neurosci. Abst. 16: 1196.
Berry-Kravis, E., and Dawson, G. (1983). Characterization of an adenylate cyclase-linked receptor in a neuroblastoma X brain explant hybrid cell line (NCB-20). J. Neurochem. 40: 977–985.
Bond, R. A., Craig, D. A., Charlton, K. G., Ornstein, A. G., and Clarke, D. E. (1989). Partial agonist activity of GR43175 at the inhibitory prejunctional 5-HT$_{1-like}$ receptor in rat kidney. J. Auto. Pharmacol. 9: 201–210.
Bradford, M. (1976). A rapid and sensitive method for the quantification of microgram quantities of protein utilizing the principle of protein-dye binding. Anal. Biochem. 72: 248–254.
Bradley, P. B., Engel, G., Feniuk, W., Fozard, J. R., Humphrey, P. P. A., Middlemiss, D. N., Mylecharane, E. J., Richardson, B. P., Saxena, P. R. (1986). Proposals for the classification and nomenclature of functional receptors for 5-hydroxytryptamine. Neuropharmacol. 25: 563–576.
Branchek, T., Mawe, G., and Gershon, M. (1988). Characterization and localization of a peripheral neural 5-hydroxytryptamine receptor subtype (5-HT$_{1P}$) with a selective agonist, ^{3}H-5-hydroxyindalpine. J. Neurosci. 8: 2582–2595.
Branchek, T., Weinshank, R., Macchi, M., Zgombick, J., and Hartig, P. (1990a). Cloning and expression of a human 5-HT$_{1D}$ receptor. Second IUPHAR Satellite Meeting on Serotonin, p. 35.
Branchek, T., Adham, N., Macchi, M., Kao, H.-T., and Hartig, P. R. (1990b). [^{3}H]-DOB (4-bromo-2,5-dimethoxyphenylisopropylamine) and [^{3}H]-ketanserin label two affinity states of the cloned human 5-hydroxytryptamine$_2$ receptor. Mol. Pharmacol. 38: 6046–609.
Buchheit, K. H., and Buhl, T. (1990). 5-HT$_{1D}$ and 5-HT$_4$ agonists stimulate the peristaltic reflex in the isolated guinea pig ileum. Second IUPHAR Satellite Meeting on Serotonin, p. 66.
Charlton, K. G., Bond, R. A., and Clarke, D. E. (1986). An inhibitory prejunctional 5-HT$_{1-like}$ receptor in the isolate perfused rat kidney. Naunyn-Schmiedeberg's Arch. Pharmacol. 332: 8–15.
Cullen, B. R. (1987). Use of eucaryotic expression techniques in the functional analysis of cloned genes. Methods in Enzymology, 152: 684–704.
Fargin, A., Raymond, J., Lohse, M., Kobilka, B., Caron, M., and Lefkowitz, R. (1988). The genomic clone G-21 which resembles the B-adrenergic receptor sequence encodes the 5-HT$_{1A}$ receptor. Nature 335: 358–360.
Feniuk, W., Humphrey, P. P. A., Perrin, M. J., and Watts, A. D. (1985). A comparison of 5-hydroxytryptamine receptors mediating contraction in rabbit isolated aorta and dog saphenous vein. Evidence for different receptor types obtained by the use of selective agonists and antagonists. Br. J. Pharmacol. 86: 697–704.
Gershon, M. D., Mawe, G., and Branchek, T. (1989). 5-Hydroxytryptamine and enteric neurons. In J. R. Fozard (ed.), The Peripheral Actions of 5-HT. U.K: Oxford Press, pp. 247–273.

Herrick-Davis, K., Titeler, M., Leonhardt, S., Struble, R., and Price, D. (1988). Serotonin 5-HT$_{1D}$ receptors in human prefrontal cortex and caudate: Interaction with a GTP binding protein. J. Neurochem. 51: 1906–1912.

Herrick-Davis, K., Maisonneuve, I., and Titeler, M. (1989). Postsynaptic localization and up-regulation of serotonin 5-HT$_{1D}$ receptros in rat brain. Brain Res. 483: 155–157.

Heuring, R. E., and Peroutka, S. J. (1987). Characterization of a novel ^{3}H-5-hydroxytryptamine binding site subtype in bovine brain membranes. J. Neurosci. 7: 894–903.

Hoyer, D. (1989). 5-Hydroxytryptamine receptors and effector coupling mechanism in peripheral tissues. In: J. R. Fozard (ed.), The Peripheral Actions of 5-HT. Oxford: Oxford University Press, pp. 72–99.

Hoyer, D., and Middlemiss, D. M. (1989). Species differences in the pharmacology of 5-HT terminal autoreceptors in mammalian brain. Trends Pharmacol. Sci. 10: 130–132.

Hoyer, D., and Schoeffter, P. (1988). 5-HT$_{1D}$ receptors inhibit forskolin stimulated adenylate cyclase activity in calf substantia nigra. Eur. J. Pharmacol. 147: 145–147.

Humphrey, P. P. A., Feniuk, W., Perrin, M. J., Connor, H. E., Oxford, A. W., Coates, I. H., and Butina, D. (1988). GR43175, a selective agonist for the 5-HT$_{1-like}$ receptor in dog isolated saphenous vein. Br. J. Pharmacol. 94: 1123–1132.

Humphrey, P. P. A., Apperley, E., Feniuk, W., and Perren, M. (1990). A rational approach to identifying a fundamentally new drug for the treatment of migraine. In: P. R. Saxena, (ed.), Cardiovascular Pharmacology of 5-Hydroxytryptamine, Dordrecht: Kluwer, pp. 417–431.

Julius, D., MacDermott, A. B., Axel, R., and Jessell, T. M. (1988). Molecular characterization of a functional CDNA encoding the serotonin 1C receptor. Science 241: 558–564.

Julius, D., Huang, K., Livelli, T., Axel, R., and Jessel, T. (1990). The 5-HT2 receptor defines a family of structurally distinct but functionally conserved serotonin receptors. Proc. Natl. Acad. Sci. (USA) 87: 928–932.

Kao, H. -T, Olsen, M. A., and Hartig, P. R. (1989). Isolation and characterization of a human 5-HT$_2$ receptor clone. Soc. Neurosci. Abst. 15, 486.

Kaumann, A. J., Sanders, L., Brown, A., Murray, K., and Brown, M. (1990). 5-HT and the human heart. Second IUPHAR Satellite Meeting on Serotonin, p. 107.

Kobilka, B. K., Frielle, T., Collins, S., et al. (1987b). An intronless gene encoding a potential member of the family of receptors coupled to guanine nucleotide regulatory proteins. Nature 329: 75–79.

Leonhardt, S., Herrick-Davis, K., and Titeler, M. (1989). Detection of a novel serotonin receptor subtype (5-HT$_{1E}$) in human brain: interaction with a GTP-binding protein. J. Neurochem. 53: 465–471.

Lübbert, H., Hoffman, B., Snutch, T., Van Dyke, T., Hartig, P., Lester, H., and Davidson, N. (1987). cDNA cloning of a serotonin 5-HT$_{1C}$ receptor by electrophysiological assays of MRNA-injected Xenopus oocytes. Proc. Natl. Acad. Sci. (USA) 84: 4332–4336.

Mawe, G., Branchek, T., and Gershon, M. D. (1986). Peripheral neural serotonin receptors: Identification and characterization with specific agonists and antagonists. Proc. Natl. Acad. Sci. (USA) 83: 9799–9803.

Mei, L., Lai, J., Yamamura, H., and Roeske, W. (1989). The relationship between agonist states of the M1 muscarinic receptor and the hydrolysis of inositol lipids in transfected murine fibroblast cells (B82). J. Pharmacol. Exp. Ther. 251: 90–97.

Middlemiss, D., Suman-Chauhan, N., Smith, S., Picton, C., Shaw, D., and Bevan, Y. (1990). Subpopulations of 5-HT$_{1D}$ recognition sites in guinea-pig, rabbit, dog, and human cortex. Second IUPHAR Satellite Meeting on Serotonin, p. 49.

Molderings, G., Engel, G., Roth, E., and Gothert, M. (1989). Characterization of an endothelial 5-hydroxytryptamine (5-HT) receptor mediating relaxation of the porcine coronary artery. Naunyn-Schmiedeberg's Arch. Pharmacol. 340: 300–308.

Pritchett, D., Bach, A., Wozny, A., Taleb, O., Dal Taso, R., Shih, J., and Seebury, P. (1988). Structure and functional expression of cloned rat serotonin 5-HT$_2$ receptor. EMBO J. 13: 4135–4140.

Salomom, Y. (1979). Adenylate cyclase assay. Anal. Biochem. 58: 541.

Schlicker, E., Fink, K., Göthert, M., Hoyer, D., Molderings, G., Roschke, I., and Schoeffter, P. (1989). The pharmacological properties of the presynaptic serotonin autoreceptor in the pig brain cortex conform to the 5-HT$_{1D}$ subtype. Naunyn-Schmiedeberg's Arch. Pharmacol. 340: 45–51.

Schoeffter, P., Waeber, C., Palacios, J. M., and Hoyer, D. (1988). The 5-hydroxytryptamine 5-HT$_{1D}$ receptor subtype is negatively coupled to adenylate cyclase in calf substantia nigra. Naunyn-Schmiedberg's Arch. Pharmacol. 337: 602–608.

Schoeffter, P., and Hoyer, D. (1989). How selective is GR 43175? Interactions with functional 5-HT$_{1A}$, 5-HT$_{1B}$, 5-HT$_{1C}$, and 5-HT$_{1D}$ receptors. Naunyn-Schmiedeberg's Arch. Pharmacol. 340, 135–138.

Schoeffter, P., and Hoyer, D. (1990). 5-Hydroxytryptamine (5-HT)-induced endothelium-dependent relaxation of pig coronary arteries is mediated by 5-HT receptors similar to the 5-HT$_{1D}$ subtype. J. Pharmacol. Exp. Ther. 252: 387–395.

Schoeffter, P., and Sahin-Erdemli, I. (1990). Further characterization of the 5-HT$_{1-like}$ receptor mediating contraction of guinea-pig iliac artery. Second IUPHAR Satellite Meeting on Serotonin, p. 81.

Sumner, M. J., Feniuk, W., and Humphrey, P. P. A. (1989) Further characterization of the 5-HT receptor mediating vascular relaxation and elevation of cyclic AMP in the porcine isolated vena cava. Br. J. Pharmacol. 97: 292–300.

Sumner, M. J., and Humphrey, P. P. A. (1989). 5-HT$_{1D}$ binding sites in porcine brain can be sub-divided by GR43175. Br. J. Pharmacol. 98: 29–31.

Sumner, M. J., and Humphrey, P. P. A. (1990). Sumatriptan (GR43175) inhibits cyclic-AMP accumulation in dog isolated saphenous vein. Br. J. Pharmacol. 99: 219–220.

Toda, N., and Okamura, T. (1990). Comparison of the response to 5-carboxamidotryptamine and serotonin in isolated human, monkey, and dog coronary arteries. J. Pharmacol. Exp. Ther. 253: 676–682.

Waeber, C., Dietl, M. M., Hoyer, D., Probst, A., and Palacios, J. M. (1988a). Visualization of a novel serotonin recognition site (5-HT$_{1D}$) in the human brain by autoradiography. Neurosci. Let. 88: 11–16.

Waeber, C., Schoeffter, P., Palacios, J. M., and Hover, D. (1988b). Molecular pharmacology of 5-HT$_{1D}$ recognition sites: radioligand binding studies in human, pig, and calf brain membranes. Naunyn-Schmiedberg's Arch. Pharmacol. 337: 595–601.

Waeber, C., Dietl, M. M., Hoyer, D., and Palacios, J. M. (1989a). 5-HT$_1$ receptors in the vertebrate brain. Regional distribution examined by autoradiography. Nauny-Schmiedberg's Arch. Pharmacol. 340: 486–494.

Waeber, C., Hoyer, D., and Palacios, J. M. (1989b). GR 43175: a preferential 5-HT$_{1D}$ agent in monkey and human brains. Synapse 4: 168–170.

Waeber, C., Schoeffter, P., Palacios, J. M., and Hoyer, D. (1989c). 5-HT1D receptors in guinea-pig and pigeon brain: radioligand binding and biochemical studies. Naunyn-Schmiedeberg's Arch. Pharmacol. 340: 479–485.

Xiong, W.-C., and Nelson, D. L. (1989). Characterization of a [3H]-5-hydroxytryptamine binding site in rabbit caudate nucleus that differs from the 5-HT$_{1A}$, 5-HT$_{1B}$, 5-HT$_{1C}$, and 5-HT$_{1D}$ subtypes. Life Sci. 45: 1433–1442.

Serotonin: Molecular Biology, Receptors and Functional Effects
ed. by J. R. Fozard/P. R. Saxena
© 1991 Birkhäuser Verlag Basel/Switzerland

Developmental and Synaptic Regulation of 5-HT$_2$ and 5-HT$_{1c}$ Serotonin Receptors

B. L. Roth, M. W. Hamblin, R. Desai, and R. D. Ciaranello

Nancy Pritzker Laboratory of Developmental and Molecular Neurobiology, Department of Psychiatry, Stanford University School of Medicine, Stanford, CA 94305, USA

Summary. 5-HT$_2$ and 5-HT$_{1c}$ receptors are regulated by a number of exogenous and endogenous factors via biochemical and molecular mechanisms which remain largely uncharacterized. To clarify some of these we investigated the effects of (1) normal perinatal development and (2) acute and chronic receptor blockade on the levels of 5-HT$_2$ and 5-HT$_{1c}$ receptor binding and mRNAs. During rat brain development, 5-HT$_2$ receptors increased 8-fold while 5-HT$_2$ receptor mRNA rise by 13-fold. 5-HT$_{1c}$ receptors displayed a 2.2-fold elevation while 5-HT$_{1c}$ mRNA increased 6-fold. On the basis of these observations, we postulate that transcriptional processes appear to be important for regulating 5-HT$_2$ receptors during development.

Mianserin, administered i.p. for 4, 10 or 21 days caused a 50–70% decrease in brain levels of 5-HT$_2$ and 5-HT$_{1c}$ receptors as assessed by quantitative receptor autoradiography or radioligand binding. A small increase in 5-HT$_{1c}$ receptor mRNA (20–25%) was seen after 10 and 21 days of continuous mianserin treatment while 5-HT$_2$ receptor mRNA levels were unchanged. These results indicate that mianserin decreases 5-HT$_2$ and 5-HT$_{1c}$ receptors without altering steady-state mRNA levels.

Introduction

5-HT$_2$ and 5-HT$_{1c}$ serotonin receptors, as recently summarized (Roth et al. 1991b), may be regulated by many factors including antipsychotic agents, antidepressants, receptor agonists and antagonists and unknown developmentally regulated substances. In general, most exogenous agents cause a decrease in 5-HT$_2$ and 5-HT$_{1c}$ receptor levels (see Table I). Thus, antipsychotics such as clozapine (Matsubara and Meltzer 1989), antidepressants like amitryptiline (Kellar et al. 1981), receptor agonists such as DOI ([1-(2,5-dimethoxy-4-indophenyl]-2-aminopropane) and DOM (1-[2,5-dimethoxy-4-methylphenyl]-2-aminopropane) (Roth et al. 1991b) and receptor antagonists like ketanserin (Gandolfi et al. 1985) all *decrease* the numbers of 5-HT$_2$ receptors. With the exception of DOI and DOM, all of the known agents which decrease the number of brain 5-HT$_2$ receptors function as receptor antagonists both *in vivo* and *in vitro*.

5-HT$_2$ and 5-HT$_{1c}$ receptors may also be regulated by other exogenous agents. For example, depleting brain 5-HT levels with parachlorophenylananine (PCPA) but not with 5,7-dihydroxytryp-

Table 1. Regulation of 5-HT$_2$ receptors compared with other receptor classes

Agent/class	Expected	Measured
Exogenous agents which decrease 5-HT$_2$ receptor binding:		
DOI, DOM/agonist	Decrease	Decrease
Ketanserin/Antagonist	Increase	Decrease**
Mianserin/Antagonist, Anitdepressant	Increase	Decrease**
Clozapine/Antagonist, Antipsychotic	Increase	Decrease**
Exogenous agents which increase 5-HT$_2$ receptor binding:		
p-chloro-phenylalanine/5-HT synthesis inhibition	Increase	Increase
Exogenous agents expected to alter 5-HT$_2$ receptor binding which do not:		
Fluoxetine/5-HT uptake inhibitor	Decrease	No change*, **
5,7-DHT/5-HT neurotoxin	Increase	No change**
Endogenous conditions which increase 5-HT$_2$ receptor binding:		
Post-natal development	Increase	Increase
Violent suicide	Unknown	Increase
Endogenous conditions which decrease 5-HT$_2$ receptor binding:		
Schizophrenia	Unknown	Decrease
Alzheimer's disease	Unknown	Decrease

These results represent a summary of many reports (reports in Roth et al. 1991b) which demonstrate the effects of various experimental manipulations on the 5-HT$_2$ receptor binding. The first column indicates in which direction we predict the 5-HT$_2$ receptor should change; the second column shows the measured alterations.
*although 5-HT$_2$ receptors are not changed, 5-HT$_2$-stimulated PI metabolism is decreased.
**these results are opposite to what is predicted from studies with other neurotransmitter receptors.

tamine (5,7-DHT) (Brunello et al. 1982; Roth et al. 1987) *increases* cortical 5-HT$_2$ receptors. Preliminary findings suggest that cortical 5-HT$_{1c}$ receptors may be increased by neonatal 5,7-DHT treatment (Pranzatelli 1990). Augmenting the synaptic levels of 5-HT using 5-HT-uptake inhibitors does not alter the number of 5-HT$_2$ receptors, but does *decrease* the ability of 5-HT to activate phosphoinositide (PI) metabolism (Conn and Sanders-Bush 1987). These results are contrary to our preconceived notions of receptor regulation and serve to highlight the differences between 5-HT$_2$ and other neurotransmitter receptors (see Table I). Thus, for instance, we might have expected that receptor antagonists would *increase* 5-HT$_2$ receptor levels, while increasing synaptic levels of 5-HT would *decrease* 5-HT$_2$ receptors.

5-HT$_2$ and 5-HT$_{1c}$ receptors are also altered by undefined endogenous factors. Brain levels of 5-HT$_2$ receptors are increased during post-natal development in rats (Roth et al. 1991a) and among individuals who suffer violent deaths from suicide (Stanley and Mann 1983). 5-HT$_2$ receptors are decreased in Alzheimer's disease and in schizophrenia (Arora and Meltzer 1991). There is also evidence for altered 5-HT$_2$ receptor sensitivity in obsessive-compulsive disorder (Bastiani et al. 1990), following clozapine treatment of schizophrenia and in certain depressed patients (see Lesch et al. 1990; Meltzer and Lowy 1987). Taken together, all of these results suggest that alterations in 5-HT$_2$, and possibly 5-HT$_{1c}$ receptors, may be important for many pharmacological, developmental, psychological and pathophysiological events. The molecular and biochemical details responsible for these changes remain largely undefined. Insights into the mechanisms by which these differences occur could shed light on a number of important processes.

Accordingly, we began a series of studies aimed at discovering the molecular and biochemical mechanisms responsible for regulating 5-HT$_2$ and 5-HT$_{1c}$ receptor levels. We initially focused on two experimental systems: (1) perinatal development and (2) the effects of receptor antagonists. We studied these two systems because each has predictable and reproducible effects and hence could serve as a model system for investigating the roles of exogenous (receptor antagonists) and endogenous (developmentally regulated) factors. We found that during development transcriptional processes can play a major part in regulating 5-HT$_2$ and 5-HT$_{1c}$ receptors. Receptor antagonists like mianserin, on the other hand, appear to alter 5-HT$_2$ and 5-HT$_{1c}$ receptors by post-transcriptional means.

Methods

Animals: Time-pregnant Sprague-Dawley rats were sacrificed at embryonic days E17 and E19; embryos were dissected out and their brains removed for receptor and mRNA quantification, and stored at $-80°C$ until assay. At several postnatal days, (P2, P5, P13, P21, P27 Adult) rats were killed by decapitation and their brains removed.

In other experiments, adult male Sprague-Dawley rats were injected with either mianserin (15 mg/kg) or vehicle (saline) for 4, 10 or 21 days and then, 18 hr after the last injection, decapitated and their brains removed for mRNA and receptor measurements.

DNA Probes and mRNA Analysis: For these studies the following probes were constructed and used (see Roth et al. 1991a for structures). Unless otherwise specified, each cDNA was subcloned into pBluescript KS−: (1) pRACHI which contains the complete cDNA for cyclophilin; (2) pSR1CΔ3′ which contains bp 1735–2900 of the 5HT1c cDNA; (3) pRACH8 which contains an AluI exonic fragment of a 5′-directed 5-HT$_2$ genomic clone (λRR13) (bp 635–913). In some cases, a synthetic oligonucleotide complementary to bp 811–862 of the 5-HT$_2$ DNA sequence was used for Northern blot analysis. cDNA probes were labelled by the random-primer technique while oligonucleotide probes were end-labelled (Roth et al. 1991a).

For Northern blots, total RNA was isolated by the guanidinium-isothiocyanate CsCl gradient technique as previously detailed (Chirgwin

36

et al. 1979; Roth et al. 1991a). Total RNA was then prepared for Northern blots and run as described by Davis (Davis et al. 1986). RNA was
then transferred to Duralon-UV membranes, cross-linked by UV irradi-
ation and pre-hybridized at 42°C (6X SSC, 2X Denhardt's, 1% SDS,
50% formamide, 10% dextran sulfate, 500 μg/ml ssDNA, 250 μg/ml
yeast tRNA). After hybridization overnight in the same buffer with
1×10^6 cpm/ml labelled probe, the membranes were washed as follows:
2X SSC/0.1% SDS (1 h; room temp) and then either 1X SSC/0.1% SDS
@ 55°C (oligonucleotide probes) or 0.1% SSC/0.1% SDS @ 55°C
(plasmid-derived probes). After exposure to X-Ray film (X-OMAT) with
Cronex screens, mRNA was quantified by scanning laser densitometry.

Radioligand Binding: 5-HT$_2$ receptors were quantified with either
[^{3}H]-ketanserin (Roth et al. 1987; Roth et al. 1991a) or [^{125}I]-LSD
under site-selective conditions. 5-HT$_{1c}$ receptors were measured with
[^{125}I]-LSD in the presence of 50 nM spiperone to block binding to D$_2$
and 5-HT$_2$ receptors. Data were analyzed using the LIGAND program
of Munson and Rodbard (1980). Protein was determined by the method
of Bradford (1976).

Results

*Developmental Regulation of 5-HT$_2$ and 5HT$_{1c}$ Receptors and
mRNAs:* Preliminary studies showed that our methods easily distinguished between 5-HT$_2$ and 5-HT$_{1c}$ receptors and mRNAs. We found
that 5-HT$_2$ receptor mRNA was enriched in cerebral cortex and striatum while 5-HT$_{1c}$ receptor mRNA was found in choroid plexus, ventral
hippocampus, pons-medulla and other brain regions (not shown).

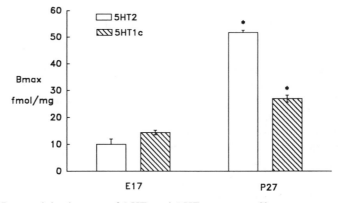

Figure 1. Postnatal development of 5-HT$_2$ and 5-HT$_{1c}$ receptors. Shown are mean $\pm$ sem of
4 independent determinations of 5-HT$_2$ (white bars) and 5-HT$_{1c}$ (cross-hatched bars) receptor
Bmax for Embryonic day 17 (E17) and postnatal day 27 (P27). *$p < 0.01$ comparing E17 and
P27.

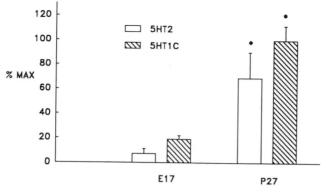

Figure 2. Postnatal development of 5-HT$_2$ and 5-HT$_{1c}$ receptor mRNAs. Shown are mean $\pm$ sem of 4 independent determinations of 5-HT$_2$ (white bars) and 5-HT$_{1c}$ (cross-hatched bars) receptor mRNAs for E17 and P27. Data is expressed as percentage maximum mRNA. *$p < 0.01$ comparing E17 and P27.

Figure 1 shows the developmental time-course of 5-HT$_2$ and 5-HT$_{1c}$ receptors. mRNA changes are shown in Figure 2. 5-HT$_2$ receptor mRNA increases 13-fold while receptor levels change by a factor of 8 during this time period. 5-HT$_{1c}$ mRNA shows a 6-fold augmentation during development while 5-HT$_{1c}$ receptors increases 2-fold. We also measured the mRNA for cyclophilin as a control since this gene has been reported not to be developmentally regulated (Milner et al. 1987). We found this to be technically true, but observed that cyclophilin mRNA rises modestly (70%) between P2 and P13 and then declines to adult levels by P27.

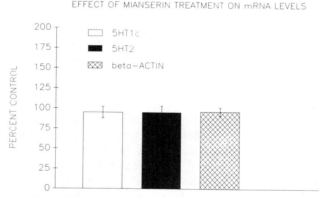

Figure 3. Effect of mianserin treatment on 5-HT$_2$ and 5-HT$_{1c}$ receptor mRNA levels. Rats were treated with either mianserin (15 mg/kg ip) or vehicle (saline) for 4 days and then 18 hr after the last injection brains removed for receptor and mRNA determinations. 5-HT$_2$ and 5-HT$_{1c}$ receptors were decreased by 50–70% ($p < 0.01$ vs control). Data shows mean $\pm$ sem of 8 independent determinations for the 5-HT$_2$ receptor mRNA, 5-HT$_{1c}$ receptor mRNA and, as a control, β-actin mRNA. No significant differences were found for mRNA levels.

Effect of Mianserin on 5-HT$_2$ and   5-HT$_{1c}$ Receptors and mRNAs:
Mianserin causes a 50–56% decrease in the levels of 5-HT$_2$ receptors as
assessed by radioligand binding after 4 days of continuous treatment
($p < 0.01$ vs control; data not shown). No changes in 5-HT$_2$, 5-HT$_{1c}$, or
β-Actin mRNAs were measured after 4 days of mianserin treatment
($p > 0.05$ vs control; Figure 3). These results demonstrate that mianserin
decreases 5-HT$_2$ receptor levels without altering steady-state 5-HT$_2$
mRNA levels.

Discussion

The major findings of these studies are that, first, developmentally
induced alterations in 5-HT$_2$ and 5-HT$_{1c}$ receptors are likely to result
from increased levels of receptor mRNAs. Secondly, mianserin, a prototypical receptor antagonist, alters 5-HT$_2$ and 5-HT$_{1c}$ levels without
altering steady-state receptor mRNAs. These results imply that a number of biochemical and molecular events might be important for regulating 5-HT$_2$ and 5-HT$_{1c}$ receptors *in vivo*. Each of these results will be
discussed in turn in relation to the existing literature and then summarized into a coherent model of 5-HT$_2$/5-HT$_{1c}$ receptor regulation.

The Significance of Developmentally Regulated 5-HT2 receptors: We
were surprised by the fact that 5-HT$_2$ receptors, which are predominantly present in the cerebral cortex (Roth et al. 1987; Pazos et al.
1985), display striking developmental expression over a relatively restricted time period. We could not reliably quantify 5-HT$_2$ receptors
with [^{3}H]-ketanserin prior to post-natal day 5 (P5). Even using [^{125}I]-
LSD under site-specific conditions, we found less than 10 fmol/mg
protein of 5-HT$_2$ receptors prior to P2. Between P2 and P13, 5-HT$_2$
receptor increased 8-fold while a concomitant 13-fold increase in receptor mRNA was measured. These results suggest that a burst of receptor
gene expression occurs during the immediate perinatal period in rat
brain. This period roughly corresponds to the period of synaptogenesis
in the rat.

Over this time period the cytoarchitecture of the cortex changes
dramatically. At embryonic day 17 (E17), at which time few 5-HT$_2$
receptors are measured, the cortex exists as a primordium with little
cellular specialization. Some ascending serotonergic axons have been
described to reach the medial and lateral aspects of the neocortex
(Lidov and Molliver 1982). By E19, the cortex has specialized into
cortical plate, subplate and marginal zones. At this stage of development, serotonergic axons reach the pallium but no axons are found in
the parietal, dorsofrontal or occipital cortices (Lidov and Molliver
1982). Within the cortical plate, a bilaminar pattern is seen though little
5-HT$_2$ receptor gene expression is noted. No terminal arborization of

the serotonin terminals is seen in cortex (Lidov and Molliver 1982) by E19. Around P2, where 5-HT$_2$ receptor mRNA shows a burst of expression, the density of serotonergic axons has greatly increased (Lidov and Molliver 1982). Between P2 and P13, the time period in which the greatest increases in 5-HT$_2$ receptors and mRNA levels occur, intense serotonergic innervation of the cortex is evident. This is also the time in which terminal arborization and final specialization of cortical layers occurs in the rat (Hicks and D'Amato 1968). Also during this time period a hyperinnervation of the parietal cortex has been described in the rat (D'Amato et al. 1987). The time-course of 5-HT$_2$ receptor expression appears, then, to roughly correlate with serotonergic innervation of the cerebral cortex.

These results suggest two possible mechanisms for regulating the levels of 5-HT$_2$ receptors during development. First, levels of serotonin and the development of functional serotonergic synapses could be important for 5-HT$_2$ gene expression. The second is that serotonergic innervation and 5-HT$_2$ receptor gene expression occur independently but that each is coordinately regulated. The first hypothesis would predict that ablation of serotonin terminals (with 5,7-DHT) or blocking 5-HT$_2$ receptors (with mianserin) should decrease subsequent 5-HT$_2$ receptor gene expression. Previous studies by Sachs and colleagues measuring [^{3}H]-LSD binding after neonatal 5,7-DHT lesions would suggest that endogenous 5-HT and/or the presence of intact 5-HT terminals does not alter subsequent maturation of 5-HT$_2$ receptor binding (Jonsson et al. 1978). Also, our preliminary findings with mianserin treatment during ontogeny suggest that occupation of 5-HT$_2$ receptors by an antagonist does not alter subsequent 5-HT$_2$ receptor gene expression. These results imply that 5-HT$_2$ receptor gene expression may be regulated by factors other than serotonergic innervation, and would support the second hypothesis. We are currently carrying out additional studies to test these hypotheses further.

All of these findings suggest that regulation of 5-HT$_2$ receptor gene expression may be important for developmentally induced changes in 5-HT$_2$ receptors. It is important, though, not to forget that processes other than transcriptional ones could be regulating the levels of 5-HT$_2$ mRNA during development. As outlined in Figure 4, these include: (1) control at the level of mRNA splicing; (2) control via post-transcriptional modifications; (3) and regulation of mRNA stability and translational efficacy. To demonstrate a change in 5-HT$_2$ receptor gene expression will require nuclear run-on experiments from cortical nuclei prepared from rats of different developmental ages.

Recent studies by Julius have suggested that both the 5-HT$_{1c}$ and the 5-HT$_2$ receptors can transduce mitogenic signals when transfected into certain cell lines. This is not particularly surprising since 5-HT has been known for many years to be a mitogen. Protein kinase C activation by

40

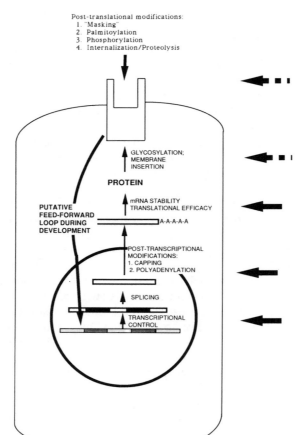

Figure 4. Possible molecular mechanisms for regulating 5-HT$_2$ and 5-HT$_{1c}$ receptor levels. This figure shows various points at which regulation could occur for the 5-HT$_2$ and 5-HT$_{1c}$ receptors. The solid arrows represent transcriptional and post-transcriptional processes; the dashed arrows represent post-translational processes. The long arrow from the receptor to the nucleus indicates a possible feedback loop from receptor to transcription. At present, we have no evidence which conclusively implicates any molecular process, though, as stated in the text, developmental regulation appears to proceed via alterations in mRNA levels.

phorbol esters and other agents has been shown to be mitogenic as well. Since both 5-HT$_2$ and 5-HT$_{1c}$ receptors, are coupled to PI metabolism and activate protein kinase C, it is expected that the 5-HT$_2$ and 5-HT$_{1c}$ receptors might serve as mitogenic stimuli in cell lines. Because the 5-HT$_2$ receptors are expressed around the time period corresponding to the "brain growth spurt", we might hypothesize that neonatal 5-HT$_2$ receptors serve to promote cortical growth during this developmental epoch.

Synaptic Regulation of 5-HT$_2$ Receptors: As has been recently reviewed (Roth et al. 1991b), a large number of agents can alter brain levels of 5-HT$_2$ and 5-HT$_{1c}$ receptors (see also Table I). Many of these agents act as receptor antagonists or agonists and serve to induce a decrease in the level of 5-HT$_2$ and 5-HT$_{1c}$ receptors. The molecular means by which this occurs is unknown but could result from transcriptional, post-transcriptional and post-translational modifications (e.g. receptor internalization, proteolytic degradation, phosphorylation, covalent modification and "masking"; see Figure 4). It is conceivable, for example, that agonists and antagonists serve to decrease transcription of 5-HT$_2$ receptor mRNA. As we have discovered, though, in the case of mianserin, no changes in the levels of 5-HT$_2$ or 5-HT$_{1c}$ mRNA occur after acute or chronic mianserin treatment. Consistent with our findings is the notion that mianserin covalently binds with the receptor to "mask" the recognition site. Other possibilities (see Figure 4) are, of course, possible. On the other hand, preliminary findings obtained in collaboration with Meltzer's group suggest that facilitating serotonergic transmission may induce a decrease in 5-HT$_2$ receptors and mRNA (B. L. Roth, C. Stockmeyer and H. Y. Meltzer, unpublished observations).

Taken together, these findings suggest that synaptic regulation of 5-HT$_2$ receptors may operate at the level of transcription, translation, and post-translational processes (see Figure 4). Sorting out the effects of various drugs on the rates of transcription and post-translational modification will require, though, that new reagents and methodologies be applied to 5-HT$_2$ receptor studies. In particular, the development of antibodies specific for the 5-HT$_2$ receptor will facilitate mechanistic studies on 5-HT$_2$ receptor regulation. Production of such antibodies is currently in progress in this laboratory. As well, the isolation of genomic clones which carry regulatory sequences for *trans*-acting factors will assist studies aimed at elucidating the tissue-specific factors important for regulating 5-HT$_2$ gene expression. We have recently isolated two 5'-directed genomic clones for the 5-HT$_2$ receptor and are characterizing them in preparation for a search for endogenous *trans*-acting factors which regulate 5-HT$_2$ receptor gene expression.

In conclusion, our studies have demonstrated potential regulation of 5-HT$_2$ and 5-HT$_{1c}$ receptors by two biochemical processes: mRNA regulation and post-translational modification. Further studies should clarify the precise molecular details responsible for these receptor alterations. These studies will enhance our knowledge of serotonin receptor regulation and could illuminate the underlying mechanisms by which certain psychoactive drugs act.

Acknowledgements

Supported by a grant from the NIMH (MH 39437) and the John Merck and Spunk Funds. BLR was supported in part by a Dana Foundation Research Fellowship. RDC is the recipient of a Research Scientist Award from the NIMH (MH 00219).

42

References

Arora, M., and Meltzer, H. Y. (1991). Altered 5HT2 receptors in schizophrenia. J Neurotrans. (in press).

Bastiani, M., Nash, J. F., and Meltzer, H. Y. (1990). Prolactin and cortisol responses to MK-212, a serotonin agonist, in obsessive-compulsive disorder. Arch. Genl. Psych. 47: 833–839.

Bradford, M. M. (1976). A rapid and sensitive method for the quantitation of microgram quantities of protein utilizing the principle of protein-dye binding. Anal. Biochem. 72: 248–254.

Brunello, N., Chaung, D-M., and Costa, E. (1982). Different synaptic location of mianserin and imipramine binding sites. Sci. 215: 1112–1115.

Chirgwin, J. M., Przybyla, A. E., MacDonald, R. J., and Rutter, W. J. (1979). Isolation of biologically active ribonucleic acid from sources enriched in ribonuclease. Biochem. 18: 5294–5299.

Conn, P. J., and Sanders-Bush, E. (1987). Central serotonin receptors: effector systems, physiological roles and regulation. Psychopharmacol. 92: 267–277.

D'Amato, R. J., Blue, M. E., Largent, B. L., Lynch, D. R., Ledbetter, D. J., Molliver, M. E., and Synder, S. H. (1987). Ontogeny of the serotonergic projection to rat neocortex: transient expression of a dense innervation to primary sensory areas. Proc. Natl. Acad. Sci. USA. 84: 4322–4326.

Davis, L. G., Dibner, M. D., and Battey, J. F. (1986). Basic Methods in Molecular Biology. New York: Elsevier Science Publishing, pp 129–152.

Gandolfi, O., Barbaccia, M. L., and Costa, E. (1985). Different effects of serotonin antagonists on [^{3}H]-mianserin and [^{3}H]-ketanserin recognition sites. Life Sci. 36: 713–721.

Hicks, S. P., and D'Amato, R. (1968). Cell migrations to the isocortex of the rat. Anat Rec. 160: 619–634.

Jonsson, G., Pollare, T., Hallman, H., and Sachs, C. (1978). Developmental plasticity of central serotonin neurons after 5,7-dihydroxytryptamine treatment. Ann. NY. Acad. Sci. 328–345.

Kellar, K. J., Cascio, C. S., Butler, J. A., and Kurtzke, R. N. (1981). Different effects of electroconvulsive shock and antidepressant drugs on serotonin receptors in rat brain. Eur. J. Pharmacol. 69: 515–518.

Lesch, K. P., Mayer, S., Disselkamp-Tietze, J., Hoh, H., Wiesmann, M., Osterheider, M., and Schulte, H. M. (1990). 5-HT1a receptor responsivity in unipolar depression: evaluation of ipsapirone-induced ACTH and cortisol secretion in patients and controls. Biol. Psych. 28: 620–628.

Lidov, H. G. W., and Molliver, M. E. (1982). An immunohistochemical study of serotonin neuron development in the rat: ascending pathways and terminal fields. Brain Res. Bull. 8: 389–430.

Matsubara, S., and Meltzer, H. Y. (1989). Effect of typical and atypical antipsychotic drugs on 5-HT2 receptor density in rat cerebral cortex. Life Sci. 45: 1397–1406.

Meltzer, H. Y., and Lowy, M. T. (1987). The Serotonin hypothesis of depression, Meltzer, H. Y., Psychopharmacology: the third generation of progress, New York.

Milner, R. J., Bloom, F. E., and Sutcliffe, J. G. (1987). Brain specific genes: Strategies and issues. Cur. Top. Dev. Bio. 21: 117–150.

Munson, P. J., and Rodbard, D. (1980). LIGAND: a versatile computerized approach for characterization of ligand binding systems. Anal. Biochem. 107: 220–239.

Pazos, A., Cortes, A., and Palacios, J. M. (1985). Quantitative autoradiographic mapping of serotonin receptors in rat brain. II. Serotonin-2 receptors. Brain Res. 34: 231–249.

Pranzatelli, M. R. (1990). Neonatal 5,7-DHT lesions alter [3H]-mesulergine-labelled 5HT1c receptors in rat brain. Soc. Neurosci. Abstr. 16.

Roth, B. L., Hamblin, M. W., and Ciaranello, R. D. (1991). Developmental regulation of 5HT2 and 5HT1c mRNA and receptor levels. Devl. Brain Res. 58: 51–58.

Roth, B. L., Hamblin, M. W., and Ciaranello, R. D. (1990). Regulation of 5HT$_2$ and 5HT$_{1c}$ Serotonin Receptor Levels: Methodology and Mechanisms. Neuropscychopharm. 3: 427–433.

Roth, B. L., McLean, S., Zhu, X. Z., and Chaung, D. M. (1987). Characterization of two [^{3}H]-ketanserin recognition sites in rat striatum. J. Neurochem. 49: 1833–1838.

Stanley, M., and Mann, J. J. (1983). Increased serotonin-2 binding sites in frontal cortex of suicide victims. Lancet 1: 214–216.

Serotonin: Molecular Biology, Receptors and Functional Effects
ed. by J. R. Fozard/P. R. Saxena
© 1991 Birkhäuser Verlag Basel/Switzerland

Serotonin Trophic Factors in Development, Plasticity and Aging

P. M. Whitaker-Azmitia[1] and E. C. Azmitia[2]

[1]Department of Psychiatry, State University of New York, Stony Brook, NY 11794, USA;
[2]Department of Biology, New York University, NY 10003, USA

Summary. Serotonin has long been known to function as a regulator of brain development, prior to assuming its role as a neurotransmitter in the mature brain. In addition to acting as a trophic factor on other neuronal populations, serotonin also regulates its own development. Serotonin neurons are also one of the most plastic neurotransmitter systems in the brain – that is after lesioning of terminals, the serotonin neuron has the capacity to regenerate them. In order for the phenomena of plasticity to take place, some of the trophic (or growth-promoting) molecules present during development, must be present even in the adult brain.

In the adult mammalian brain, serotonergic fibers are the most widely distributed single neurotransmitter system. Obviously, the mechanisms by which serotonin develops from a small collection of cell bodies in a relatively restricted region in the brainstem, into such an expansive distribution network must be highly sophisticated and specific.

These developmental signals may be present in the adult brain. This can be surmised from the fact that serotonin neurons, when lesioned, are one of the most plastic neurotransmitter systems known. Simply stated, this means that severed serotonin terminals will, in time, regrow and make appropriate functional connections. In order for the phenomena of plasticity to take place, therefore, there must be some trophic (or growth-promoting) molecules present, even in the adult brain.

Since serotonin is itself a developmental signal, regulating the maturation and differentiation of many other types of neurons, any alterations in the serotonin system in aging or development would also result in changes in other systems.

This chapter will summarize our work on factors which we have found to influence the development or the adult plasticity of serotonin neurons. It has recently been proposed that aging is, in a sense, development in reverse – development is a process by which positive and negative factors are in a balance such that growth takes place. Aging is a process in which the balance is altered such that the negative or toxic factors predominate over the trophic factors. With this hypothesis in mind, the final section will summarize what our findings may imply for models of serotonin in aging.

Trophic Factors

S-100β and Other Neuronal Growth Factors: Astroglial cells have long been known to be a source of factors which are trophic to the development of neurons. In both tissue culture and whole animals studies, we found evidence indicating that stimulation of a high affinity serotonin receptor was trophic to the development of serotonin neurons (Whitaker-Azmitia et al. 1990a). To test the hypotheses that the trophism may be through an astroglial-derived factor, we exposed immature astroglial cells in culture to various agonists of high affinity serotonin receptors. This glial-conditioned media (GCM) was then screened for serotonergic growth factors by adding it to the culture media of serotonin neurons derived from 14-day fetal rat brain. Using this approach, we found that the $5-HT_1$ receptor was responsible for the release and/or production of astroglial specific protein S-100β (Whitaker-Azmitia et al. 1990b).

The exact function of S-100β in the central nervous system has been unclear, but recently there have been several important new discoveries. S-100β can stimulate neurite growth of serotonergic and cortical cells (Azmitia et al. 1990a; Kligman and Marshak 1985). The developmental pattern of S-100β is such that its expression peaks at the right time and place to have an influence on developing serotonin neurons (Van Hartsveldt et al. 1986). The gene for S-100β was found to be localized to chromosome 21 in the region considered obligate for Down's Syndrome (Allore et al. 1988) – a disease long thought to be associated with altered development of the serotonin system.

The receptor which releases S-100β, $5-HT_{1A}$, is one of the receptors which is transiently expressed during development at very high levels and then decreases as the brain matures (Daval et al. 1987) consistent with our findings that this receptor disappears from astrocytes as they mature (Whitaker-Azmitia and Azmitia 1986).

In studies of adult plasticity, we have found that selective 5,7-dihydroxytryptamine (5,7-DHT) lesions of the serotonergic fibers that innervate the hippocampus result in sprouting of undamaged serotonergic fibers, while neither noradrenergic nor cholinergic fibers sprout (see Azmitia et al. 1990b). Similarly, transplantation of fetal serotonergic or noradrenergic neurons into the 5-HT denervated hippocampus results in a 400% increase in serotonin levels but no change in noradrenaline neurons. This suggests that the trophic signal is specific to the growth of serotonin neurons, that it is present in fetal as well as adult tissue, and that both fetal and adult neurons respond to it. In subsequent experiments, we have found that the 5,7-DHT lesion results initially in a decrease in S-100β in the hippocampus, but that stimulation with a 5-HT$_{1A}$ agonist, such as ipsapirone, causes an increase in S-100β. This increase itself has no effect on release of S-100β in control animals (Azmitia et al. 1990c). Therefore, our *in vivo* results indicate that 5-HT fibers autoregulate the secretion of their own growth factor by activating the 5-HT$_{1A}$ on astrocytes which in turn produce S-100β.

This finding offers a unique example of a neurotransmitter system regulating its own development and plasticity by receptor activation. However, the story appears to be more complex since S-100β is a known cortical growth factor with potent regulatory effects on cortical plasticity in neonatal kittens. The many observations of 5-HT having a function as a developmental factor on forebrain maturation would be consistent with the observed 5-HT/astrocyte interaction.

We have also tried a number of other known trophic factors and found them to have no effect on the development of serotonin neurons in culture. Amongst those tested and found to have no effect were nerve growth factor (NGF), insulin, epidermal growth (EGF) and calmodulin. NGF is also ineffective in promoting adult plasticity of serotonergic neurons.

Laminin: Laminin is a component of the extracellular matrix, which is released by immature astrocytes. In tissue culture studies of developing neurons, it was found that laminin served to strongly attach neurons to the surface and permit neurite extension (Rogers et al. 1983). Laminin is synthesized by astrocytes and transposed to external membrane. Here it serves to anchor neurites and both facilitate and guide growth. Laminin plays an important role during early development, but appears to disappear from the brain during maturation. This absence of an adhesive molecule has been proposed to explain the decreased sprouting seen in the mature brain (Liesi 1985). In the adult brain, microinjections of purified laminin serves to stimulate and guide the outgrowth of fetal serotonergic neurons transplanted into a number of sites (Zhou and Azmitia 1988).

Thus astrocyte products both promote (by S-100β) and guide (by laminin) the growth of serotonergic fibers in the adult brain.

Other Neurotransmitters and Neuromodulators: The effects of adrenal steroids and adrenocorticotrophin hormone (ACTH) on serotonin development and plasticity, has been a focus of our work for some time (see Azmitia et al. 1990b). If rat pups are treated prenatally with ACTH, there is an increase in serotonin terminal density in the hippocampus up to 21 days postnatally. On the other hand, if adrenal steroids are removed by bilateral adrenalectomy, the normal developmental rise of tryptophan hydroxylase is suppressed (Sze 1980). In our tissue culture model ACTH, and its fragment $ACTH_{4-10}$ are stimulatory to growth.

Adrenal steroids also play an important role in regulation of serotonin synthesis and turnover in the adult brain (DeKloet et al. 1982) and in adult plasticity. If rats are adrenalectomized prior to receiving a 5,7-DHT lesion, the normal sprouting does not take place.

Inhibitory Factors

Dopamine: One of the best known interactions between developing neurotransmitter systems is that which has been reported for the interactions between dopamine and serotonin. If neonatal rat pups are treated with 6-hydroxydopamine to remove dopamine terminals, the serotonin terminals overgrow, principally in the caudate. Since D_1 dopamine receptors have been shown to be inhibitory to growth (Lankford et al. 1988) and since serotonin terminals have D_1 receptors, we proposed that during development dopamine is tonically inhibitory to the development of serotonin neurons, possibly through the D_1 receptor. To test this hypotheses, we tested the response of serotonin cultures to the dopamine agonist SKF 38393, as well as treating rats prenatally with this agonist. In both the tissue culture model and the whole animals, SKF 38393 was found to decrease serotonin outgrowth. In the whole animal studies, even as adults these rats were permanently impaired in their serotonin neurochemistry and in their responses to the non-selective serotonin agonist quipazine (Whitaker-Azmitia et al. 1990c).

Serotonin Release: In tissue culture models, both MDMA and fenfluramine are inhibitory to development of the serotonin neuron. This effect can be inhibited by the calcium channel antagonist nimodipine, suggesting a role for calcium in the toxicity (Azmitia et al. 1990d). Immature serotonin neurons have been shown to have an active calcium channel, which remains but becomes inactive as the neuron matures. In adult animals, MDMA and possibly fenfluramine are also toxic to the serotonin terminal. The exact mechanism of this toxicity is unknown but it is interesting to speculate on whether or not it is through the sodium/calcium channels becoming re-activated.

$5-HT_3$ Receptors: In our original investigations of the role of serotonin receptors in regulating development, we found that two high affinity serotonin receptors were involved in the development of serotonin neurons themselves. One receptor stimulated outgrowth while the other inhibited it. We termed this phenomena autoregulation of development. We now believe that the serotonin receptor responsible for the trophic actions of serotonin is the $5-HT_{1A}$ receptor, through the release of S-100β, as described above.

The receptor responsible for the inhibitory effect has not been so easily characterized, possibly because in many cases immature receptors do not necessarily have the same pharmacological profile as their adult counterparts. We have tested a number of drugs in both the tissue culture model and the whole animal model, and found them to have limited effects. This included serotonin uptake inhibitors (fluoxetine), $5-HT_{1B}$ receptor agonists (TFMPP, mCPP) and $5-HT_2$ agonists (DOI). However, we did find interesting results using the $5-HT_3$ receptor active drugs 2-methylserotonin, phenylbiguanide and MDL 72222.

Our results with $5-HT_3$ receptor active drugs are the first studies we have done which show a change in serotonin innervation densities into the spinal cord. In the tissue culture studies of ascending raphe nuclei, 2-methylserotonin was toxic to the developing serotonin neurons. Using immunocytochemistry, the same appears to be true in the hippocampus of aminals treated prenatally with phenylbiguanide. However, the effect in the spinal cord appears to be the opposite – an increase in outgrowth after prenatal treatment with phenylbiguanide. This was observable both with a test of nocioception (tail flick latency) and in our preliminary findings with ^{3}H-paroxetine binding. This may be an example of a developmental phenomena referred to as "pruning" – if terminals are prevented from growing into one area they will overgrow elsewhere.

Significance of Findings to Aging

Since aging may in part arise from lack of trophic factors, it may be important to evaluate our findings in development and plasticity as a model in aging.

In Alzheimer's patients, there is a selective loss of serotonin terminals in cortex and hippocampus, but not elsewhere. This loss is correlated with senile plaque density. These regions also lack $5-HT_{1A}$ receptor (Cross 1990). Interestingly, hippocampus and cortex are two regions which contain high levels of S-100, presumably under the regulation of a $5-HT_{1A}$ receptor. Thus alteration in the $5-HT_{1A}$ receptor in Alzheimer's disease would lead to a decrease in the release of S-100 (thus explaining the post-mortem results of increased S-100 within cells (Griffin et al. 1989)), degeneration of serotonin terminal and of the hippocampal and cortical neurons. This hypothesis is a focus of our current studies.

48

Acknowledgements

The authors gratefully acknowledge the input of their many collaborators including, Ann Shemer, Randy Murphy, Hoomayoon Akbari, Xiao Ping Hou and James Bell. The authors are also deeply indebted to the National Institute for Child Health and Human Development (PMW-A), the National Institute for Neurological Disease and Stroke (PMW-A), National Institute on Drug Abuse (ECA) and the National Science Foundation (ECA) for financial support.

References

Allore, R., O'Hanlon, D., Price, R., Nielsen, K., Willard, H. F., Cox, D. R., Marks, A., and Dunn, R. J. (1988). Gene encoding the beta subunit of S-100 protein is on chromosome 21: Implications for Down Syndrome. Science 239: 1311–1313.

Azmitia, E. C., Dolan, K., and Whitaker-Azmitia, P. M. (1990a). S-100$_B$ but not NGF, EGF, insulin or calmodulin functions as a CNS serotonergic growth factor. Brain Res. 516: 354–356.

Azmitia, E. C., Frankfurt, M., Davila, M., Whitaker-Azmitia, P. M., and Zhou, F. (1990b). Plasticity of adult CNS neurons: role of growth-regulating factors. Ann. N. Y. Acad. Sci. 600: 343–365.

Azmitia, E. C., Hou, X. P., Chen, Y., Marshak, D., and Whitaker-Azmitia, P. M. (1990c). 5-HT$_{1A}$ receptors regulate hippocampal S-100, a 5-HT and cortical growth factor. Soc. Neurosci. Abs. 16: 74.

Azmitia, E. C., Murphy, R., and Whitaker-Azmitia, P. M. (1990d). Nimodipine reversal of MDMA (Ecstasy)-induced degeneration of serotonergic neurons in culture. Brain Res. 510: 97–103.

Cross, A. J. (1990). Serotonin in Alzheimer-type dementia and other dementing illnesses. Ann. New York Acad. Sci. 600: 405–417.

Daval, G., Verge, D., Becerril, A., Gozlan, H., Spampinato, U., and Hamon, M. (1987). Transient expression of 5-HT$_{1A}$ receptor binding sites in some areas of the rat CNS during postnatal development. International J. Develop. Neurosci. 5: 171–180.

De Kloet, E. R., Kovacs, G. L., Szabo, G., Telegfy, G., Bohus, B., and Versteeg, D. H. G. (1982). Decreased serotonin turnover in the dorsal hippocampus of rat brain shortly after adrenalectomy: selective normalization after corticosterone substitution. Brain Res. 239: 659–663.

Griffin, W. S. T., Stanley, L. C., Ling, C., White, L., MacLeod, V., Perrot, L. J., White, C. L., and Araoz, C. (1989). Brain interleukin 1 and S-100 immunoreactivity are elevated in Down Syndrome and Alzheimer Disease. Proc. Natl. Acad. Sci. 86: 7611–7615.

Kligman, D., and Marshak, D. (1985). Purification and characterization of a neurite extension factor from bovine brain. Proc. Natl. Acad. Sci. 82: 7136–7139.

Lankford, K. L., Fernando, F. G., and Klein, W. L. (1988). D$_1$ type dopamine receptors inhibit growth cone motility in cultured retina neurons: evidence that neurotransmitters act as morphogenic growth regulators in the developing central nervous system. Proc. Natl. Acad. Sci. 85: 4567–4571.

Liesi, P. (1985). Laminin-immunoreactive glia distinguish regenerative adult CNS systems from non-regenerative ones. EMBO J. 4: 2505–2511.

Rogers, S. L., Letourneau, P. C., Palm, S. L., McCarthy, J., and Furcht, L. T. (1983). Neurite extension by peripheral and central nervous system neurons in response to substratum bound fibronectin and laminin. Dev. Biol. 98: 212–220.

Sze, P. Y. (1980). Glucocorticoids as a regulatory factor for brain tryptophan hydroxylase during development. Devel. Neurosci. 3: 217–223.

Van Hartesveldt, C., Moore, B., and Hartman, B. K. (1986). Transient midline raphe glial structure in the developing rat. J. Comp. Neurol. 253: 175–184.

Whitaker-Azmitia, P. M., and Azmitia, E. C. (1986). 3H-5-Hydroxytryptamine binding to brain astroglial cells: Differences between intact and homogenized preparations and mature and immature cultures. J. Neurochem, 46(4): 1186–1189.

Whitaker-Azmitia, P. M., Shemer, A. V., Caruso, J., Molino, L., and Azmitia, E. C. (1990). Role of High Affinity Serotonin Receptors in Neuronal Growth. Ann. N.Y. Acad. Sci. 600: 315–330.

Whitaker-Azmitia, P. M., Murphy, R., and Azmitia, E. C. (1990). S-100 protein is released from astroglial cells by stimulation of 5-HT$_{1a}$ receptors and regulates development of serotonin neurons. Brain Res. 528: 155–158.

Whitaker-Azmitia, P. M., Quartermain, D., and Shemer, A. V. (1990). Prenatal treatment with SKF, a selective D-1 receptor agonist: Longterm consequences on ^{3}H-paroxetine binding and on dopamine and serotonin receptor sensitivity. Dev. Brain Res. 57: 181–185.

Zhou, F. C., and Azmitia, E. C. (1988). Laminin facilitates and guides fiber outgrowth of transplanted neurons in adult brain. J. Chem. Neuroantat. 1: 133–146.

Serotonin: Molecular Biology, Receptors and Functional Effects
ed. by J. R. Fozard/P. R. Saxena
© 1991 Birkhäuser Verlag Basel/Switzerland

A Neurotrophic Factor – SNTF – for Serotonergic Neurons

F. C. Zhou[1] and E. C. Azmitia[2]

[1]Department of Anatomy, Indiana University, Indianapolis, IN 46202, USA;
[2]Department of Biology, New York University, NY 10003, USA

Summary. We have found that the 5-HT target tissue of denervated adult rat brain contains high levels of a serotonergic neurons related-trophic factor(s) [SNTF]. Specific denervation of 5-HT fibers from the hippocampus [by 5,7-dihydroxytryptamine injection in the afferent pathway] induces 5-HT homotypic collateral sprouting in a month. The 5-HT-denervated hippocampus was shown to be a rich environment for the growth of grafted fetal 5-HT neurons. Grafted fetal 5-HT neurons exhibited higher 5-HT levels and 5-HT synaptosomal high-affinity uptake, more dense fibers and larger cell bodies in the 5-HT-denervated hippocampus than in normal hippocampus. Extracts [hypotonic solution in high-speed supernatant fraction] obtained from the denervated hippocampus were found to be trophic in vitro and in vivo. Extracts placed in raphe cell culture [rich in 5-HT neurons] increased the 5-HT high-affinity uptake. Extracts when added to grafted 5-HT neurons increased the 5-HT content of fetal 5-HT neurons in the normal hippocampus. The grafted 5-HT neurons and trophic extracts were also tested in the cerebellum, a brain region with sparse 5-HT innervation. We found that grafted 5-HT neurons seldom survived in this brain region. The denervated hippocampal extracts greatly increased the survival rate of the grafted 5-HT neurons, and significantly increased the density of the fibers in the cerebellum. These observations indicate that brain derived SNTF is trophic to 5-HT neurons at three levels – survival, neurite extension and transmitter maturation.

Introduction

It is now generally accepted that NGF does not affect all types of neurons (Levi-Montalcini and Angeletti 1963; Harper and Thoenen 1980; Hefti et al. 1989). Rather, its influence on nerve growth is largely limited to two parts of the nervous system, cholinergic neurons in the central nervous system and the sympathetic and sensory ganglionic neurons in the peripheral nervous system (recent review: Hefti et al. 1989). Its effects on other central neurons has largely been undefined. It is intriguing to suppose that neurons in other parts of the nervous system, which show little or no sensitivity to NGF, depend on other trophic factors that serve the same general purpose.

The pursuit of other neuronal trophic factors (NTFs) has been quite limited. Evidence of NTFs other than NGF was not obtained until the late 1970's, when an NTF for the parasympathetic system was found in avian ciliary ganglia. The isolated ciliary ganglion cells die in culture unless they are supported by their normal targets or extracts of target tissues (Nishi and Berg 1977, 1979, 1981; Adler et al. 1979; Manthorpe et al. 1980, 1982; Bonyhady et al. 1980; Barde et al. 1983). Neuronal death during development was hypothesized to be the result of failure of the developing neurons to compete for the supply of a specific neurotrophic factor present in limited amounts in their target field (Thoenen and Barde 1980). The trophic signals locally produced in target areas are thought to regulate survival and maturation of neurons innervating that area (Hefti et al. 1989; Thoenen and Edgar 1985; Hamburger 1980).

Injury Derived NTF

Nerve Growth Factor would not have been so thoroughly studied if it had not been present in extremely high concentrations in snake venom, mouse tumor, and male mouse submaxillary glands. Other NTFs have not been found in such large abundance. It has been shown only recently that endogenous NTFs with demonstrable activity have been observed in injured nerve tissue. These injury induced NTFs have been shown to occur in the fluid that surrounds a regenerating peripheral nerve (Caday et al. 1989; Ignatius et al. 1987; Lundborg et al. 1982), and in the fluid that fills a wound cavity in the brain (Nieto-Sampedro et al. 1982, 1983: Manthorpe et al. 1983; Gage et al. 1984; Bjorklund and Stenevi 1981). These NTFs are generally not present or are present in very low concentrations in normal nervous and non-neuronal target tissue. A dramatic increase of NTF activity, of the order of 10-to-100-fold, was observed 1 to 3 weeks post-injury or lesion. Injury induced NTF to peripheral neurons (dorsal root ganglia, ciliary ganglia, sympathetic ganglia) as well as central neurons (spinal cord, corpus striatum, cortical) was proposed in the mid 1980's. Injury induced high NTF activity has become another major breakthrough in the pursuit of putative endogenous NTFs since the discovery of exogenous Nerve Growth Factor in the salivary gland and target-derived NTFs.

Injury induced high activity NTF has enabled the collection of fluid from a limited source of defined target tissue for the assay of a specific type of neuron. Recently, fluid with induced NTF has been effectively collected using a silicon chamber in peripheral nerves (Longo et al. 1982, 1983; Ignatius et al. 1987), gel foam in brain wound cavities, and extracts of homogenized target brain tissue (Heacock et al. 1986; Caday et al. 1989; Needels et al. 1987; Nieto-Sampedro et al. 1983; Manthrope et al. 1983; Gage et al. 1984; Nakagawa and Ishihara 1988; Zhou and Azmitia 1990; Zhou et al. 1987b). A number of novel NTFs for CNS neurons has been proposed since the concept of "injury induced NTF" was introduced.

Specificity of NTF

The specificity of target-derived NTF seems to depend on the neuronal type innervating the target tissue. For example, the iris muscles are innervated by two sets of autonomic nerves. The dilator muscle is innervated largely by sympathetic ganglion cells, whereas the sphincter is innervated by parasympathetic ciliary ganglion cells. Evidently each of these neuronal populations is selectivity maintained by a different NTF family: the sympathetic axons by Nerve Growth Factor derived from the dilator muscle, and the parasympathetic axons by an NTF family derived from sphincter muscle. The concept of specific target tissue derived NTF to innervating neurons has led to the identification of the chick eye ciliary neurotrophic factor (CNTF) which was subsequently purified and characterized (Barbin et al. 1984; Ebendal 1987). A new family of target derived NTFs – Brain Derived Neurotrophic Factor, (BDNF) specific to sensory neurons, was also identified and molecularly cloned (Thoenen and Barde 1980; Barde et al. 1980; Edgar et al. 1981; Barde et al. 1982; and Ebendal et al. 1979, 1983; Leibrock et al. 1989). Cultured chick sensory neurons and retinal ganglion cells can survive and grow in the absence of nerve growth factor when this NTF is present (Turner et al. 1983; Johnson et al. 1986).

5,7-DHT Lesion-Induced NTF for Serotonergic Neurons – SNTF
(see Figure 1)

In the past ten years, we have developed a unique model to study plasticity in a single transmitter system. Specific injury to serotonergic (5-HT) fibers can be made using the neurotoxin, 5,7-dihydroxytryptamine (5,7-DHT) when catecholamine neurons are protected with a selective uptake inhibitor (recent review: see Baumgarten et al. 1981). 5,7-DHT lesion in a hippocampal afferent path (cingulum bundle) caused a specific partial denervation in the hippocampus (Zhou and Azmitia 1984). A month after the lesion, serotonergic (5-HT) axons in another afferent path (fornix-fimbria, FF) were triggered to reinnervate the hippocampus (a homotypic collateral sprouting, sprouting of non-damaged fibers – which carry the same transmitter as the denervated ones) (Azmitia et al. 1978; Clewan and Azmitia 1984; Zhou and Azmitia 1984, 1985, 1986). The norepinephrine (NE) fibers, which travel in the same pathways and terminate in an overlapping area in the hippocampus, do not seem to respond to the denervation of 5-HT fibers. A number of examples of homotypic sprouting were observed after lesions in one hemisphere of bilaterally projecting neurons (Loesche and Stewart 1977; Lynch et al. 1976; Scheff and Cotman 1977; Steward et al. 1976; Nakamura et al. 1974). Similar homotypic regrowths were reported in cholinergic and norepinephrinergic systems in the hippocampus (David and Van Deusen 1984; Gage et al. 1983), the visual system (Cowen et al.

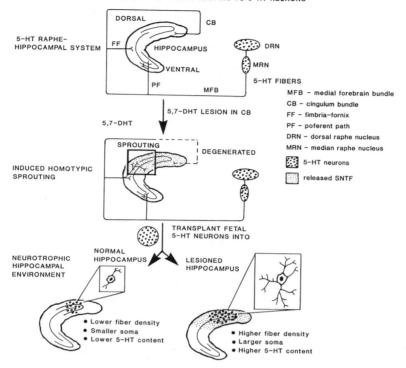

5-HT DENERVATED HIPPOCAMPUS IS TROPHIC TO 5-HT NEURONS

Figure 1. Serotonergic neurons in the raphe innervate the hippocampus through CB, FF, and PF pathways. Specific lesion of 5-HT fibers in CB using 5,7-DHT partially denervates hippocampus. One month later, a homotypic sprouting is induced through the FF pathway. This denervated hippocampus was found to be an enriched environment which supported grafted fetal 5-HT neuronal growth in a significantly higher degree. An extract from the lesioned hippocampus was obtained for the proposed study.

1984; Lund and Lund 1976; Stenevi et al. 1972), and the entorhinal cortex (Steward and Loesche 1977).

Specific injury, therefore, appears to trigger a trophic signal to facilitate regeneration or sprouting of a homotypic neuronal type. Such a regrowth-triggering signal may be initiated within the 5-HT neuron, or extrinsically, from the hippocampus. The latter speculation was favored by the results of an *in vivo* assay on transplanted fetal 5-HT neurons in the lesioned hippocampus, and an *in vitro* assay on cultured fetal 5-HT neurons. The grafted 5-HT neurons had higher 5-HT levels, denser 5-HT fibers, and larger cell-bodies in the hippocampus with a previous lesion rather than in the lesion-free hippocampus (Zhou et al. 1987a); and the supernatant from lesioned-hippocampus increased 5-HT high-affinity synaptic uptake of the cultured 5-HT neurons (Azmitia and Zhou 1986). Thus, a 5,7-DHT lesion of 5-HT fibers engendered a trophic environment in the hippocampus. Similarly, a mechanical lesion in the FF, which involved injury of cholinergic fibers, facilitated the growth of grafted cholinergic neurons (Bjorklund and Gage 1987). Production of NGF-like neurotrophic factor was proposed in the hippocampus (Crutcher and Collins 1982). A lesion made by aspiration in the FF resulting in the loss of stained cholinergic neurons in the septum can be prevented by infusion of NGF to the septum (Williams et al. 1986). This evidence indicated that injury in a brain region can trigger a trophic signal in that brain area to encourage neuronal growth and survival.

Bioassay

A major limitation to the investigation of putative NTF was the lack of a sensitive and appropriate bioassay for quantifying neuronal survival, growth and/or transmitter development. Currently we are using a transplantation method to measure the survival and neurite growth of fetal neurons implanted into a brain wound cavity or denervated target brain area where putative NTFs have accumulated, or into a non-target brain area with supplements of extracted NTFs (Bjorklund and Stenevi 1981; Nieto-Sampedro et al. 1984; Needles et al. 1987 and Gage et al. 1984; Zhou and Azmitia 1990; Zhou et al. 1987a, b). The number, size and transmitter content of the neurochemically identified neurons can also be quantified.

This method has the advantage of directly assaying 5-HT neurons *in vivo* which avoids any possible complication of culture assay when projecting its effect on the brain, and the advantage of assaying for fetal as well as adult 5-HT neurons. They are highly sensitive and they can be done in a manageable time frame.

In the present work, we have prepared a humoral extract from the 5,7-DHT lesioned hippocampus. The trophic effect was examined by a unique bioassay which determines the survival and growth of transplanted fetal 5-HT neurons in two brain regions – cerebellum and hippocampus. Cerebellum normally receives the least 5-HT innervation of all areas in the CNS. This study has shown that the cerebellum supports only minimal survival of grafted 5-HT neurons. The SNTF may be needed to maintain the survival. On the other hand, hippocampus normally receives abundant 5-HT innervation. Grafted 5-HT neurons largely survive and grow to a fixed degree. The SNTF was added to test if additional growth can be achieved.

Materials and Methods

Bioassays of extracted SNTF were done *in vitro* (cell culture) and *in vivo* (transplant). Transplant assays were performed in two areas – the target area (hippocampus) and non-target area (cerebellum). The survival, neurite extensions, and transmission maturation of the 5-HT neurons were measured using the methods described below.

5,7-Dihydroxytryptamine (5,7-DHT) Lesion

Methods were detailed in our earlier reports (Zhou et al. 1987a, b; Zhou and Azmitia, 1984, 1986). In brief, Sprague-Dawley rats were injected intraperitoneally with the NE uptake blocker, desipramine (10 mg/kg), or nomifensine (25 mg/kg), 45 minutes before they were anesthetized with chloropent (1 mg/100 g body-weight, i.p.). The animals with 5,7-DHT lesions were all treated with the NE uptake-blocker unless otherwise noted. The microinjection of 0.4 ul of ascorbic saline (0.01% ascorbic acid in physiological saline) containing 4 ug of the free base of 5,7-DHT was made through a glass micropipette into the fornix-fimbria at a rate of 30 nl/minutes. The tip diameter of the pipette was between 50 and 80 um to limit mechanical damage and ensure a smooth delivery of the solution.

Extraction of SNTF

The extraction of a 5-HT-promoting substance was made from the hippocampus with a prior 5,7-DHT lesion. Two weeks after lesioning, the animals were decapitated, their hippocampi were collected, pooled, weighed, and homogenized in 1:10 (w/v) sterile phosphate buffer (25 mM, pH 7.2) with saline (0.14 M). The homogenates were then centrifuged at 10,000 g at 4°C for 10 minutes. The supernatant, derived by serotonin-denervation and functional as a neurotrophic factor to the growth and survival of serotonergic neurons, was designated as Serotonergic Neurons related-Trophic Factor (SNTF). The SNTF was subsequently collected and stored at 70°C until use.

Transplantation

The fetal mesencephalic raphe tissue, rich in 5-HT neurons was used for transplant and cell culture assay. The transplantation procedures were performed as previously described (Azmitia et al. 1981; Zhou et al. 1987a; b; Zhou et al. 1988; Auerbach et al. 1985). In brief, fetuses were removed from pregnant Sprague-Dawley rats (on day 14 of gestation) and placed in Hank's solution. A midsagittal strip ventral to the cerebral aqueduct, and between mesencephalic and pontine flexures was dissected from mesencephalic raphe tissue. The blocked tissues for each group were then pooled and dissociated and adjusted to a concentration of 0.3×10^6 cell/ul suspensions. The cell suspension was mixed with SNTF (1:1 dilution in Hank's solution, 5 animals; or 1:200 dilution, 15 animals, 5 of them for HPLC, 9 for immunocytochemistry) or with Hank's solution as a control (20 animals, half of them for HPLC, the other half for immunocytochemistry), before transplant through a glass micropipette into serotonergic target area (hippocampus) or non-target area (cerebellum) of the same-strain of young-adult rats. Each animal received 0.6×10^6 cells in 4 ul of mixed cocktail. All the animals were processed for analysis 4–5 weeks after transplantation.

Immunocytochemistry

Four to five weeks after transplantation, all animals were pretreated i.p. with pargyline, 200 mg/kg, 80 minutes prior to perfusion and L-tryptophan, 200 mg/kg, 60 minutes prior to perfusion. These animals were then perfused with formaldehyde intracardially under deep anesthesia and the brains sectioned at 50 um for immunocytochemical staining. The 5-HT antiserum used to stain 5-HT neurons was produced against 5-HT-Limulus hemocyanin-conjugate in rabbit. The Sternberger's peroxidase-anti-peroxidase (PAP) (Sternberger 1979) indirect-enzyme method was used for staining.

Neuronal Count and Image-analysis of Innervation Density

The 5-HT immunoreactive neurons were counted through the cerebellum in every section. The numbers were corrected with Abercrombie's formula (Abercrombie 1946), according to the thickness of the section and the average size of the neurons.

The innervation density of the transplant in the cerebellum was analyzed by measuring the density of 5-HT immunoreactive fibers in boxed-areas selected at region between 0.3 and 0.5 mm medial to the transplant. The fibers in the above areas were digitized under a Leitz Ortho II microscope with a AIC3000 image-analyzer.

HPLC

The animals for high-performance liquid-chromatography (HPLC) measurement were sacrificed by neck-dislocation. The cerebellum was removed, weighed, and homogenized to make a 10% (w/v) solution in 0.1 M perchloric acid containing 100 mM EDTA. Homogenates were centrifuged at 15,000 g for 15 minutes and the supernatants analyzed for

5-HT and 5-HIAA by HPLC with electrochemical detection (Zhou et al. 1987a; b; Zhou et al. 1988). The above quantitative data were compared between experimental and control groups with the two-tailed Student's t-test, or among groups with ANOVA.

Raphe Cell Culture and ^{3}H-5-HT High Affinity Uptake

The cell suspension was obtained as described in "Transplant", and the pellet resuspended in complete medium (MEM with 1% glucose non-essential amino acids, and 5% fetal calf serum), and the cell density is determined using a hemocytometer. The cells were plated onto 96-well Linbro plates previously coated with poly-D-lysine (25 ug/ml). The wells were washed twice with distilled water and once with Hank's Balanced salt solution. The cells were plated at initial plating densities ranging between 0.25 and 3.0×10^6 cells/um^2 by adding 200 ul of the cell suspension to each well (area of 0.28 mm^2). The media was changed after 3 days and once a week thereafter in long-term studies. No antibiotics were used in the complete media.

Cultures of raphe cells alone and co-cultures with SNTF with serial dilution were compared. The cultures were incubated with HBSS-G with 6×10^{-8} M [^{3}H]5-HT (26 Ci/mmol. New England Nuclear) for 20 minutes at 37°C. This concentration of [^{3}H]5-HT has been shown to be selectively retained by neurons as demonstrated by radioautography pharmacology. Non-specific accumulation was determined by incubating as described with the addition of 10^{-5} M Fluoxetine. At the end of 20 minutes the incubation medium was removed and the cultures were washed twice with PBS. The cells were allowed to air-dry before adding 200 ul of absolute ethanol for 1 hour. Then 150 ul of the alcohol was removed to scintillation vials and counted on a Beckman liquid scintillation counter (counting efficiency 40%). The specific accumulation of [^{3}H]5-HT is the difference between the total and the non-specific retention. This uptake and storage system has been employed by several workers to monitor neuronal maturation and innervation density. Protein determinations were done in individual wells when poly-D-lysine was used as substrate.

Results

(a) Bioassay in Cell Culture

The SNTF placed in cultured raphe neurons increased the serotonin high affinity uptake of the 5-HT neurons. A serial dilution profile had shown that SNTF was potent up to 1:5000 dilution (Figure 2).

58

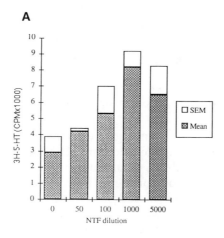

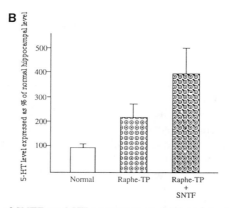

Figure 2. The effects of SNTF on 5-HT neurons *in vitro* (A) and *in vivo* (B). (A) shows that SNTF increased high-affinity uptake up to 1:5000 dilution. (B) shows that transplanted 5-HT neurons increase their 5-HT content to a given level in the hippocampus, and SNTF can further increase the 5-HT content of transplanted 5-HT neurons.

(b) Bioassay in the 5-HT Target Brain Region – Hippocampus

Fetal 5-HT neurons grafted into the intact hippocampus with SNTF supplement had a dramatically higher 5-HT content (Figure 2) and fiber density than the graft without SNTF (Zhou et al. 1987b; 1988). This indicates that (a) the target tissue has a minimal amount of SNTF to support the survival of the grafted 5-HT or to maintain a degree of intrinsic innervation, but (b) individual grafted 5-HT neurons contain a higher 5-HT content and number of fibers when supplemented with SNTF. This notion was further supported by the fact that the SNTF supported neurons were significantly larger than the ones without SNTF (Zhou et al. 1987a).

(c) Bioassay in the Non-Target Brain Region – Cerebellum

The cerebellum is the brain region receiving the least 5-HT innervation, regardless of its proximity to 5-HT neurons in the dorsal raphe. Fetal 5-HT neurons grafted into the cerebellum seldom survived. HPLC showed that regardless of raphe graft containing rich 5-HT neurons, the 5-HT content in the cerebellum did not increase. Immunocytochemistry showed that few, if any, 5-HT neurons survive in this non-target brain region (Zhou and Azmitia 1990). This indicates that the cerebellum contains little SNTF. It supports a very mimimal amount of innervation from intrinsic 5-HT neurons, and minimal degree of survival of grafted fetal 5-HT neurons. However, the fetal 5-HT neurons survived and grew in the cerebellum when supplemented with SNTF (1:200) (Figure 3). The injury induced SNTF supplemented 5-HT neurons sent out fibers to a distance of a millimeter and had significantly higher contents of 5-HT and its metabolite 5-HIAA in the cerebellum (Zhou and Azmitia 1990).

Discussion

The bioassay *in vitro* showed that SNTF can increase the high affinity uptake of serotonin of cultured 5-HT neurons. The dilution profile of the SNTF indicates that the SNTF is effective in serial dilution up to 1:5000. We also found, in *in vivo* assays (Zhou and Azmitia 1990) that in extremely high concentrations (1:1) SNTF is detrimental to the 5-HT neurons. A neurotoxic factor may exist in the injury factor which is lost by dilution. Our previous culture study also show that the hippocampal extracted factor is heat-liable and tryspin-denatured (Azmitia et al. 1990). This suggests that the SNTF is a protein or peptide-like soluble factor.

Assay in the cerebellum shows that the unsuitable condition of the cerebellum to 5-HT nerve fiber-growth during development remained unfavorable into the adult stage, as indicated by (a) the limited survival of transplanted fetal 5-HT neurons, (b) limited fiber extension into cerebellum of a small number of surviving 5-HT neurons, (c) low levels of 5-HT/5-HIAA after raphe graft (Figures 3, 4). Since the brain 5-HT system reaches forebrain regions in early gestation, the extremely limited innervation of 5-HT intrinsic as well as grafted fibers in the cerebellum suggests that this brain region either contains a specific neurotoxic factor that inhibits the growth of 5-HT fibers or there is a lack of specific trophic factor for 5-HT neurons. However, supplementing the grafted fetal 5-HT neurons with denervation-induced SNTF, greatly enhanced the survival of 5-HT neurons in the cerebellum, and significantly increased the density of 5-HT fibers in the cerebellum. A noticeable increase in the size of transplant which contained a large

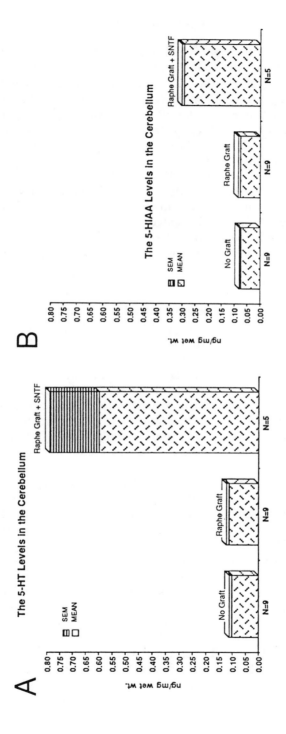

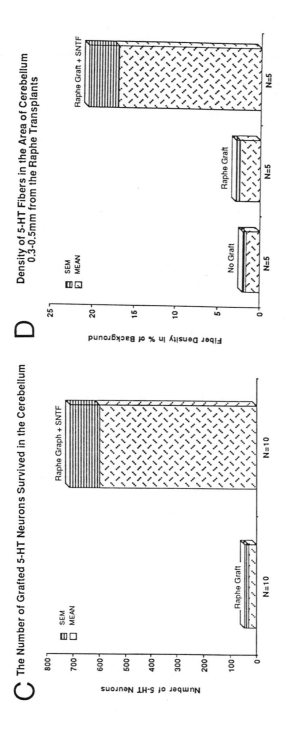

Figure 3. The normal cerebellum has very low levels of 5-HT and 5-HIAA (A, B). Grafting fetal raphe containing rich 5-HT neurons in the cerebellum does not significantly increase the 5-HT and 5-HIAA levels. However, grafting SNTF-supplemented fetal raphe in the cerebellum significantly increased the 5-HT and 5-HIAA levels (ANOVA). The number of grafted 5-HT neurons in the cerebellum is dramatically higher in the graft supplemented with SNTF than the graft without (C) (two-tailed Student's t-test, $p < 0.016$). The density of 5-HT fibers (D) measured in the area in cerebellum was the highest in the group with SNTF-supplemented graft than in the group without graft and group with graft but without supplement. There is no significant difference between the last two groups (ANOVA). The fiber density was measured with an AIC image-analyzer as % area of 5-HT fibers in an area 0.3 to 0.5 mm medial to the edge of the transplant (see insert) or in a comparable area in the contralateral side of the cerebellum.

62

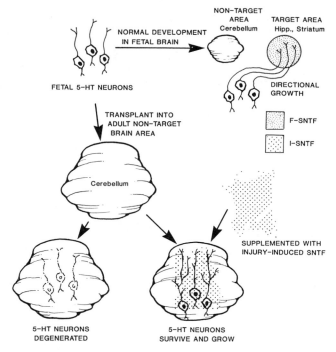

Figure 4. During development, 5-HT neurons grow discriminately toward selected target areas. The cerebellum, although adjacent to dorsal raphe where largest population of 5-HT neurons reside, receives little innervation. Similarly, grafted fetal 5-HT neurons do not survive in the cerebellum. However, when provided with injury induced SNTF(I-SNTF), the fetal 5-HT neurons survive and grow in the unfavorable cerebellar environment.

portion of non-5-HT neurons in the humoral factor-supplemented raphe graft [data not shown] indicates that this humoral factor may also support non-5-HT neurons in the raphe, or that the non-5-HT neurons could be supported by the increased 5-HT neurons which provide the trophic interaction (Lauder and Bloom 1974).

The hippocampus is a normal brain target region of 5-HT neurons. During development, the hippocampus receives 5-HT innervation with a specific pattern and given amount of 5-HT fibers. It indicates that there is a supporting factor, and the amount of supporting factor may permit a given amount of 5-HT innervation. Thus, the maximal growth of these fetal 5-HT neurons in the host target brain region seems to be tailored to a given degree due to many unknown host supporting factors – nutrition and oxygen supplies, availibility of receptor, and neurotrophic factor. The assay in the adult hippocampus indicates that the injury induced SNTF is not only able to increase the survival, neurite extension and transmitter maturation of fetal 5-HT neurons in the unfavorable environment where neurotrophic factor may be absent or limited, but also can increase 5-HT content of 5-HT neurons above normal maximal levels in a favorable environment where neurotrophic support is seemingly available. The SNTF supplement is not really nutritional but rather trophical as is indicated by the fact that fetal 5-HT neurons can survive normally in the hippocampus or striatum on a long term basis (Zhou et al. 1987, Zhou and Azmitia 1990), that SNTF was supplied in the cell pellet before the cells were grafted in the host brain but not continuously infused into the brain and that the specificity of SNTF to 5-HT neurons but not norepinephrine neurons (Zhou et al. 1987a, b).

The bioassays in the cell culture, in the cerebellum and the hippocampus, taken together, provide a number of pieces of information: (a) SNTF is capable of increasing high affinity uptake of 5-HT neurons and it is potent up to 1:5000 dilution, (b) the supernatant from the 5-HT-denervated hippocampus contains some signal released in the hippocampus after 5-HT denervation (Zhou and Azmitia 1986); (c) this humoral signal increases maturation of the 5-HT neuronal transmitter in the environment suitable for the survival of 5-HT neurons [hippocampus] (Zhou et al. 1987); (d) in the environment unsuitable for the survival of 5-HT neurons [cerebellum], it also enhances survival rate; (e) this denervation factor may affect non-5-HT neurons; (f) this denervation factor is probably not an NGF, since NGF has no known effect on 5-HT neurons; (g) the trophic factor to 5-HT neurons is injury-inducible.

Injury to the brain involves many changes in the local environment. Among them, five extrinsic factors are closely associated with the regrowth of nerve fibers in the injured area: mobilization of glial cells, reorganization of vasculature, changes in extracellular matrices, exposure to blood-bound factors, and induction of neurotrophic factors. Each has been studied to various extents in order to obtain the optimal growing condition. Of these five, the last two have been shown to have the potential to "stimulate" the new growth of nerve fibers, which has ceased in the mature brain. Platelet-derived growth factor and NGF are examples of the two categories of the blood-bound and brain-derived neurotrophic factors respectively. These identified factors provided evidence that soluble peptide or protein factor can stimulate the growth or enhance the survival of neurons *in vitro*. These factors accumulating in the injured brain area may be favorable for regeneration and/or sprouting of local nerve fibers. Recently documented wound-cavity factors which accumulated in the injured brain cavity enhanced the survival of transplanted neurons and can be extracted and used to enhance neuronal growth (Manthrope et al. 1983; Needels et al. 1982; Neito-Sampedro et al. 1982; Whittemore et al. 1985; Zhou 1988).

The origin of the trophic signal in our extracted supernatant could be a blood-bound factor or it could be a product of hippocampal cells, a brain-derived factor. For instance, wound-cavity factor could be either or both of the two cases, since blood vessels were severely damaged and the local blood-brain barrier might not be mended during the extraction of the factor. For our trophic factor, evidence favors the latter case – a brain-derived factor, because, the 5,7-DHT lesion was made at one location [fornix-fimbria] and the extract was obtained from an area [hippocampus] some distance away. The entrance of large molecules across the ruptured blood-brain barrier would be more likely to occur in the fornix-fimbria rather than in the hippocampus.

The brain derived trophic factor in response to injury is not an isolated case. Recently the phenomenon has been reported in quite a number of neural regions (Bjorklund and Stenevi 1981; Crutcher and Collins 1986; Gage et al. 1984; Hadani et al. 1984; Heacock et al. 1986; Kanakis et al. 1985; Muller et al. 1987; Nieto-Sampedro et al. 1984; Schwartz et al. 1985; Skene and Shooter 1983). The humoral factor which we extracted from the denervated hippocampus is unique in that (a) it was triggered by selective lesion of 5-HT fibers; (b) it was extracted from a brain region where sprouting of 5-HT fibers occurred; (c) it has effects on the survival, the transmitter maturation, and the neurite outgrowth of the fetal 5-HT neurons. However, two important connections remain to be investigated: (a) is this extracted factor responsible for the sprouting or regeneration of 5-HT fibers after injury? (b) what is the effect of the SNTF on "normal" adult 5-HT neurons in the brain? With respect to these two issues, we have previously demonstrated that transplanted fetal striatal tissue (Zhou and Buchwald 1989) and hippocampus (Azmitia et al. 1981) serve as very strong stimulants to the regrowth of adult (host) 5-HT fibers. The comparison between fetal striatum- and hippocampal-derived serotonergic trophic factors and the hippocampal denervation-derived serotonergic trophic factors is currently under investigation in our laboratory.

Acknowledgements

This study is supported by NIH Grant NS23027 to F.C.Z. The authors wish to thank Ms. Karen Fitzgerald and Sharon Bledsoe for assistance in editing this manuscript, Ms. Sharon Teal and Mr. John Nixon for graphics drawing and Mr. John Le for assistance in surgery and histology.

References

Abercrombie, M. (1946). Estimation of nuclear population from microtome sections. Anat. Rec. 94: 239–247.

Adler, R., Landa, K. B., Manthrope, M., and Varon, S. (1979). Cholinergic neuronotrophic factors. II. Selective intraocular distribution of soluble trophic activity for ciliary ganglionic neurons. Science 204: 1434–1436.

Auerbach, S., Zhou, F. C., Jacobs, B. L., and Azmitia, E. C. (1985). Serotonin turnover in raphe neurons transplanted into rat hippocampus. Neurosci. Lett. 61: 147–152.

Azmitia, E. C., and Zhou, F. C. (1986a). Induced homotypic collateral sprouting of hippocampal serotonergic fibers. In: Gilad G. M., Gorio, A., and Kreutzberg, G. W. (eds), Processes of recovery from Neuronal Trauma. Amsterdam: Elsevier, Exp. Brain Res. (Suppl 13), 129–141.

Azmitia, E. C., and Zhou, F. C. (1986b). A specific serotonergic growth factor from 5,7-DHT lesioned hippocampus: *in vitro* evidence from dissociated cultures of raphe and locus ceruleus. Soc. Neurosci. Abst. 11: 1048.

Azmitia, E. C., Buchan, A. M., and Williams, J. H. (1978). Structural and functional restoration by collateral sprouting of hippocampal 5-HT axons. Nature 274: 374–376.

Azmitia, E. C., Lama, P., Segal, M., Whitaker-Azmitia, P. M., Murphy, R. B., and Zhou, F. C. (1991). Effects of hippocampal supernatant extract on the development of serotonergic neurons in dissociated microcultures. Int. J. Develop. Neurosci. (In Press).

Azmitia, E. C., Perlow, M. J., Brennan, M. J., and Lauder, J. M. (1981). Fetal raphe and hippocampal transplants into adult and aged C57BL/6N mice: A preliminary immunocytochemical study. Brain Res. Bull. 7: 703–710.

Barbin, G., Selak, I., Manthorpe, M., and Varon, S. (1984). Use of central neuronal cultures for detection of neuronotrophic agents. Neurosci. 12: 33–43.

Barde, Y.-A., Edgar, D., and Thoenen, H. (1980). Sensory neurons in culture: changing requirements for survival factors during embryonic development. Proc. Natl. Acad. Sci. USA 77: 1199–1203.

Barde, Y.-A., Edgar, D., and Thoenen, H. (1982). Purification of a new neurotrophic factor from mammalian brain. EMBO J. 1: 549–553.

Barde, Y.-A., Edgar, D., and Thoenen, H. (1983). New neurotrophic factors. Annu. Rev. Physiol. 45: 601–612.

Baumgarten, H. G., Jenner, S., and Klemin, H. P. (1981). Serotonin neurotoxins: Recent advances in the mode of administration and molecular mechanism of action. J. Physio. 77: 309–314.

Bjorklund, A., and Gage, F. H. (1987). Grafts of fetal septal cholinergic neurons to the hippocampal formation in aged or Fimbria-Fornix-Lesioned rats. In: Azmitia, E. C., and Bjorklund, A. (eds), Cells and tissue transplantation into the adult brain. Annals of the New York Academy of Science. 495: 120–137.

Bjorklund, A., and Stenevi, U. (1981). *In vivo* evidence for a hippocampal adrenergic neuronotrophic factor specifically released on septal deafferentation. Brain Res. 229: 403–428.

Bonyhady, R. E., Hendry, I. A., Hill, C. E., and McLennan, I. S. (1980). Characterization of a cardiac muscle factor required for the survival of cultured parasympathetic neurons. Neurosci. Lett. 18: 197–201.

Caday, C. G., Apostolides, P. J., Benowitz, L. I., Perrone-Bizzozero, N. I., and Finklestein, S. P. (1989). Partial purification and characterization of a neurite-promoting factor from the injured goldfish optic nerve. Molec. Brain Res. 5: 45–50.

Clewans, C., and Azmitia, E. C. (1984). Tryptophan hydroxylase in the hippocampus and midbrain following unilateral injection of 5,7-dihydroxytryptamine. Brain Res. 307: 125–133.

Cowan, W. M., Fawcett, J. W., Oleary, D. D. M., and Stanfield, B. B. (1984). Regressive events in neurogenesis. Science 225: 1258–1265.

Crutcher, K. A., and Collins, F. (1982). *In vitro* evidence for two distinct hippocampal growth factors: Basis of neuronal plasticity? Science 217: 67–68.

Crutcher, K. A., and Collins, F. (1986). Entorhinal lesions result in increased nerve growth factor-like growth-promoting activity in medium conditioned by hippocampal slices. Brain Res. 399: 383–389.

David, A. R., and Van Deusen, E. B. (1984). Recovery of enzyme markers for cholinergic terminals in septo-temporal regions of the hippocampus following selective fimbrial lesions in adult rat. Brain Res. 324: 119–128.

Ebendal, T. (1987). Comparative screening for ciliary neurotrophic activity in organs of the rat and chicken. J. Neurosci. Res. 17: 19–24.

Ebendal, T., Belew, M., Jacaboson, P.-O., and Porath, J. (1979). Neurite outgrowth elicited by embryonic chick heart: partial purification of the active factor. Neurosci. Lett. 14: 91–95.

Ebandal, T., Norrgren, G., and Hedlund, K.-O. (1983). Nerve growth-promoting activity in the chick embryo: quantitative aspects. Med. Biol. 61: 65–72.

Edgar, D., Barde, T.-A., and Thoenen, H. (1981). Subpopulations of cultured chick sympathetic neurons differ in their requirements for survival factor. Nature, Lond. 289: 294–295.

Gage, F. H., Bjorklund, A., and Stenevi, U. (1983). Reinnervation of the partially deafferented hippocampus by compensatory collateral sprouting from spared cholineergic and noradrenergic afferents. Brain Res. 268: 27–37.

Gage, F. H., Bjorklund, A., and Stenevi, U. (1984). Denervation releases a neuronal survival factor in adult rat hippocampus. Nature 308: 637–639.

Hadani, M., Harel, A., Solomon, A., Belkin, M., Lavie, V., and Schwartz, M. (1984). Substances originating from the optic nerve of neonatal rabbit induce regeneration-associated response in the injured optic nerve of adult rabbit. Proc. Natl. Acad. Sci. USA. 81: 7965–7969.

Hamburger, V. (1980). Trophic interactions in neurogenesis: a personal historical account. Ann. Rev. Neurosci. 3: 269–278.

Harper, G. P., and Thoenen, H. (1980). Nerve growth factor: Biological significance, measurement, and distribution. J. Neurochem. 34: 5–16.

Heacock, A. M., Schonfeld, A. R., and Katzman, R. (1986). Hippocampal neurotrophic factor: Characterization and response to denervation. Brain Res. 363: 299–306.

Hefti, F., Hartikka, J., and Knusel, B. (1989). Function of neurotrophic factors in the adult and aging brain and their possible use in the treatment of neurodegenerative diseases. Neurobiol. Aging. 10: 515–533.

Ignatius, M. J., Skene, J. H. P., Muller, H. W., and Shooter, E. M. (1987). Examination of a nerve injury-induced, 37 kDa protein: purification and characterization. Neurochem. Res. 12: 967–976.

Johnson, J. E., Barde, Y. A., Schwab, M., and Thoenen, H. (1986). Brain-derived neurotrophic factor supports the survival of cultured rat retinal ganglion cells. J. Neurosci. 6: 3031–3038.

Kanakis, S. J., Hill, C. E., Hendry, I. A., and Watters, D. J. (1985). Sympathetic neuronal survival factors change after denervation. Developmental Brain Res. 20: 197–202.

Lauder, J. M., and Bloom, F. E. (1974). Ontogeny of monoamine neurons in the locus coeruleus, raphe nuclei and substantia nigra. I. Cell Differentiation, J. Comp. Neurol. 155: 469–481.

Leibrock, J., Lottspeich, F., Hohn, A., Hofer, M., Hengerer, B., Masiakowski, P., Thoenen, H., and Barde, Y.-Al. (1989). Molecular cloning and expression of brain-derived neurotrophic factor. Nature 341: 149–152.

Levi-Montalcini, R., and Angeletti, P. U. (1963). Essential role of the nerve growth factor in the survival and maintenance of dissociated sendory and sympathetic nerve cells in vitro. Dev. Biol. 7: 655–659.

Longo, F. M., Manthrope, M., Skaper, S. D., Lundborg, G., and Varon, S. (1983). Neurotrophic activities in fluid accumulated in vivo within silicone nerve regeneration chambers. Brain Res. 261: 1099–1117.

Longo, F. M., Skaper, S. D., Manthorpe, M., Lundborg, G., and Varon, S. (1982). Further characterization of neuronotrophic factors accumulating in vivo within nerve stump-containing silicone chambers. Soc. Neurosci. Abst. 8: 861.

Loesche, J., and Stewart, O. (1977). Behavioral correlents of denervation and reinnervation of the hippocampal formation of rat: Recovery of alternation performance following unilateral entorhinal cortex lesion. Brain Res. Bull. 2: 31–39.

Lund, R. D., and Lund, J. S. (1976). Plasticity in the developing visual system: the effects of retinal lesion made in young rats. J. Comp. Neurol. 169: 133–154.

Lundborg, G., Longo, F. M., and Varon, S. (1982). Nerve regeneration model and trophic factors in vivo. Brain Res. 232: 157–161.

Lynch, G., Gall, C., Rose, G., and Ctoman, C. W. (1976). Changes in the distribution of the dentate gyrus associational system following unilateral or bilateral entorhinal lesion in the adult rat. Brain Res. 110: 57–61.

Manthrope, M., Nieto-Samoedro, M., Skaper, S. D., Lewis, E. R., Barbin, G., Longo, F. M., Cotman, C. W., and Varon, S. (1983). Neuronotrophic activity in brain wounds of the developing rat. Correlation with implant survival in the wound cavity. Brain Res. 267: 47–56.

Manthrope, M., Skaper, S. D., Adler, R., Landa, K. B., and Varon, S. (1980). Cholinergic neuronotrophic factors: IV Fractionation properties of an extract from selected chick embytonic eye tissue. J. Neurochem. 34: 69–75.

Manthrope, M., Skaper, S. D., Barbin, G., and Varon, S. (1982). Cholinergic neuronotrophic factors (CNTFs): VII Concurrent activities on certain nerve growth factor-responsive neurons. J. Neurochem. 38: 415–421.

Muller, H. W., Geibicke-Harter, P. J., Hangen, D. H., and Shooter, E. M. (1987). A specific 37,000-dalton protein that accumulates in regenerating but not in nonregenerating mammalian nerves. Science 228: 499–501.

Nakagawa, Y., and Ishihara, T. (1988). Enhancement of neurotrophic activity in cholinergic cells by hippocampal extract prepared from colchicine-lesioned rats. Brain Res. 439: 11–18.

Nakamura, Y., Mizuno, N., Konishi, A., and Sato, M. (1974). Synatpic reorganization of the red nucleus after chronic deafferentation from cerebellorubral fibers: an electron microscope study in the cat. Brain Res. 82: 298–301.

Needels, D. L., Nieto-Sampedro, M., and Cotman, C. W. (1987). Long-term support by injured brain extract of a subpopulation of ciliary ganglion neurons purified by differential adhesion. Neurochem. Res. 12: 901–907.

Nieto-Sampedro, M., Lewis, E. R., Cotman, C. W., Manthorpe, M., Skaper, S. D., Barbin, G., Longo, F. M., and Varon, S. (1982). Brain injury causes a time-dependent increase in neuronotrophic activity at the lesion site. Science 217: 860–861.

Nieto-Sampedro, M., Manthorpe, M., Barbin, G., Varon, S., and Cotman, C. W. (1983). Injury-induced neuronotrophic activity in adult rat brain: Correlation with survival of delayed implants in the wound cavity. Neurosci. 3: 2219–2229.

Nieto-Sampedro, M., Whittemore, Scott R., Needels, D. L., Larson, J., and Cotman, C. W. (1984). The survival of brain transplants is enhanced by extracts from injured brain. Proc. Natl. Acad. Sci. USA 81: 6250–6254.

Nishi, R., and Berg, D. K. (1977). Dissociated ciliary ganglion neurons in vitro: survival and synapse formation. Proc. Natl. Acad. Sci. U.S.A. 74: 5174–5175.

Nishi, R., and Berg, D. K. (1979). Survival and development of ciliary ganglion neurons grown alone in cell culture. Nature. 277: 232–234.

Nishi, R., and Berg, D. K. (1981). Two components from eye tissue that differentially stimulate the growth and development of ciliary ganglion neurons in cell culture. J. Neurosci. 1: 505–513.

Scheff, S. W., and Cotman, C. W. (1977). Recovery of spontaneous alteration following lesion of entorhinal cortex in adult rats: Possible correlation to axon sprouting. Behav. Biol. 21: 286–293.

Schwartz, M., Belkin, M., Harel, A., Solomon, A., Valie, V., Hadani, M., Rachailovich, I., and Stein-Izsak, C. (1985). Regenerating fish optic nerves and a regeneration-like response in injured optic nerves of adult rabbits. Science 228: 600–603.

Skene, J. H. P., and Shooter, E. M. (1983). Denervated sheath cells secret a new protein after nerve injury. Proc. Natl. Acad. Sci. USA 80: 4169–4173.

Stenevi, U., Bjorklund, A., and Moore, R. Y. (1972). Growth of intact central adrenergic axons in the denervated lateral geniculate body. Exp. Neurol. 35: 29–299.

Sternburger, L. A. (1979). Immunocytochemistry, 2nd edn., Wiley, New York, 104–169.

Steward, O., and Loesche, J. (1977). Quantitative autoradiographic analysis of the time course of proliferation of contralateral entorhinal efferents in the dentate gyrus denervated by ipsilateral entorhinal lesions. Brain Res. 125: 11–21.

Steward, O., Cotman, C. W., and Lynch, G. A. (1976). A quantitative autoradiographic and electrophysiological study of the reinnervation of the dentate gyrus by contralateral entorhinal cortex following ipsilateral entorhinal lesions. Brain Res. 114: 181–200.

Thoenen, H., and Barde, Y.-A. (1980). Physiology of nerve growth factor. Physiol. Rev. 60: 1284–1335.

Thoenen, H., and Edgar, D. (1985). Neurotrophic factors. Science 229: 238–242.

Turner, J. E., Barde, Y.-A., Schwab, M., and Thoenen, H. (1983). Extract from brain stimulates neurite outgrowth from fetal rat retinal explants. Dev Brain Res. 6: 77–84.

Watters, D., Belford, D., Hill, C., and Hendyr, I. (1989). Monoclonal antibody that inhibits biological activity of a mammalian cilaiary neurotrophic factor. J. Neurosci. Res. 22: 60–64.

Whittemore, S. R., Nieto-Sampedro, M., Needels, D. L., and Cotman, C. W. (1985). Neuronotrophic factors for mammalian brain neurons: Injury induction in neonatal, adult and aged rat brain. Develop. Brain Res. 20: 169–178.

Williams, L. R., Varon, S., Peterson, G. M., Wictorin, K., Fischer, W., Bjorklund, A., and Gage, F. H. (1986). Continuous infusion of nerve growth factor prevents basal forebrain neuronal death after fimbria fornix transaction. Proc. Natl. Acad. Sci. USA 83: 9231–9235.

Zhou, F. C. (1988). Mechanical or chemical injury in the brain induces laminin production by astrocyte, Anat. Rec. 220: 108A.

Zhou, F. C., Auerbach, S., and Azmitia, E. (1987a). Denervation of serotonergic fibers in the hippocampus induces a trophic factor which enhances the maturation of transplanted serotonergic neurons but not norepinephrinergic neurons. J. Neurosci. Res. 17: 235–246.

Zhou, F. C., Auerbach, S., Azmitia, E. C. (1987b). Injury of serotonergic fiber induces factor which promotes the maturation of serotonergic neurons but not norepinephrine neurons. In: Cell and Tissue transplantation into the Adult Brain. Azmitia, E. C., and Bjorklund, A. (eds), Ann. NY Acad. Sci. 495: 138–152.

Zhou, F. C., Auerbch, S., Azmitia, E. C. (1988). Transplanted raphe and hippocampal fetal neurons do not displace the afferent inputs to the dorsal hippocampus from serotonergic neurons on the median raphe nucleus of the rat. Brain Res. 450: 51–59.

Zhou, F. C., and Azmitia, E. C. (1984). Induced homotypic collateral sprouting of serotonergic fibers in hippocampus. Brain Res. 308: 53–62.

Zhou, F. C., and Azmitia, E. C. (1985). The effect of adrenalectomy and corticosterone on homotypic collateral sprouting of serotonergic fibers in hippocampus. Neurosci. Lett. 54: 111–116.

Zhou, F. C., and Azmitia, E. C. (1986). Induced homotypic sprouting of serotonergic fibers in hippocampus: II, An immunocytochemical study. Brain Res. 373: 337–348.

Zhou, F. C., and Azmitia, E. C. (1990). Neurotrophic factor for serotonergic neurons prevent degeneration of grafted raphe neurons in the cerebellum. Brain Res. 507: 301–308.

Zhou, F. C., and Buchwald, N. (1989). Connectivities of the striatal grafts in adult rat brain: a rich afference and scant striato-nigral efference. Brain Res. 504: 15–30.

Serotonin: Molecular Biology, Receptors and Functional Effects
ed. by J. R. Fozard/P. R. Saxena
© 1991 Birkhäuser Verlag Basel/Switzerland

Molecular Mechanisms of Action Induced by 5-HT$_3$ Receptors in a Neuronal Cell Line and by 5-HT$_2$ Receptors in a Glial Cell Line

G. Reiser

Physiologisch-chemisches Institut der Universität Tübingen, Hoppe Seyler Str. 4, 7400 Tübingen, FRG

Summary. The molecular mechanisms of action of serotonin were investigated in a neuronal cell line expressing 5-HT$_3$ receptors (neuroblastoma × glioma hybrid cells) and in a glioma cell line with 5-HT$_2$ receptors. In both cell lines, serotonin induces a transient rise of cytosolic Ca^{2+} activity. Ca^{2+} channel blockers (Ni^{2+} and La^{3+}) suppress the Ca^{2+} response to serotonin in the neuronal cells but not in the glial cells. When internal Ca^{2+} stores are depleted and short-circuited by applying Ca^{2+} ionophores (ionomycin and A23187), serotonin still induces the normal Ca^{2+} response in the neuronal hybrid cells but not in the glioma cells. Thus, in the neuronal cell line cytosolic Ca^{2+} activity is enhanced through stimulation of Ca^{2+} entry into the cells from the extracellular environment via 5-HT$_3$ receptors. The depolarization response caused by serotonin in these cells is due to activation of a cation conductance, and consequent entry of extracellular Ca^{2+}. In the neuronal cell line, serotonin induces a rise of the cyclic GMP level, which depends on the rise of cytosolic Ca^{2+} activity. This conclusion is derived from the following findings: The serotonin-stimulated rise of cyclic GMP level is inhibited by i) reduced extracellular Ca^{2+} concentration (half-maximal stimulation at 0.3 mM Ca^{2+}); ii) addition of inorganic (La^{3+}, Mn^{2+}, Co^{2+}, Ni^{2+}) or organic blockers (diltiazem, methoxyverapamil) of Ca^{2+} permeable channels; and iii) intracellular application of the Ca^{2+} chelator BAPTA. The suppression of the cyclic GMP effect of serotonin by the arginine analogues (N^G-monomethyl-L-arginine, N^G-nitro-L-arginine and canavanine) and by incubation in media containing oxyhemoglobin indicates that after stimulation with serotonin nitric oxide released from arginine acts as an intercellular stimulant of soluble guanylate cyclase. The rise of cytosolic Ca^{2+} activity appears to be a prerequisite for the formation of nitric oxide as an activator of guanylate cyclase. In the glial cell line, however, ketanserin-sensitive 5-HT$_2$ receptors mainly cause liberation of Ca^{2+} from internal stores. In the glioma cells, Ca^{2+} released from internal stores opens Ca^{2+}-dependent K$^+$ channels which results in the hyperpolarizing response.

Introduction

Serotonin has been found to induce a range of electrophysiological responses (see Reiser et al. 1989) via the several receptor types for serotonin (5-HT) occurring in the mammalian central nervous system (Peroutka 1988). A pivotal question in the understanding of the molecular mechanism of action of serotonin is the elucidation of the cellular effector systems coupled to the various receptor types. Subtypes of 5-HT$_1$ receptors are reported to activate adenylate cyclase (Peroutka 1988) and phospholipase C (Jankowsky et al. 1984) in rat hippocampus and to mediate presynaptic inhibition of evoked transmitter (5-HT) release (Richardson and Engel 1986). Serotonin 5-HT$_{1c}$ receptors in Xenopus oocytes injected with rat brain mRNA (Lübbert et al. 1987) mediate a rise of cytosolic Ca^{2+} activity apparently through inositol 1,4,5-trisphosphate formation (Gundersen et al. 1984). 5-HT$_2$ receptors which are linked to neuronal depolarization are suggested to be associated with a phosphoinositide-specific phospholipase C in rat frontal cortex (Conn and Sanders-Bush 1985).

In the study presented here two neural cell lines were used to further investigate the mechanisms of action of serotonin at different types of 5-HT receptors. Firstly, a polyploid rat glioma cell line (Heumann et al. 1982) derived from C6 glioma cells was studied. Biochemical (Ananth et al. 1987) and electrophysiological (Ogura and Amano 1984) responses to serotonin have been reported for C6 cells. Secondly, as a neuronal cell line we employed mouse neuroblastoma × rat glioma hybrid cells (Hamprecht et al. 1985) with a depolarization response to serotonin (Christian et al. 1978). In the hybrid cells 5-HT$_3$ recognition sites have been identified by radioligand binding studies (Hoyer and Neijt, 1987). In this report the influence of serotonin on cytosolic Ca^{2+} activity, on ^{45}Ca^{2+} uptake, on membrane potential, and on levels of inositolphosphates and of cyclic GMP is compared in both cell lines, summarizing some of the results published previously (Reiser et al. 1989; Reiser 1990).

Methods and Materials

Cell Cultures: Mouse neuroblastoma × rat glioma hybrid cells, clone 108CC15 and clone 108CC25 (for review see Hamprecht et al. 1985), and polyploid rat glioma cells, clone C6-4-2 (Heumann et al. 1982) of passage numbers between 14 and 32 were cultured as described (Hamprecht et al. 1985). Methods and materials used, as described in detail for Ca^{2+} determination (Reiser et al. 1989), for electrophysiology (Reiser et al. 1990b), for measurement of cyclic GMP (Reiser 1990) and of inositolphosphates (Donié and Reiser, 1989) are briefly outlined.

Determination of Cytosolic Ca^{2+} Activity: Cells detached from their substrate by treatment with trypsin (viability more than 96%) were suspended in Dulbecco's Modified Eagle's Medium without bicarbonate, but buffered with 20 mM HEPES adjusted to pH 7.4 with Tris. The cell suspension (density 3 to 4×10^6 cells/ml) was continuously stirred at 37°C and supplemented with 3 to 4 μM fura-2 pentaacetoxymethylester (Grynkiewicz et al. 1985). After dye loading, between 5 and 10×10^6 cells were centrifuged, washed once with 2 ml incubation medium, and resuspended in 2 ml incubation medium which contained (mM) NaCl (145), KCl (5.4), CaCl$_2$ (1.8), MgCl$_2$ (1.0), glucose (20), HEPES (20) adjusted to pH 7.4 with Tris, osmolarity 320–350 mOsmol/l. The cells were placed into a quartz cuvette in an Aminco SPF-500 spectrofluorimeter (Silver Spring, Maryland, USA) kept at 37°C. Fura-2 fluorescence with excitation at 340 nm (2 nm slit width) and emission at 500 nm (5 nm slit width), was calibrated in terms of Ca^{2+} activity, as described by Grynkiewicz et al. (1985).

Electrophysiology: For electrophysiological experiments, neuroblastoma × glioma hybrid cells (clone 108CC25) or polyploid rat glioma cells were grown on glass coverslips, transferred into a perfusion chamber, and superfused constantly with recording medium which had the same composition as the incubation medium given above. Membrane potential was recorded as previously described (Reiser et al. 1990b). Serotonin was applied iontophoretically.

Measurement of Cyclic GMP Level: Hybrid cells (clone 108CC15, 3 to 4×10^5 cells in plastic Petri dishes with diameter 50 mm) were grown for 2 to 3 days. To start the experiment, the growth medium was removed and the cells were washed twice with 2 ml incubation medium which in addition contained 2 mM Na$_2$HPO$_4$. Cells were preincubated at 37°C in 2 ml incubation medium for 30 min allowing the cells to equilibrate. Stimulations were started by adding 20 μl of a concentrated stock solution of the compounds to be tested, and were stopped by adding 1 ml ethanol. Cellular content of cyclic GMP was quantified by radioimmunoassay as described by Reiser et al. (1984) and referred to cellular protein. Cellular protein was determined according to the Lowry method using bovine serum albumin as standard.

Determination of Cellular Levels of Ins(1,4,5)P$_3$ and Ins(1,3,4,5)P$_4$: For the radioligand binding assay of Ins(1,3,4,5)P$_4$ and Ins(1,4,5)P$_3$ membranes prepared from pig cerebellum and from bovine liver, respectively, were taken. Incubation and separation of bound from free label were carried out as described (Donié and Reiser, 1989). A calibration curve obtained by using a series of different concentrations of Ins(1,3,4,5)P$_4$ supplemented with 1 mM trichloroacetic acid adjusted to pH 5.0 with NaOH was used to calculate the amount of Ins(1,3,4,5)P$_4$ in tissue samples. In the experiments cells were preincubated for 30 min in incubation medium. The challenge incubation was started by addition of 20 μl stock solution of the test compounds. Thereafter the medium was aspirated and 1 ml ice-cold trichloroacetic acid (10%) was added. The supernatant was extracted with diethylether for removing trichloroacetic acid. The supernatant was used for measuring Ins(1,3,4,5)P$_4$. In samples neutralized with the appropriate volume of 1 M NaHCO$_3$ the contents of Ins(1,4,5)P$_3$ was determined.

72

Results

Regulation of Cytosolic Ca²⁺ Activity by Serotonin

In the neuronal hybrid cells loaded with fura-2, serotonin caused a transient increase in fluorescence intensity (Figure 1A) which is due to a rise of cytosolic Ca^{2+} activity (Grynkiewicz et al. 1985). The basal level of Ca^{2+} activity was reached again about 40 s after addition of serotonin. In the hybrid cells the EC_{50} value for serotonin was 0.6 μM and for 2-methyl-serotonin 4.5 μM. The maximum, obtainable at about 2 μM serotonin was, however, in general smaller than the signal induced by adding the peptide bradykinin (Figure 1A). In the glioma cells, serotonin caused a comparable increase of the cytosolic Ca^{2+} activity

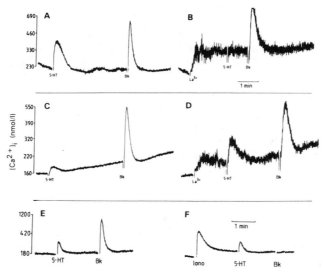

Figure 1. Serotonin (5-HT) increases cytosolic Ca^{2+} activity, derived from fura-2 fluorescence, in the neuronal hybrid cells and the glioma cells. A-D: Influence of La^{3+} on the rise of cytosolic Ca^{2+} activity induced by serotonin (1 μM) or bradykinin (Bk; 20 nM) in the hybrid cells (A, B) or in the glioma cells (C, D). Responses after adding 0.5 mM La^{3+} are given in B and D, respectively. The increase of fluorescence intensity after addition of La^{3+} is discussed in the text. E–F: Influence of the Ca^{2+} ionophore ionomycin (Iono) on the Ca^{2+} response to serotonin (2.5 μM) and to bradykinin (400 nM) in the hybrid cells; (E) responses under control conditions and (F) after addition of 0.5 μM ionomycin. Experiments A–B, C–D, E–F respectively, were carried out with samples from the same cell suspension. The upper (lower) time bar applies to A–D (E–F). The test agents were added as aliquots of 100-fold concentrated solutions to the cell suspension at the time points given by an interruption of the fluorescence trace due to opening the chamber of the spectrofluorimeter. Figures 1 and 2 show results which are representative of at least 3 experiments.

(Figure 1C) but 2-methyl-serotonin was ineffective at concentrations up to 50 μM.

The pharmacology of the receptors was characterized by using specific antagonists. In the glioma cells the Ca^{2+} response was blocked by ketanserin, a 5-HT_2 receptor antagonist (Leysen et al. 1982), at concentrations between 1 and 10 nM. In the hybrid cells, however, ketanserin did not affect the stimulation by serotonin of cytosolic Ca^{2+} activity. Antagonists of 5-HT_3 receptors, ICS 205-930 (Richardson and Engel 1986), MDL 72222 (Fozard 1984) and ondansetron (Kilpatrick et al. 1987), at nanomolar concentrations, inhibited the Ca^{2+} response to serotonin in the hybrid cells. In the glioma cells, however, ICS 205-930 in concentrations up to 0.2 μM had no effect on the rise in Ca^{2+} activity induced by serotonin. Cyproheptadine, a serotonin antagonist with limited specificity, suppressed the Ca^{2+} response to serotonin in both the hybrid cells and in the glioma cells at concentrations from 2 to 20 μM.

A series of experiments was made to elucidate the mechanisms by which serotonin causes the rise of cytosolic Ca^{2+} activity in the two different cell lines. After addition of 0.5 mM La^{3+} to a suspension of fura-2-loaded hybrid cells (Figure 1B) the fluorescence signal increased which is explainable by the affinity of La^{3+} for fura-2. The larger fluctuations after addition of La^{3+} are attributed to some aggregation of the cells. In the presence of La^{3+} (Figure 1B), Ca^{2+} activity in the hybrid cells remained constant after addition of serotonin, whereas the cells still responded to bradykinin as seen before under control conditions (Figure 1A). With the glioma cells (Figures 1C and D), addition of 0.5 mM La^{3+} also caused an increase of the fluorescence intensity (Figure 1D) similar to that observed in the hybrid cells. On the other hand, the response to serotonin did not disappear in the glioma cells. In the experiment shown the rise in Ca^{2+} activity was even enhanced in the presence of La^{3+}. Further experiments using Ni^{2+}, which also blocks Ca^{2+} channels, demonstrated that Ni^{2+} (5 mM) suppressed the rise of Ca^{2+} activity upon addition of serotonin in the hybrid cells but not in glioma cells. These results suggest that Ca^{2+} entry is the underlying mechanism of Ca^{2+} activation by serotonin in the hybrid cells and that in the glioma cells Ca^{2+} is released from internal stores.

Ca^{2+} ionophores are useful tools to further elucidate the role of intracellular Ca^{2+} stores in generating the Ca^{2+} signal after serotonin application. Ionomycin, at 0.5 μM, caused a fast rise of cytosolic Ca^{2+} activity in the neuronal hybrid cells (Figure 1F), comparable in size to that elicited by bradykinin, but with a longer duration. After the ionomycin-induced rise, Ca^{2+} activity returned to baseline within 50–60 s. Thereafter the size of the Ca^{2+} signal caused by serotonin (Figure 1F) was not significantly different from the corresponding control value

(Figure 1E). The Ca^{2+} response to bradykinin, however, was suppressed after pretreatment with ionomycin. After addition of 4-Br-A23187, another Ca^{2+} ionophore with low intrinsic fluorescence, the hybrid cells showed a substantially reduced response to bradykinin, but the effect of serotonin was not affected. In the glioma cells, ionomycin caused a comparable transient rise of cellular Ca^{2+} activity. The sizes of subsequent responses of the cells to both serotonin and to bradykinin were greatly reduced as compared to the effects seen in control cells (not shown). This suggests that in the glioma cells both hormones cause release of Ca^{2+} from internal stores. Furthermore, we found that serotonin enhanced the rate of $^{45}Ca^{2+}$ uptake of the hybrid cells, by a factor of 2.7 at 2.5 μM serotonin. In the glioma cells, however, 5 μM serotonin caused an increase of only (maximally) 20% above control (Reiser et al. 1990a; b).

The following experiment sheds some light on the mechanism of Ca^{2+} removal after the increase induced by serotonin. In the hybrid cells, the Ca^{2+} response to ionomycin was augmented when a serotonin response had been elicited before adding ionomycin. At the concentrations used, ionomycin primarily acted on internal Ca^{2+} stores. The higher amplitude of the response to ionomycin after addition of serotonin seems to be due to an extra load of Ca^{2+} in the stores. Therefore, the excess cytosolic Ca^{2+} after the serotonin stimulus in the neuronal cells is most likely taken up into internal stores.

Influence of Serotonin on Levels of Inositolphosphates

The levels of $Ins(1,4,5)P_3$ and $Ins(1,3,4,5)P_4$ were determined in neuroblastoma $\times$ glioma hybrid cells and in glioma cells stimulated with serotonin $(1 - 10 \mu M)$. In the neuronal cells serotinin did not significantly change the concentrations of either inositolphosphate.

In the glioma cells, serotonin induced a slight increase in the level of $Ins(1,4,5)P_3$. The stimulation reached maximally twofold the basal level (Donié & Reiser, unpublished results). However, the variability of the degree of stimulation between different experiments warrants further detailed analysis.

Mechanism of Activation of Guanylate Cyclase by Serotonin in the Neuronal Cell Line

Serotonin raises the levels of cyclic GMP in neuroblastoma $\times$ glioma hybrid cells, half-maximally at a concentration of 1 μM. The time course of stimulation which shows a peak at around 20 s after addition of serotonin was not affected by the addition of $1\,\mu M$ tetrodotoxin (Figure 2A).

The stimulation of cyclic GMP levels by serotonin depends on the extracellular Ca^{2+} concentration. The lowest Ca^{2+} concentration shown in the test in Figure 2B was $10\,\mu M$. The concentration-response curve, which reaches half-maximal stimulation at 0.3 mM Ca^{2+}, can be fitted with a hyperbolic curve. Inorganic blockers of Ca^{2+} channels were added to evaluate a contribution of extracellular Ca^{2+} to the serotonin-induced cyclic GMP response. The cells were exposed only briefly to Ni^{2+}, Co^{2+}, Mn^{2+} or La^{3+} before being challenged with serotonin. Figure 2C shows that under these conditions the effect of serotonin was blocked half-maximally at $40\,\mu M$ La^{3+}, 0.4 mM Mn^{2+}, 1.2 mM Ni^{2+} or 0.9 mM Co^{2+}. The cyclic GMP response to serotonin was also suppressed by organic Ca^{2+} channel blockers, i.e. desmethoxyverapamil (D888) with an IC_{50} of $0.5\,\mu M$ for the S-enantiomer and diltiazem showing an IC_{50} of $6\,\mu M$. As Figure 2D shows, tubocurarine, a substance blocking the ion channel of the $5\text{-}HT_3$ receptor (Peters et al. 1990), suppresses the cyclic GMP respose to serotonin half-maximally at 30 nM.

In some experiments the neuronal cells were preloaded with the acetoxymethyl ester of the Ca^{2+} chelator BAPTA. The ester permeates the cell membrane and is cleaved by intracellular hydrolytic enzymes. Thus BAPTA accumulates in the cells and clamps the concentration of cytosolic free Ca^{2+}. With increasing concentrations of BAPTA, the cyclic GMP response to serotonin was suppressed, half-maximally at about $2\,\mu M$ BAPTA/acetoxymethylester in the incubation medium. The reduced capacity of the cells to raise cytosolic Ca^{2+} levels is apparently the reason for the diminished cyclic GMP synthesis.

Ca^{2+} ionophores were used to study the contribution of intracellular Ca^{2+} stores to the activation of guanylate cyclase by serotonin. At low concentrations, the Ca^{2+} ionophores primarily release Ca^{2+} from intracellular stores and prevent their refilling. Preincubation of the hybrid cells with ionomycin or A23187 did not greatly affect the cyclic GMP response to serotonin, but substantially depressed the effect of bradykinin on the level of cyclic GMP (Figure 2E).

Mepacrine, an inhibitor of phospholipase A_2, blocked the rise of the cyclic GMP level induced by serotonin. Methylene blue, an agent known to block soluble guanylate cyclase (Waldman and Murad 1987), inhibited the response to both serotonin and to bradykinin with a half-maximal inhibitory concentration (IC_{50}) of around $7\mu M$.

Hemoglobin which inhibits the activity of the endothelium-derived relaxing-factor (EDRF) partially suppressed the rise of cyclic GMP level caused by serotonin. Oxyhemoglobin prepared as described (Reiser 1990) reduced the stimulation of cyclic GMP synthesis induced by serotonin or bradykinin with half-maximal activity at $0.5\,\mu M$; maximal inhibition was 90%.

76

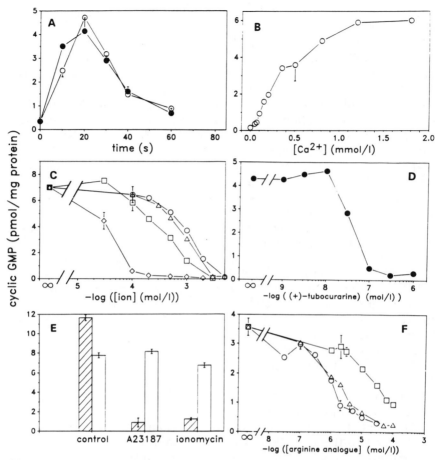

Figure 2. Regulation of cyclic GMP levels by serotonin in the neuronal hybrid cells. (A) Time course of cellular contents of cyclic GMP after addition of 5 μM serotonin in the absence ($\bigcirc$) or presence ($\bullet$) of 1 μM tetrodotoxin. Influence of (B) extracellular Ca^{2+} concentration, (C) the inorganic blockers of Ca^{2+} channels Ni^{2+} ($\bigcirc$), Mn^{2+} ($\square$), Co^{2+} ($\triangle$) and La^{3+} ($\diamond$), (D) tubocurarine, (E) Ca^{2+} ionophores, and (F) the arginine analogues N^G-monomethyl-L-arginine ($\triangle$), N^G-nitro-L-arginine ($\bigcirc$), and canavanine ($\square$), on the serotonin-induced rise of cyclic GMP level. Cells were challenged for 20 s with 5 μM serotonin after preincubation for 30 min in incubation medium containing various concentrations of $CaCl_2$ (B), or in incubation medium supplemented with 1 μM A23187 or ionomycin (E), or without addition (A, C, D, and F). In C, 1.5 min before the challenge incubation with serotonin, the Ca^{2+} channel blockers were added. Tetrodotoxin (A), (+)-tubocurarine (D) or arginine analogues (F) were added 2.5 min before the end of the 30 min preincubation period. Data represent mean values from duplicate incubations, whose cyclic GMP content was calculated from the average of duplicate determinations in the radioimmunoassay. Deviations of the experimental values from the mean values are shown only when they exceed the size of the symbol used. Basal content of cyclic GMP was between 0.1 and 0.3 pmol/mg protein in the experiments shown.

Some guanidino analogues of arginine inhibit the formation of nitric oxide and NO-containing compounds (Schmidt et al. 1988). Preincubating the hybrid cells with canavanine, a guanidinooxy analogue, abolished the ability of either serotonin or bradykinin to raise the cyclic GMP level with IC_{50} values between 10 and 50 μM. N^G-monomethyl-L-arginine and N^G-nitro-L-arginine suppressed the rise in cyclic GMP level caused by serotonin with an IC_{50} of around 1 μM (Figure 2F).

Electrophysiological Responses to Serotonin

Iontophoretic application of serotonin elicited a fast depolarization response in the hybrid cells (Reiser et al. 1988; 1989). The depolarization was associated with a conductance increase. The dependence of the amplitude on the extracellular Na^+ concentration has been taken as evidence for activation of a cation conductance in the hybrid cells. In some cells the depolarization response to serotonin was associated with one or several action potentials which could be suppressed by tetrodotoxin (Reiser et al. 1988). The response to serotonin in the hybrid cells was blocked reversibly by ICS 205-930, MDL 72222 and by ondansetron, but was not sensitive to ketanserin (up to 2 μM). The kinetics and the single channel properties of the ion channel activated by 5-HT$_3$ receptors have been characterized in various neuronal cell lines (Lambert et al. 1989; Neijt et al. 1989) and in cultured hippocampal cells (Yakel and Jackson 1988).

In the glioma cells, however, iontophoretic pulses of serotonin induced a transient hyperpolarization with a delay of a few seconds after the current pulse. Ketanserin or methysergid blocked the effect of serotonin, which reappeared after washing out the antagonist. Ketanserin completely blocked the membrane potential response to serotonin at a range of concentrations from 5 to 100 nM in different cells. ICS 205-930 and ondansetron, however, even at 10 μM concentrations had no effect on the responsiveness of the same cells to serotonin (not shown).

Discussion

The sequence of reactions triggered by 5-HT$_3$ receptors in the neuronal cells or by 5-HT$_2$ receptors in the non-excitable glioma cells is shown schematically in Figure 3. The graph indicates the interactions of the receptors with the respective second messengers and effector systems. In the neuronal cells, 5-HT$_3$ receptors are involved since inhibitors, such as ICS 205-930, MDL 72222 and ondansetron, blocked the responses at nanomolar concentrations. Binding studies using radiolabelled ICS 205-930 have demonstrated the existence of 5-HT$_3$ recep-

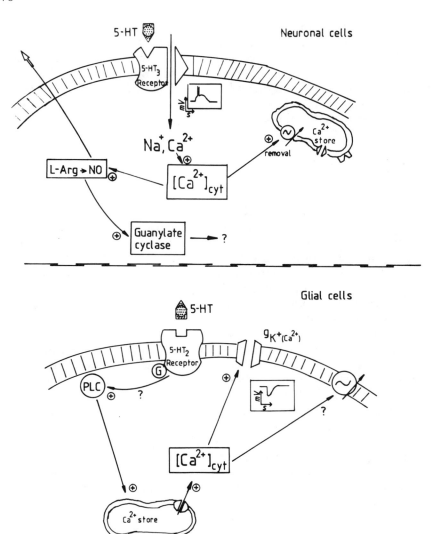

Figure 3. Schematic representation of the interaction of serotonin receptors with the respective second messengers and effector systems. 5-HT$_3$ receptors in the neuronal cell line induce a fast depolarization (inset picture shows schematic depolarization response) leading to entry of Na$^+$ into the cell and a concomitant rise in cytosolic Ca^{2+} activity ([Ca^{2+}]$_{cyt}$) through Ca$^+$ entry. The rise in [Ca^{2+}]$_{cyt}$ activates formation of NO from L-arginine (L-Arg) which stimulates guanylate cyclase. Ca^{2+} is removed by uptake into internal stores. 5-HT$_2$ receptors in the glial cell line activate (via a G-protein and phospholipase C?) release of Ca^{2+} from internal stores. The ensuing rise in [Ca^{2+}]$_{cyt}$ leads to opening of Ca^{2+} -dependent K$^+$ channels ($g_{K(Ca2+)}$), inset picture shows hyperpolarization response. Ca^{2+} is probably removed by extrusion. '+' denotes stimulating activity.

tors in the hybrid cells (Hoyer and Neijt 1987). In the glioma cells, 5-HT_3 receptor antagonists even at micromolar concentrations did not influence the effects of serotonin, whereas ketanserin blocked the responses in the nanomolar range suggesting the presence of 5-HT_2 receptors on the glioma cells.

In both the hybrid and in the glioma cells, serotonin induced similar, concentration-dependent, transient rises in cytosolic Ca^{2+} activity. However, the mechanisms of activation by serotonin in the two cell lines are obviously different. Ni^{2+} and La^{3+}, known to block Ca^{2+} channels, suppressed the response to serotonin in the neuronal cells but not in the glioma cells. This indicates that serotonin causes the entry of extracellular Ca^{2+} through the plasma membrane of the neuronal hybrid cells, whereas in the glioma cells serotonin mainly releases Ca^{2+} from intracellular stores. It still has to be elucidated whether a small component of entry of external Ca^{2+} is required for activation of internal Ca^{2+} release by serotonin in the glioma cells. Similarly in a previous study we have found that the Ca^{2+} response elicited by the neuropeptide bradykinin in the hybrid cells (e.g. Figure 1) seems to be largely due to release of Ca^{2+} from intracellular stores (Reiser et al. 1990a).

The Ca^{2+} ionophores ionomycin and A23187 were employed to establish the route of Ca^{2+} activation by serotonin more clearly. It has been reported that treatment of platelets with $1\ \mu M$ ionomycin prevents resequestration of internally released Ca^{2+} (Pollock et al. 1987). Similarly A23187 and ionomycin at micromolar concentrations functionally eliminate intracellular stores in the neural cell lines (Reiser et al. 1990a). Thus, in the continued presence of low concentrations of the Ca^{2+} ionophores, the internal stores are short-circuited, and recovery of an elevated cytosolic Ca^{2+} activity is almost entirely due to Ca^{2+} extrusion (Pollock et al. 1987). The glioma cells, after being challenged with $1\ \mu M$ ionomycin no longer responded to serotonin. This is consistent with the interpretation that serotonin in the glioma cells induces the release of Ca^{2+} from internal stores. The activation of the Ca^{2+} stores does not involve a pertussis toxin-sensitive G-protein, but is possibly mediated by inositol(1,4,5)trisphosphate. Also in the related rat glioma cell line C6, serotonin receptors have been demonstrated to slightly stimulate inositolphosphate metabolism (Ananth et al. 1987).

In the hybrid cells, however, after treatment with ionomycin, serotonin still caused a transient rise of Ca^{2+} activity which was comparable to the signal seen in cells without previous exposure to the Ca^{2+} ionophore. Furthermore, in the hybrid cells serotonin largely enhanced the uptake of $^{45}Ca^{2+}$, but only slightly in the glioma cells. These observations lend support to the conclusion that in the hybrid cells following addition of serotonin, Ca^{2+} enters the cells mainly from the extracellular space.

The experiments using Ca^{2+} ionophores shed light on the mechanism which brings the cytosolic Ca^{2+} level back to baseline after a serotonin-induced rise. In the hybrid cells, after the addition of serotonin the amount of Ca^{2+} releasable by ionophores from internal Ca^{2+} stores is increased. This supports the conclusion that a considerable part of Ca^{2+} entering the cells after serotonin receptor activation is taken up into intracellular stores.

Electrophysiological recordings showed that in the polyploid glioma cells, iontophoretic application of serotonin caused a hyperpolarization response, as already reported for rat glioma cells, clone C6-BU-1 (Ogura and Amano 1984). The hyperpolarizing response in the polyploid glioma cells closely resembles the response to bradykinin which has been shown to involve activation of Ca^{2+}-dependent K^+ channels (Reiser et al. 1990b). In the neuronal hybrid cells, however, the 5-HT_3 receptors enhance without delay and directly, a cation conductance with limited selectivity leading to depolarization of the cell (Jakel and Jackson 1988), concomitantly augmenting the entry of extracellular Ca^{2+}. The regulation of the permeability of this channel for Ca^{2+} has still to be investigated.

The sensitivity of the serotonin-induced cyclic GMP response in the hybrid cells to drugs like methylene blue indicates that serotonin activates the soluble guanylate cyclase (for refs. see Waldman and Murad 1987). In the neuronal cells, some of the properties of the transducer between the serotonin receptor and guanylate cyclase resemble those of endothelium-derived relaxing-factor (EDRF). EDRF, a diffusible agent released from the endothelial cells induces relaxation of the smooth muscle cells acting via stimulation of cyclic GMP synthesis (Furchgott and Vanhoutte 1989). Recently one EDRF has been identified as nitric oxide, NO (Palmer et al. 1987). The activity of EDRF is inhibited by hemoglobin and some redox compounds (for review see Moncada et al. 1989). Since hemoglobin cannot penetrate the cells, this large protein could act by scavenging an activator substance which accumulates in the extracellular space upon exposure to serotonin. Thus, part of the stimulation of cyclic GMP synthesis by serotonin observed in the homogeneous cell population studied here could arise from a diffusible factor-possibly NO-which is released from the cells and acts on neighbouring cells.

Nitric oxide or a NO-containing compound has been reported to be formed enzymatically from the terminal guanidino nitrogen of L-arginine (Schmidt et al. 1988; Moncada et al. 1989). The synthesis of nitric oxide can be suppressed by structural analogues of L-arginine such as N^G-monomethyl-L-arginine, N^G-nitro-L-arginine and canavanine. Since the analogues of arginine effectively block the rise of cyclic GMP level caused by serotonin, the same mechanism, i.e. formation of nitric oxide from arginine, seems to apply to the action of serotonin in the hybrid cells.

A rise in cytosolic Ca^{2+} activity is the crucial step for activation of guanylate cyclase in the neuronal hybrid cells, since buffering intracellular Ca^{2+} by the chelator BAPTA (Grynkiewicz et al. 1985) suppressed the stimulation of cyclic GMP level caused by serotonin. Inorganic and organic blockers of Ca^{2+} channels completely suppressed the cyclic GMP response to serotonin. Pretreatment of the hybrid cells with the Ca^{2+} ionophores ionomycin and A23187 did not affect responsiveness to serotonin. These findings corroborate the conclusion that the rise of cytosolic Ca^{2+} activity, which is necessary for activation of guanylate cyclase, is caused by serotonin-induced entry of extracellular Ca^{2+}.

Tetrodotoxin which blocks action potentials associated with the depolarizing activity of serotonin in the hybrid cells (Reiser et al. 1988) did not alter the time course of cyclic GMP formation induced by serotonin. Moreover, (+)-tubocurarine which blocks the ion channels activated by 5-HT_3 receptors (Peters et al. 1990) potently inhibited the cyclic GMP formation. These findings make an indirect stimulation of Ca^{2+} entry through voltage dependent Ca^{2+} channels very unlikely and are consistent with the conclusion that serotonin stimulates Ca^{2+} entry through 5-HT_3 receptors in the neuronal cells.

The summary in Fig. 3 shows that in the glioma cells 5-HT_2 receptors cause, probably via inositol(1,4,5)trisphosphate, release of Ca^{2+} from internal stores. The rise in cytosolic Ca^{2+} activity leads to opening of Ca^{2+}-dependent K^+ channels causing a hyperpolarization of the plasma membrane. The excess Ca^{2+} is most likely extruded through the plasma membrane. In the neuronal cells, activation of 5-HT_3 receptors is associated with a rapid increase of a cation conductance and rise of cytosolic Ca^{2+} activity which is due to entry of Ca^{2+} through the 5-HT_3 receptor channel. The increase in cytosolic Ca^{2+} activity is a prerequisite for the cascade of nitric oxide formation from L-arginine and subsequent activation of guanylate cyclase. The cascade can be interrupted either by chelating cytosolic Ca^{2+} or by blocking nitric oxide formation. Ca^{2+} is removed through uptake into internal stores.

Acknowledgements

This work was supported by a project grant from the Deutsche Forschungsgemeinschaft (Re 563/2-1 and 2). The skilful technical assistance of Britta Baumann is gratefully acknowledged. I thank Frédéric Donié for his most valuable contributions to parts of the work presented here and his helpful comments on the manuscript.

References

Ananth, U. S., Leli, U., and Hauser, G. (1987). Stimulation of phosphoinositide hydrolysis by serotonin in C6 glioma cells. J. Neurochem. 48: 253–261.

Christian, C. N., Nelson, P. G., Bullock, P., Mullinax, D., and Nirenberg, M. (1978). Pharmacological responses of cells of a neuroblastoma × glioma hybrid clone and modulation of synapses between hybrid cells and mouse myotubes. Brain Res. 147: 261–276.

Conn, P. J., and Sanders-Bush, E. (1985). Serotonin-stimulated phosphoinositide turnover: mediation by the S_2 binding site in rat cerebral cortex but not in subcortical regions. J. Pharmacol. Exp. Ther. 234: 195–203.

Donié F., and Reiser G. (1989). A novel, specific binding protein assay for quantitation of intracellular inositol 1,3,4,5-tetrakisphosphate ($InsP_4$) using a high-affinity $InsP_4$ receptor from cerebellum. FEBS Lett. 254: 155–158.

Fozard, J. R. (1984). MDL 72222: a potent and highly selective antagonist at neuronal 5-hydroxytryptamine receptors. Naunyn Schmiedeberg's Arch. Pharmacol. 326: 36–44.

Furchgott, R. F. & Vanhoutte, P. M. (1989). Endothelium-derived relaxing and contracting factors. FASEB J. 3: 2007–2018.

Grynkiewicz, G., Poenie, M., and Tsien, R. Y. (1985). A new generation of Ca^{2+} indicators with greatly improved fluorescence properties. J. Biol. Chem. 260: 3440–3450.

Gundersen, C. B., Miledi, R., and Parker, I. (1984). Messenger RNA from human brain induces drug- and voltage-operated channels in Xenopus oocytes. Nature (Lond.) 308: 421–424.

Hamprecht, B., Glaser, T., Reiser, G., Bayer, E., and Propst, F. (1985). Culture and characteristics of hormone-responsive neuroblastoma × glioma hybrid cells. Meth. Enzymol. 109: 316–341.

Heumann, R., Reiser, G., van Calker, D., and Hamprecht, B. (1982). Polyploid rat glioma cells: production, oscillations of membrane potential and response to neurohormones. Exp. Cell Res. 139: 117–126.

Hoyer, D., and Neijt, H. C. (1987). Identification of serotonin 5-HT$_3$ recognition sites by radioligand binding in NG108-15 neuroblastoma-glioma cells. Eur. J. Pharmacol. 142: 291–292.

Jankowsky, A., Labarca, R., and Paul, S. A. (1984). Characterization of neurotransmitter receptor mediated phosphatidylinositol hydrolysis in the rat hippocampus. Life Sci. 35: 1953–1961.

Kilpatrick, G. J., Jones, B. J., and Tyers, M. B. (1987). Identification and distribution of 5-HT$_3$ receptors in rat brain using radioligand binding. Nature (Lond.) 330: 746–748.

Lambert, J. J., Peters, J. A., Hales, T. G., and Dempster, J. (1989). The properties of 5-HT$_3$ receptors in clonal cell lines studied by patch-clamp techniques. Br. J. Pharmacol. 97: 27–40.

Leysen, J. E., Niemegeer, C. J. E., Van Neuten, J. M., and Laduron, P. M. (1982). [^{3}H]Ketanserin (R41 468), as selective ^{3}H-ligand for serotonin$_2$ receptor binding sites: binding properties, brain distribution and functional role. Mol. Pharmacol. 21: 304–314.

Lübbert, H., Hoffman, B. J., Snutch, T. P., van Dyke, T., Levine, A. J., Hartig, P. R., Lester, H. A., and Davidson, N. (1987). cDNA cloning of serotonin 5-HT$_{1C}$ receptor by electrophysiological assays of mRNA-injected Xenopus oocytes, Proc. Natl. Acad. Sci. USA. 84: 4332–4336.

Moncada, S., Palmer, R. M. J., and Higgs, E. A. (1989). Biosynthesis of nitric oxide from L-arginine. Biochem. Pharmacol. 38: 1709–1715.

Neijt, H. C., Plomb, J. J., and Vijverberg, H. P. M. (1989). Kinetics of the membrane current mediated by serotonin 5-HT$_3$ receptors in cultured mouse neuroblastoma cells. J. Physiol. 411: 257–269.

Ogura, A., and Amano, T. (1984). Serotonin-receptor coupled with membrane electrogenesis in a rat glioma clone. Brain Res. 297: 387–391.

Palmer, R. M. J., Ferrige, A. G., and Moncada, S. (1987). Nitric oxide release accounts for the biological activity of endothelium-derived relaxing factor. Nature (Lond.) 327: 524–526.

Peroutka, S. J. (1988). 5-Hydroxytryptamine receptor subtypes. Annu. Rev. Neurosci. 11: 45–60.

Peters, J. A., Malone, H. M., and Lambert, J. J. (1990). Antagonism of 5-HT$_3$ receptor mediated currents in murine N1E-115 neuroblastoma cells by (+)-tubocurarine. Neurosci. Lett. 110: 107–112.

Pollock, W. K., Sage, S. O., and Rink, T. J. (1987). Stimulation of Ca^{2+} efflux from fura-2-loaded platelets activated by thrombin or phorbol myristate acetate. FEBS Lett. 210: 132–136.

Reiser, G. (1990). Mechanism of stimulation of cyclic GMP level in a neuronal cell line mediated by serotonin (5-HT$_3$) receptors: involvement of nitric oxide, arachidonic-acid metabolism and cytosolic Ca^{2+}. Eur. J. Biochem. 189: 547–552.

Reiser, G., Walter, U., and Hamprecht, B. (1984). Bradykinin regulates the level of guanosine 3′, 5′-cyclic monophosphate (cyclic GMP) in neural cell lines. Brain Res. 290: 367–371.

Reiser, G., Binmöller, F.-J., and Koch, R. (1988). Memantine (1-amino-3,5-dimethyladamantane) blocks the serotonin-induced depolarization response in a neuronal cell line. Brain Res. 443: 338–344.

Reiser, G., Donié, F., and Binmöller, F.-J. (1989). Serotonin regulates cytosolic Ca^{2+} activity and membrane potential in a neuronal and in a glial cell line via 5-HT$_3$- and 5-HT$_2$-receptors by different mechanisms. J. Cell. Sci. 93: 545–555.

Reiser, G., Binmöller, F.-J., and Donié, F. (1990a). Mechanisms for activation and subsequent removal of cytosolic Ca^{2+} in bradykinin-stimulated neuronal and glial cell lines. Exp. Cell. Res. 186: 47–53.

Reiser G., Binmöller F.-J., Strong P. N., and Hamprecht B. (1990b). Activation of a K$^+$ conductance by bradykinin and by inositol-1,4,5-trisphosphate in rat glioma cells: involvement of intracellular and extracellular Ca^{2+}. Brain Res. 506: 205–214.

Richardson, B. P., and Engel, G. (1986). The pharmacology and function of 5-HT$_3$ receptors. Trends Neurosci. 9: 424–428.

Schmidt, H. H. H. W., Nau, H., Wittfoht, W., Gerlach, J., Prescher, K. E., Klein, M. M., Niroomand, F., and Böhme, E. (1988). Arginine is a physiological precursor of endothelium-derived nitric oxide. Eur. J. Pharmcol. 154: 213–216.

Waldman, S. A., and Murad, F. (1987). Cyclic GMP synthesis and function. Pharmacol. Rev. 39: 163–196.

Yakel, J. L., and Jackson, M. B. (1988). 5-HT$_3$ receptors mediate rapid responses in cultured hippocampus and a clonal cell line. Neuron 1: 615–621.

Serotonin: Molecular Biology, Receptors and Functional Effects
ed. by J. R. Fozard/P. R. Saxena

Characterization of 5HT₃ Receptor Mediated Electrical Responses in Nodose Ganglion Neurones and Clonal Neuroblastoma Cells Maintained in Culture

J. A. Peters, H. M. Malone, and J. J. Lambert

Department of Pharmacology and Clinical Pharmacology, Ninewells Hospital & Medical School, The University, Dundee, DD1 9SY, Scotland

Summary. Voltage-clamped adult rabbit nodose ganglion (NG) neurones and murine N1E-115 neuroblastoma cells respond to locally applied 5-HT with an inward current response mediated by the opening of 5-HT₃ receptor-linked cation selective ion channels. 5-HT-induced currents reversed in sign at a potential close to 0 mV and desensitized rapidly. The conductance of the 5-HT₃ receptor ion channel complex in N1E-115 cells (~ 0.3 pS) was estimated to be approximately 155-fold lower than that found in rabbit NG neurones (~ 17.0 pS). Ondansetron and metoclopramide demonstrated similar IC_{50} values in blocking 5-HT-induced currents in the two cell types, but the antagonist potencies of (+)-tubocurarine and cocaine in N1E-115 cells (IC_{50} values 0.85 nM and 7.9 μM respectively) were different to those determined in rabbit NG neurones (160 nM and 80 nM) under similar recording conditions. The relative contributions of species and tissue differences to these discrepancies are discussed, together with preliminary data obtained from guinea-pig nodose ganglion neurones, which suggests the former to be an important variable.

Introduction

In peripheral autonomic and sensory neurones, and several neuronal clonal cell lines, 5-HT₃ receptor activation elicits a depolarizing response associated with an increase in membrane conductance (Peters and Lambert 1989, Wallis 1989). Recent electrophysiological studies employing voltage-clamp and single-channel recording techniques have demonstrated that the 5-HT-induced depolarization is due to a transient inward current response contingent upon the opening of cation selective ion channels (Higashi and Nishi 1982, Lambert et al. 1989, Derkach et al. 1989). Published estimates of the conductance of the channel vary considerably. Recording from N1E-115 neuroblastoma cells, Lambert et al. (1989) inferred a value of 0.31 pS from fluctuation analysis of whole-cell currents, whereas Derkach et al. (1989) reported conductances of 15 and 9 pS by the direct observation of single-channel events in outside-out membrane patches excised from guinea-pig submucous plexus neurones. These discrepancies, coupled with suggestions of heterogeneity in the pharmacological properties of 5-HT₃ receptors

(Richardson et al. 1985) prompted the present comparative study of 5-HT$_3$ receptor mediated whole-cell- and single-channel-currents in N1E-115 neuroblastoma cells and nodose ganglion (NG) neurones isolated from rabbit and guinea-pig.

Materials and Methods

Cell Isolation and Culture

N1E-115 neuroblastoma cells were maintained in culture as previously described (Peters et al. 1988). Nodose ganglia excised from adult New Zealand rabbits and guinea-pigs were enzymatically dispersed into a single cell suspension using the method of Ikeda et al. (1986) with minor modifications. NG neurones were maintained in culture for 1–7 days prior to use in electrophysiological experiments.

Electrophysiology

Agonist-evoked whole-cell and single-channel currents were recorded using standard patch-clamp techniques (Hamill et al. 1981). In all single channel, and the majority of whole-cell recordings, a List Electronics L/M EPC 7 converter headstage and amplifier was used. Voltage errors were minimised by the use of low resistance (2–5 MΩ) patch pipettes in conjunction with series resistance compensation (30–80%). Whole-cell currents were also recorded with an Axoclamp 2A in discontinuous single electrode voltage-clamp mode (switching frequency 8–14 KHz). The results obtained with two recording systems were consistent. Unless stated otherwise, cells were continually superfused (2–4 ml min^{-1}) with a recording medium comprising (in mM):NaCl 140, KCl 2.8, MgCl$_2$ 2.0, CaCl$_2$ 1.0 and HEPES 10 (pH 7.2). The pipette solution employed to dialyse the cell interior consisted of: KCl 140, MgCl$_2$ 2, CaCl$_2$ 0.1, EGTA 1.1 and HEPES 10 (pH 7.2). All recordings were performed at room temperature (17–23°C). Quantitative data are given as the arithmetic mean $\pm$ the standard error of the mean.

Results

Ionic Basis of 5-HT$_3$ Receptor Mediated Currents

Under constant-current recording conditions, rabbit NG neurones displayed a resting potential of -55.7 ± 1.5 mV ($n = 12$) and an input resistance of 338.3 ± 55.6 MΩ ($n = 12$). In these unclamped neurones,

5-HT (10 μM), applied by pressure ejection from a modified patch pipette, evoked a rapid membrane depolarization associated with action potential discharge and a fall in input resistance (Figure 1A). Such responses were frequently succeeded by a slow hyperpolarization as described by others (e.g. Higashi & Nishi 1982). When rabbit NG neurones were voltaged-clamped at a holding potential of -60 mV,

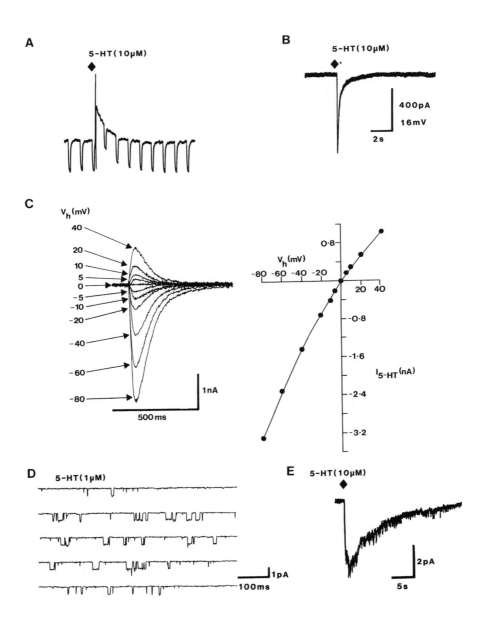

similar applications of 5-HT elicited a transient monophasic inward current associated with an increase in membrane conductance (Figure 1B). The amplitude of the current decreased with membrane depolarization, and using the recording solutions detailed above, reversed in sign at a potential of -1.6 ± 1.0 mV ($n = 5$) (Figure 1C). The total replacement of KCl in the pipette solution by CsCl had no significant effect upon the reversal potential of the response to 5-HT ($E_{5\text{-HT}}$), whereas reducing the internal concentration of K^+ to 20 mM, by partial substitution of KCl with tetraethylammonium. Cl, resulted in a positive shift in $E_{5\text{-HT}}$ to 32.5 ± 3.3 mV ($n = 3$) (Table 1). When the concentration of NaCl in the extracellular medium was reduced to 35 mM by isoosmotic replacement with sucrose, $E_{5\text{-HT}}$ was negatively displaced to -33.7 ± 1.6 mV ($n = 3$). Replacement of extracellular Cl alone, by the poorly permeant isethionate anion, had little effect upon $E_{5\text{-HT}}$ (Table 1). The above results are consistent with the proposal by Higashi and Nishi (1982), that an increase in membrane Na and K permeability underlies the response to 5-HT in rabbit NG neurones. 5-HT_3 receptor-induced currents in N1E-115 cells have a similar ionic basis; indeed the ratio of Na and K permeabilities (P_{Na}/P_K) reported in N1E-115 cells (0.92, Lambert et al. 1989) is in excellent agreement with that of 0.94 calculated from the Goldman-Hodgkin-Katz voltage equation (Hille 1984) and the measurements of $E_{5\text{-HT}}$ in rabbit NG neurones reported above.

Properties of 5-HT$_3$ Receptor-Gated Single-Channel Currents

Acting upon outside-out membrane patches excised from rabbit NG neurones, 5-HT (1 μM) elicited well resolved single-channel events

Figure 1. Properties of 5-HT_3 receptor mediated electrical responses in rabbit nodose ganglion neurones (A-D) and N1E-115 neuroblastoma cells (E). (A): Depolarizing response to pressure applied 5-HT (10 μM, 20 ms, 1.4×10^5 Pa) recorded in current-clamp mode from a rabbit nodose ganglion neurone. Note that during the response to 5-HT, the anelectrotonic potential elicited by hyperpolarizing constant current pulses decreases in size, indicating an apparent increase in membrane conductance. Cell resting potential was -68 mV. (B): Inward current response to an identical application of 5-HT recorded from the same neurone as in (A) under voltage-clamp conditions (holding potential $= -68$ mV). (C): Current responses to pressure applied 5-HT (10 μM, 10 ms, 1.4×10^5 Pa) recorded from a rabbit nodose ganglion neurone. Currents were recorded at holding potentials (V_h) in the range -80 to $+40$ mV as indicated. Each trace is the computer generated average of 4 responses to 5-HT. A plot of 5-HT-induced current amplitude ($I_{5\text{-HT}}$) as a function of holding potential yields a current-voltage relationship demonstrating slight inward rectification and a reversal potential of 0 mV in this example. Control extracellular solution, Cs^+ replacing K^+ in the internal solution (see Table 1). (D): Single-channel currents, recorded from an outside-out membrane patch excised from a rabbit nodose ganglion neurone, in response to bath applied 5-HT (1 μM). Channel currents were recorded at a holding potential of -80 mV. The records are sequential. (E): Response of an outside-out membrane patch excised from an N1E-115 neuroblastoma cell to pressure applied 5-HT (10 μM, 1 s, 1.4×10^5 Pa). Note the absence of resolved single-channel activity. Patch current recorded at a holding potential of -60 mV. Currents in (B), (C) and (E) were low-pass filtered at 0.5 KHz, those in (D) at 1.0 KHz.

Table 1. The effects of ion substitutions upon the reversal potential ($E_{5\text{-}HT}$) of 5-HT$_3$ receptor-induced currents in rabbit nodose ganglion neurones

External solution	Internal solution	$E_{5\text{-}HT}$(mV)	$\Delta E_{5\text{-}HT}$(mV)
Control ($Na^+ = 143$ $K^+ = 2.8$, $Cl^- = 149$)	Control ($Na^+ = 6$, $K^+ = 140$, $Cl^- = 144$)	-1.6 ± 1.0 ($n = 5$)	—
Control ($Na^+ = 143$, $K^+ = 2.8$, $Cl^- = 149$)	Cs^+ replaces K^+ ($Na^+ = 6$, $Cs^+ = 140$, $Cl^- = 144$)	-2.1 ± 0.7 ($n = 8$)	-0.5
Control ($Na^+ = 143$, $K^+ = 2.8$, $Cl^- = 149$)	TEA replaces K^+ ($Na^+ = 6$, $K^+ = 20$ TEA $= 120$, $Cl^- = 144$)	$+32.5 \pm 3.3$ ($n = 3$)	34.1
Reduced Na^+ and Cl^- ($Na^+ = 35$, $K^+ = 2.8$, $Cl^- = 41$)	Cs^+ replaces K^+ ($Na^+ = 6$, $Cs^+ = 140$, $Cl^- = 144$)	-33.7 ± 1.6 ($n = 3$)	-31.6
Reduced Cl^- ($Na^+ = 140$, $K^+ = 2.8$, $Cl^- = 6$ mM)	Cs^+ replaces K^+ ($Na^+ = 6$, $Cs^+ = 140$, $Cl^- = 144$)	-3.2 ± 1.1 ($n = 3$)	-1.1

Ion substitutions were performed as described in the text. All external solutions contained Ca^{2+} 1 mM, Mg^{2+} 2 mM, HEPES 10 mM (pH 7.2) in addition to ions listed in the table. Internal solutions contained Mg^{2+} 2 mM, Ca^{2+} 0.1 mM, EGTA 1.1 mM and HEPES 10 mM (pH 7.2) as standard. $E_{5\text{-}HT}$ has been corrected for liquid junction potentials as described previously (Lambert et al. (1989)). Results expressed as mean $\pm$ s.e.m. (n) $\cdot$ $\Delta E_{5\text{-}HT}$ = change in reversal potential. Figures in brackets give ion concentrations in mM.

which declined in frequency in the continued presence of the agonist (Figure 1D). At a holding potential of -70 mV, single-channel currents had a mean amplitude of -1.07 ± 0.04 pA ($n = 4$), which assuming $E_{5\text{-}HT}$ to be ~ -2.0 mV (see above), corresponds to a single channel conductance of 16.6 ± 0.7 pS ($n = 4$). A similar value of 18 pS was determined directly from a detailed examination of the single channel current-voltage relationship in one membrane patch. The probability of channel opening was reversibly reduced when metoclopramide (1 μM) was co-applied with 5-HT (1 μM). Metoclopramide had no effect upon single-channel current amplitude. Outside-out membrane patches excised from N1E-115 cells, voltage clamped at a holding potential of -60 mV, responded to 5-HT with a relatively smoothly rising and decaying inward current response (Figure 1E). Discrete single-channel events within such signals were not resolved. The currents recorded from such patches displayed desensitization, reversed in sign at a potential of ~ 0 mV, and were reversibly suppressed by ondansetron (1 nM). Fluctuation analysis of whole-cell currents evoked by 1 μM 5-HT suggests a single-channel current amplitude of 18 fA at a holding potential of -60 mV, corresponding to a channel conductance of 0.31 pS (Lambert et al. 1989).

Table 2. The potencies of 5-HT$_3$ antagonists in several species and neurone types

Species	Mouse	Rabbit	Guinea-pig		
Preparation	N1E-115 neuroblastoma	Nodose ganglion neurone	Nodose ganglion neurone	Submucous plexus neurone[1]	Coeliac ganglion neurone[2]
Antagonist					
ICS 205-930	ND	10.3	ND	7.9	ND
Ondansetron	9.6	10.1	7.6	7.0	ND
MDL-72222	ND	9.4	6.5	5.5	5.8
Metoclopramide	7.6	7.9	5.9	ND	ND
Cocaine	5.1	7.1	5.7	5.5	ND
(+)-Tubocurarine	9.1	6.8	5.0	4.7	4.4

The table lists antagonist pIC$_{50}$ values [i.e. -log$_{10}$IC$_{50}$(M)] determined for the compounds acting against 5-HT-induced inward current responses recorded under voltage-clamp (this study), or depolarizing responses recorded with intracellular electrodes[1,2]. Data for guinea-pig submucous and coeliac ganglion neurones were taken from Vanner and Surprenant (1990), and Wallis and Dun (1988) respectively. ND = Not determined, or data unavailable[1,2].

Pharmacological Characterization

Inward currents recorded from rabbit NG neurones in response to locally applied 5-HT (10 μM) were resistant to antagonism by bath applied methysergide (1 μM, $n = 3$) or ketanserin (1 μM, $n = 4$). In contrast, the selective 5-HT$_3$ receptor antagonists ICS 205-930, ondansetron and MDL-72222 potently and concentration-dependently blocked the response with IC$_{50}$ values of 49, 78 and 380 pM respectively. The less selective compounds, cocaine, (+)-tubocurarine [(+)-Tc], and metoclopramide were also effective as antagonists (Figure 2 and Table 2). IC$_{50}$ values determined in rabbit NG neurones agreed well with those obtained, under similar recording conditions, from N1E-115 cells in the case of ondansetron and metoclopramide, but large discrepancies in the potencies of cocaine and (+)-tubocurarine were evident (Table 2). The rank order of antagonist potencies in rabbit NG neurones (ondansetron > metoclopramide > cocaine > (+)-Tc) was consequently

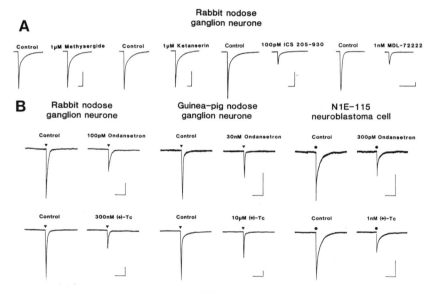

Figure 2. Antagonist pharmacology of 5-HT-induced currents in rabbit nodose ganglion neurones (A), and a comparison of the effects of ondansetron and (+)-Tc upon responses to 5-HT recorded from rabbit and guinea-pig nodose ganglion neurones and N1E-115 neuroblastoma cells (B). (A): Currents evoked by pressure applied 5-HT (10 μM, 10–30 ms, 10–30 ms, 1.4×10^5 Pa) in rabbit nodose ganglion neurones are unaffected by methysergide (1 μM), and ketanserin (1 μM), but antagonised by ICS 205-930 (100 pM) and MDL-72222 (1 nM). (B): The influence of ondansetron and (+)-Tc upon 5-HT-evoked currents recorded from rabbit and guinea-pig nodose ganglion neurones and N1E-115 cells. 5-HT was applied either by pressure ejection (10 μM, 10–30 ms, 1.4×10^5 Pa), or in the case of N1E-115 cells by ionophoresis (40 nA, 40 ms). All currents were recorded at a holding potential of -60 mV and low-pass filtered at 0.5 KHz. Calibration: vertical bar 0.2 nA all traces; horizontal bar 1s, except N1E-115 cells, 10 s.

different to that found in N1E-115 cells (ondansetron $>(+)$-Tc $>$ metoclopramide $>$ cocaine). Preliminary studies on guinea-pig NG cells indicate the potencies of all antagonists studied to be considerably less than in the corresponding neurones from the rabbit, but their rank order of effectiveness is nonetheless the same (Table 2).

Discussion

The results of this study indicate that 5-HT$_3$ receptors in rabbit NG neurones are linked to a cation selective channel with a permeability ratio P_{Na}/P_K of 0.92. This estimate is within the range of P_{Na}/P_K values (0.91–1.12) calculated from measurements of E_{5-HT} in the neuronal clonal cell lines N1E-115 (Lambert et al. 1989), NG 108-15 (Yakel et al. 1990) and N18 (Yang 1990), but differs from that of 2.3 reported by Higashi and Nishi (1982) in a voltage clamp study of 5-HT induced currents in undissociated rabbit nodose ganglion neurones. Here, patch pipettes giving rise to known intracellular concentrations of ions were used, whereas Higashi and Nishi employed intracellular microelectrodes and were compelled to estimate the Na and K equilibrium potentials from the peak of the action potential and spike after hyperpolarization respectively. Errors in these estimates, or the determination of E_{5-HT}, may well explain the discrepancy in P_{Na}/P_K values. However, it should be noted that 5-HT$_3$ receptor-linked ion channels in guinea-pig submucous neurones (Derkach et al. 1989) demonstrate a P_{Na}/P_K ratio identical to that reported by Higashi and Nishi (1982).

The conductance of 5-HT$_3$ receptor channels has recently been examined by several groups using fluctuation analysis and single-channel recording techniques. In outside-out membrane patches excised from rabbit NG neurones, 5-HT activated an apparently homogeneous population of channels with a conductance of 17 pS. This value is close to that of 15 pS reported for 5-HT$_3$ receptors mediating the opening of large amplitude single-channel currents in guinea-pig submucous plexus neurones. However, lower amplitude currents, corresponding to a conductance of 9 pS, are additionally observed in the latter preparation (Derkach et al. 1989). In undifferentiated N1E-115 cells, fluctuation analysis of whole-cell currents suggests a 5-HT$_3$ receptor channel with a conductance of only 0.31 pS (Lambert et al. 1989). Using similar techniques, Yang (1990) has reported a channel conductance of 0.59 pS in N18 neuroblastoma cells, whereas Yakel et al. (1990) calculated a value of 4.4 pS from their study of 5-HT$_3$ receptor mediated currents in differentiated NG 108-15 neuroblastoma × glioma hybrid cells. The results obtained with fluctuation analysis techniques give no information upon the possible existence of multiple conductance channels, or subconductance states of a homogeneous channel population, but instead provide a weighted-mean estimate of channel conductance that is influenced by the relative frequency of all conductance states. The presence of two channel conductances associated with 5-HT$_3$ receptors in guinea-pig submucous neurones reinforces the need for caution in interpreting the results of fluctuation analysis in terms of a channel of unitary conductance. Even so, the available evidence suggests single channel conductance(s) in neuronal cell lines to be considerably less than in either rabbit NG neurones or guinea-pig submucous neurones. Whether the small conductance(s) of 5-HT$_3$ channels in clonal cell lines reflects their murine origin, or is a characteristic of the transformed cell, warrants further investigation. In this regard, it is interesting to note that in outside-out membrane patches excised from adult rat dorsal root ganglion neurones, 5-HT$_3$ receptor mediated currents cannot be resolved into discrete single channel events, again suggesting a channel of low, but as yet undetermined conductance (Robertson and Bevan 1991).

In this study, an attempt was made to compare the pharmacological properties of 5-HT$_3$ receptors in N1E-115 cells, rabbit NG neurones and latterly, guinea-pig NG neurones. (+)-Tubocurarine was identified as a compound of particular interest, since it had previously been demonstrated to block 5-HT$_3$ receptor mediated currents in the clonal cell lines N1E-115 and NG108-15, and mouse hippocampal neurones in culture, at very low concentrations (IC$_{50}$ 0.8–1.5 nM; Yakel and Jackson 1988, Peters et al. 1990). Compared to its potency in the above cell types, (+)-Tc was found to be approximately 200-fold and 1200-fold less effective as an antagonist of 5-HT in rabbit and guinea-pig neurones respectively (Table 2). In a recent comparative study of the extracellularly recorded depolarizing response to 5-HT$_3$ receptor activation in mouse, rat and guinea-pig superior cervical ganglia (SCG), Newberry et al. (1991) also noted a marked species dependence in the antagonist potency of (+)-Tc. Given its apparent ability to discriminate between 5-HT$_3$ receptors in several species, the mechanism of action of (+)-Tc is of considerable importance. In mouse and rat SGG, (+)-Tc produces a parallel dextral shift in the dose response curve to 2-methyl-5-HT, a 5-HT$_3$ receptor agonist (Newberry et al. 1991). Similar findings have been made in rat vagus nerve (unpublished observations reported by Kilpatrick et al. 1990). In N1E-115 cells, blockade by (+)-Tc is voltage- and use-independent (Peters et al. 1990) and preliminary studies suggest this to be the case in rabbit NG neurones also (Peters et al. unpublished observations). All of these findings are consistent with a competitive action of (+)-Tc at the 5-HT$_3$ receptor. However, in rabbit NG neurones a transition between an apparently competitive to a non-competitive action may occur with increasing concentrations of (+)-Tc (Higashi and Nishi 1982), and in the guinea-pig SCG, both the slope and maximum of the dose response curve to 2-methyl-5-HT are reduced by (+)-Tc (Newberry et al. 1991). By analogy to the nicotinic cholinoceptor at the neuromuscular junction, an obvious possibility is that (+)-Tc exerts both receptor- and ion channel-blocking actions (Colquhoun et al. 1979). The mechanism which predominates at the 5-HT$_3$ receptor-channel complex may well vary between species.

The proposal that species differences exist in the properties of 5-HT$_3$ receptors is supported by several other observations. The results presented in Table 2 suggest a 100-fold differential in the potency of cocaine acting against 5-HT-induced currents in rabbit NG neurones and N1E-115 cells. When antagonist potencies determined in rabbit and guinea-pig NG neurones are compared, it is clear that all compounds tested are considerably less effective in the latter species. MDL-72222 in particular demonstrates an almost 1000-fold difference in potency. Consistent with these findings, Lattimer *et al.* (1989) and Burridge et al. (1989) have reported several 5-HT$_3$ receptor antagonists to display lower affinities in guinea-pig vagus nerve compared to rat vagus nerve. In contrast, a comparison of antagonist affinities in guinea-pig ileum and guinea-pig vagus nerve gave little evidence of receptor heterogeneity (Burridge et al. 1989). Electrophysiological data emerging from single cell studies, albeit limited, reinforce the view that tissue differences in the properties of 5-HT$_3$ receptors are not as pronounced as originally suggested in the 5-HT$_{3A}$, 5-HT$_{3B}$, 5-HT$_{3C}$ subclassification proposed by Richardson et al. (1985) and Richardson and Engel (1986). An examination of Table 2 reveals the potencies of several antagonists to be similar in nodose ganglion, coeliac ganglion and submucous plexus neurones of the guinea-pig. Clearly, further work is necessary to elucidate the relative contributions of species and tissue differences to the purported heterogenity of 5-HT$_3$ receptors, and whilst it is certainly premature to rule out the possibility of receptor subtypes within a particular species, it is now clear that the situation is not as simple as A, B, C.

Acknowledgements

This work was supported by the Wellcome Trust (JJL) and a Dundee University Research Initiative grant (JAP). We are grateful to Drs. M. Jackson, N. Newberry, B. Robertson, J. Yakel, and J. Yang who provided preprints of their papers in press at the time of writing.

References

Burridge, J., Butler, A., and Kilpatrick, G. J. (1989). 5-HT$_3$ receptors mediate depolarization of the guinea-pig isolated vagus nerve. Br. J. Pharmacol. 96: 269P.

Colquhoun, D., Dreyer, F., and Sheridan, R. E. (1979) The actions of tubocurarine at the frog neuromuscular junction. J. Physiol. 293: 247–284.

Derkach, V., Surprenant, A., and North, R. A. (1989). 5-HT$_3$ receptors are membrane ion channels. Nature 339: 706–709.

Hamill, O. P., Marty, A., Neher, E., Sakmann, B., and Sigworth, F. J. (1981). Improved patch-clamp techniques for high resolution current recordings from cells and cell-free membrane patches. Pflügers Arch. 391: 85–100.

Higashi, H., and Nishi, S. (1982). 5-Hydroxytryptamine receptors of visceral primary afferent neurones on rabbit nodose ganglia. J. Physiol. 323: 543–567.

Hille, B. (1984). Ionic channels of excitable membranes. Sunderland, Massachusetts: Sinauer Associates, pp. 227–248.

Ikeda, S. R., Scholfield, G. G., and Weight, F. F. (1986). Na^+ and Ca^{2+} currents of acutely isolated adult rat nodose ganglion cells. J. Neurophysiol. 55: 527–539.

Kilpatrick, G. J., Butler, A., Hagan, R. M., Jones, B. J., and Tyers, M. B. (1990). [^{3}H] GR67330, a very high affinity ligand for 5-HT$_3$ receptors. Naunyn Schiedeberg's Arch. Pharmacol. 342: 22–30.

Lambert, J. J., Peters, J. A., Hales, T. G., and Dempster, J. (1989). The properties of 5-HT$_3$ receptors in clonal cell lines studied by patch-clamp techniques. Br. J. Pharmacol. 97: 27–40.

Lattimer, N., Rhodes, K. F., and Saville, V. L. (1989). Possible differences in 5-HT$_3$-like receptors in the rat and guinea-pig. Br. J. Pharmacol. 96: 270P.

Newberry, N. R., Cheshire, S. H., and Gilbert, M. J. (1991). Evidence that the 5-HT$_3$ receptors of the rat, mouse and guinea-pig superior cervical ganglion may be different. Br. J. Pharmacol. 102: 615–620.

Peters, J. A., Hales, T. G., and Lambert, J. J. (1988). Divalent cations modulate 5-HT$_3$ receptor-induced currents in N1E-115 neuroblastoma cells. Eur. J. Pharmacol. 151: 491–495.

Peters, J. A., and Lambert, J. J. (1989). Electrophysiology of 5-HT$_3$ receptors in neuronal cell lines. Trends Pharmacol. Sci. 10: 172–175.

Peters, J. A., Malone, H. M., and Lambert, J. J. (1990). Antagonism of 5-HT$_3$ receptor mediated currents in murine N1E-115 neuroblastoma cells by (+)-tubocurarine. Neurosci. Lett. 110: 107–112.

Richardson, B. P., Engel, G., Donatsch, P., and Stadler, P. A. (1985). Identification of serotonin M-receptor subtypes and their specific blockade by a new class of drugs. Nature, 316: 126–131.

Richardson, B. P., Engel, G. (1986). The pharmacology and function of 5-HT$_3$ receptors. Trends Neurosci. 9: 424–428.

Robertson, B., and Bevan, S. (1991). Properties of 5-HT$_3$ receptor-gated currents in adult rat dorsal root ganglion neurones. Br. J. Pharmcol. 102: 272–276.

Vanner, S., and Surprenant, A. (1990). Effects of 5-HT$_3$ receptor antagonists on 5-HT and nicotinic depolarizations in guinea-pig submucosal neurones. Br. J. Pharmcol. 99: 840–844.

Wallis, D. I. (1989). Interaction of 5-hydroxytryptamine with autonomic and sensory neurones. In: Fozard, J. R. (ed.), The peripheral actions of 5-hydroxytryptamine. Oxford University Press, pp. 220–246.

Wallis, D. I., and Dun, N. J. (1988). A comparison of fast and slow depolarizations evoked by 5-HT in guinea-pig coeliac ganglion cells in vitro. Br. J. Pharmacol. 93: 110–120.

Yakel, J. L., and Jackson, M. B. (1988). 5-HT$_3$ receptors mediate rapid responses in cultured hippocampus and a clonal cell line. Neuron. 1: 615–621.

Yakel, J. L., Shao, X. M., and Jackson, M. B. (1990). The selectivity of the channel coupled to the 5-HT$_3$ receptor. Brain Res. 533: 46–52.

Yang, J. (1990). Ion permeation through 5-HT-gated channels in neuroblastoma N18 cells. J. Gen. Physiol. 96: 1177–1198.

Serotonin: Molecular Biology, Receptors and Functional Effects
ed. by J. R. Fozard/P. R. Saxena
© 1991 Birkhäuser Verlag Basel/Switzerland

Specific Antibodies as New Tools for Studies of Central 5-HT$_{1A}$ Receptors

M. Hamon[1], S. El Mestikawy[1], M. Riad[1], H. Gozlan[1], G. Daval[2], and D. Vergé[2]

[1]INSERM U288, Neurobiologie Cellulaire et Fonctionnelle, CHU Pitié-Salpêtrière, 75634 Paris cedex 13; [2]Institut des Neurosciences, CNRS URA 1199, Université Pierre et Marie Curie, 75005 Paris, France

Summary. Polyclonal antibodies were raised by the repeated injection of rabbits with a synthetic peptide corresponding to a highly selective portion (amino-acid residues 243 to 268) of the amino-acid sequence of the rat 5-HT$_{1A}$ receptor. The anti-peptide antiserum allowed the immunoprecipitation of 5-HT$_{1A}$ receptors but not of other 5-HT receptor types and adrenergic receptors solubilized from rat hippocampal membranes. Immuno-autoradiographic labelling of rat brain sections with the anti-peptide antiserum exhibited the same regional distribution as 5-HT$_{1A}$ sites labelled by selective radioligands such as [^{3}H]8-OH-DPAT and [^{125}I]BH-8-MeO-N-PAT. However, regional differences apparently existed between the respective intensity of labelling by the agonist radioligands and the antiserum, which might be explained by variations in the degree of coupling of 5-HT$_{1A}$ receptor binding subunits with G proteins from one brain area to another.

Introduction

Among the multiple classes of serotonin (5-HT) receptors which have been identified in the central nervous system (CNS) to date, the 5-HT$_{1A}$ type is one of the best known mainly because selective ligands have been available for *in vitro* and *in vivo* studies during the last years. In particular, several tritiated ([^{3}H]8-OH-DPAT, [^{3}H]5-MeO-DPAC, [^{3}H]ipsapirone, [^{3}H]5-Me-urapidil, etc.) and radioiodinated ([^{125}I]BH-8-MeO-N-PAT) ligands have allowed the specific labelling of 5-HT$_{1A}$ receptor binding sites in brain membranes and sections. Such studies led to the extensive description of the pharmacological properties and regional distribution of 5-HT$_{1A}$ receptors in the brain of various species (see Hamon et al. 1990b for a review). It thus became apparent that 5-HT$_{1A}$ receptors are the molecular targets of new anxiolytic/antidepressant drugs (ipsapirone, buspirone and gepirone). Furthermore, as expected from their possible involvement in mood control, these receptors are particularly abundant in limbic structures notably the hippocampus, septum, frontal and entorhinal cortex and amygdala (Pazos and Palacios 1985; Vergé et al. 1986). Binding studies with brain membranes also demonstrated that guanine nucleotides (GTP and its analogues

GppNHp and GTPγS) interfere with the specific binding of selective radioligands to 5-HT$_{1A}$ receptors, suggesting that the latter belong to the G protein-coupled receptor superfamily (Gozlan et al. 1983; Emerit et al. 1990). Elegant confirmation of this inference was provided by Albert et al. (1990) who cloned and sequenced the 5-HT$_{1A}$ receptor from the rat brain. Like that for other receptors of the same superfamily, the ligand binding site of the 5-HT$_{1A}$ receptor is located on a single polypeptide with seven transmembrane hydrophobic domains. This functional coupling to G (i.e. Gi and Go) proteins allows 5-HT$_{1A}$ receptors to exert either a negative control on adenylate cyclase activity or a stimulatory control on K$^+$ channels when they are occupied by agonists (Hamon et al. 1990b).

In the case of 5-HT$_{1A}$ receptors, as for any G protein-coupled receptor, only the receptor binding subunit (R 5-HT$_{1A}$) physically associated with G exhibits a high affinity for agonists. As all the 5-HT$_{1A}$ selective radioligands presently available are agonists, this implies that only the R 5-HT$_{1A}$-G complexes can be labelled by these molecules in brain membranes and sections. By contrast, free R 5-HT$_{1A}$ which are not coupled with G, and therefore exhibit a low affinity for agonists, cannot be detected by these radioligands. This limitation has important consequences notably for the absolute quantification of R 5-HT$_{1A}$ when assessing possible up and down regulation after various treatments. Indeed, only the fraction of R 5-HT$_{1A}$ coupled to G can be quantified by measurement of the specific binding of agonist radioligands.

In order to label all R 5-HT$_{1A}$, independently of their possible coupling with G, two approaches can be used. First, binding studies may be performed with an antagonist radioligand as its affinity for 5-HT$_{1A}$ sites should not be affected by the coupling or uncoupling of these sites with G proteins. Unfortunately, no antagonist radioligand has yet been developed for the labelling of 5-HT$_{1A}$ receptors. The second approach consists of raising specific antibodies which would recognize all R 5-HT$_{1A}$ in brain membranes or sections. This strategy was chosen in our laboratory. This paper reports the successful use of an anti-5-HT$_{1A}$ receptor antiserum for the immunoprecipitation, visualisation and quantification of 5-HT$_{1A}$ receptors in the rat brain.

Materials and Methods

By comparing the aminoacid sequences of all the G protein-coupled receptors which have been cloned to date with that of the rat 5-HT$_{1A}$ receptor (Albert et al. 1990), it appeared that minimal homology occurred in the third intracellular loop. This led us to select a portion of 26 amino-acids within this loop (amino-acid residues 243 to 268: GTSS-APPPKKSLNGQPGSGDWRRCAE) for making a synthetic peptide to be injected into rabbits. Five mg of the peptide were coupled to 15 mg of bovine serum albumin (BSA) by 1-ethyl-3-(3-dimethylaminopropyl) carbodiimide (see Cesselin et al. 1981), and the conjugate was extensively dialyzed against 0.9% NaCl. Aliquots (0.2 ml corresponding to ~0.2 mg of the peptide) were emulsified in Freund's adjuvant (Difco) to be injected every 4 weeks to male New Zealand rabbits. Blood was taken each time, and the serum heated at 56°C for 30 min, dialyzed extensively against 0.9% NaCl (in order to remove endogenous 5-HT), and finally mixed with an equal volume of glycerol to be kept at −30°C. The results reported herein were obtained with a serum collected four weeks after the 8th booster injection.

Immunoprecipitation of soluble $5\text{-}HT_{1A}$ receptors by the rabbit antiserum was tested as follows: Aliquots (0.3 ml corresponding to ~1.6 mg protein) of the 100,000 g supernatant of a suspension of rat hippocampal membranes preincubated for 60 min with 10 mM 3-[3-(cholamidopropyl)dimethylammonio]-1-propane sulfonate (CHAPS, to solubilize $5\text{-}HT_{1A}$ sites, El Mestikawy et al. 1989) were mixed with 30 μl of the antiserum diluted (1:200) in 0.05 M Tris-HCl, pH 7.4. After 30 min at 4°C, the mixture was supplemented with 70 μl of protein A-Sepharose CL-4B (Pharmacia) diluted by half in the same buffer, and the incubation proceeded for two hours. Samples were then centrifuged (2,500 g, 10 min, 4°C), the supernatants were saved and the pellets were washed twice with 1.5 ml of 0.05 M Tris-HCl, pH 7.4, before their final suspension in 0.33 ml of this buffer. Binding assays were performed on 80 μl aliquots of each supernatant and corresponding pellet suspension using the following radioligands: [³H]8-OH-DPAT (1.20 nM, 100 Ci/mmol, Service des Molécules Marquées, CEA) for the specific labelling of $5\text{-}HT_{1A}$ sites (El Mestikawy et al. 1989); [³H]5-HT (1.40 nM, 12.5 Ci/mmol, Amersham) in the presence of 0.1 μM 8-OH-DPAT and 0.1 μM mesulergine for the labelling of $5\text{-}HT_{1B}$ sites (Emerit et al. 1986); [³H]mesulergine (2.40 nM, 75.8 Ci/mmol, Amersham) for $5\text{-}HT_{1C}$ sites (Hoyer et al. 1985); [³H]ketanserin (0.65 nM, 64 Ci/mmol, New England Nuclear) for $5\text{-}HT_2$ sites (Leysen et al. 1982); [³H]zacopride (0.70 nM, 82 Ci/mmol, Laboratories Delalande, Rueil-Malmaison, France) for $5\text{-}HT_3$ sites (Bolaños et al. 1990); [³H]prazosin (0.56 nM, 25.4 Ci/mmol, Amersham) for α_1 adrenoceptors (Morrow and Creese 1986) and [³H]dihydroalprenolol (1.10 nM, 49.4 Ci/mmol, Amersham) in the presence of 10 μM 5-HT (see Hoyer et al. 1985) for the specific labelling of β-adrenoceptors (Bylund and Snyder 1976). Control experiments consisted of incubating the CHAPS soluble extract from hippocampal membranes with either the pre-immune serum (dilution: 1:200) or the antiserum preincubated for 2 h with 50 μg/ml of the synthetic peptide instead of the native antiserum.

Forskolin-activated adenylate cyclase was assayed in rat hippocampal homogenates by following the conversion of $\alpha\text{-}[^{32}P]ATP$ (0.5 mM) into

[^{32}P]cyclic AMP in the presence of 0.1 M NaCl and 10 μM GTP as described in detail elsewhere (Hamon et al. 1988). The inhibitory effect of 5-HT$_{1A}$ receptor stimulation by 8-OH-DPAT (0.1 μM) on the enzyme activity was tested on untreated homogenates and on those which had been preincubated for 1 h at 4°C with the antiserum or the pre-immune serum (1:50 and 1:500 final dilutions).

For the immuno-autoradiography experiments, adult male Sprague-Dawley rats (250–300 g body weight) were anaesthetized with chloral hydrate (350 mg/kg, i.p.), and perfused via the ascending aorta with 200 ml of 0.9% NaCl containing 1 g/l of sodium nitrite. Animals were decapitated, and the brain was quickly removed and frozen in isopentane at -30°C. Coronal and horizontal sections (thickness: 20 μm) were cut at -15°C and thaw-mounted onto gelatin-coated slides. After storage at -20°C for 2 weeks, sections were preincubated for 30 min in phosphate-saline buffer (PBS: 50 mM NaH$_2$PO$_4$/Na$_2$HPO$_4$, 154 mM NaCl, pH 7.4) containing BSA (3%, w/v) at room temperature (as for all the following steps) and washed with PBS alone for 5 min. Incubation then proceeded for 30 min in PBS containing 1% BSA and the antiserum or pre-immune serum at a final dilution of 1:1000. Slides were washed in PBS (3 × 5 min), and dipped in the same buffer supplemented with [^{35}S]lgG-anti rabbit lgG (1.0 μCi/ml, 570 Ci/mmol, Amersham) for 2 h. They were then washed 3 more times in PBS for 10 min each, once in distilled water for 15 s, dried with cold air, and exposed to βmax films (Amersham) at -80°C for 48 h. Autoradiograms were developed in Kodak Microdol (10 min at 20°C), and their quantitative analysis was performed with an IMSTAR densitometer.

The autoradiographic labelling of 5-HT$_{1A}$ sites by [^{3}H]8-OH-DPAT (2.0 nM) and subsequent quantitative analysis were carried out as described by Vergé et al. (1986). Data are expressed in optical density (O.D.) units.

Results

Selective Immunoprecipitation of 5-HT$_{1A}$ Sites by the Antiserum in a Soluble Extract from Rat Hippocampal Membranes: Control experiments showed that under the conditions used for immuno-precipitating 5-HT$_{1A}$ sites, neither the pre-immune serum (dilution: 1:200) nor the antiserum which had been preincubated with the synthetic peptide affected the specific binding of [^{3}H]8-OH-DPAT in the soluble hippocampal extract. By contrast, a significant decrease in [^{3}H]8-OH-DPAT specific binding capacity was found in the supernatant from samples which were incubated with the native antiserum (1:200) and protein A-Sepharose CL4B (Table 1). Concomitantly, [^{3}H]8-OH-DPAT specific binding could be measured in the corresponding pellets, and the

Table 1. Specific binding of various radioligands to serotoninergic and adrenergic receptors in hippocampal extracts which had been incubated with the preimmune serum or the anti-5-HT_{1A} antiserum

Receptor type	Radioligand	[^{3}H]ligand specifically bound (fmol/ml)		
		Preimmune serum	Anti-5-HT_{1A} antiserum	%
5-HT_{1A}	[^{3}H]8-OH-DPAT	267.1 ± 9.0	151.4 ± 8.2*	57
5-HT_{1B}	[^{3}H]5-HT	209.8 ± 11.3	215.4 ± 9.7	103
5-HT_{1C}	[^{3}H]Mesulergine	25.5 ± 2.2	27.0 ± 1.9	106
5-HT_2	[^{3}H]Ketanserin	12.2 ± 0.4	12.3 ± 0.3	101
5-HT_3	[^{3}H]Zacopride	27.9 ± 1.4	29.4 ± 1.9	105
α1	[^{3}H]Prazosin	47.2 ± 3.1	45.4 ± 2.0	96
β	[^{3}H]Dihydroalprenolol	13.5 ± 1.0	13.0 ± 0.4	96

Soluble extracts from hippocampal membranes were preincubated for 30 min at 4°C with the preimmune serum or the anti-5-HT_{1A} antiserum at a final dilution of 1:200, and then incubated for 2 h at 4°C with protein A-Sepharose CL-4B. Samples were centrifuged and binding assays were performed on 0.08 ml aliquots of each clear supernatant (see Materials and Methods). The specific binding of each radioligand is expressed in fmol. per ml of the starting soluble extract. Each value is the mean ± S.E.M. of triplicate determinations in two separate experiments. Figures in the right column are the percentages of specific binding in the extract treated by the anti-5-HT_{1A} antiserum compared to that found after exposure to the preimmune serum.
*$p < 0.05$ when compared to the respective control value ("preimmune serum-treated extract").

sum of binding capacities in each pellet and supernatant was not significantly different from that measured in the starting CHAPS extract (El Mestikawy et al. 1990).

Data in Table 1 indicated that the procedure used for solubilizing 5-HT_{1A} sites from hippocampal membranes (El Mestikawy et al. 1989) also yielded the solubilization of other types of 5-HT receptors and adrenergic receptors. However none of these receptors was removed from the soluble fraction under the conditions for immunoprecipitating [^{3}H]8-OH-DPAT specific binding sites from this fraction to the protein A-Sepharose CL4B pellet (Table 1).

Effects of the Antiserum on [^{3}H]8-OH-DPAT Binding to- and Adenylate Cyclase Activity Associated with- 5-HT_{1A} Receptors: No change in the specific binding of [^{3}H]8-OH-DPAT to solubilized 5-HT_{1A} sites was observed when assays were carried out on the soluble CHAPS hippocampal extract supplemented with either the pre-immune serum or the antiserum at various dilutions (1:50 – 1:1000). This indicated that extensive dialysis of the serums allowed the elimination of endogenous 5-HT, which otherwise would have competed with [^{3}H]8-OH-DPAT for 5-HT_{1A} sites, and accordingly decrease its specific binding. Furthermore, the concentration-dependent inhibition of [^{3}H]8-OH-DPAT specific binding by GTP ($IC_{50} = 0.5 \mu M$) was not significantly different in the native soluble hippocampal extract and in those which were supplemented with the preimmune serum or the antiserum (1:200). However, similar experiments performed on the pellets after centrifugation of soluble extracts incubated with the antiserum and protein A-Sepharose CL4B revealed a greater potency of GTP to inhibit the specific binding of [^{3}H]8-OH-DPAT to immunoprecipitated 5-HT$_{1A}$ sites (IC$_{50}$ = 0.1 μM).

As expected from the negative coupling of 5-HT$_{1A}$ receptors with adenylate cyclase, the 5-HT$_{1A}$ agonist 8-OH-DPAT (0.1 μM) significantly reduced the production of [^{32}P]cyclic AMP (-27%) by hippocampal homogenates exposed to 10 μM forskolin (Table 2). No change in forskolin-stimulated adenylate cyclase activity was noted after preincubation of these homogenates with the pre-immune serum or the antiserum at two different dilutions: 1:50 and 1:500 (Table 2). In addition, preincubation with the pre-immune serum did not significantly alter the reduction in [^{32}P]cyclic AMP production due to 0.1 μM 8-OH-DPAT (Table 2). By contrast, exposure of hippocampal homogenates to the antiserum at 1:50 (but not 1:500) prevented to some extent the inhibitory effect of the 5-HT$_{1A}$ receptor agonist of forskolin-stimulated adenylate cyclase activity (Table 2).

Immuno-Autoradiographic Labelling of 5-HT$_{1A}$ Sites by the Anti-Serum: The autoradiograms obtained from sections incubated first with the antiserum and second with [^{35}S]lgG-anti rabbit lgG revealed that the structures with the most intense labelling were the septum, hippocampus (mainly CA1 and dentate gyrus), entorhinal cortex, amygdala, dorsal

Table 2. Effects of preimmune serum and anti-5-HT$_{1A}$ antiserum on 5-HT$_{1A}$-coupled adeny-late cyclase activity in hippocampal homogenates

Addition	[^{32}P]cyclic AMP (nmol/mg prot)		
	Preimmune serum		Anti-5-HT$_{1A}$ antiserum
None	5.98 $\pm$ 0.18		
8-OH-DPAT (0.1 μM)	4.37 $\pm$ 0.11* (-27%)		
1:50	5.61 $\pm$ 0.16		6.34 $\pm$ 0.12
1:50 + 8-OH-DPAT	4.36 $\pm$ 0.08* (-22%)		5.35 $\pm$ 0.09* (-16%)
1:500	6.24 $\pm$ 0.19		5.76 $\pm$ 0.17
1:500 + 8-OH-DPAT	4.67 $\pm$ 0.13* (-25%)		4.42 $\pm$ 0.11* (-23%)

Hippocampal homogenates from adult rats were preincubated for 1 h at 0°C with two different dilutions (1:50 and 1:500, left column) of the preimmune serum or the anti-5-HT$_{1A}$ antiserum or none. Assays were performed on 50 μl aliquots of each homogenate with 10 μM forskolin, 10 μM GTP, 0.1 M NaCl (see Hamon et al. 1988), and in the presence or absence of 0.1 μM 8-OH-DPAT, as indicated. Adenylate cyclase activity is expressed as nmoles [^{32}P]cyclic AMP synthesized per mg protein and per 20 min at 30°C. Each value is the mean $\pm$ S.E.M. of triplicate determinations in two separate experiments. The percent reduction due to 8-OH-DPAT is indicated in parentheses.
*$p < 0.05$ when compared to forskolin-stimulated adenylate cyclase activity in the absence of 8-OH-DPAT.

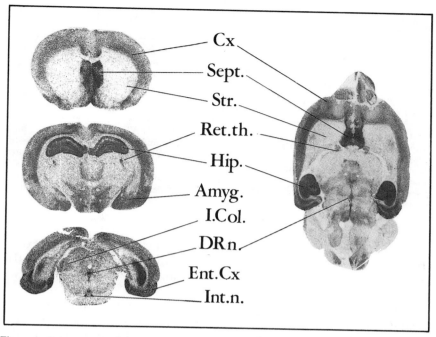

Figure 1. *Immunoautoradiograms of brain sections exposed to the anti-5-HT$_{1A}$ receptor antiserum*. Left part: coronal sections; Right part: horizontal section. Cx: cerebral cortex; Sept.: septum; Str.: striatum; Ret. th.: reticular nucleus of the thalamus; Hip.: hippocampus; Amyg.: amygdala; I. Col.: inferior colliculi; DRn.: dorsal raphe nucleus; Ent. Cx: entorhinal cortex; Int. n.: interpeduncular nucleus.

raphe nucleus and interpeduncular nucleus (Figure 1). Moderate labelling was seen in the cerebral cortex, reticular nucleus of the thalamus and inferior colliculi. By contrast, the substantia nigra (not shown) and the cerebellum remained unlabelled. Within the striatum, only a discrete lateral zone exhibited some labelling by [^{35}S]IgG-anti rabbit IgG bound to antibodies (Figure 1). Parallel experiments with the pre-immune serum yielded no labelling at any level of the brain.

Comparison of the distribution of this immuno-autoradiographic labelling with that found with selective 5-HT$_{1A}$ radioligands such as [^{3}H]8-OH-DPAT (Vergé et al. 1986) or [^{125}I]BH-8-MeO-N-PAT (Gozlan et al. 1988) indicated a perfectly similar localization, as would be expected from the recognition of 5-HT$_{1A}$ sites by the anti-peptide antiserum. Thus, the large structures enriched in 5-HT$_{1A}$ sites such as the septum, hippocampus and entorhinal cortex (see Pazos and Palacios 1985) exhibited a high intensity of labelling by both the antiserum plus [^{35}S]IgG-anti rabbit IgG and the radioligands. In addition, the similarity between the labelling by both procedures also involved more discrete zones. For instance, the lateral part of the striatum and the reticular

Table 3. Regional differences in the respective labelling of 5-HT$_{1A}$ receptors by [^{3}H]8-OH-DPAT and the anti-5-HT$_{1A}$ receptor antiserum

	Optical density	
	[^{3}H]8-OH-DPAT	Anti-5-HT$_{1A}$-antiserum
Dentate gyrus	0.293 ± 0.030	0.253 ± 0.021
Dorsal raphe nucleus	0.261 ± 0.024	0.444 ± 0.036*

Brain sections were incubated with either 2 nM [^{3}H]8-OH-DPAT or the anti-5-HT$_{1A}$-antiserum (1/1000 final dilution) then [^{35}S]lgG-anti rabbit lgG, and exposed to sensitive films for 2 days (5-HT$_{1A}$-antiserum) or 2 months ([^{3}H]8-OH-DPAT). Optical density was measured within the dentate gyrus area and the dorsal raphe nucleus on autoradiographic films. Each value is the mean ± S.E.M. of 24–30 measurements in 4 rats for each assay condition.
*$p < 0.05$ when compared to corresponding value for the dentate gyrus in the same sections labelled by the anti-5-HT$_{1A}$-receptor antiserum.

nucleus of the thalamus where the antiserum bound to some extent (Figure 1) have previously been found to specifically bind [^{3}H]8-OH-DPAT and [^{125}l]BH-8-MeO-N-PAT (Gozlan et al. 1988; El Mestikawy et al. 1990).

Further comparison between the labelling by the antiserum plus [^{35}S]lgG-anti rabbit lgG and [^{3}H]8-OH-DPAT was then made by measuring the O.D. on the respective autoradiograms. In the case of 5-HT$_{1A}$ labelling by the radioligand, similar O.D. values were found in the dentate gyrus and the dorsal raphe nucleus, suggesting a similar density of 5-HT$_{1A}$ receptors coupled to G proteins (see Introduction) in these two areas. By contrast, quantification of the immuno-autoradiographic labelling revealed marked differences between these regions since O.D. measured within the dorsal raphe nucleus was 75% higher than that found in the dentate gyrus (Table 3).

Discussion

As previously reported for other marker proteins (see for instance Lee et al. 1989; Roth et al. 1989), the present data confirmed that specific antibodies against a neurotransmitter receptor could be raised from a synthetic peptide corresponding to a selective portion of its amino-acid sequence. Indeed, the antiserum presently obtained was able to immunoprecipitate selectively 5-HT$_{1A}$ receptor binding sites in a soluble extract from rat hippocampal membranes, and to label these sites in brain sections. Furthermore, immunoblot experiments (not shown) have demonstrated that this antiserum recognized a protein of 63 kDa in a soluble fraction enriched in 5-HT$_{1A}$ receptors (i.e. the eluate of a wheat germ agglutinin-agarose column loaded with a CHAPS soluble hippocampal extract, see El Mestikawy et al. 1989), and previous studies have shown that this molecular weight corresponded exactly to that of the native 5-HT$_{1A}$ receptor binding subunit (Emerit et al. 1987).

Interestingly, the antiserum alone did not affect the specific binding of [^{3}H]8-OH-DPAT to solubilized 5-HT$_{1A}$ sites, and the sum of [^{3}H]8-OH-DPAT specific binding in the supernatant and the pellet after adsorption of the immune complexes to protein A-Sepharose beads was in the same range as that found in the starting soluble extract. Clearly the anti-5-HT$_{1A}$ receptor antiserum did not interfere with the specific binding of [^{3}H]8-OH-DPAT, which was not unexpected since antibodies were raised against a portion of the third intracellular loop, whereas the agonist binding site is probably located in the transmembrane domains of the 5-HT$_{1A}$ receptor molecule (Hartig 1989). However the third intracellular loop is involved in the interaction of the receptor molecule with G protein, and some alterations in G protein-dependent events should have occurred when the antibodies occupied at least part of the loop onto the 5-HT$_{1A}$ receptor binding subunit. Thus, a greater sensitivity of [^{3}H]8-OH-DPAT specific binding to GTP was noted in the immuno-precipitate where 5-HT$_{1A}$ receptor-antibody complexes were adsorbed onto protein A-Sepharose beads. In addition, the efficacy of 8-OH-DPAT to inhibit forskolin-stimulated adenylate cyclase was reduced in membrane preparations which had been exposed to the antiserum. These two series of data indicated that the 5-HT$_{1A}$ receptor – G protein interaction was not prevented, but only weakened, by the antibodies. Previous studies on the β_2-adrenoceptor have shown that only the N and C terminal (including $\sim$ 10 amino-acid residues each), but not the medial portions of the third intracellular loop are required for the interaction with G protein (Strader et al. 1989). It bears emphasis that the peptide sequence that we selected for raising anti-5-HT$_{1A}$ receptor antibodies was located in the medial part of the third intracellular loop of the 5-HT$_{1A}$ receptor molecule. Therefore, the apparent weakness of the 5-HT$_{1A}$ receptor-G protein association in the presence of antibodies could not result from some competition between the latter and G protein for the same binding site onto the receptor molecule. Instead, the binding of antibodies to the medial part of the third intracellular loop might have rendered the proximal and distal parts of this zone less accessible to the G protein. Studies with antibodies directed against other zones of the 5-HT$_{1A}$ receptor molecule are currently being performed in order to validate this interpretation.

Of particular interest is the fact that the antiserum can be used for the visualization of 5-HT$_{1A}$ receptors in brain sections. In addition to the present data which illustrate the use of the antiserum for the quantitative immuno-autoradiography of 5-HT$_{1A}$ receptors, we have recently applied classical immunohistochemical procedures for the visualization of 5-HT$_{1A}$ receptors at the light microscope level on cell bodies and dendrites in various brain regions (Sotelo et al. 1990). At all levels examined the antiserum mainly labels patches in the plasmic membrane, with little or no intracellular staining. Of particular interest was the demonstration of the location of 5-HT$_{1A}$ receptors on the somas and dendrites of serotonin neurones (identified with anti-5-HT antibodies) within the dorsal and median raphe nuclei (Sotelo et al. 1990). This provides the first direct evidence of the presynaptic location of 5-HT$_{1A}$ receptors as expected for their role as autoreceptors controlling the electrical and metabolic activity of serotonin neurones (Innis et al. 1988; Hamon et al. 1988). The next goal will be to use the antiserum for the visualization of 5-HT$_{1A}$ receptors at the electron microscope level. Because of the possibility that serotonin neurotransmission occurs outside differentiated synapses (Séguéla et al. 1989), extrasynaptic location of 5-HT$_{1A}$ receptors might be expected. Ultrastructural investigations with anti 5-HT$_{1A}$ receptor antibodies should be a valuable approach to confirm this inference.

Another promising use of antibodies is the quantification of total 5-HT$_{1A}$ receptor binding subunits independently of their coupling with G proteins (see Introduction). The first series of data reported herein indicated that the ratio of 5-HT$_{1A}$ receptors associated with G protein (which could be quantified by the specific binding of an agonist radioligand such as [^{3}H]8-OH-DPAT) over total 5-HT$_{1A}$ receptor binding subunits (labelled by the antiserum plus [^{35}S]lgG-anti rabbit lgG) exhibited regional differences. Thus the proportion of 5-HT$_{1A}$ receptors involved in complexes with G proteins was apparently less in the dorsal raphe nucleus than in the hippocampus. This observation is interesting as Meller et al. (1990) recently reported that a 5-HT$_{1A}$ receptor reserve might exist in the dorsal raphe nucleus (but not in the hippocampus). More generally, variations in the degree of coupling with G proteins might explain at least some of the regional differences in the functional properties of 5-HT$_{1A}$ receptors (see Hamon et al. 1990a). Studies on other G protein-coupled receptors have shown that the association of the receptor binding subunit with G is controlled by the phosphorylation of specific amino-acid residues onto the receptor molecule (Harada et al. 1990; Strader et al. 1989). As the 5-HT$_{1A}$ receptor molecule also contains consensus amino-acid sequences to be phosphorylated (Albert et al. 1990), one possible explanation of the regional differences noted above is that the degree of 5-HT$_{1A}$ receptor phosphorylation may vary from one brain area to another. More generally, changes in the degree of 5-HT$_{1A}$ receptor phosphorylation might explain some of the adaptive regulation of the functional properties of 5-HT$_{1A}$ receptors under a given experimental (notably pharmacological, see Schechter et al. 1990) circumstance. With the availability of specific anti-5-HT$_{1A}$ receptor antibodies, such hypotheses would be soon tested by measuring directly the incorporation of ^{32}Pi into immunoprecipated 5-HT$_{1A}$ receptor molecules.

In conclusion, the present study has shown that specific anti-5-HT$_{1A}$ receptor antibodies can be raised from a synthetic peptide corresponding to a selective portion of the receptor molecule. The same approach should be equally fruitful for raising antibodies against 5-HT$_{1C}$ and 5-HT$_2$ receptors whose amino-acid sequences have also been reported (Hartig 1989).

Acknowledgements

This research has been supported by grants from INSERM and BAYER-PHARMA. We are very grateful to Dr O. Civelli for the kind communication of the amino-acid sequence of the rat 5-HT$_{1A}$ receptor prior to publication.

References

Albert, P. R., Zhou, Q. Y., Van Tol, H. H. M., Bunzow, J. R., and Civelli, O. (1990). Cloning, functional expression and mRNA tissue distribution of the rat 5-hydroxy-tryptamine$_{1A}$ receptor gene. J. Biol. Chem. 265: 5825–5832.

Bolaños, F. J., Schechter, L. E., Miquel, M. C., Emerit, M. B., Rumigny, J. F., Hamon, M., and Gozlan, H. (1990). Common pharmacological and physico-chemical properties of 5-HT$_3$ binding sites in the rat cerebral cortex and NG 108–15 clonal cells. Biochem. Pharmacol. 40: 1541–1550.

Bylund, D. B., and Snyder, S. H. (1976). Beta-adrenergic receptor binding in membrane preparations from mammalian brain. Mol. Pharmacol. 12: 568–580.

Cesselin, F., Soubrié, P., Bourgoin, S., Artaud, F., Reisine, T. D., Michelot, R., Glowinski, J., and Hamon, M. (1981). In vivo release of met-enkephalin in the cat brain. Neuroscience, 6: 301–313.

El Mestikawy, S., Riad, M., Laporte, A. M., Vergé, D., Daval, G., Gozlan, H., and Hamon, M. (1990). Production of specific anti-rat 5-HT$_{1A}$ receptor antibodies in rabbits injected with a synthetic peptide. Neurosci. Lett. 118: 189–192.

El Mestikawy, S., Taussig, D., Gozlan, H., Emerit, M. B., Ponchant, M., and Hamon, M. (1989). Chromatographic analyses of the serotonin 5-HT$_{1A}$ receptor solubilized from the rat hippocampus. J. Neurochem. 53: 1555–1566.

Emerit, M. B., El Mestikawy, S., Gozlan, H., Cossery, J. M., Besselievre, R., Marquet, A., and Hamon, M. (1987). Identification of the 5-HT$_{1A}$ receptor binding subunit in rat brain membranes using the photoaffinity probe [^{3}H]8-methoxy-2-[N,n-propyl,N-3-(2-nitro-4-azi-dophenyl) aminopropyl]aminotetralin. J. Neurochem. 49: 373–380.

Emerit, M. B., El Mestikawy, S., Gozlan, H., Rouot, B., and Hamon, M. (1990). Physical evidence of the coupling of solubilized 5-HT$_{1A}$ binding sites with G regulatory proteins. Biochem. Pharmacol. 39: 7–18.

Emerit, M. B., Gozlan, H., Marquet, A., and Hamon, M. (1986). Irreversible blockade of central 5-HT$_{1A}$ receptor binding sites by the photoaffinity probe 8-methoxy-3'-NAP-amino-PAT. Eur. J. Pharmacol. 127: 67–81.

Gozlan, H., El Mestikawy, S., Pichat, L., Glowinski, J., and Hamon, M. (1983). Identification of presynaptic serotonin autoreceptors using a new ligand: ^{3}H-PAT. Nature (Lond.) 305: 140–142.

Gozlan, H., Ponchant, M., Daval, G., Vergé, D., Ménard, F., Vanhove, A., Beaucourt, J. P., and Hamon, M. (1988). [^{125}I]Bolton-Hunter-8-methoxy-2-[N-propyl-N-propylamino] tetralin as a new selective radioligand of 5-HT$_{1A}$ sites in the rat brain. In vitro binding and autoradiographic studies. J. Pharmacol. Exp. Ther. 244: 751–759.

Hamon, M., Emerit, M. B., El Mestikawy, S., Gallissot, M. C., and Gozlan, H. (1990a). Regional differences in the transduction mechanisms of serotonin receptors in the mammalian brain. In: Saxena, P. R., Wallis, D., Wouters, W., and Bevan, P. (eds), "Cardiovascular Pharmacology of 5-hydroxytryptamine: prospective therapeutic applications". Kluwer: Dordrecht, pp. 41–59.

Hamon, M., Fattaccini, C. M., Adrien, J., Gallissot, M. C., Martin, P., and Gozlan, H. (1988). Alterations of central serotonin and dopamine turnover in rats treated with ipsapirone and other 5-hydroxytryptamine$_{1A}$ agonists with potential anxiolytic properties. J. Pharmacol. Exp. Ther. 246; 745–752.

Hamon, M., Gozlan, H., El Mestikawy, S., Emerit, M. B., Bolaños, F., and Schechter, L. (1990b). The central 5-HT$_{1A}$ receptors: pharmacological, biochemical, functional and regulatory properties. Ann. N.Y. Acad. Sci. 600: 114–131.

Harada, H., Ueda, H., Katada, T., Ui, M., and Satoh, M. (1990). Phosphorylated μ-opioid receptor purified from rat brains lacks functional coupling with Gi1, a GTP-binding protein in reconstituted lipid vesicles. Neurosci. Lett. 113: 47–49.

Hartig, P. R. (1989). Molecular biology of 5-HT receptors. Trends Pharmacol. Sci. 10: 64–69.

Hoyer, D., Engel, G., and Kalkman, H. O. (1985). Molecular pharmacology of 5-HT$_1$ and 5-HT$_2$ recognition sites in rat and pig brain membranes: radioligand binding studies with [^{3}H]5-HT, [^{3}H]8-OH-DPAT, $(-)$[^{125}I]iodocyanopindolol, [^{3}H]mesulergine and [^{3}H]ketanserin. Eur. J. Pharmacol. 118: 13–23.

Innis, R. B., Nestler, E. J., and Aghajanian, G. K. (1988). Evidence for G protein mediation of serotonin- and GABA$_B$- induced hyperpolarization of rat dorsal raphe neurons. Brain Res. 459: 27–36.

Lee, K. Y., Lew, J. Y., Tang, D., Schlesinger, D. H., Deutch, A. Y., and Goldstein, M. (1989). Antibodies to a synthetic peptide corresponding to a ser-40-containing segment of tyrosine hydroxylase: activation and immunohistochemical localization of tyrosine hydroxylase. J. Neurochem. 53: 1238–1244.

Leysen, J. E., Niemegeers, C. J. E., Van Nueten, J. M., and Laduron, P. M. (1982). [^{3}H]ketanserin (R 41468), a selective [^{3}H]ligand for serotonin$_2$ receptor binding sites. Binding properties, brain distribution and functional role. Mol. Pharmacol. 21: 301–314.

Meller, E., Goldstein, M., and Bohmaker, K. (1990). Receptor reserve for 5-hydroxytryptamine$_{1A}$-mediated inhibition of serotonin synthesis: possible relationship to anxiolytic properties of 5-hydroxytryptamine$_{1A}$ agonists. Mol. Pharmacol. 37: 231–237.

Morrow, A. L., and Creese, I. (1986). Characterization of α_1-adrenergic receptor subtypes in rat brain: a reevaluation of [^{3}H]WB4101 and [^{3}H]prazosin binding. Mol. Pharmacol. 29: 321–330.

Pazos, A., and Palacios, J. M. (1985). Quantitative autoradiographic mapping of serotonin receptors in the rat brain. I. Serotonin-1 receptors. Brain Res. 346: 205–230.

Roth, B. L., Iadarola, M. J., Mehegan, J. P., and Jacobowitz, D. M. (1989). Immunohistochemical distribution of β-protein kinase C in rat hippocampus determined with an antibody against a synthetic peptide sequence. Brain Res. Bull. 22: 893–897.

Schechter, L. E., Bolaños, F. J., Gozlan, H., Lanfumey, L., Haj-Dahmane, S., Laporte, A. M., Fattaccini, C. M., and Hamon, M. (1990). Alterations of central serotoninergic and dopaminergic neurotransmission in rats chronically treated with ipsapirone – Biochemical and electrophysiological studies. J. Pharmacol. Exp. Ther. 255: 1335–1347.

Séguéla, P., Watkins, K. C., and Descarries, L. (1989). Ultrastructural relationships of serotonin axon terminals in the cerebral cortex of the adult rat. J. Comp. Neurol. 289: 129–142.

Sotelo, C., Cholley, M., El Mestikawy, S., Gozlan, H., and Hamon, M. (1990). Direct immunohistochemical evidence of the existence of 5-HT$_{1A}$ autoreceptors on serotoninergic neurones in the midbrain raphe nuclei. Eur. J. Neurosci. 2: 1144–1154.

Strader, C. D., Sigal, I. S., and Dixon, R. A. F. (1989). Mapping of functional domains of the β-adrenergic receptor. Am. J. Respir. Cell. Mol. Biol. 1: 81–86.

Vergé, D., Daval, G., Marcinkiewicz, M., Patey, A., El Mestikawy, S., Gozlan, H. and Hamon, M. (1986). Quantitative autoradiography of multiple 5-HT$_1$ receptor subtypes in the brain of control or 5,7-dihydroxytryptamine-treated rats. J. Neurosci. 6: 3474–3482.

Serotonin: Molecular Biology, Receptors and Functional Effects
ed. by J. R. Fozard/P. R. Saxena
© 1991 Birkhäuser Verlag Basel/Switzerland

5-HT$_{1C}$, 5-HT$_{1D}$ and 5-HT$_2$ Receptors in Mammalian Brain: Multiple Affinity States with a Different Regional Distribution

C. Waeber and J. M. Palacios

Preclinical Research, Sandoz Pharma Ltd., CH-4002 Basel, Switzerland

Summary. In vitro ligand binding autoradiography was used to:
- investigate regional differences in the sensitivity of 5-HT$_{1D}$/5-HT$_{1B}$ binding sites to the agonist 5-carboxamidotryptamine (5-CT),
- compare the distribution of 5-HT$_{1C}$ sites labelled by [^{3}H]5-HT, [^{125}I]DOI or by [^{3}H]mesulergine,
- compare the distribution of 5-HT$_2$ sites labelled by [^{125}I]DOI, [^{3}H]LSD or [^{3}H]ketanserin.
  The data show that 5-CT competes for 5-HT$_{1D}$ binding sites with a biphasic profile. In rat, guinea-pig and human brain, the proportion of sites with a high affinity for 5-CT is higher in the substantia nigra and globus pallidus than in the caudate/putamen and the cortex. These studies also demonstrate that antagonist ligands reveal a more widespread distribution of 5-HT$_{1C}$ and 5-HT$_2$ sites than agonist ligands, indicating that these receptors might exist in multiple affinity states with different regional distributions.

Introduction

Biochemical and molecular genetic evidence reveals that 5-HT$_1$ and 5-HT$_2$ receptors belong to the family of receptors coupled to GTP binding proteins (Hartig 1989; Branchek et al. 1990). Such receptors have been found to exist in at least two interconvertible states and to present a higher affinity for agonist ligands when they form a complex with the GTP binding protein (DeLean et al. 1980). Based on radioligand binding data with homogenates, the existence of subtypes of 5-HT$_{1D}$ (Leonhardt et al. 1989) and 5-HT$_2$ (McKenna and Peroutka 1989) receptors has recently been proposed. However, adequate ligands (in particular antagonists for 5-HT$_1$ receptors and agonists for 5-HT$_2$ receptors) are lacking, so that binding techniques alone do not allow discrimination between the presence of multiple receptor subtypes or multiple affinity states of the same protein. In order to provide further information on this subject, the regional distribution of heterogeneous receptor populations has been investigated.

Materials and Methods

Human brains were obtained at autopsy from 5 subjects deceased without history of neurological or psychiatric disease. Guinea-pigs and rats were killed by decapitation. 10 μm thick sections were cut from different brain regions with a microtome-cryostat, mounted on gelatin-coated slides and stored at $-20°$C until used. Incubations were made according to the following procedure. Slides were brought to room temperature (RT) 30 min before preincubation (RT, 30' for [^{3}H]5-HT and [^{3}H]mesulergine, 15' for the other ligands) in Tris · HCl buffer pH 7.5 (containing 4 mM $CaCl_2$ for agonist ligands). Slides were then incubated at RT in the same buffer containing the radioligand ([^{3}H]5-HT: 2 nM; [^{3}H]mesulergine: 5 nM; [^{3}H]ketanserin: 1 nM; [^{3}H]LSD: 2 nM; [^{125}I]DOI: 0.1 nM), in the presence or the absence of competing drugs, and then washed by successive immersions in ice-cold buffer (1 × 5' for [^{3}H]5-HT, 2 × 10' for the other ligands). Finally, slides were dipped in ice-cold water to remove the salts and dried quickly under a stream of cold air. They were apposed to radiation sensitive films (Amersham, UK). After exposure (60 days for [^{3}H]5-HT and [^{3}H]LSD; 30 days for [^{3}H]mesulergine and [^{3}H]ketanserin; 5 days for [^{125}I]DOI) and development of the films, the densities of autoradiographic grains were quantified using a computer-aided image analysis system (MCID, Ontario, Canada).

Results

5-HT$_{1D}$ Receptors

In human, guinea-pig and rat brains, increasing concentrations of 5-CT displaced [^{3}H]5-HT (in the presence of 100 nM 8-OH-DPAT and 100 nM mesulergine) homogeneously in all brain areas and in particular the caudate/putamen, the substantia nigra and the cortex. In guinea-pig and human brains, 5-CT competed for 5-HT$_{1D}$ binding sites in a biphasic manner in all regions examined (Figure 1). In rat brain, this pattern was less marked, nevertheless a biphasic model significantly improved the curve fitting. In all species, the proportion of sites exhibiting a high affinity for 5-CT was dependent on the brain region under consideration, being higher in the globus pallidus and substantia nigra than in the caudate/putamen. In the guinea-pig, the superficial grey layer of the superior colliculus contained a majority of sites displaying a high affinity for 5-CT. A similar observation was made in the layer IVcβ of the human visual cortex (A17 of Brodman), whereas in the other cortical regions of the human brain, all layers contained mostly sites with a low affinity for 5-CT.

5-HT$_{1C}$ Receptors

Agonist ([^{3}H]5-HT and [^{125}I]DOI) and antagonist ([^{3}H]mesulergine) radioligands label high densities of 5-HT$_{1C}$ receptors in the human choroid plexus (Figure 2). However, 5-HT$_{1C}$ sites are labelled only by the antagonist [^{3}H]mesulergine in the globus pallidus and substantia nigra (Figure 3). In the latter regions, competition studies reveal that 5-HT displays a slightly lower affinity for [^{3}H]mesulergine binding sites than for those located in the choroid plexus. Species differences also appear to exist, as no significant densities of [^{3}H]mesulergine binding sites have been detected in the substantia nigra and globus pallidus of rat and guinea-pig.

5-HT$_2$ Receptors

In the human striatum, the selective antagonist [^{3}H]ketanserin labels 5-HT$_2$ sites which are homogeneously distributed (not shown). This constrasts with the heterogeneous distribution of 5-HT$_2$ sites labelled using the agonist [^{125}I]DOI (Figure 4). The mixed agonist/antagonist ligand [^{3}H]LSD (in the presence of 100 nM 5-CT and 30 μM sulpiride to block receptors other than 5-HT$_{1C}$ and 5-HT$_2$) also displays an heterogeneous labelling pattern, although less marked than that obtained with [^{125}I]DOI (data not shown).

The distribution of [^{125}I]DOI labelled patches corresponds rather well with that of patches obtained with the benzodiazepine receptor ligand [^{3}H]flunitrazepam. Moreover, striatal areas rich in [^{125}I]DOI binding sites possess a localization similar to that of areas with low acetyl-cholinesterase activity.

Discussion

Using 'classical' ligands (i.e. [^{3}H]5-HT, [^{3}H]mesulergine and [^{3}H]ketanserin), the distributions of 5-HT$_{1D}$, 5-HT$_{1C}$ and 5-HT$_2$ receptors described in this study are fully comparable with those previously reported (for reviews, see Palacios and Dietl 1988; Radja et al. 1990). However the results obtained using newly developed ligands or the analysis of competition profiles by selective compounds point to the heterogeneity of these receptor populations.

[^{3}H]5-HT, in the presence of 100 nM 8-OH-DPAT and 100 nM mesulergine, can be used to label 5-HT$_{1B}$ sites in the rat and mouse brain and 5-HT$_{1D}$ sites in the brains of other vertebrate species (Waeber et al. 1989a; b). Several studies have shown that 5-CT displaces [^{3}H]5-HT from 5-HT$_{1D}$ sites with a biphasic pattern (Waeber et al. 1988;

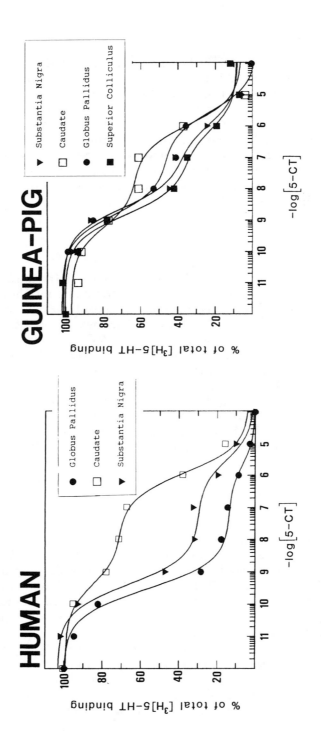

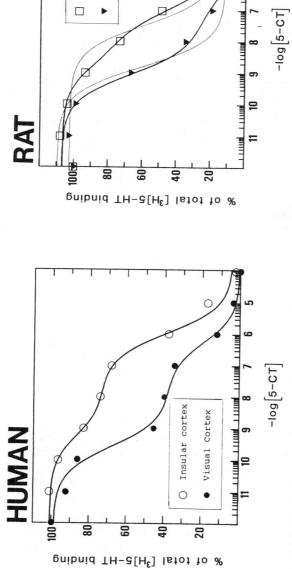

Figure 1. Analysis of the competition for [³H]5-HT binding sites obtained by increasing concentrations of 5-CT (performed in the presence of 100 nM 8-OH-DPAT and 100 nM mesulergine). Results have been obtained by microdensitometric quantification of autoradiograms generated as described in the text. Data have been fitted using two site model for all tissues. For comparison, monophasic curves are displayed for the rat.

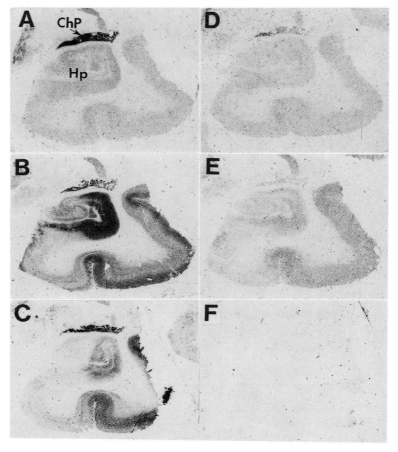

Figure 2. Autoradiograms obtained with human brain sections at the level of the hippocampus (Hp). Consecutive sections have been labelled using [³H]mesulergine (A: total binding; D: in the presence of 100 nM 5-HT), [³H]5-HT (B: total; E: in the presence of 100 nM 8-OH-DPAT and mesulergine) and [¹²⁵I]DOI (C: total; F: in the presence of 100 nM 5-HT). All three ligands densely label the choroid plexus (ChP).

Waeber et al, 1989a; Sumner and Humphrey 1989). The existence of a new subtype of 5-HT$_1$ receptor or of multiple affinity states of 5-HT$_{1B}$/ 5-HT$_{1D}$ receptors could be proposed as an explanation for the displacement profile of 5-CT. Supporting the first assumption, 5-HT$_{1E}$ sites, with drug binding properties distinct from those of 5-HT$_{1D}$ sites, have been characterized in the human brain (Leonhardt et al. 1989). Interestingly, [³H]5-HT binding to these sites was sensitive to GTP analogs, indicating a possible interaction with a G-protein. On the other hand, the fact that the populations of sites displaying a high and low affinity for 5-CT are present in the same brain regions and present only variations in their relative proportions in different areas could suggest

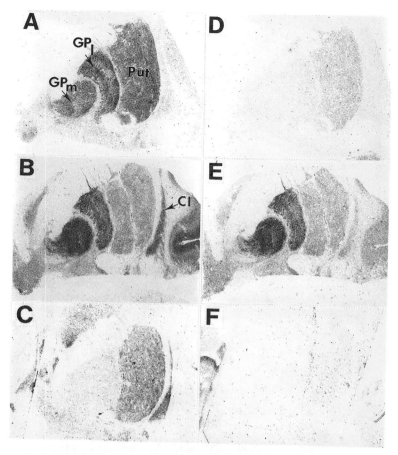

Figure 3. Autoradiograms obtained with human brain sections at the level of the nucleus lentiformis. Ligands and conditions are the same as in Fig. 2. [³H]mesulergine labeling in the globus pallidus (lateral: GP_l; medial: GP_m) is readily displaced by 100 nM 5-HT, whereas [³H]5-HT labeling in this regions is not displaced by 100 nM mesulergine. The level of [¹²⁵I]DOI specific binding is low in the globus pallidus. Abbreviations are putamen (Put), claustrum (Cl).

the existence of different conformations, or slight modifications, of the same receptor protein.

5-HT$_{1C}$ receptors, labelled by [³H]mesulergine, are commonly reported to be enriched in the choroid plexus. However, in the human brain, high densities of binding sites for this ligand are also found in the substantia nigra and the globus pallidus. In these regions [³H]5-HT labelling is accounted for exclusively by 5-HT$_{1D}$ receptors, which display an intermediate affinity for mesulergine. In addition, 5-HT possess a slightly lower affinity for nigral and striatal [³H]mesulergine binding sites than for those located in the choroid plexus. This difference of

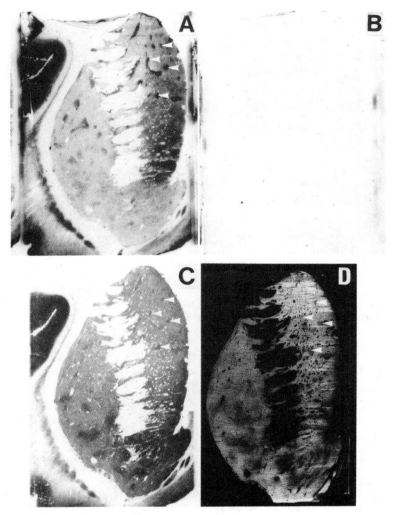

Figure 4. Pattern of [^{125}I]DOI labeling on sections of the human striatum (A: total binding; B: in the presence of 10 μM methysergide). Consecutive sections have been labeled using [^{3}]flunitrazepam (C) or stained for acetylcholinesterase activity (dark areas in D contain a low enzymatic activity).

potency could point to the existence of multiple affinity states of 5-HT$_{1C}$ receptors and hence explain why only the antagonist ligand (in contrast to [^{3}H]5-HT and [^{125}I]DOI) labels significant densities of 5-HT$_{1C}$ sites in the globus pallidus and the substantia nigra (where only receptors in a low affinity state would exist).

The different labelling patterns obtained using the 5-HT$_2$ receptor antagonist, [^{3}H]ketanserin, the mixed agonist/antagonist [^{3}H]LSD and the agonist [^{125}I]DOI could be explained by the fact that the majority of 5-HT$_2$ receptors in the human striatum are in a low affinity state (recognized only by the antagonist), whereas sites in a high affinity state exist exclusively in well defined patches in this region. Although the functional significance of these patches is still unknown, it is interesting to note that their localization corresponds to that of areas rich in benzodiazepine binding sites, which also contain a low acetyl-cholinesterase activity. This indicates that high affinity 5-HT$_2$ receptors are present in a striatal compartment previously described as striosomes (for review, see Graybiel 1990) and would be involved in the specific functions of this division, which presents a chemical cytoarchitecture and a connectivity pattern different from that of the surrounding striatal tissue.

Conclusion

For each of the receptors investigated, our results could be accounted for by the existence of affinity states or of multiple receptor subtypes. Development of new tools, such as antibodies, and the increasing use of molecular biology should provide powerful approaches to help resolve the controversy. However, at the present time, comparing the distributions of receptors as revealed by agonist and antagonists ligands can provide interesting insights. However, antagonist ligands for 5-HT$_{1A}$ and 5-HT$_{1D}$ receptors are still lacking. The availability of such tools would allow to detect these receptors in a possible low affinity state and would help to further define the heterogeneities reported so far for these receptors.

References

Branchek, T. A., Zgombick, J. M., Macchi, M. J., Hartig, P. R., and Weinshank, R. L. (1991). Cloning and expression of a human 5-HT$_{1D}$ receptor. This volume, pp. 21–32.
DeLean, A., Stadel, J. M., and Lefkowitz, R. J. (1980). A ternary complex explains the agonist-specific binding properties of the adenylate cyclase-coupled ß-adrenergic receptor. J. Biol. Chem. 255: 7108–7117.
Graybiel, A. M. (1990). Neurotransmitters and neuromodulators in the basal ganglia. Trends Neurosci. 13: 244–254.
Hartig, P. (1989). Molecular biology of 5-HT receptors. Trends Pharmacol. 10: 64–69.
Leonhardt, S., Herrick-Davis, K., and Titeler, M. (1989). Detection of a novel serotonin receptor subtype (5-HT$_{1E}$) in human brain: interaction with a GTP-binding protein. J. Neurochem. 53: 465–471.
McKenna, D. J., and Peroutka, S. J. (1989). Differentiation of 5-hydroxytryptamine$_2$ receptor subtypes using ^{125}I-R-(-)2,5-dimethoxy-4-iodo-phenylisopropylamine and ^{3}H-ketanserin. J. Neurosci. 9: 3482–3490.
Palacios, J. M., and Dietl, M. M. (1988). Autoradiographic studies on 5-HT receptors. In Sanders-Bush, E. (ed.), The Serotonin Receptors. Clifton: Humana Press, pp. 89–138.

Radja, F., Laporte, A.-M., Daval, G., Vergé, D., Gozlan, H., and Hamon, M. (1990). Autoradiography of serotonin receptor subtypes in the central nervous system. Neurochem. Int. 18: 1–15.

Sumner, M. J., and Humphrey, P. P. A. (1989). Heterogeneous 5-HT$_{1D}$ binding sites in porcine brain can be differentiated by GR 43175. Br. J. Pharmacol. 98: 29–31.

Waeber, C., Schoeffter, P., Palacios, J. M., and Hoyer, D. (1988). Molecular pharmacology of 5-HT$_{1D}$ recognition sites: radioligand binding studies in human, pig and calf brain membranes. Naunyn-Schmiedeberg's Arch. Pharmacol. 337: 595–601.

Waeber, C., Schoeffter, P., Palacios, J. M., and Hoyer, D. (1989a). 5-HT$_{1D}$ receptors in guinea-pig and pigeon brain: radioligand binding and biochemical studies. Naunyn-Schmiedeberg's Arch. Pharmacol. 340: 479–485.

Waeber, C., Dietl, M. M., Hoyer, D., and Palacios, J. M. (1989b). 5-HT$_1$ receptors in the vertebrate brain: regional distribution examined by autoradiography. Naunyn-Schmiedeberg's Arch. Pharmacol. 340: 486–494.

Serotonin: Molecular Biology, Receptors and Functional Effects
ed. by J. R. Fozard/P. R. Saxena
© 1991 Birkhäuser Verlag Basel/Switzerland

Second Messengers in the Definition of 5-HT Receptors

D. Hoyer, H. Boddeke, and P. Schoeffter

Preclinical Research, Sandoz Pharma Ltd, CH-4002 Basel, Switzerland

Summary. Radioligand binding studies have been instrumental in the discovery of 5-HT receptor subtypes, but have clear limitations. By contrast, second messenger studies are elegant tools, although somewhat overlooked for classifying *functional* receptors. In the particular case of 5-HT$_1$ receptor subtypes, few "classical" isolated tissue preparations have been described, whereas second messenger systems are available. 5-HT$_{1A}$, 5-HT$_{1B}$ and 5-HT$_{1D}$ receptors are negatively coupled to adenylate cyclase, whereas 5-HT$_{1C}$ receptors (like 5-HT$_2$ receptors) stimulate phospholipase C. The rank orders of potency of agonists and antagonists established in second messenger studies correlated significantly with rank orders of affinity determined in specific radioligand binding assays. Subsequently, a variety of functional 5-HT$_1$ receptor subtypes-mediated effects have been described in both the CNS and periphery. The profile established for 5-HT$_{1C}$ receptor-mediated stimulation of phospholipase C will be especially useful in the definition of functional 5-HT$_{1C}$ receptors, although 5-HT$_{1C}$ and 5-HT$_2$ receptors have similar pharmacological profiles and second messenger systems. 5-HT$_3$ receptors form ligand gated channels and thus, do not directly activate second messenger systems. Even more interesting is the case of the 5-HT$_4$ receptor which stimulates adenylate cyclase in colicullus cell cultures, and for which no binding site has been as yet described. Second messenger studies have permitted the identification of similar functional receptors in various peripheral models, and have contributed significantly to the definition of new functional 5-HT receptor subtypes.

Introduction

The heterogeneity of 5-HT receptors is now a commonly accepted fact which has been substantiated in radioligand binding, autoradiographical, biochemical, electrophysiological, behavioral and what is usually termed functional studies (see Bradley et al. 1986; Peroutka 1988; Hoyer 1988b; 1989). Most recently the cloning of at least four genes coding for 5-HT$_{1A}$, 5-HT$_{1C}$, 5-HT$_{1D}$ and 5-HT$_2$ receptor proteins has provided even more convincing evidence for the heterogeneous nature of 5-HT receptors at the molecular level. It is evident that developments in medicinal chemistry have provided crucial tools (i.e. agonists and antagonists) for the identification of receptor subtypes. Thus, the availability of compounds with high affinity for 5-HT receptors has permitted the development of radioligands which made possible the characterisation of binding sites with different pharmacological profiles and distributions. However, in several cases, these binding sites could not readily be equated with the 5-HT receptors known from functional studies.

Conversely, certain functional receptors have not yet been identified in binding studies (e.g. 5-HT$_4$ receptors). Clearly, a binding site, however well characterised with respect to pharmacological selectivity and/or tissue distribution, is not sufficient proof for the existence of a functional receptor. When Bradley et al. (1986) proposed a scheme for the subdivision of 5-HT receptors into three families (5-HT$_1$-like, 5-HT$_2$ and 5-HT$_3$), the definition of 5-HT$_2$ and 5-HT$_3$ receptors appeared to be fairly straightforward, in contrast to that of 5-HT$_1$ receptors. Thus, 5-HT$_2$ and 5-HT$_3$ receptors were the equivalent of the 5-HT-D and 5-HT-M receptors of Gaddum and Picarelli (1957), respectively. Furthermore, it had been clearly established that 5-HT-D receptors corresponded to the 5-HT$_2$ sites originally defined by Peroutka and Snyder (1979) (see Engel et al. 1984). By contrast, the definition of 5-HT$_1$-like receptors was less exact. In fact, at the time of the Bradley et al. classification, there was little or no correlation between subtypes of 5-HT$_1$ sites (Pedigo et al. 1981; Gozlan et al. 1983; Pazos et al. 1984; Hoyer et al. 1985a; b) and certain functional effects of 5-HT; thus, the term '5-HT$_1$-like' was adopted.

Although most of the binding sites identified in the mid- to late 1980's are now recognised as true receptors (5-HT$_{1A}$, 5-HT$_{1B}$, 5-HT$_{1C}$, 5-HT$_{1D}$, 5-HT$_2$ and 5-HT$_3$), the contribution of radioligand binding studies to receptor characterisation has met serious criticism (Green 1987; Laduron 1987). Indeed, the use of radioligand binding has clear limitations and pitfalls: a radioligand should have sufficient high affinity and selectivity, the binding should be homogeneous, saturable, reversible, stereospecific and display the appropriate pharmacology. Further, radioligands may label a variety of sites (in addition to receptors) such as uptake sites, enzymes, transporters, and so called non-specific binding sites. Another limitation of binding studies, is the lack of prediction for the agonist or antagonist nature of the ligands interacting with that site. Finally, it is clear that a 'cultural' bias exists towards radioligand binding, the level of acceptance of this type of studies being usually greater among molecular as opposed to classical pharmacologists.

There is one approach in the 5-HT receptor field, which in recent years has proven invaluable in the definition of new receptors, that is second messenger studies. For several of the newly identified 5-HT recognition sites, functional correlates were initially lacking (e.g. 5-HT$_{1A}$, 5-HT$_{1B}$, 5-HT$_{1C}$, and 5-HT$_{1D}$ sites). For each of these sites, the second messenger system was characterised before functional correlates were found. Studies with second messengers allowed definition of whether ·drugs are agonists or antagonists. Subsequently, the results obtained in second messenger tests were confirmed in classical functional studies and for some of these receptors by the cloning and expression of the receptor protein. The second messenger approach was the basis of the discovery of the $5\text{-}HT_4$ receptor, for which no binding equivalent has yet been reported; $5\text{-}HT_4$ receptors have since been identified in several peripheral tissues, which were previously classified as containing atypical 5-HT receptors (see for example, Bom et al. 1988; Villalon et al. 1990; 1991). Here, we review the 5-HT receptor second messenger approach, show how it can and has contributed to the definition of receptor subtypes, and how it may explain apparent discrepancies which have been observed with respect to agonist potencies in various functional and/or behavioural models.

Radioligand Binding Studies and the Subclassification of 5-HT Receptors

All the recognised $5\text{-}HT_1$ receptor subtypes have been characterised using high affinity radioligands and usually tissue preparations enriched in one or other of the subtypes: $5\text{-}HT_{1A}$, $5\text{-}HT_{1B}$, $5\text{-}HT_{1C}$ and $5\text{-}HT_{1D}$ receptors are labelled respectively, with [^{3}]8-OH-DPAT (Gozlan et al. 1983), [^{125}I]cyanopindolol (Hoyer et al. 1985a), [^{3}H]mesulergine (Pazos et al. 1984) and [^{3}H]5-HT under appropriate conditions (Heuring and Peroutka 1987). For each of the receptor subtypes, except for $5\text{-}HT_{1D}$, other radioligands have been described. $5\text{-}HT_{1B}$ receptors have only been found in some rodents (rat, mouse, hamster), whereas $5\text{-}HT_{1D}$ receptors are found in most other mammals, including man (Hoyer et al. 1985b; 1986; Waeber et al. 1988a; b; 1989a; b). [^{3}H]Ketanserin is the most commonly used radioligand for $5\text{-}HT_2$ receptors (Leysen et al. 1982). Other $5\text{-}HT_2$ receptor ligands have become available, but most of them (except ketanserin and spiperone), label both $5\text{-}HT_{1C}$ and $5\text{-}HT_2$ receptors (Hoyer 1988a; 1989); this also holds true for high affinity agonist ligands, such as [^{3}H]DOB and [^{125}I]DOI. Two subtypes of the $5\text{-}HT_2$ receptor have been proposed (Peroutka et al. 1988), labelled by [^{3}H]DOB or [^{125}I]DOI ($5\text{-}HT_{2A}$) and [^{3}H]ketanserin ($5\text{-}HT_{2A}$ and $5\text{-}HT_{2B}$); however, this issue has recently been clarified as agonists appear to label the high affinity state of the $5\text{-}HT_2$ receptor (Branchek et al. 1990; Teitler et al. 1990). $5\text{-}HT_3$ receptors have been labelled using [^{3}H]ICS 205-930, [^{3}H]GR 65630, [^{3}H]quipazine, [^{3}H]quaternised-ICS 205-930, [^{3}H]zacopride and [^{3}H]BRL 43694 (Hoyer and Neijt 1987; 1988; Kilpatrick et al. 1987; Watling et al. 1988; Peroutka and Hamik 1988; Barnes et al. 1988). These ligands allowed $5\text{-}HT_3$ sites to be detected in the brain and in the periphery. Subtypes of functional $5\text{-}HT_3$ receptors have been proposed (Fozard 1983; Richardson and Engel 1986), but there is little convincing evidence from either radioligand binding or functional studies for $5\text{-}HT_3$ receptor subtypes. Species differences may be the basis of the differences between $5\text{-}HT_3$ receptor subtypes described up to now (Hoyer 1990). $5\text{-}HT_4$ receptors have not yet been identified using radioligand binding, and no high affinity agonist or antagonist adequate to be radiolabelled has yet been described.

5-HT Receptors and Second Messenger Studies

5-HT₁ Receptors

5-HT₁ₐ Receptors: It is well known that 5-HT$_{1A}$ sites are particularly enriched in the hippocampus of several species (Gozlan et al. 1983; Pazos and Palacios 1985); thus, it is not surprising that most 5-HT$_{1A}$ receptor second messenger studies have been carried out in hippocampal preparations. 5-HT-stimulated adenylate cyclase has been reported in guinea pig and rat hippocampus (Shenker et al. 1985; 1987; Markstein et al. 1986). However, 5-HT receptor mediated inhibition of forskolin- or vasoactive intestinal polypeptide-stimulated adenylate cyclase has also been described in rat, guinea pig, mouse and calf hippocampus preparations with the pharmacological profile of a 5-HT$_{1A}$ receptor-mediated effect (DeVivo and Maayani 1986; Weiss et al. 1986; Bockaert et al. 1987; Schoeffter and Hoyer 1988; 1989b; Dumuis et al. 1988c). Fargin et al. (1988) reported on the cloning of the human 5-HT$_{1A}$ receptor, which couples negatively to adenylate cyclase when expressed in various cells. These authors also provided evidence that 5-HT$_{1A}$ receptors transfected in Hela cells may couple to phospholipase C (Fargin et al. 1989).

Table 1 compares 5-HT$_{1A}$ receptor-mediated inhibition of forskolin-stimulated activity in calf hippocampus, with 5-HT$_{1A}$ receptor mediated calcium mobilisation (measured with Fura 2) in HeLa cells transfected with 5-HT$_{1A}$ receptors. It has previously been documented that 5-HT$_{1A}$ binding in these cells was fully comparable with 5-HT$_{1A}$ binding performed in mammalian brain tissue (Fargin et al. 1988) both with respect to rank order of affinity and actual affinity values. In these HeLa cells, which express about 500 fmol/mg 5-HT$_{1A}$ sites, 5-HT and other agonists produce calcium mobilisation with a pharmacological profile compatible with that of a 5-HT$_{1A}$ receptor. In addition, the potency of NAN 190 and pindolol as antagonists, was similar in the Fura-2 test in HeLa cells and in the cyclase response in hippocampus preparations. It is striking that 5-HT is significantly more potent with respect to cyclase inhibition ($pEC_{50} = 20$ nM) in HeLa cells than compared to phospholipase C activation ($pEC_{50} = 3.2\ \mu$M) in these same cells (Fargin et al. 1989). Further, it is clear that not all drugs acting as agonists at the cyclase do so in the calcium response. For instance, sumatriptan behaves as a full agonist in the hippocampus, but as a low efficacy partial agonist in HeLa cells. Even more striking is the behaviour displayed by MDL 73005, ipsapirone, buspirone and spiroxatrine. These drugs have been described as agonists, partial agonists or even antagonists depending on the test model studied (Dumuis et al. 1988c; Schoeffter and Hoyer 1988; Moser et al. 1990; Rydelek-Fitzgerald et al. 1990). In our studies, all four compounds act as agonists at the cyclase, but as antagonists of the

Table 1. Comparison of 5-HT$_{1A}$ receptors negatively coupled to adenylate cyclase in hippocampal or cortical preparation and receptor-mediated calcium mobilisation in HeLa cells transfected with human 5-HT$_{1A}$ receptors

Drug	Calf hippocampus homogenates		Mouse hippocampus neurones		Mouse cortex neurones		HeLa cells	
	pK	IA(%)	pK	IA(%)	pK	IA(%)	pK	IA(%)
5-HT	7.83	100	7.30	100	7.00	100	6.01	100
5-CT	8.59	108	7.70	100	7.50	100	6.71	93
8-OH-DPAT	8.22	96	8.10	100	6.60	50	6.52	106
RU 24969	7.81	105	7.16	100	6.95	100		
flesinoxan	7.68	106					6.22	96
sumatriptan	5.57	102					5.60	36
ipsapirone	7.48	77	7.10	77	6.40	0	7.84	0
buspirone	7.32	77	6.90	66	inactive as agonist	inactive as agonist	7.33	0
TFMPP	6.72	67	6.93	35	inactive as agonist	inactive as agonist		
mCPP	5.91	40	6.95	55	inactive as agonist	inactive as agonist		
metergoline	7.58	111	6.90	94	7.90	0		
methysergide	6.40	105	6.35	92	7.00	0		
spiroxatrine	7.75	91						
MDL 73005	7.32	92					8.16	0
cyanopindolol	8.16	43	7.80	0	6.30	0	7.52	0
WB 4101	7.85	49	7.55	0	6.50	0		
NAN 190	8.60	0	7.55	0	7.20	0	8.42	0
methiothepin	7.73	0	7.90	0	7.24	0	8.23	0
pindolol	7.87	0	6.40	0	7.40	0	7.74	0
propranolol	6.60	0						

pK = pEC$_{50}$ values (agonists) or pK$_B$ values (antagonists). Efficacy (IA) = % of the maximal effect produced by 5-HT. Calf hippocampus: inhibition of forskolin-stimulated adenylate cyclase (data are from Schoeffter and Hoyer, 1988, and from the author's laboratory). HeLa cells: (transfected with human 5-HT$_{1A}$ receptors, provided by Dr. A. Fargin, Howard Hughes Medical Institute, Duke University, Durham, N. C.), calcium mobilisation measured using Fura 2 (Data are from the author's laboratory). Mouse hippocampal and cortical neurones (data are from Dumuis et al. 1988c).

calcium mobilisation. These differences may be explained in several ways. Either the receptor reserve is much higher in the hippocampus than in HeLa cells, or the coupling efficacy is lower in HeLa cells, or the receptor is able to couple to different G proteins, one of which acts preferentially on adenylate cyclase. Whatever the explanation, it seems clear that depending on the model studied, agonists do not necessarily show the same intrinsic activity at different $5\text{-}HT_{1A}$ receptors.

A further illustration of this comes from the comparison of the effects of drugs on adenylate cyclase in mouse hippocampus and cortex. In the mouse hippocampus, with a few exceptions, the potency and efficacy of agonists is comparable to that observed in calf hippocampus. By contrast, in mouse cortex, most agonists show lower potency and efficacy, and in several cases, compounds acting as full or potent partial agonists in the hippocampus, turn out to display antagonist activity in the cortex (e.g. metergoline, methysergide, ipsapirone, buspirone, TFMPP and mCPP). These differences led Dumuis et al. (1988c), to suggest that the cortex receptor might represent a subtype different from the $5\text{-}HT_{1A}$ receptor present in the hippocampus. Although this cannot be ruled out unequivocally, the results obtained in cortical neurones are very similar to those in HeLa cells, which are known to express exclusively $5\text{-}HT_{1A}$ receptors. Thus, an alternative suggestion for the discrepancy between cortex and hippocampus would be poor coupling or low receptor reserve of the cortical receptor. We propose that variations in intrinsic activity of a variety of compounds acting at $5\text{-}HT_{1A}$ receptors, may not reflect differences in receptor subtypes, but conditions prevailing in the various test models. In other words, drugs such as MDL 73005, buspirone or ipsapirone may behave as agonists or antagonists depending on the properties of the receptors (e.g. pre *vs* post-synaptic, hippocampus *vs* cortex or dorsal raphe).

$5\text{-}HT_{1B}$ Receptors: $5\text{-}HT_{1B}$ sites in the rat brain are particularly concentrated in the nigro-striatal system, especially in the substantia nigra (Pazos and Palacios 1985). Thus, it was reasonable to search for the second messenger system coupled to $5\text{-}HT_{1B}$ receptors in this tissue. $5\text{-}HT_{1B}$ receptors were indeed found to be negatively coupled to adenylate cyclase (Bouhelal et al. 1988; Schoeffter and Hoyer 1989a; b) in homogenates of rat substantia nigra, but also in hamster lung (Seuwen et al. 1988) and opossum kidney cell lines (Murphy and Bylund 1988). Second messenger studies performed in substantia nigra of rat, guinea-pig and calf, have clearly demonstrated that the pharmacological profile of the $5\text{-}HT_{1B}$ receptor in rat was different from that of the $5\text{-}HT_{1D}$ receptor observed in calf or guinea-pig, as previously suggested by radioligand and autoradiographical studies (see below).

$5\text{-}HT_{1D}$ Receptors: As in the case of $5\text{-}HT_{1B}$ receptors, the substantia nigra is the tissue which contains the highest density of $5\text{-}HT_{1D}$ sites

in higher mammals (Waeber et al. 1988a; b). Activation of 5-HT_{1D} receptors leads to inhibition of forskolin stimulated adenylate cyclase activity in calf and guinea-pig substantia nigra (Hoyer and Schoeffter 1988; Schoeffter et al. 1988; Waeber et al. 1989b). The pharmacology of the inhibition of adenylate cyclase in rat substantia nigra (5-HT_{1B}), was clearly different from that of the cyclase activity in guinea-pig and calf substantia nigra (both 5-HT_{1D}) (Schoeffter and Hoyer 1989a; Waeber et al. 1989b) thus confirming the species differences first observed in radioligand binding and autoradiographic studies. Adenylate cyclase inhibition was the first functional test reported to be mediated by 5-HT_{1D} receptors. Subsequently, several other functional 5-HT_{1D} receptor models have been reported, e.g. terminal autoreceptors in guinea-pig and pig cortex (Schlicker et al. 1989; Hoyer and Middlemiss 1989); relaxation of pig coronary artery (Schoeffter and Hoyer 1990) or more recently, contraction of human pial arterioles (Hamel and Bouchard 1990; this volume). In each case, the results from second messenger studies correlated with those from functional studies. Finally, a human 5-HT_{1D} receptor has been cloned whose pharmacology agrees with that defined in second messenger and binding studies (Branchek et al. 1990; this volume).

Table 2 compares the effects of drugs on 5-HT_{1D} receptor-mediated inhibition of adenylate cyclase in calf and guinea-pig substantia nigra, with inhibition of 5-HT release in pig cortex and relaxation of pig coronary artery. As with 5-HT_{1A} receptors, clear differences can be observed with respect to agonist potency and intrinsic activity. SDZ 21009, cyanopindolol, yohimbine and CGS 12066 for instance, are agonists both in the substantia nigra and the cortex, but behave essentially as antagonists in the coronary artery. Metergoline is an agonist at the cyclase, but an antagonist in both the release and relaxation models. On the other hand, the pK_B values of antagonists are very similar in all four models. Thus, as for 5-HT_{1A} receptors, there are 5-HT_{1D} receptor agonists which display marked variations in potency and intrinsic activity depending on the test model. Again, these observations may shed new light, on the apparent lack of correlation between 5-HT_{1D} binding and the receptor-mediated effects in various vascular tissues, termed "5-HT_1 like", which may well belong to the 5-HT_{1D} receptor category (e.g. saphenous vein, basilar artery), as reported in human pial vessels (Hamel and Bouchard, this volume).

5-HT₂ Receptors

5-HT₁C Receptors: The choroid plexus contains the highest density of 5-HT_{1C} sites in all species studied (Pazos et al. 1984; Pazos and Palacios 1985), and thus was the ideal preparation to look for the 5-HT_{1C} second

Table 2. Comparison of 5-HT$_{1D}$ receptors negatively coupled to adenylate cyclase in substantia nigra (calf and guinea-pig), 5-HT$_{1D}$ receptor mediated inhibition of 5-HT release (pig cortex) and 5-HT$_{1D}$ receptor mediated relaxation (pig coronary artery)

Drug	Calf subs. nigra homogenates		Guinea-pig subs. nigra homogenates		Pig cortex release		Pig coronary artery	
	pK	IA(%)	pK	IA(%)	pK	IA(%)	pK	IA(%)
5-HT	7.62	100	7.62	100	7.08	ago	6.66	100
5-CT	8.05	111	7.96	107	6.60	ago	6.53	121
RU 24969	6.82	81	6.69	84	6.22	ago	5.52	49
5-MeOT	7.31	98			6.61	ago	6.79	125
sumatriptan	6.28	102					5.09	48
SDZ 21009	5.91	61	5.27	55	5.86	ago	5.05	0
yohimbine	6.79	56	7.26	49	5.77	ago	6.37	0
cyanopindolol	6.80	92			5.69	ago	5.32	0
CGS 12066	7.11	83			5.37	ago	6.73	0
8-OH-DPAT	5.84	76	6.30	73	5.47	ago	<5.0	0
metergoline	7.53	83			7.05	0	6.86	0
mianserin	6.49	0			5.64	0	6.60	0
methiothepin	7.04	0			7.47	0	7.30	0
isamoltane	4.55	0	5.22	0	<5.5		4.33	0
propranolol	5.23	0	4.34	0	<5.5		4.43	0
spiperone	4.75	0					4.44	0

pK = pEC$_{50}$ values (agonist) or pK$_B$ values (antagonists). Efficacy (IA) = % of the maximal effect produced by 5-HT. Calf and guinea-pig substantia nigra: inhibition of forskolin-stimulated adenylate cyclase (data are from Schoeffter et al, 1988, Schoeffter and Hoyer, 1989b, Waeber et al, 1989c and from the author's laboratory), pig cortex: inhibition of 5-HT release (data from Schlicker et al, 1989), pig coronary artery: inhibition of contraction (data from Schoeffter and Hoyer, 1990).

messenger system. 5-HT$_{1C}$ receptor activation in rat, pig and mouse choroid plexus led to the activation of phospholipase C and accumulation of inositol phosphates (Conn and Sanders-Bush 1986; Conn et al. 1986; Hoyer 1988a; Hoyer et al. 1989). Antagonists acting selectively at 5-HT$_2$ receptors (ketanserin, cinanserin, spiperone, pirenperone), although acting as competititve antagonists in the choroid plexus, were weak, ruling out a 5-HT$_2$ receptor-mediated event. Further, both in radioligand binding and autoradiographical studies the presence of 5-HT$_2$ sites could not be detected in the choroid plexus. As for other 5-HT receptors, second messenger studies provided the first functional evidence indicating that 5-HT$_{1C}$ sites represent true receptors. Due to the analogy between 5-HT$_{1C}$ and 5-HT$_2$ receptors, with respect to pharmacology, second messenger coupling and structure, it has been proposed that the 5-HT$_{1C}$ receptor should be classified as a member of the 5-HT$_2$ receptor family (Hoyer 1988a; 1989).

5-HT$_2$ Receptors: 5-HT$_2$ receptor-stimulated hydrolysis of inositol lipids and/or calcium mobilisation has been reported in brain (Conn and Sanders-Bush 1984; 1985), C$_6$ glioma cells (Ananth et al. 1987), platelets (De Chaffoy de Courcelles et al. 1985) and smooth muscle (Nakaki et al. 1985; Doyle et al. 1986; Cory et al. 1986). The coupling of the 5-HT$_2$ receptor to phospholipase C was confirmed when this receptor was cloned and expressed in cells (Pritchett et al. 1988).

5-HT$_3$ Receptors

Binding studies indicate that 5-HT$_3$ receptors are not linked to G-proteins (Kilpatrick et al. 1987; Hoyer and Neijt 1988). Evidence from electrophysiological studies show that, like nicotine receptors, 5-HT$_3$ receptors activate fast ion channels (Neijt et al. 1988; Yakel et al. 1988; Peters and Lambert 1989; see also Wallis, this volume). Thus, it has been suggested that 5-HT$_3$ receptors are members of the ion channel gated family (like nicotine, glycine or GABA$_A$ receptors). The recent work of Derkach et al. (1989), using patch clamp techniques, has clearly demonstrated this to be the case.

5-HT$_4$ Receptors

5-HT stimulated adenylate cyclase activity resistant to blockade by classical 5-HT receptor antagonists was first reported in various brain tissues, such as hippocampus and colliculus (Bockaert et al. 1981). The changes in cyclase activity in the hippocampus were complex, with various agonists (e.g. 5-CT) producing biphasic concentration response

Table 3. Comparison of 5-HT$_4$ receptor-mediated stimulation of adenylate cyclase in mouse colliculus neurones, contraction of guinea-pig ileum and relaxation of rat oesophagus

Drug	Mouse colliculus neurones pK	Guinea-pig ileum contraction pK	Rat oesophagus relaxation pK
5-HT	7.03	8.55	8.38
5-MeOT	7.00	7.62	8.08
renzapride	6.90	7.13	7.08
α-Me-5-HT		6.76	7.15
cisapride	7.14	6.46	7.53
zacopride	5.95	6.48	6.36
5-CT	5.50	6.46	6.42
metoclopramide	5.34		6.30
2-Me-5-HT	< 5	< 5	5.25
ICS 205-930	6.01-6.27	6.4-6.6	6.7

pK = pEC$_{50}$ values (agonists) or pK$_B$ or pA$_2$ values (ICS 205-930). Mouse colliculus: stimulation of adenylate cyclase (data from Dumuis et al. 1988, 1989, Bockaert et al. 1990), guinea-pig ileum (cholinergically mediated twitch response, data from Craig and Clarke 1990), rat oesophagus: relaxation (data adapted from Baxter and Clarke 1990).

curves (Shenker et al. 1985). A high affinity component was identified as a 5-HT$_{1A}$ effect, whereas a low affinity component could not be defined using a variety of 5-HT receptor antagonists. Recently, Bockaert and collaborators (Dumuis et al. 1988a; b; 1989a; b; Bockaert et al., this volume) have characterised pharmacologically this activity in neurones of mouse colliculi and guinea-pig hippocampus. The rank order of potency of agonists shown for mouse colliculi in table 3, does not fit to any of the known receptor subtypes. The only competitive antagonist described to date is ICS 205-930, which blocked the effects of various agonists, but at concentrations much higher than those affecting 5-HT$_3$ receptors. The authors proposed that this receptor be named 5-HT$_4$. Significantly, the low affinity component adenylate cyclase stimulation in hippocampus was antagonized by ICS 205-930, at concentrations similar to those acting in the colliculi preparation.

The 5-HT$_4$ receptor is also present in the guinea-pig ileum (Craig and Clarke 1990; Clarke et al. 1989), where it mediates the prokinetic activities of compounds such as cisapride, BRL 24924 and metoclopramide (Sanger 1987). During the past year, 5-HT$_4$ receptors have been identified in functional tests e.g. in the human heart (Kaumann et al. 1990; this volume), pig heart (Villalon et al. 1990; 1991; this volume), in the ascending colon (Elswood et al. 1990) and rat oesophagus (Baxter and Clarke, this volume). Table 3 compares the potency of a series of drugs at the colliculus 5-HT$_4$ receptors, and at receptors in two preparations (guinea-pig ileum and rat oesophagus), which until recently remained poorly defined. Overall, the rank order of potency is similar in all three models. It should be added that a whole variety of drugs known to interact with high affinity at the various subtypes of $5\text{-}HT_1$, $5\text{-}HT_2$ and $5\text{-}HT_3$ receptors, are largely devoid of activity in all three models (Dumuis et al. 1988; 1989; Bockaert et al. 1990; Clarke et al. 1989; Craig and Clarke 1990; Craig et al. 1990; Baxter and Clarke 1990). Nevertheless, ICS 205-930 acts as a surmountable antagonist, with low but very similar potency in all three models. Generally, agonists are somewhat more potent in the functional as opposed to second messenger models, which suggests a smaller receptor reserve and/or poor coupling in the latter. The $5\text{-}HT_4$ receptor is probably *the* example illustrating the value of second messenger models for the definition of a new receptor.

Conclusion

It is now clear that multiple 5-HT receptors subtypes exist and that biochemical approaches such as radioligand binding, autoradiography and second messenger studies have been of fundamental importance in the identification and characterisation of several of the new receptors.

Second messenger studies represent an appropriate, although somewhat underestimated approach, for classifying *functional* receptors. In the particular case of $5\text{-}HT_1$ receptor subtypes, few "classical" functional models have been described; in contrast, second messenger systems are known for each of these receptor subtypes. For each receptor, rank orders of potency (second messengers studies) and affinity (binding studies) of agonists and antagonists correlate significantly. Where functional tests have appeared (e.g. for $5\text{-}HT_{1A}$, $5\text{-}HT_{1B}$ and $5\text{-}HT_{1D}$ receptors), results from functional studies have correlated significantly with those from second messenger studies. Of particular interest, is the $5\text{-}HT_4$ receptor, which has no binding equivalent, yet which was thoroughly and convincingly demonstrated using the positive link to adenylate cyclase. For those sites where functional tests do not yet exist (e.g. $5\text{-}HT_{1C}$ receptors), the lead given by second messenger studies has proven helpful in defining the pharmacology of the receptor.

Finally, considerations of second messenger and structural information ($5\text{-}HT_{1A}$, $5\text{-}HT_{1C}$, $5\text{-}HT_{1D}$ and $5\text{-}HT_2$ receptors have been cloned and the $5\text{-}HT_3$ receptor forms part of a fast cation channel), rather than pharmacological considerations (presently, the $5\text{-}HT_4$ receptor does not fit into any of the recognised classes), would allow 5-HT receptors to be classified, like acetylcholine receptors, into two large families: receptors coupled to G-proteins ($5\text{-}HT_1$, $5\text{-}HT_2$ and $5\text{-}HT_4$) and those forming ligand-gated ion channels ($5\text{-}HT_3$), which are probably members of the nicotine /$GABA_A$/ glycine channel "super-family".

128

Acknowledgements

We wish to express our sincere thanks to John Fozard for constant support and critically reading the manuscript.

References

Ananth, U.S., Leli, U., and Hauser, G. (1987). Stimulation of phosphoinositide hydrolysis by serotonin in C6 glioma cells. J. Neurochem. 48: 253–261.

Barnes, N. M., Costall, B., and Naylor, R. J. (1988). [^{3}H]Zacopride: Ligand for identification of 5-HT$_3$ recognition sites. J. Pharm. Pharmacol. 40: 548–551.

Baxter, G. S., and Clarke, D. E. (1990). Putative 5-HT$_4$ receptors mediate relaxation of rat oesophagus. The Second IUPHAR Satellite Meeting on Serotonin, Basel July 11-13, 1990. Abstr. 85: 78.

Bockaert, J., Dumuis, A., Bouhelal, R., Sebben, M., and Cory, R. N. (1987). Piperazine derivatives including the putative anxiolytic drugs, buspirone and ipsapirone, are agonists at 5-HT$_{1A}$ receptors negatively coupled with adenylate cyclase in hippocampal neurons. Naunyn-Schmiedeberg's Arch. Pharmacol. 335: 588–592.

Bockaert, J., Nelson, D. L., Herbet, A., Adrien, J., Enjalbert, A., and Hamon, M. (1981). Serotonin receptors coupled with an adenylate cyclase in the rat brain: non identity with [^{3}H]-5-HT binding sites. Adv. Exp. Med. Biol. 133: 327–345.

Bockaert, J., Sebben, M., and Dumuis, A. (1990). Pharmacological characterization of 5-hydroxytryptamine-4 (5-HT$_4$) receptors positively coupled to adenylate cyclase in adult guinea-pig hippocampal membranes effect of substituted benzamide derivatives. Mol. Pharmacol. 37: 408–411.

Bom, A. H., Duncker, D. J., Saxena, P. R., and Verdouw, P. D. (1988). 5-Hydroxy-tryptamine-induced tachycardia in the pig: possible involvement of a new type of 5-hydroxytryptamine receptor. Br. J. Pharmacol. 93: 663–671.

Bouhelal, R., Smounya, L., and Bockaert, J. (1988). 5-HT$_{1B}$ receptors are negatively coupled with adenylate cyclase in rate substantia nigra. Eur. J. Pharmacol. 151: 189–196.

Bradley, P. B., Engel, G., Fenuik, W., Fozard, J. R., Humphrey, P. P. A., Middlemiss, D. N., Mylecharane, E. J., Richardson, B. P., and Saxena, P. R. (1986). Proposals for the classification and nomenclature of functional receptors for 5-hydroxytryptamine. Neuropharmacol. 25: 563–576.

Branchek, R. L., Adham, N., Macchi, M. J., Kao, H. T., and Hartig, P. R. (1990). [^{3}H]-DOB (4-bromo-2,5-dimethoxyphenyliopropylamine) and [^{3}H]ketanserin label two affinity states of the cloned human 5-hydroxytryptamine$_2$ receptor. Mol. Pharmacol. 38: 604–609.

Clarke, D. E., Craig, D. A., and Fozard, J. R. (1989), The 5-HT$_4$ receptor: naughty but nice. TiPS. 10: 385–386.

Conn, P. J., and Sanders-Bush, E. (1984). Selective 5-HT antagonists inhibit serotonin stimulated phosphatidylinositol metabolism in cerebral cortex. Neuropharmacol. 8: 993–996.

Conn, P. J., and Sanders-Bush, E. (1985). Serotonin stimulated phosphoinositide turnover: mediation by the S2 binding site in rat cerebral cortex but not in subcortical regions. J. Pharmacol. Exp. Ther. 234: 195–203.

Conn, P. J., and Sanders-Bush, E. (1986). Agonist-induced phosphoinositide hydrolysis in choroid plexus. J. Neurochem. 47: 1754–1760.

Conn, P. J., Sanders-Bush, E., Hoffman, B. J., and Hartig, P. R. (1986). A unique serotonin receptor in choroid plexus is linked to phosphatidylinositol turnover. Proc. Natl. Acad. Sci. USA 83: 4086–4088.

Cory, R. N., Berta, P., Haiech, J., and Bockaert, J. (1986). 5-HT$_2$ receptor-stimulated inositol phosphate formation in rat aortic myocytes. Eur. J. Pharmacol. 131: 153–157.

Craig, D. A., and Clarke, D. E. (1990). Pharmacological characterization of a neuronal receptor for 5-hydroxytryptamine in guinea-pig ileum with properties similar to the 5-hydroxtryptamine-4 receptor. J. Pharmacol Exp. Ther. 252: 1378–1386.

Craig, D. A., Eglen, R. M., Walsh, L. K. M., Perkins, L. A., Whiting, R. L., and Clarke, D. E. (1990). 5-Methoxytryptamine and 2-methyl-5-hydroxytryptamine-induced desensitization as a discriminative tool for the 5-HT3 and putative 5-HT4 receptors in guinea pig ileum. Naunyn-Schmeideberg's Arch. Pharmacol. 342: 9–16.

De Chaffoy de Courcelles, D., Leysen, J. E., De Clerck, F., Van Belle, H., and Janssen, P. A. J. (1985) Phospholipid turnover is the signal transducing system coupled to serotonin-S_2 receptor sites. J. Biol. Chem. 260: 7603–7608.

Derkach, V., Surprenant, A. M., and North, R. A. (1989). 5-HT_3 receptors are membrane ion channels. Nature (London) 339: 706–709.

De Vivo, M., and Maayani, S. (1986). Characterization of 5-hydroxytryptamine$_{1A}$-receptor-mediated inhibition of forskolin-stimulated adenylate cyclase activity in guinea-pig and rat hippocampal membranes. J. Pharmacol. Exp. Ther. 238: 248–253.

Doyle, V. M., Creba, J. A., Rüegg, U. T., and Hoyer, D. (1986). Serotonin increases the production of inositol phosphates and mobilises calcium via the 5-HT_2 receptor in A_7r_5 smooth muscle cells. Naunyn-Schmiederberg's Arch. Pharmacol. 333: 98–107.

Dumuis, A., Bouhelal, R., Sebben, M., and Bockaert, J. (1988a). A 5-HT receptor in the central nervous system, positvely coupled with adenylate cyclase, is antagonized by ICS 205 930. Eur. J. Pharmacol. 146: 187–188.

Dumuis, A., Bouhelal, R., Sebben, M., Cory, R., and Bockaert, J. (1988b). A non-classical 5-hydroxytryptamine receptor positively coupled with adenylate cyclase in the central nervous system. Mol. Pharmacol. 34: 880–887.

Dumuis, A., Sebben, M., and Bockaert, J. (1988c). Pharmacology of 5-Hydroxytryptamine-1A-receptors which inhibit cAMP production in hippocampal and cortical neurons in primary culture. Molecular Pharmacology 33: 178–186.

Dumuis, A., Sebben, M., and Bockaert, J. (1989a). BRL 24924: A potent agonist at a non-classical 5-HT receptor positively coupled with adenylate cyclase in colliculi neurons. Eur. J. Pharmacol. 162: 381–384.

Dumuis, A., Sebben, M., and Bockaert, J. (1989b). The gastrointestinal prokinetic benza-mide derivatives are agonists at the non-classical 5-HT receptor 5-HT-4 positively coupled to adenylate cyclase in neurons. Naunyn-Schmeideberg's Arch Pharmacol. 340: 403–410.

Elswood, C. J., Bunce, K. T., and Humphrey, P. P. A. (1990). Identification of 5-HT_4 receptors in guinea-pig ascending colon. The Second IUPHAR Satellite Meeting on Serotonin, Basel July 11–13, 1990. Abstr. 86, 78.

Engel, G., Hoyer, D., Kalkman, H. O., and Wick, M. B. (1984). Identification of 5-HT_2 receptors on longitudinal muscle of the guinea-pig ileum. J. Rec. Res. 4: 113–126.

Fargin, A., Raymond, J. R., Lohse, M. J., Kobilka, B. K., Caron, M. G., and Lefkowitz, R. J. (1988). The genomic clone G-21 which resembles a β-adrenergic receptor sequence encodes the 5-HT_{1A} receptor. Nature 335: 358–360.

Fargin, A., Raymond, J. R., Regan, J. W., Cotecchia, S., Lefkowitz, R. J., and Caron, M. G. (1989). Effector coupling mechanisms of the cloned 5-HT_{1A} receptor. J. Biol. Chem. 264: 14848–52.

Fozard, J. R. (1983) Differences between receptors for 5-hydroxytryptamine on autonomic neurones revealed by nor-(-)-cocaine. J. Auton. Pharmac. 3: 21–26.

Gaddum, J. H., and Picarelli, Z. P. (1957). Two kinds of tryptamine receptor. Br. J. Pharmacol. Chemother. 12: 323–328.

Gozlan, H., El Mestikawy, S., Pichat, L., Glowinski, J., and Hamon, M. (1983). Identification of presynaptic serotonin autoreceptors by a new ligand: ^{3}H-PAT. Nature (London) 305: 140–142.

Green, J. P. (1987). Nomenclature and classification of receptors and binding sites: the need for harmony. Trends Pharmacol. Sci. 8: 90–94.

Hamel, E., and Bouchard, D. (1990). Contractile 5-HT_1 receptors in human pial arterioles: correlation with 5-HT_{1D} binding sites. The Second IUPHAR Satellite Meeting on Serotonin, Basel July 11-13, 1990. Abstr. 68.

Heuring, R. E., and Peroutka, S. J. (1987). Characterization of a novel ^{3}H-5-hydroxytryptamine binding site subtype in bovine brain membranes. J. Neurosci. 7: 894–903.

Hoyer, D. (1988a). Molecular pharmacology and biology of 5-HT_{1C} receptors. TIPS 9: 89–94.

Hoyer, D. (1988b). Functional correlates of serotonin 5-HT_1 recognition sites. J. Rec. Res. 8: 59–81.

Hoyer, D. (1989). Biochemical mechanisms of 5-HT receptor-effector coupling in peripheral tissues, In: Fozard, J. R. (Ed) The peripheral actions of 5-hydroxytryptamine. Oxford University Press, pp 72-99.

Hoyer, D. (1990). 5-HT_3, 5-HT_4 and 5-HT-M receptors. Neuropsychopharmacology, 3: 371–383.

130

Hoyer, D., Engel, G., and Kalkman, H. O. (1985a). Characterization of the 5-HT$_{1B}$ recognition site in rat brain: binding studies with [^{125}I]iodocyanopindolol. Eur. J. Pharmacol. 118: 1–12.

Hoyer, D., Engel, G., and Kalkman, H. O. (1985b). Molecular pharmacology of 5-HT$_1$ and 5-HT$_2$ recognition sites in rat and pig brain membranes: radioligand binding studies with [^{3}H]5-HT, [^{3}H]8-OH-DPAT,(-)[^{125}I]iodocyanopindolol, [^{3}H]mesulergine and [^{3}H]ketanserin. Eur. J. Pharmacol. 118: 13–23.

Hoyer, D., and Middlemiss, D. N. (1989). The pharmacology of the terminal 5-HT autoreceptors in mammalian brain: evidence for species differences. Trends Pharmacol. Sci. 10: 130–132.

Hoyer, D., and Neijt, H. C. (1987). Identification of serotonin 5-HT$_3$ recognition sites by radioligand binding in NG 108-15 neuroblastoma-glioma cells. Eur. J. Pharmacol. 143: 191–192.

Hoyer, D., and Neijt, H. C. (1988). Identification of serotonin 5-HT$_3$ recognition sites in membranes of N1E-115 neuroblastoma cells by radioligand binding. Mol. Pharmacol. 33: 303–309.

Hoyer, D., Pazos, A., Probst, A., and Palacios, J. M. (1986). Serotonin receptors in the human brain. I: Characterization and autoradiographic localization of 5-HT$_{1A}$ recognition sites. Apparent absence of 5-HT$_{1B}$ recognition sites. Brain Res. 376: 85–96.

Hoyer, D., and Schoeffter, P. (1988). 5-HT$_{1D}$ receptor-mediated inhibition of forskolin-stimulated adenylate cyclase activity in calf substantia nigra. Eur. J. Pharmacol. 147: 145–147.

Hoyer, D., Schoeffter, P., Waeber, C., Palacios, J. M., and Dravid, A. (1989). 5-HT$_{1C}$ receptor-mediated stimulation of inositol phosphate production in pig choroid plexus; a pharmacological characterization. Naunyn Schmiedeberg's Arch. Pharmacol. 339: 252–258.

Kaumann, A. J., Sanders, L., Brown, A. M., Murray, K. J., and Brown, M. J. (1990). A 5-hydroxytryptamine receptor in human atrium. Br. J. Pharmacol. 100: 879–885.

Kilpatrick, G. J., Jones, B. J., and Tyers, M. B. (1987). The identification and distribution of 5-HT$_3$ receptors in rat brain using radioligand binding. Nature 330: 746–748.

Laduron, P. M. (1987). Limitations of binding studies for receptor classification, in Perspectives on receptor classification. Alan R. Liss, Inc. 71–79.

Leysen, J. E., Niemegeers, C. J. E., Van Nueten, J. M., and Laduron, P. M. (1982). [^{3}H]ketanserin (R41 468), a selective ^{3}H ligand for serotonin 2 receptor binding sites. Mol. Pharmacol. 21: 301–314.

Markstein, R., Hoyer, D., and Engel, G. (1986). 5-HT$_{1A}$-receptors mediate stimulation of adenylate cyclase in rat hippocampus. Naunyn-Schmiedeberg's Arch. Pharmacol. 333: 335–341.

Moser, P. C., Tricklebank, M. D., Middlemiss, D. N., Mir, A. K., Hibert, M. F., and Fozard, J. R. (1990). Characterization of MDL-73005EF as a 5-HT-1A selective ligand and its effects in animal models of anxiety, comparison with buspirone, 8-hydroxy-dpat and diazepam. Br. J. Pharmacol. 99: 343–349.

Murphy, T. J., and Bylund, D. B. (1988). Oxymetazoline inhibits adenylate cyclase by activation of serotonin-1 receptors in the OK cell, an established renal epithelial cell line. Mol. Pharmacol. 34: 1–7.

Nakaki, T., Roth, B. L., Chuang, D., and Costa, E. (1985). Phasic and tonic components in 5-HT$_2$ receptor-mediated rat aorta contraction: participation of Ca++ channels and phospholipase C. J. Pharmacol. Exp. Ther. 234: 442–446.

Neijt, H. C., te Duits, I. J., and Vijverberg, H. P. M. (1988). Pharmacological characterization of serotonin 5-HT$_3$ receptor-mediated electrical response in cultured mouse neuroblastoma cells. Neuropharmacol. 27: 301–307.

Pazos, A., Hoyer, D., and Palacios, J. M. (1984). The binding of serotonergic ligand to the porcine choroid plexus: characterization of a new type of serotonin recognition site. Eur. J. Pharmacol. 106: 539–546.

Pazos, A., and Palacios, J. M. (1985). Quantitative autoradiographic mapping of serotonin receptors in the rat brain. I. Serotonin-I receptors. Brain Res. 346: 205–230.

Pedigo, N. W., Yamamura, H. I., and Nelson, D. L. (1981). Discrimination of multiple [^{3}H]5-hydroxytryptamine-binding sites by the neuroleptic spiperone in rat brain. J. Neurochem. 36: 220–226.

Peroutka, S. J. (1988). 5-Hydroxytryptamine receptor subtypes. Ann. Rev. Neurosci. 11: 45–60.

Peroutka, S. J., and Hamik, A. (1988). [^{3}H]Quipazine labels 5-HT$_3$ recognition sites in rat cortical membranes. Eur. J. Pharmacol. 148: 297–299.

Peroutka, S. J., and Snyder, S. H. (1979). Multiple serotonin receptors: differential binding of [^{3}H]5-hydroxytryptamine, [^{3}H]lysergic acid diethylamide and [^{3}H]spiroperidol. Mol. Pharmacol. 16: 687–699.

Peroutka, S. J., Hamik, A., Harrington, M. A., Hoffman, A. J., Mathis, C. A., Pierce, P. A., and Wang, S. S. H. (1988). (R)-(-)-[^{77}Br]4-bromo-2, 5-dimethoxyamphetamine labels a novel 5-hydroxytryptamine binding site in brain membranes. Mol. Pharmacol. 34: 537–542.

Peters, J. A., and Lambert, J. L. (1989). Electrophysiology of 5-HT$_3$ receptors in neuronal cell lines. TIPS 10: 172-175.

Pritchett, D. B., Bach, A. W. J., Wozny, M., Taleb, O., Dal Toso, R., Shih, J. C., and Seeburg, P. H. (1988). Structure and functional expression of cloned rat serotonin 5-HT$_2$ receptor. EMBO J. 7: 4135–4140.

Richardson, B. P., and Engel, G. (1986). The pharmacology and function of the 5-HT$_3$ receptors. Trends Neurosci. 9: 424–428.

Rydelek-Fitzgerald, L., Teitler, M., Fletcher, P. W., Ismaiel, A. M., and Glennon, R. A. (1990). NAN-190: agonist and antagonist interactions with brain 5-HT$_{1A}$ receptors. Brain Res. 532: 191–196.

Sanger, G. J. (1987). Increased gut cholinergic activity and antagonism of 5-hydroxytryptamine M-receptors by BRL 24924: potential clinical importance of BRL 24924. Br. J. Pharmacol. 91: 77–87.

Schlicker, E., Fink, K., Göthert, M., Hoyer, D., Molderings, G., Roschke, I., and Schoeffter, P. (1989). The pharmacological properties of the presynaptic 5-HT autoreceptor in the pig brain cortex conform to the 5-HT$_{1D}$ receptor subtype. Naunyn-Schmiedeberg's Arch. Pharmacol. 340: 45–51.

Schoeffter, P., and Hoyer, D. (1988). Centrally acting hypotensive agents with affinity for 5-HT$_{1A}$ binding sites inhibit forskolin-stimulated adenylate cyclase activity in calf hippocampus. Br. J. Pharmacol. 95: 975–985.

Schoeffter, P., and Hoyer, D. (1989a). 5-Hydroxytryptamine 5HT$_{1B}$ and 5-HT$_{1D}$ receptors mediating inhibition of adenylate cyclase activity. Pharmacological comparison with special reference to the effects of yohimbine, rauwolscine and some β-adrenoceptor antagonists. Naunyn-Schmiedeberg's Arch. Pharmacol. 340: 285–292.

Schoeffter, P., and Hoyer, D. (1989b). Interactions of arylpiperazines with 5-HT$_{1A}$, 5-HT$_{1B}$, 5-HT$_{1C}$ and 5-HT$_{1D}$ receptors: do discriminatory 5-HT$_{1B}$ ligands exist? Naunyn Schmiedeberg's Arch. Pharmacol. 339: 675–683.

Schoeffter, P., and Hoyer, D. (1990). 5-hydroxytryptamine (5-HT) induced endothelium-dependent relaxation of pig coronary arteries is mediated by 5-HT receptors similar to the 5-HT$_{1D}$ receptor subtype. J. Pharmacol. Exp. Therap. 252: 387–395.

Schoeffter, P., Waeber, C., Palacios, J. M., and Hoyer, D. (1988). The serotonin 5-HT$_{1D}$ receptor subtype is negatively coupled to adenylate cyclase in calf substantia nigra. Naunyn Schmiedeberg's Arch. Pharmacol. 337: 602–608.

Seuwen, K., Magnaldo, I., and Pouysségur, J. (1988). Serotonin stimulates DNA synthesis in fibroblasts acting through 5-HT$_{1B}$ receptors coupled to a G$_i$-protein. Nature 335: 254–256.

Shenker, A., Maayani, S., Weinstein, H., and Green, J. P. (1985). Two 5-HT receptors linked to adenylate cyclase in guinea pig hippocampus are discriminated by 5-carboxamidotryptamine and spiperone. Eur. J. Pharmacol. 109: 427–429.

Shenker, A., Maayani, S., Weinstein, H., and Green, J. P. (1987). Pharmacological characterization of two 5-hydroxytryptamine receptors coupled to adenylate cyclase in guinea pig hippocampal membranes. Mol. Pharmacol. 31: 357–367.

Teitler, M., Leonhardt, S., Weisberg, E., and Hoffman, B. J. (1990). 4-[^{125}I]iodo-(2, 5-dimethoxy)phenylisopropylamine and [^{3}H]ketanserin labeling of 5-hydroxytryptamine$_2$ (5-HT$_2$) receptors in mammalian cells transfected with a rat 5-HT$_2$ cDNA: evidence for multiple states and not multiple 5-HT$_2$ receptor subtypes. Mol. Pharmacol. 38: 594–598.

Villalon, C. M., Den Boer, M. O., Heiligers, J. P. C., and Saxena, P. R. (1990). Mediation of 5-hydroxytryptamine-induced tachycardia in the pig by the putative 5-HT$_4$ receptor. Br. J. Pharmacol. 100: 665–667.

Villalon, C. M., Den Boer, M. O., Heiligers, J. P. C., and Saxena, P. R. (1991). Further characterization, by use of tryptamine and benzamide derivatives, of the putative 5-HT$_4$ receptor mediating tachycardia in pig. Br. J. Pharmacol. 102: 107–112.

Waeber, C., Dietl, M. M., Hoyer, D., Probst, A., and Palacios, J. M. (1988a). Visualization of a novel serotonin recognition site (5-HT$_{1D}$) in the human brain by autoradiography. Neurosci. Lett. 88: 11–16.

Waeber, C., Schoeffter, P., Palacios, J. M., and Hoyer, D. (1988b). Molecular pharmacology of 5-HT$_{1D}$ recognition sites: radioligand binding studies in human, pig and calf brain membranes. Naunyn Schmiedeberg's Arch. Pharmacol. 337: 595–601.

Waeber, C., Dietl, M. M., Hoyer, D., and Palacios, J. M. (1989a). 5-HT$_1$ receptors in the vertebrate brain: regional distribution examined by autoradiography. Naunyn Schmiedeberg's Arch. Pharmacol. 340: 486–494.

Waeber, C., Schoeffter, P., Palacios, J. M., and Hoyer, D. (1989b). 5-HT$_{1D}$ receptors in the guinea-pig and pigeon brain: radioligand binding and biochemical studies, Naunyn-Schmiedeberg's Arch. Pharmacol. 340: 479–485.

Watling, K. J., Aspley, S., Swain, C. J., and Saunders, J. (1988). [^{3}H]-Quaternised ICS 205-930 labels 5-HT$_3$ receptor binding sites in rat brain. Eur. J. Pharmacol. 149: 397–398.

Weiss, S., Sebben, M., Kemp, D. E., and Bockaert, J. (1986). Serotonin 5-HT$_1$ receptors mediate inhibition of cyclic AMP production in neurons. Eur. J. Pharmacol. 120: 227–230.

Yakel, J. L., and Jackson, M. B. (1988). 5-HT$_3$ receptors mediate rapid responses in cultured hippocampus and a clonal cell line. Neuron. 1: 615–621.

Serotonin: Molecular Biology, Receptors and Functional Effects
ed. by J. R. Fozard/P. R. Saxena
© 1991 Birkhäuser Verlag Basel/Switzerland

5-HT$_{1P}$ Receptors in the Bowel: G Protein Coupling, Localization, and Function

M. D. Gershon, P. R. Wade, E. Fiorica-Howells, A. L. Kirchgessner, and H. Tamir

Department of Anatomy and Cell Biology, Columbia University College of Physicians and Surgeons, New York, NY 10032, USA

Summary. Although serotinin (5-HT) has been demonstrated to mediate slow exicitatory postsynaptic potentials in neurons of the myenteric plexus of the bowel, its role in the physiology of gastrointestinal motility and secretion is not well understood. Five subtypes of 5-HT receptor have been reported in the gut. The 5-HT$_2$ receptor appears to be present on smooth muscle, while the others, 5-HT$_{1A}$, 5-HT$_{1P}$, 5-HT$_3$, and 5-HT$_4$ are neuronal. The present experiments were undertaken in order to obtain more information about the location, action, and function of 5-HT$_{1P}$ receptors. GTP-γ-S, was found to inhibit the binding of ^{3}H-5-HT by the 5-HT$_{1P}$ receptor (the IC$_{50}$ for GTP-γ-S was $1.8 \pm 0.4 \mu$M). GTP-γ-S was more potent than GTP, while ATP and GMP (100 μM) were without effect. These observations are compatible with the hypothesis that the 5-HT$_{1P}$ receptor is coupled to a G protein. 5-HT (0.1–10 μM) was also found to elevate levels of cAMP in preparations of isolated myenteric ganglia. Renzapride also increased cAMP, but the response neither to 5-HT nor to renzapride was inhibited by the 5-HT$_{1P}$ antagonist, N-acetyl-5-hydroxytryptophyl-5-hydroxytryptophan amide (5-HTP-DP) or the 5-HT$_3$/5-HT$_4$ antagonist, ICS 205–930, even at 10 μM. The action of 5-HT on ganglionic cAMP, therefore, cannot now be attributed to 5-HT$_{1P}$ or 5-HT$_4$ receptors. Polyclonal anti-idiotypic antibodies were raised that recognize some, but not all, subtypes of 5-HT receptor. These antibodies were demonstrated to bind to 5-HT$_2$, 5-HT$_{1C}$, 5-HT$_{1P}$, and 5-HT$_3$, but not 5-HT$_{1A}$ receptors. When applied to enteric neurons, the anti-idiotypic antibodies transiently mimicked and then blocked the actions of 5-HT at 5-HT$_3$ and 5-HT$_{1P}$ receptors. The blockage of responses to 5-HT that followed application of the antibody was specific, since neurons continued to respond as before to carbachol or substance P. The 5-HT$_{1P}$ agonist-like actions of the antibodies were prevented by desensitization of 5-HT receptors and by renzapride (1 μM). The anti-idiotypic antibodies were found to be useful for the immunocytochemical localization of specific 5-HT receptor subtypes, if they were applied to tissues in the presence of appropriate antagonists. Use of these antibodies indicated that 5-HT$_{1P}$ receptors are located, not only on neurons in myenteric and submucosal ganglia, but also on a subepithelial plexus of nerve fibers. Finally, 5-HT$_{1P}$ receptor antagonists (5-HTP-DP, renzapride, and anti-idiotypic antibodies), were found to prevent excitation of submucosal or myenteric neurons by mucosal application of cholera toxin. This observation, and the localization of 5-HT$_{1P}$ receptors, is consistent with the idea that they play a role in the excitation of enteric neurons during the peristaltic reflex.

Introduction

"Functional" diseases of the gastrointestinal tract are among the most common and least effectively treated maladies facing modern medicine. These conditions cause a great deal of morbidity but they are poorly understood and little effective therapy is available for them. Clearly, a pharmacological approach designed to paralyze neuromuscular transmission leaves a great deal to be desired, as would the use of curare in the treatment of muscle cramps. The enteric nervous system (ENS) is large and complex (Furness and Costa 1987; Gershon 1981). Unlike the ganglia found in other organs, the ENS is able to mediate reflex activity in the absence of input from the brain or spinal cord. This ability depends on the action of intrinsic primary afferent neurons as well as on a variety of interneurons. The fact that interneurons modulate synaptic transmission in the ENS suggests that some of the actions of the ENS are subtle and susceptible to being influenced by drugs in ways that do not impair the motility of the bowel.

Serotonin (5-hydroxytryptamine; 5-HT) is present both in the ENS and in the gastrointestinal epithelium (Gershon et al. 1989). Serotonergic neurons are restricted to the myenteric plexus of the guinea pig gut (Furness and Costa 1982), but are located in both myenteric and submucosal plexuses in other species, such as the pig (Timmermans et al. 1990). Epithelial 5-HT is found in specialized enterochromaffin cells (Erspamer 1966). In the nervous system, 5-HT has been shown to be one of the transmitters responsible for a slow excitatory postsynaptic potential (EPSP) (Gershon et al. 1989). The 5-HT of the enterochromaffin cells can be released by pressure and has been proposed to be involved in the activation of primary afferent neurons to initiate the peristaltic reflex (Bülbring and Crema 1959). These actions of 5-HT are of a type that suggests that compounds that modify its enteric effects might modulate, but not stop gastrointestinal motility. Compounds that modulate motility without stopping it might well be expected to be useful in treating "functional" gastrointestinal disorders. It is therefore important to define enteric 5-HT receptors and to determine the physiological processes mediated by them.

Five types of 5-HT receptor have been reported to be present in the bowel. 5-HT_2 receptors are located on muscle (Richardson et al. 1985), while the other 4 receptors, 5-HT_{1A} (Galligan and North 1988; Galligan et al. 1988; Surprenant and Crist 1988), 5-HT_{1P} (Branchek et al. 1984; Mawe et al. 1986a), 5-HT_3 (Galligan and North 1988; Galligan et al. 1988; Mawe et al. 1986a; Mawe et al. 1989), and 5-HT_4 (Craig and Clarke 1990), all appear to be neuronal. 5-HT_{1A} receptors are responsible for inhibitory actions, mediating presynaptic inhibition of the release of acetylcholine at nicotinic synapses and a postsynaptic hyperpolarization (Galligan and North 1988; Galligan et al. 1988; Surprenant and Crist 1988). 5-HT_{1P} receptors are responsible both for presynaptic inhibition (an effect similar to that mediated by 5-HT_{1A} receptors) (Takaki et al. 1985) and for postsynaptic excitation (Mawe et al. 1986a). The excitatory response mediated by 5-HT_{1P} receptors is a slow one. The depolarization is associated with inhibition of a

Ca^{2+}-activated potassium conductance and is thus accompanied by an increase in input resistance. The 5-HT$_3$ receptor, like the 5-HT$_{1P}$, mediates a postsynaptic excitatory response; however, the ionic bases of the two responses differ. The 5-HT$_3$ response is fast, not slow, and is accompanied by a decrease in input resistance (Galligan and North 1988; Galligan et al. 1988; Mawe et al. 1986a; Mawe et al. 1989). The kinetics of the 5-HT$_{1P}$ response are typical of those of a receptor coupled to a second messenger, while the 5-HT$_3$ receptor is a ligand-gated ion channel (Derkach et al. 1989). The 5-HT$_4$ receptor has not been characterized in terms of a neuronal response. Thus far, it has only been identified on the basis of mechanical responses of the intestinal musculature. The 5-HT$_4$ receptor is defined as a receptor responsible for the augmentation by 5-HT, 5-methoxytryptamine, and substituted benzamides of electrically-driven twitches of the intestinal smooth muscle that are antagonized by higher concentrations of ICS 205-930 than are necessary to block responses mediated by 5-HT$_3$ receptors (Craig and Clarke 1990). Of the neuronal receptors, only the 5-HT$_{1P}$ has been established to be responsible for a physiological effect of 5-HT, the slow EPSP.

5-HT$_{1P}$ receptors have been studied by radioligand binding techniques as well as by electrophysiology. ^{3}H-5-HT binds to the isolated membranes of enteric neurons reversibly and with high affinity ($K_D \sim 2-3$ nM). The binding of ^{3}H-5-HT is not inhibited by drugs that are known to be antagonists at other subtypes of 5-HT receptor, but it is blocked by hydroxylated indoles (Branchek et al. 1984), hydroxylated indalpines (Branchek et al. 1988a), and by dipeptides of 5-hydroxytryptophan, such as N-acetyl-5-hydroxytryptophyl-5-hydroxy-tryptophan (5-HTP-DP) (Takaki et al. 1985). The hydroxylated indalpines are agonists (Branchek et al. 1988a) and 5-HTP-DP is an antagonist (Takaki et al. 1985) at 5-HT$_{1P}$ receptors. Radioautographic studies have revealed the presence of 5-HT$_{1P}$ receptors in both the submucosal and myenteric plexuses as well as on nerve fibers in the lamina propria of the mucosa (Branchek et al. 1988b; Branchek et al. 1984; Mawe et al. 1986b). The effects of drugs on the binding of ^{3}H-5-HT, analyzed radiographically, are the same as when binding is studied by rapid filtration. The current investigation was undertaken in order to characterize further 5-HT$_{1P}$ receptors. The hypothesis that the receptor is coupled to a G protein was tested by determining the effects on the binding of ^{3}H-5-HT of the non-hydrolyzable derivative of GTP, GTP-γ-S. In addition, the effects on enteric neurons of anti-idiotypic antibodies that recognize 5-HT receptors were studied as was also the distribution in the bowel of sites to which these antibodies bind. Finally, the possibility that 5-HT$_{1P}$ receptors play a role in the excitation of enteric neurons during the peristaltic reflex was evaluated.

Materials and Methods

Binding of ^{3}H-5-HT to membranes isolated from dissected preparations of rabbit, guinea pig and mouse longitudinal muscle with adherent myenteric plexus (LM-MP) were investigated as in previous studies (Branchek et al. 1988b; Branchek et al. 1984; Mawe et al. 1986b). Radioautography was carried out with unfixed, frozen sections of tissue, dried, and applied to tritium sensitive film as previously described (Branchek et al. 1988b; Branchek et al. 1984; Mawe et al. 1986b). For electrophysiological recording from myenteric neurons, the methods were identical to those used by Mawe et al. (1986). The method used to raise anti-idiotypic antibodies and their characterization have been described in a prior report (Tamir et al. 1989). Immunocytochemistry was carried out on dissected laminar preparations of bowel or on frozen sections as described by Kirchgessner et al. (1988). Ganglia were isolated from the myenteric plexus of the bowel by digesting LM-MP preparations with collagenase (Yau et al. 1989) and collecting ganglia by filtration through Nucleopore™ filters with an average pore size of 8 μm. These preparations were cultured overnight in minimal essential media and used for the measurement of cAMP under a variety of experimental conditions. cAMP was measured by using a commercially available kit (New England Nuclear Corp., Boston MA). Cholera toxin (CTX) was employed to evoke the peristaltic reflex. It is known that peristalsis is massively stimulated when CTX is applied to the luminal surface of the bowel (Cassuto et al. 1983). CTX has been proposed to evoke the peristaltic reflex by releasing 5-HT from mucosal enteroendocrine cells (Cassuto et al. 1982). The released 5-HT, in turn, has been postulated to simulate the mucosal endings of primary afferent nerves, which activate the reflex. Sacs of guinea pig small intestine ($\cong 3.0$ cm in length) were filled with 250 μl of Krebs solution containing 10.0 μg of CTX. Control sacs were filled with a similar solution from which CTX was omitted. Alternatively, sacs were filled with Krebs solution containing both CTX (10.0 μg) and experimental drugs. The sacs were then tied at both ends with silk threads and incubated in oxygenated Krebs solution at 37°C for 3 hr. At the end of the incubation period the sacs were opened along the mesenteric border, pinned flat mucosal side up, and fixed for the histochemical demonstration of cytochrome oxidase activity in enteric neurons according to the method of Mawe and Gershon (Mawe and Gershon 1986). Cytochrome oxidase activity was measured by computer-assisted video microdensitometry. Cytochrome oxidase activity in neurons increases rapidly as the cells are stimulated to fire action potentials and thus reflects the degree to which neurons have been activated at the time of fixation.

Results

Anti-idiotypic Antibodies: Anti-idiotypic antibodies that recognize 5-HT receptors were raised by immunizing rabbits with affinity purified antibodies to 5-HT (Tamir et al. 1989). When this was done, the resulting polyclonal sera contained both anti-idiotypic and anti-anti-idiotypic antibodies. The anti-anti-idiotypic antibodies, which recognized 5-HT itself, were removed by passing the crude serum through an anti-5-HT column and the sera were further purified by affinity chromatography on columns constructed with antibodies to 5-HT. The anti-idiotypic antibodies were demonstrated by immunocytochemistry to be able to detect some, but not all subtypes of 5-HT receptor. For example, purified anti-idiotypic antibodies immunostained fibroblasts transfected *in vitro* with cDNA encoding the $5-HT_{1C}$ or the $5-HT_2$ receptor, but not HeLa cells transfected with human genomic DNA encoding the $5-HT_{1A}$ receptor. Immunostaining of cells by the anti-idiotypic antibodies was inhibited by appropriate pharmacological agents; that of cells expressing $5-HT_{1C}$ receptors was blocked by $1.0\,\mu M$ mesulergine (but not $10.0\,\mu M$ ketanserin or $10.0\,\mu M$ 8-hydroxy-2-[di-*n*-propylamino]tetralin [8-OH-DPAT]), while that of cells expressing $5-HT_2$ receptors was blocked by $1.0\,\mu M$ ketanserin or spiperone (but not $10.0\,\mu M$ mesulergine or 8-OH-DPAT). In the gut the anti-idiotypic antibodies immunostained neurons in the myenteric and submucosal plexuses and subepithelial nerve fibers in the lamina propria. Immunofluorescence was reduced by $10\,\mu M$ 5-HTP-DP or ICS 205-930 and blocked by a combination of these antagonists. This pattern of immunostaining was virtually identical to that seen in radioautographs prepared with ^3H-5-HT to label $5-HT_{1P}$ receptors. When microejected onto the surface of myenteric type 2/AH neurons (see Wood, 1987 for criteria used in the classification of myenteric neurons), the antibodies mimicked both the $5-HT_{1P}$- and $5-HT_3$-mediated responses to 5-HT; that is, the antibodies first elicited a fast depolarization associated with a decrease in input resistance, followed by a prolonged depolarization, during which input resistance was markedly increased. Cross-desensitization was observed between responses to the antibodies and responses to 5-HT; however, following application of anti-idiotypic antibodies, $5-HT_3$-mediated responses rapidly recovered and could be elicited again, while $5-HT_{1P}$-mediated responses remained blocked for extended periods of time. Moreover, desensitization of receptors with the antibodies sufficient to completely abolish slow responses to 5-HT failed to affect either the slow (muscarinic) response to carbachol or the slow response to substance P. When $5-HT_{1P}$ receptors are desensitized by anti-idiotypic antibodies, 5-HT-mediated slow EPSPs are also blocked. The anti-idiotypic antibodies thus recognize both $5-HT_{1P}$ and $5-HT_3$ receptors and act as an agonist at each, although the antibodies evidently dissociate more slowly from 5-HT$_{1P}$ receptors. It should be noted that neither the slow response to 5-HT, nor the 5-HT-mediated slow EPSP could be blocked by the 5-HT$_3$/5-HT$_4$ antagonist, ICS 205-930 (0.1–10.0 μM).

Effect of GTP-γ-S on the Binding of ^{3}H-5-HT: GTP-γ-S was found to inhibit the binding of ^{3}H-5-HT. When the concentration of ^{3}H-5-HT was 10.0 nM, the maximum inhibition of binding was 89.3 $\pm$ 4.5%. The IC$_{50}$ for GTP-γ-S was 1.8 $\pm$ 0.4 μM for mouse intestine and 2.8 $\pm$ 0.1 μM for rabbit intestine. The binding of ^{3}H-5-HT to enteric neuronal membranes was also potently inhibited by GDP-β-S; the IC$_{50}$ in mice was 2.1 $\pm$ 0.6 μM and in rabbits was 2.8 $\pm$ 0.1 μM. GTP (100 μM) itself reduced binding of ^{3}H-5-HT to mouse intestinal membranes by 52.5 $\pm$ 5.7%. In contrast to GTP and its non-hydrolyzable analogues, GMP and ATP failed to inhibit the binding of ^{3}H-5-HT. At 1 μM, GTP-γ-S was found to significantly reduce the number of high affinity binding sites for ^{3}H-5-HT; the B$_{max}$ changed from the control value of 21.8 $\pm$ 3.7 to 10.1 $\pm$ 3.2 pmol/mg/protein in the presence of GTP-γ-S. No significant change was noted in the K$_D$ of the ^{3}H-5-HT binding site (control: 3.6 $\pm$ 0.9 nM; GTP-γ-S: 2.7 $\pm$ 0.6 nM). It was not possible to measure binding of ^{3}H-5-HT to a lower affinity site. It is thus likely that GTP-γ-S shifts the receptor to an affinity state too low for the measurement of the binding of ^{3}H-5-HT. Similar effects of non-hydrolyzable analogues of GTP were observed when the binding of ^{3}H-5-HT was assayed by radiography. Binding of ^{3}H-5-HT in both the ganglia and subepithelial nerve fibers in mouse, rabbit or guinea pig small intestine was blocked by GTP-γ-S > GDP-β-S $\gg$ ATP-γ-S.

Effect of 5-HT on the Biosynthesis of cAMP by Isolated Myenteric Ganglia: In contrast to peripheral nerve in most regions of the body collagen is absent from the interior of the ganglia of the myenteric plexus (Gabella 1987). Instead of collagen, glial cells support the neural elements of the ENS. As a result, digestion of LM-MP preparations with collagenase yields suspensions of single muscle cells and fibroblasts, while ganglia remain intact. In previous studies, ganglia have been collected mechanically (Yau et al. 1989); however it was found that they could be obtained on a preparative scale by filtration with minimal contamination by other cells. Accordingly, ganglia were prepared, cultured overnight to permit recovery from the isolation procedure, and used to study the ability of 5-HT to promote the generation of cAMP. Ganglia were incubated for 10 min in Ca^{2+}-free Krebs solution containing EDTA (4.0 mM) and the phosphodiesterase inhibitor, isobutyl-methylxanthine (IBMX; 10.0 μM) in the presence or absence of 5-HT (1.0–10.0 μM) or renzapride (BRL 24924; 10.0 μM). When ICS 205-930 was tested as a potential antagonist, ganglia were exposed to the agent for 10 min prior to the addition of IBMX and agonist.

5-HT (10.0 μM) significantly increased the generation of cAMP (to 227 $\pm$ 9.0% of control; $p < 0.001$). The substituted benzamide, renzapride (10.0 μM) also increased the production of cAMP (to 201.8 $\pm$ 19.0% of control; $p < 0.001$). ICS 205-930 (10.0 μM), which at the concentration tested is an antagonist at both 5-HT$_3$ and 5-HT$_4$ receptors, failed to inhibit the elevation of cAMP seen in response to the addition either of 5-HT or of renzapride.

5-HT and the Simulation of Enteric Neurons by CTX: Previous studies (Cassuto et al. 1982; Cassuto et al. 1983) have suggested that 5-HT and enteric neurons might participate in the mediation of the action of CTX on the gut. We tested this hypothesis by investigating the effects of intraluminal CTX on neuronal activity (assessed by the cytochrome oxidase technique) *in vitro* in sacs of guinea pig small intestine. Results were analyzed by one way ANOVA (differences between means were considered significant at $p < 0.05$). CTX increased cytochrome oxidase activity in neurons of both plexuses (to 168 $\pm$ 4% of control in the myenteric plexus and to 148 $\pm$ 4% of control in the submucosal plexus). This effect was blocked by lidocaine (10 μM), TTX (1.0 μM), and 5-HT receptor desensitization (with 10 μM 5-HT). CTX-stimulation of enteric neuronal cytochrome oxidase activity was also antagonized by the anti-idiotypic antibodies discussed above, 5-HTP-DP (10 μM), and ICS 202-930 (0.1 μM) (Table 1). Other compounds that inhibited the activation of enteric neurons by CTX were the substituted benzamides, renzapride (1 μM), racemic (1.0 μM), and S-zacopride (1.0 μM). These observations suggest that intraluminal CTX activates nerve fibers in the lamina propria through a mechanism that involves both 5-HT$_3$ and 5-HT$_{1P}$ receptors since blockade of either subtype prevents the CTX-stimulated increase in enteric neuronal activity. The participation of 5-HT in the activation of mucosal nerve fibers is consistent with the idea that mucosal release of 5-HT may be involved in initiating the peristaltic reflex.

Table 1. Relative activity of cytochrome oxidase in enteric neurons

Conditions	Submucosal Plexus (% control)	Myenteric Plexus (% control)
Control	98.7 $\pm$ 3.1	100.0 $\pm$ 3.0
CTX	147.8 $\pm$ 3.9	168.0 $\pm$ 4.0
CTX + anti-id Ab	98.5 $\pm$ 4.5	82.3 $\pm$ 2.9
CTX + N-acetyl-5-HTP-DP	78.0 $\pm$ 4.0	91.7 $\pm$ 3.5
CTX + ICS 205-930	101.6 $\pm$ 3.1	85.3 $\pm$ 3.0
CTX + renzapride	82.3 $\pm$ 4.3	110.4 $\pm$ 3.8
CTX + S-zacopride	91.8 $\pm$ 4.4	123 $\pm$ 3.9
CTX + racemic zacopride	133.6 $\pm$ 6	118.8 $\pm$ 5.2

Discussion

The current experiments have provided new information about 5-HT_{1P} receptors. These receptors are among those that are recognized by 5-HT receptor-specific polyclonal anti-idiotypic antibodies. Interestingly, the anti-idiotypic antibodies, like 5-HT, act initially as an agonist at these receptors; however, the receptors rapidly desensitize. The antibodies evidently dissociate very slowly from 5-HT_{1P} sites; consequently, following trials with the antibodies, neurons fail to manifest 5-HT_{1P}-mediated responses for long periods of time. In contrast, although the anti-idiotypic antibodies also recognize 5-HT_3 receptors, their effects on this site are transient. 5-HT_3-mediated responses recover soon after the antibodies are washed out. These properties of the anti-idiotypic antibodies make them useful for the study of 5-HT receptor subtypes in the ENS. The ENS contains 5-HT_{1A} (Galligan and North 1988; Galligan et al. 1988; Suprenant and Crist 1988), 5-HT_{1P}, and 5-HT_3 (Galligan and North 1988; Galligan et al. 1988; Gershon et al. 1989; Mawe et al. 1986b; Suprenant and Crist 1988) receptors. The antibodies fail to recognize 5-HT_{1A} receptors and rapidly dissociate from 5-HT_3 sites; therefore, long-term antagonism of neurally-mediated responses following exposure of the bowel to the antibodies is likely to be due to 5-HT_{1P} receptor blockade. The anti-idiotypic antibodies thus block the 5-HT-mediated slow EPSPs, an effect shared by 5-HTP-DP and renzapride which also antagonize 5-HT_{1P}-mediated effects (Mawe et al. 1989; Takaki et al. 1985), but not by even high concentrations of the $5\text{-HT}_3/5\text{-HT}_4$ antagonist, ICS 205-930. Anti-idiotypic antibodies, like 5-HTP-DP and renzapride, also inhibit activation of enteric neurons by intraluminal application of CTX. It seems likely that this effect too is related to the ability of the antibodies to bind to 5-HT_{1P} receptors.

In addition to their physiological utility, the anti-idiotypic antibodies are valuable for the localization of 5-HT receptors. Despite the fact that the antibodies recognize more than one 5-HT receptor subtype, they can be made subtype-specific when employed as a localizing reagent by coupling their use with an appropriate antagonist. Anti-idiotypic antibodies combine with the 5-HT recognition site of receptors; therefore, binding of the antibodies to the receptors is prevented when the receptors are occupied by an antagonist. The sites revealed immunocytochemically by applying anti-idiotypic antibodies to sections correspond to 5-HT_{1P} sites revealed by radioautography. Not only are receptors located in the enteric plexuses, but they are also present on nerve fibers of the lamina propria. This location is consistent with the idea that 5-HT derived from enteroendocrine cells of the gastrointestinal mucosa participates in initiating the peristaltic reflex. Such an action of mucosal 5-HT is supported by the ability of agents that recognize 5-HT receptors to block the activation of enteric neurons by intraluminal application of CTX.

A question that has not been adequately answered is what is the relationship between 5-HT$_{1P}$ and 5-HT$_4$ receptors. In collicular neurons 5-HT and substituted benzamides are positively coupled to the generation of cAMP (Dumuis et al. 1989). These actions, blocked by high concentrations of ICS 205-930, are attributed to activation of 5-HT$_4$ receptors. Both 5-HT and substituted benzamides also increase cAMP in isolated myenteric ganglia and the 5-HT$_{1P}$ receptor, like other receptors of the 5-HT$_1$ class is linked to a G protein, which might potentially couple its actions to adenyl cyclase; however, in contrast to the effects of 5-HT and renzapride on collicular neurons, the increase in cAMP seen following their application to myenteric ganglia is not blocked by ICS 205-930, even at 10.0 μM. Similarly, neither the slow response to 5-HT, nor the similar slow EPSP elicited by the release of endogeneous 5-HT from stimulated enteric nerves is antagonized by ICS 205-930. Moreover, contractile responses of the bowel that have been attributed to 5-HT$_4$ receptors are not minicked by hydroxylated indalpines, which are 5-HT$_{1P}$ agonists, nor are they inhibited by 5-HTP-DP, which is a 5-HT$_{1P}$ antagonist. If one assumes that 5-HT$_{1P}$ and 5-HT$_4$ receptors are different, as the existing pharmacology suggests, then the 5-HT$_4$ receptor cannot be associated with any known physiological response to 5-HT. All of the known responses are accounted for by other subtypes of 5-HT receptor. On the other hand, neither the compounds used to characterize 5-HT$_{1P}$ receptors, nor those used to identify 5-HT$_4$ sites are adequate. High concentrations of these drugs must be used. More effective agonists and antagonists are thus necessary to analyze these receptors in a definitive manner.

References

Branchek, T., Mawe G., and Gershon, M. D.(1988a). Characterization and localization of a peripheral neural 5-hydroxytryptamine receptor subtype with a selective agonist, ^{3}H-5-hydroxyindalpine. J. Neuroscience. 8: 2582–2595.

Branchek, T., Mawe G., and Gershon, M. D. (1988b). Characterization and localization of a peripheral neural 5-hydroxytryptamine receptor subtype with a selective agonist, ^{3}H-5-hydroxyindalpine. J. Neuroscience. 8: 2582–2595.

Branchek, T., Rothman T., and Gershon M. D. (1984). Serotonin receptors on the processes of intrinsic enteric neurons: Reduction in the aganglionic bowel of the ls/ls mouse. Soc. Neurosci. Abstr. 10: 1097.

Branchek, T. A., Kates M., and Gershon, M. D. (1984). Enteric receptors for 5-hydroxytryptamine. Brain Res. 324: 107–118.

Bulbring, E., and Crema A. (1959). The action of 5-hydroxytryptamine, 5-hydroxytryptophan and reserpine on intestinal peristalsis in anaesthetized guinea-pigs. J. Physiol. (Lond.). 146: 29–53.

Cassuto, J., Jodal, M., Tuttle, R., and Lundgren, O. (1982). 5-Hydroxytryptamine and cholera secretion. Scand. J. Gastroenterol. 17: 695–703.

Cassuto, J., Siewert, A., Jodal, M., and Lundgren, O. (1983). The involvement of intraluminal nerves in cholera toxin induced intestinal secretion. Acta Physiol. Scand. 117: 195–202.

142

Craig, D. A., and Clarke, D. E. (1990). Pharmacological characterization of a neuronal receptor for 5-hydroxytryptamine in guinea pig ileum with properties similar to the 5-hydroxytryptamine 4 receptor. J. Pharmacol. Exp. Ther. 252: 1378–1386.

Derkach, V., Surprenant, A., and North, R. A. (1989). 5-HT$_3$ receptors are membrane ion channels. Nature. 339: 706–709.

Dumuis, A., Sebben M., and Bockaert, J. (1989). BRL24924: a potent agonist at a non-classical 5-HT receptor positively coupled with adenylate cyclase in colliculi neurons. Eur. J. Pharmacol. 162: 381–384.

Erspamer, V. (1966). Occurrence of indolealkylamines in nature. In Erspamer, V. (ed.) Handbook of Experimental Pharmacology: 5-Hydroxytryptamine and Related Indolealkylamines. New York: Springer-Verlag, pp 132–181.

Furness, J. B., and Costa, M. (1982). Neurons with 5-hydroxytryptamine-like immunoreactivity in the enteric nervous system: Their projections in the guinea pig small intestine. Neurosci. 7: 341–350.

Furness, J. B., and Costa, M. (1987). The Enteric Nervous System. New York: Churchill Livingstone.

Gabella, G. (1987). Structure of muscles and nerves of the gastrointestinal tract. In Johnson, L. R., Christensen, J., Jackson, M. J., Jacobson, E. D., and Walsh, J. H. (eds.) Physiology of the gastrointestinal tract. New York: Raven Press, pp 335–382.

Galligan, J. J., and North, R. A. (1988). Drug receptors on single enteric neurons. Life Sci. 43: 2183–2192.

Galligan, J. J., Suprenant, A., Tonini, M., and North, R. A. (1988). Differential localization of 5-HT$_1$ receptors on myenteric and submucosal neurons. Am. J. Physiol. (Gastrointest. Liver Physiol. 18). 255: G603–G611.

Gershon, M. D. (1981). The enteric nervous system. Ann. Rev. Neurosci. 4: 227–272.

Gershon, M. D., Mawe, G. M., and Branchek, T. A. (1989). 5-Hydroxytryptamine and enteric neurons. In Foxard J. A. (ed.) Peripheral actions of 5-Hydroxytryptamine. Oxford: Oxford University Press, pp. 247–264.

Kirchgessner, A. L., Dodd J., and Gershon, M. D. (1988). Markers shared between dorsal root and enteric ganglia. J. Comp. Neurol. 276: 607–621.

Mawe, G. M., Branchek, T., and Gershon, M. D. (1986a). Peripheral neural serotonin receptors: Identification and characterization with specific agonists and antagonists. Proc. Natl. Acad. Sci. (USA). 83: 9799–9803.

Mawe, G. M., Branchek, T., and Gershon, M. D. (1986b). Peripheral neural serotonin receptors: Identification and characterization with specific agonists and antagonists. Proc. Natl. Acad. Sci. (USA). 83: 9799–9803.

Mawe, G. M., Branchek, T., and Gershon, M. D. (1989). Blockade of 5-HT mediated enteric slow EPSPs by BRL 24924: Gastrokinetic effects. Am. J. Physiol. (Gastrointest. Liver Physiol.). 257: G386–396.

Mawe, G. M., and Gershon, M. D. (1986). Functional heterogeneity in the myenteric plexus: demonstration using cytochrome oxidase as a verified cytochemical probe of the activity of individual enteric neurons. J. Comp. Neurol. 249: 381–391.

Richardson, B. P., Engel, G., Donatsch, P., and Stadler, P. A. (1985). Identification of serotonin M-receptor subtypes and their specific blockade by a new class of drugs. Nature. 316: 216–231.

Surprenant, A., and Crist, J. (1988). Electrophysiological characterization of functionally distinct 5-HT receptors on guinea-pig submucous plexus. Neurosci. 24: 283–295.

Takaki, M., Branchek, T., Tamir, H., and Gershon, M. D. (1985). Specific antagonism of enteric neural serotonin receptors by dipeptides of 5-hydroxytryptophan: evidence that serotonin is a mediator of slow synaptic excitation in the myenteric plexus. J. Neurosci. 5: 1769–1780.

Tamir, H., Liu, K. P., Yu, P. Y., Gershon, M. D., and Kirchgessner, A. L. (1989). Characterization and use of anti-idiotypic antibodies for the localization of serotonin receptors. Neurosci. Absts. 15: 423.

Timmermans, J.-P., Scheuermann, D. W., Stach, W., Adriaensen, D., and De Groodt-Lasseel, M. H. A. (1990). Distinct distribution of CGRP-, enkephalin-. galanin-, neuromedin U-, neuropeptide Y-, somatostatin-, substence P-, VIP- and serotonin-containing neurons in the two submucosal ganglionic neural networks of the porcine small intestine. Cell Tissue. Res. 260: 367–379.

Wood, J. D. (1987). Physiology of enteric neurons. In Johnson, L. R., Christensen, J., Jackson, M. J., Jacobson, E. D., and Walsh, J. H. (eds.) Physiology of the gastrointestinal tract. New York: Raven Press, pp. 67–110.

Yau, W. M., Dorsett, J. A., and P. E. L. (1989). Characterization of acetylcholine release from enzyme-dissociated myenteric ganglia. Am. J. Physiol. (Gastrointest. Liver Physiol.). 19: G233–G239.

Serotonin: Molecular Biology, Receptors and Functional Effects
ed. by J. R. Fozard/P. R. Saxena
© 1991 Birkhäuser Verlag Basel/Switzerland

Is Contraction to Serotonin Mediated Via 5-HT$_{1C}$ Receptor Activation in Rat Stomach Fundus?

M. Baez, L. Yu, and L. Cohen

Lilly Research Laboratories, Eli Lilly and Company, Lilly Corporate Center, Indianapolis, IN 46285, USA

Summary. Although serotonin potently contracts the rat stomach fundus, characterization of the receptor responsible for this contraction has not yet been established. Several lines of pharmacological evidence suggest that the receptor may resemble the 5-HT$_{1C}$ receptor, although not all pharmacological data is consistent with this supposition. We report here our studies using recent advances in molecular biology that have permitted the cloning of 5-HT$_{1C}$ and 5-HT$_2$ receptor cDNA. Using the 5-HT$_{1C}$ receptor cDNA, we were unable to obtain any evidence for hybridization of this sequence to the mRNA obtained from rat stomach fundus. PCR amplification of cDNA synthesized from rat stomach fundus RNA with primer sequences for the rat 5-HT$_{1C}$ messenger RNA did not reveal 5-HT$_{1C}$ receptor specific mRNA in the stomach fundus. However, using such highly sensitive PCR techniques we were able to detect a modest signal in the rat stomach fundus with primers specific for the 5-HT$_2$ receptor sequence. Thus, consistent with previous pharmacological studies, we did obtain evidence for a modest population of 5-HT$_2$ receptors in the rat stomach fundus without any evidence for the presence of 5-HT$_{1C}$ receptor in this tissue. This information is consistent with the previous contention that the contractile response to serotonin in the rat stomach fundus may be mediated predominantly by a receptor that is different from any of the known, well characterized serotonin receptors.

Introduction

Serotonin is an excellent contractile agonist in many smooth muscle preparations (Cohen 1989). In particular, serotonin potently contracts the rat stomach fundus. Characterization of the receptor(s) mediating such contraction has formed the basis for several investigations. It is clear that the predominant response to serotonin is not mediated by activation of 5-HT$_2$, 5-HT$_3$, 5-HT$_{1A}$, or 5-HT$_{1B}$ receptors (Table 1). This information has been briefly summarized (Cohen 1989). More recently, we have shown that the 5-HT$_{1D}$ selective ligand, sumatriptan, did not potently contract the rat stomach fundus suggesting that the receptor mediating contraction in this tissue is not the 5-HT$_{1D}$ site as defined by sumatriptan and its potency in radioligand binding studies or the canine saphenous vein (Cohen and Robertson, unpublished observation).

The possibility that the contractile response to serotonin in the rat stomach fundus is mediated by activation of 5-HT$_{1C}$ receptors has been considered (see Cohen 1989; Table 2). In fact, an excellent correlation has emerged for the ability of 10 serotonin receptor agonists to contract the

Table 1. Rat stomach fundus: pharmacological characteristics serotonin contractile receptor is not:

● $5\text{-}HT_{1A}$
- $5\text{-}HT_{1A}$ selective ligand (LY165163) did not interact in stomach fundus with affinity consistent with its $5\text{-}HT_{1A}$ affinity
- Potency for a series of agonists in stomach fundus did not correlate with affinity at the $5\text{-}HT_{1A}$ receptor
  (Cohen and Wittenauer 1986)
  (Cohen 1989)

● $5\text{-}HT_{1B}$
- Potency for a series of agonists in stomach fundus only weakly and nonquantitatively correlated with affinity at the $5\text{-}HT_{1B}$ receptor
- $5\text{-}HT_{1A}$, $5\text{-}HT_{1B}$ ligand (cyanopindolol) did not block 5-HT-induced contractions
  (Cohen and Fludzinski 1987)
  (Cohen 1989)

● $5\text{-}HT_{1D}$
- Sumatriptan is only a weak agonist at high concentrations
  (Cohen, unpublished)

● $5\text{-}HT_2$
- Contraction to 5-HT was not competitively blocked by $5\text{-}HT_2$ receptor antagonists
  (Cohen et al. 1985)
  (Clineschmidt et al. 1985)

● $5\text{-}HT_3$
- ICS 205-930 did not block 5-HT-induced contractions
- $2\text{-}CH_3\text{-}5\text{-}HT$ is only a weak agonist
  (Cohen and Fludzinski 1987)

---

rat stomach fundus and to bind to $5\text{-}HT_{1C}$ receptors (Cohen 1989). However, the use of antagonists did not reveal a good correlation for blockade of serotonin-induced contraction to binding to the $5\text{-}HT_{1C}$ site (Cohen, unpublished observation). In fact, ligands with high affinity at $5\text{-}HT_{1C}$ receptor sites such as LY53857, ritanserin and SCH23390, (Hoyer 1989), did not inhibit contractile responses to serotonin in the rat stomach fundus with potencies consistent with their affinities at $5\text{-}HT_{1C}$ receptors (Baez et al. 1990). Furthermore, we have been unable to detect

Table 2. Rat stomach fundus: pharmacological relationship of the serotonin contractile receptor to the $5\text{-}HT_{1C}$ receptor

---

● Affinities at $5\text{-}HT_{1C}$ receptor correlated with potency of agonists in stomach fundus
  (Cohen 1989)

● $5\text{-}HT_{1C}$ receptor ligand affinities (LY53857, ritanserin and SCH23390) did not correlate with block of 5-HT-induced contractions in stomach fundus
  (Baez et al. 1990)

● 5-HT did not increase phosphoinositide turnover in stomach fundus
  (Cohen and Wittenauer 1987)
  (Secrest et al. 1989)

---

any increase in phosphoinositide turnover following serotonergic activation of receptors in the rat stomach fundus (Cohen and Wittenhauer 1987; Secrest et al. 1989), an action associated with activation of 5-HT$_{1C}$ receptors in the choroid plexus and brain (Conn et al. 1986).

To determine more conclusively the role of 5-HT$_{1C}$ receptors in serotonin-induced contraction of the rat stomach fundus, we have turned to an approach utilizing recent advances in molecular biology, capitalizing on the recent cloning of the 5-HT$_{1C}$ receptor cDNA (Julius et al. 1988; Lübbert et al. 1987). We reasoned that if the rat stomach fundus contained the 5-HT$_{1C}$ receptor, then the known cDNA clone for this receptor should hybridize to the mRNA found in rat stomach fundus. In addition, we used primer sequences for the rat 5-HT$_{1C}$ mRNA to analyze the rat fundus RNA by PCR amplification techniques. Thus, even low levels of the 5-HT$_{1C}$ receptor messenger RNA, if present in the rat stomach fundus would be detected by these procedures.

Although 5-HT$_2$ receptor antagonists do not markedly shift the concentration response curve to serotonin in the rat stomach fundus, we and others have noted a modest effect of agents such as ketanserin to produce a small inhibition (approximately 2 fold) of the contractile response to serotonin (Clineschmidt et al. 1985; Cohen – unpublished observations). This raised the possibility that we might be able to detect a small population of 5-HT$_2$ receptors in the rat stomach fundus with the use of highly sensitive PCR techniques in amplification experiments with primers synthesized from the published 5-HT$_2$ receptor cDNA sequence.

Materials and Methods

RNA Blot Analysis

Total cellular RNA was purified from rat brain, stomach fundus and liver and processed for RNA hybridization or PCR analysis as previously described (Baez et al. 1990). The mouse cDNA clone used in the RNA hybridization analysis contains 868 bases of coding sequence which is 96% homologous to the rat 5-HT$_{1C}$ receptor sequence (Lübbert et al. 1987 and Yu et al. unpublished data). A 1.0 kb fragment of a human cathepsin-D cDNA clone (Faust et al. 1985) was also used in the RNA hybridization analysis. Both probes were labeled by nick translation with [α-^{32}P]dCTP to a specific activity of $1-2 \times 10^8$ cpm/mg of DNA.

PCR and DNA Blot Analysis

cDNA was synthesized from 50 mg of total cellular RNA with a first-strand reaction mix from a cDNA synthesis kit (Pharmacia LKB Biotechnology, Piscataway, NJ). An aliquot of the cDNA was used in PCR reaction as described in a Gene AMP kit (Perkin-Elmer Cetus, Norwalk, CT). The amplification reaction consisted of 30 cycles of 96° for 15 sec, 69° for 30 sec and 72° for 1 min. The products were analyzed by agarose gel electrophoresis and DNA hybridization analysis. The probe for the DNA hybridization analysis was an oligonucleotide, 5'-GATGAAAAACGGGCACCACATGATCAGAAACACAAAGAAT-ACAATGC-3', synthesized from the sixth transmembrane region of the rat 5-HT_{1C} receptor sequence (Julius et al. 1988) and was end labeled with $[\gamma\text{-}^{32}P]ATP$ to a specific activity of approximately 10^6 cpm/pmol of DNA. The PCR primers specific for the rat 5-HT_{1C} receptor sequence (Julius et al. 1988), the mouse DHFR sequence (Crouse et al. 1982) and rat 5-HT_2 receptor sequence (Pritchett et al. 1988) were synthesized from published sequences. The primers specific for the rat 5-HT_{1C} receptor sequence were 5'-ACACCGAGGAGGAACTGGCTAATAT-3' and 5'-GACCACATTAGAGGGGTTGACTGGC-3' and define the borders of a 601-base-long DNA fragment. The 5-HT_2 receptor primers were 5'-GGGTACCGGTGGCCTTTGCCTAGCA-3' and 5'-TCCTTGTAC-TGACACTGAATGTACC-3' and define the borders of a 791-base-long DNA fragment. The DHFR primers were 5'-CTCAGGGCTGCGAT-TTCGCGCCAAACT-3' and 5'-CTGGTAAACAGAACTGCCTCCG-ACTATC-3' and define the borders of a 446–base-long DNA fragment.

Results

A mouse 5-HT_{1C} receptor cDNA clone containing 868 bases of coding sequence that is 96% homologous to the rat 5-HT_{1C} receptor sequence and encompasses 5 of 7 transmembrane region domains (see Figure 1A) was hybridized to poly $(A)^+$RNA isolated from rat brain, stomach fundus and liver (Figure 1B). Only the brain mRNA shows a 5-HT_{1C} receptor specific band of approximately 5.2 kb. The presence and size of this band in the brain sample corroborates a previous report in which a rat 5-HT_{1C} clone was used to examine the RNA from several rat tissues (Julius et al. 1988). Those investigators also noted the lack of 5-HT_{1C} receptor sequences in rat liver RNA. A second probe specific for a ubiquitously expressed gene, cathepsin-D, was also hybridized to the same samples. The presence of the expected 2.2 kb band in all three RNA preparations verifies their integrity.

Figure 2A shows the PCR products resulting from amplification with the 5-HT_{1C} receptor-specific primers. Note that only the brain RNA

A

RAT 5HT₁c RECEPTOR CODING SEQUENCE

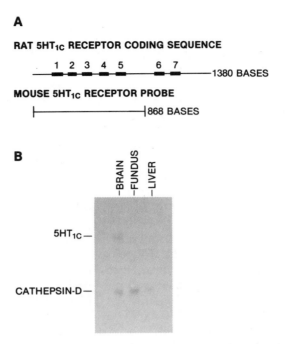

MOUSE 5HT₁c RECEPTOR PROBE

B

Figure 1. RNA hybridization analysis of rat fundus with a 5-HT$_{1C}$ receptor cDNA clone. Poly (A) $^{+}$RNA isolated from rat brain, fundus and liver was fractionated in a 1.5% agarose gel under denaturing conditions and blotted onto a nitrocellulose filter that was probed with a [α-^{32}P] dCTP labelled 5-HT$_{1C}$ receptor cDNA fragment. (B). The upper line drawing in A shows the rat 5-HT$_{1C}$ receptor coding sequence. Filled rectangles symbolize the position and relative size of the seven transmembrane domains of the encoded protein. The lower line depicts the length of the mouse 5-HT$_{1C}$ receptor probe and is aligned with the homologous region of the rat 5-HT$_{1C}$ receptor-coding sequence.

sample shows the expected 601-base-long fragment after a minimum amplification of the target sequence of 10^5 fold. The PCR products resulting from amplification of cDNAs with the 5-HT$_{1C}$ receptor-specific primers were also hybridized with a probe specific for sixth transmembrane region of the 5-HT$_{1C}$ receptor sequence and the autoradiogram in Figure 2B shows a single 601-base-long fragment in the brain sample. No signal was seen in the fundus or liver samples. However, all three samples show the 446-base-long fragment expected from the ubiquitously expressed DHFR mRNA (Figure 2A) thus indicating that the RNA preparations were intact. Furthermore, in preliminary experiments, when primers specific for the rat 5-HT$_2$ receptor were used to amplify the same sample of cDNA used in the above analysis, both the brain and fundus samples yielded DNA fragments of the appropriate size although the quantity of amplified product from the fundus was considerably lower than that of the brain (data not shown).

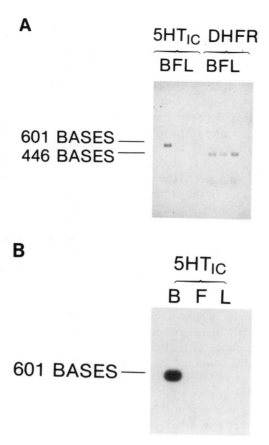

Figure 2. PCR examination of rat fundus for low levels of 5-HT$_{1C}$ receptor mRNA. cDNA synthesized from RNA isolated from rat brain (B), fundus (F) and liver (L) was amplified by PCR with oligonucleotide primers specific for 5-HT$_{1C}$ receptor and DHFR sequences. The PCR products were examined by agarose gel electrophoresis and ethidium bromide staining. A negative of the photograph of the gel is depicted in panel A. The first three lanes in A were further analyzed by DNA hybridization (B) with a [γ-^{32}P]ATP labeled oligonucleotide probe specific for the sixth transmembrane domain of the rat 5-HT$_{1C}$ receptor sequence.

Discussion

Based on previous data, it is apparent that the predominant contractile receptor responsible for serotonin-induced contractions in the rat stomach fundus is not the 5-HT$_2$, 5-HT$_3$, 5-HT$_{1A}$, 5-HT$_{1B}$ or 5-HT$_{1D}$ receptor (Table 1). The present study was prompted by pharmacological experiments showing some similarities (based on agonist potency) between the serotonin contractile response in the rat stomach fundus and binding to the 5-HT$_{1C}$ receptor (Cohen 1989). However, not all information was consistent with the presence of a 5-HT$_{1C}$ receptor mediating contraction in the stomach fundus. For example, data obtained with antagonists did not reveal a good correlation between antagonist affinities in the rat stomach fundus and affinities at the 5-HT$_{1C}$ receptor based on ligand binding studies. In fact, if one utilized high affinity 5-HT$_{1C}$ receptor ligands, such as LY53857, ritanserin and SCH23390, relatively little correlation occurred between the ability of these compounds to block serotonin-induced contraction in the stomach fundus and their 5-HT$_{1C}$ receptor affinities (Baez et al. 1990). Furthermore, at 5-HT$_{1C}$ sites in brain and choroid plexus, serotonin clearly produces an increase in phosphoinositide turnover. However, serotonin did not increase phosphoinositide turnover in the stomach fundus, suggesting that the second messenger in the stomach fundus for serotonin-induced contractions is different from that utilized in the choroid plexus and brain (Cohen and Wittenauer 1987; Secrest et al. 1989). These observations prompted us to probe more thoroughly using mRNA hybridization and PCR amplification assays, the possible presence of the 5-HT$_{1C}$ receptor mRNA in the rat stomach fundus.

Although the 5-HT$_{1C}$ receptor sequence was capable of hybridizing with brain messenger RNA, no signal was detected in the rat stomach fundus. The readily detected mRNA for the ubiquitously expressed cathepsin-D gene suggests that the absence of 5-HT$_{1C}$ receptor mRNA in fundus is not due to an artifact of sample preparation. Even with the use of highly sensitive PCR techniques, in which a minimum amplification of the target sequence of 10^5-fold was expected, we were unable to detect the 5-HT$_{1C}$ receptor mRNA in rat stomach fundus. To increase our ability to detect low levels of receptor expression we further examined the PCR amplification products by DNA hybridization analysis (Figure 2B) and again were unable to find 5-HT$_{1C}$ receptor mRNA in the fundus sample after a standard, 16 hour exposure of the autoradiogram. It should be noted however, that after prolonged exposure of the autoradiogram (6 days) we detected a weak signal in the fundus preparation and a weaker signal in the liver preparation. These extremely low levels of 5-HT$_{1C}$ receptor mRNA in rat stomach fundus and liver are probably not biologically relevant but more likely result from illegitimate transcription (Chelly et al. 1989) as has been reported for the expression of several other genes when examined by PCR amplification (Saiki et al. 1989). Consequently, our findings strongly indicate that the 5-HT$_{1C}$ receptor gene is not expressed in the rat stomach fundus at levels which could account for the potent serotonin-induced contractions of this tissue.

However, in contrast to our inability to detect the 5-HT$_{1C}$ receptor mRNA in the rat stomach fundus, the use of the 5-HT$_2$ receptor-specific primers did permit a faint but significant signal to be detected in the rat stomach fundus after PCR amplification (data not shown). Thus, we were able to demonstrate a small amount of the 5-HT$_2$ receptor sequence within this tissue. This information is consistent with previous pharmacological studies documenting a modest shift (that did not appear to be concentration dependent) in the contractile response to serotonin in the rat stomach fundus by 5-HT$_2$ receptor antagonists such as ketanserin (Clineschmidt et al. 1985). The presence of minor amounts of 5-HT$_2$ receptor in the rat stomach fundus has now been confirmed by additional PCR amplification experiments utilizing the 5-HT$_2$ receptor sequence.

In summary, the predominant receptor mediating the contractile response to serotonin in the rat stomach fundus does not appear to be any one of the presently well characterized serotonin receptors. Based on the available pharmacology, the predominant contractile receptor for serotonin in this preparation may well be a novel serotonin receptor different from any of the previously well characterized 5-HT receptors.

References

Baez, M., Yu, L., and Cohen, M. L. (1990). Pharmacological and molecular evidence that the contractile response to serotonin in rat stomach fundus is not mediated by activation of the 5-hydroxytryptamine$_{1C}$ receptor. Mol. Pharmacol. 38: 31–37.

Chelly, J., Concordet, J.-P., Kaplan, J.-C., and Kahn, A. (1989). Illegitimate transcription: Transcription of any gene in any cell type. Proc. Natl. Acad. Sci. USA 86: 2617–2621.

Clineschmidt, B. V., Reiss, D. R., Pettibone, D. J., and Robinson, J. L. (1985). Characterization of 5-hydroxytryptamine receptors in rat stomach fundus. J. Pharmacol. Exp. Ther. 235: 696–708.

Cohen, M. L. (1989). 5-hydroxytryptamine and non-vascular smooth muscle contraction and relaxation. In: Fozard, J. R. (ed.), The Peripheral Actions of 5-Hydroxytryptamine. Oxford: Oxford University Press, pp. 201–218.

Cohen, M. L., and Fludzinski, L. A. (1987). Contractile serotonergic receptor in rat stomach fundus. J. Pharmacol. Exp. Ther. 243: 264–269.

Cohen, M. L., Schenck, K. W., Colbert, W., and Wittenauer, L. (1985). Role of 5-HT$_2$ receptors in serotonin-induced contractions of nonvascular smooth muscle. J. Pharmacol. Exp. Ther. 232: 770–774.

Cohen, M. L., and Wittenauer, L. (1986). Further evidence that the serotonin receptor in the rat stomach fundus is not 5-HT$_{1A}$ or 5-HT$_{1B}$. Life Sci. 38: 1–5.

Cohen, M. L., and Wittenauer, L. (1987). Serotonin receptor activation of phosphoinositide turnover in uterine, fundal, vascular, and tracheal smooth muscle. J. Cardiovasc. Pharmacol. 10: 176–181.

Conn, P. J., Sanders-Bush, E., Hoffman, B. J., and Hartig, P. R. (1986). A unique serotonin receptor in choroid plexus is linked to phosphatidylinositol turnover. Proc. Natl. Acad. Sci. 83: 4086–4088.

Crouse, G. F., Simonsen, C. C., McEwan, R. N., and Schimke, R. T. (1982). Structure of amplified normal and variant dihydrofolate reductase genes in mouse sarcoma S180 cells. J. Biol. Chem. 257: 7887–7897.

Faust, P. L., Kornfield, S., and Chirgwin, J. M. (1985). Cloning and sequence analysis of cDNA for human cathespsin-D. Proc. Natl. Acad. Sci. USA 82: 4910–4914.

Hoyer, D. (1989). 5-Hydroxytryptamine receptors and effector coupling mechanism in peripheral tissues. In Fozard, J. R. (ed), The Peripheral Actions of 5-Hydroxytryptamine. Oxford: Oxford University Press, pp. 72–88.

Julius, D., MacDermott, A. B., Axel, R., and Jessell, T. M. (1988). Molecular Characterization of a functional cDNA encoding the serotonin 1C receptor. Science (Wash, D. C.) 241: 558–564.

Lübbert, H., Hoffman, B. J., Snutch, T. P., van Dyke, T., Levine, A. J., Hartig, P. R., Lester, H. A., and Davidson, N. (1987). cDNA cloning of a serotonin 5-HT$_{1C}$ receptor by electrophysiological assays of mRNA-injected Xenopus oocytes. Proc. Natl. Acad. Sci. USA 84: 4332–4336.

Pritchett, D. B., Bach, A. W. J., Wozny, M., Taleb, O., Dal Toso, R., Shih, J. C., and Seebury, P. H. (1988). Structure and functional expression of cloned rat serotonin 5-HT$_2$ receptor. EMBO J. 7: 4135–4140.

Saiki, R. K., Gelfand, D. H., Stoffel, S., Scharf, S. J., Higuchi, R., Hom, G. T., Mullis, K. B., and Erlich, H. A. (1989). Primer-directed enzymatic amplification of DNA with a thermostable DNA polymerase. Science 239: 487–491.

Secrest, R. J., Schoeep, D. D., and Cohen, M. L. (1989). Comparison of contractions of serotonin, carbamycholine, and prostaglandin F$_2$ alpha in rat stomach fundus. J. Pharmacol. Exp. Ther. 250: 971–978, 1989.

Serotonin: Molecular Biology, Receptors and Functional Effects
ed. by J. R. Fozard/P. R. Saxena
© 1991 Birkhäuser Verlag Basel/Switzerland

Further Definition of the 5-HT Receptor Mediating Contraction of Rat Stomach Fundus: Relation to 5-HT$_{1D}$ Recognition Sites

H. O. Kalkman and J. R. Fozard

Preclinical Research, Sandoz Pharma Ltd., CH-4002 Basel, Switzerland

Summary. The 5-HT receptor mediating contraction of rat stomach fundus is designated 5-HT$_1$-like but has defied classification in terms of the recognized subtypes of the high affinity ^{3}H-5-HT binding sites. We have now identified 5 compounds to add to the several already characterised which are silent competitive antagonists of 5-HT on rat stomach fundus. These are Org GC 94, pizotifen, MDL 73005, (−)-propranolol and (+)-propranolol. Of particular note is the high affinity of Org GC 94 and the poor stereoselectivity of the enantiomers of propranolol. Further data, certain of which confirm the literature findings, permit tentative definition of the fundus 5-HT receptor as a site where the agonist order of potency is 5-HT = (+)-α-Me-5-HT = 5-MeO-T > 5-CT > 2-Me-5-HT > sumatriptan > 8-OH-DPAT, where Org GC 94 is a particularly potent silent antagonist, where yohimbine is a potent, competitive and stereoselective antagonist and where the enantiomers of propranolol are competitive but poorly stereoselective antagonists. Although pharmacological similarities between the fundus receptor and the 5-HT$_{1D}$ recognition site are evident, there are also major differences. However, since the high affinity ^{3}H-5-HT binding site(s) labelled under "5-HT$_{1D}$ conditions" may not be homogeneous, it may be premature to dismiss categorically a link between the fundus 5-HT receptor and the 5-HT$_{1D}$ recognition site as currently defined.

Introduction

The 5-HT receptor mediating contraction of the rat stomach fundus has been designated as 5-HT$_1$-like (Bradley et al. 1986). It is, however, not a 'typical' 5-HT$_1$-like receptor, since 5-CT is clearly less active than 5-HT (Bradley et al. 1986, Buchheit et al. 1986). The 5-HT receptor of the rat fundus also defies classification in terms of the recognized subtypes of the high affinity [^{3}H]-5-HT binding sites. Phenylpiperazines such as mCPP, TFMPP and PAPP antagonize 5-HT-induced contractions; however, their antagonist potency does not correlate with their affinities for either the 5-HT$_{1A}$ or the 5-HT$_{1B}$ binding sites (Cohen and Wittenauer 1986). The fact that 5-CT is approximately 10-fold less active than 5-HT, would be consistent with the 5-HT receptor in the fundus being a 5-HT$_{1C}$ receptor. Moreover, a highly significant correlation was observed between pK$_D$ values for 5-HT$_{1C}$ binding sites and pD$_2$ values for the contractile responses of 25 agonists (Buchheit et al. 1986). However, the involvement of 5-HT$_{1C}$ receptors can be excluded on the basis of antagonist experiments; compounds like mianserin, ketanserin,

LY 53857 or pirenperone, are inactive at concentrations which theoretically would occupy 5-HT_{1C} receptors (Cohen and Fludzinski 1987; Clineschmidt et al. 1985). Yohimbine and rauwolscine are among the few compounds that are potent, competitive and stereoselective antagonists of 5-HT in the rat stomach fundus (Clineschmidt et al. 1985). Significantly, these compounds display high affinity and stereoselectivity at the 5-HT_{1D} receptor subtype and show selectivity for this site over the other high affinity $^3\text{H-5-HT}$ binding sites (Hoyer 1989). The present experiments were carried out to define further the 5-HT receptor which mediates contraction of the rat stomach fundus and in particular to explore its relationship to the 5-HT_{1D} binding site.

Materials and Methods

Agonist Studies: Experiments were carried out on fundus strips from male Sprague-Dawley rats weighing 190–230 g, killed by a sharp blow on the head. The stomach was dissected and four strips of 2 mm width and 2 cm length were cut parallel to the greater curvature. Strips were set up in organ baths of 20 ml containing Krebs solution constantly bubbled with 5% CO_2 in oxygen. The composition of the Krebs solution was (mmol/l): NaCl 118, KCl 4.7, $CaCl_2$ 1.25, KH_2PO_4 1.2, $MgSO_4$ 1.2, glucose 11 and $NaHCO_3$ 25.0. In these studies, pargyline 5×10^{-5} mol/l was present throughout the experiments. Contractions were measured isotonically under a resting tension of 1 g. Prior to testing, the strips were allowed to equilibrate for 1.5–2.5 h, during which the bath fluid was replaced every 15 min. The strips were then challenged with carbachol $(10^{-6}$ M) and 30 min later, after washout, the agonist was added according to a cumulative concentration schedule. Only a single agonist dose-response curve was established on each preparation.

Antagonist Studies: Fundic strips were prepared from male Sprague-Dawley rats as described above except that a modified Tyrode solution (composition in mmol/l: NaCl 137, KCl 2.7, $CaCl_2$ 1.8, $MgCl_2$ 1.05, $NaHCO_3$ 11.9, NaH_2PO_4 0.42, glucose 5.6, ascorbic acid 0.06) was used and contractions were measured isometrically. After an equilibration period of 0.5–1 h, during which the fundus was "primed" by eliciting 2–3 contractions to 5-HT, 10^{-5} mol/l, a concentration-response curve to 5-HT was generated cumulatively. The tissue was rinsed repeatedly until fully relaxed, then incubated with the first concentration of antagonist to be tested. Fifteen min later a second concentration-response was generated. Up to 4 concentrations of antagonist were evaluated on a single tissue. Tissues incubated only with Tyrode solution served as time controls; no significant change in sensitivity occurred during the course of the experiment. pA_2 values were obtained by the method of Arunlakshana and Schild (1959).

Drugs Used: Org GC 94 (Organon, Oss, Netherlands), pizotifen (Sandoz, Basle, Switzerland), MDL 73005 (Merrell Dow Research Institute, Strasbourg, France), ($-$) and ($+$) propranolol (ICI, Macclesfield, U.K.) were kind gifts of the respective manufacturers. Yohimbine, corynanthine, pargyline, 5-hydroxytryptamine (5-HT) and 5-methoxytryptamine (5-MeO-T) were purchased from Sigma (St Louis, U.S.A.). ($+$)-alpha-methyl-5-hydroxytryptamine (($+$)-α-Me-5-HT), 5-carboxamidotryptamine (5-CT), 2-methyl-5-hydroxytryptamine (2-Me-5-HT), sumatriptan and 8-hydroxy-2-(di-n-propylamino)tetralin (8-OH-DPAT) were synthesised at Sandoz (Basle, Switzerland).

Results

The results from the agonist study expressed as the negative logarithm of the EC_{50} (pD_2) are presented in Table 1. The order of potency was 5-HT = ($+$)-α-Me-5-HT = 5-MeO-T > 5-CT > 2-Me-5-HT > sumatriptan > 8-OH-DPAT. Comparison with the affinities for the $5\text{-}HT_{1C}$ and $5\text{-}HT_{1D}$ binding sites taken from the literature, reveals closer similarity to the $5\text{-}HT_{1C}$ site than the $5\text{-}HT_{1D}$ site. The data obtained with the antagonists are shown in Figures 1–3 and Table 1. Each antagonist induced concentration-dependent, broadly parallel shifts to the right of the curves to 5-HT with minimal changes in the slopes. Org

Table 1. Potencies of agonists and antagonists at the 5-HT receptor in rat fundus: comparison with the receptor affinities from radioligand binding studies

		Fundus		Binding data*	
	pA_2	Schild plot slope	pD_2	$5\text{-}HT_{1C}$ (pK_D)	$5\text{-}HT_{1D}$ (pK_D)
5-HT	—	—	8.0 ± 0.1	7.5	8.4
($+$)-α-Me-5-HT	—	—	8.0 ± 0.1	7.2	6.2
5-MeO-T	—	—	7.8 ± 0.2	7.6	8.4
5-CT	—	—	6.9 ± 0.1	6.2	8.6
2-Me-5-HT	—	—	6.2 ± 0.1	5.9	6.4
Sumatriptan	—	—	5.2 ± 0.2	4.1	7.5
8-OH-DPAT	—	—	4.6 ± 0.2	5.2	5.9
Org GC 94	8.4 ± 0.1	1.1 ± 0.1	—	8.5	6.0
Pizotifen	7.7 ± 0.1	1.3 ± 0.1	—	8.1	5.6
MDL 73005	7.3 ± 0.1	0.9 ± 0.1	—	6.2	6.0
($-$)-Propranolol	7.0 ± 0.3	0.9 ± 0.1	—	6.8	5.5
($+$)-Propranolol	6.6 ± 0.2	0.9 ± 0.1	—	5.3	4.0
Yohimbine	7.9 ± 0.1	1.0 ± 0.1	—	4.4	7.1
Corynanthine	5.1 ± 0.1	—	—	5.5	5.3

Means ($\pm$ SEM) of ≥ 4 individual values.
*Taken from Hoyer (1989) and (Hoyer) unpublished observations.

156

GC 94 proved a particularly potent surmountable antagonist and yohimbine blocked responses to 5-HT both potently and stereoselectively. In contrast, both ($-$) and ($+$) propranolol behaved as reasonably potent competitive antagonists but were minimally stereoselective. There were major discrepancies between the antagonist affinities defined on the fundus and those obtained in the 5-HT_{1C} and 5-HT_{1D} binding assays (Table 1).

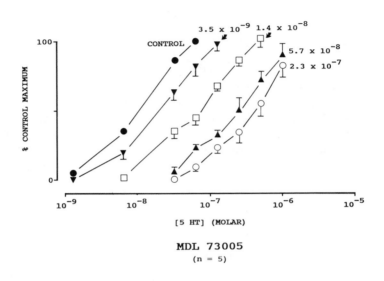

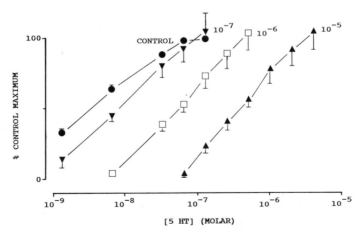

PIZOTIFEN

(n = 4)

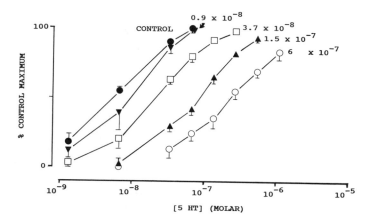

Figure 1. Blockade of responses to 5-HT on rat isolated stomach fundus by ORG GC 94, MDL 73005 and pizotifen. The molar concentrations of antagonists are given adjacent to each curve.

Discussion

As previously noted (Buchheit et al. 1986), the majority of pD_2 values for agonist responses on the rat fundus agree well with the pK_D values for the 5-HT$_{1C}$ recognition site obtained from radioligand binding studies. A notable exception is, however, sumatriptan which is more than 1 log unit more active in the functional test than in the binding assay. Reinforcement for the view that the fundus receptor and the 5-HT$_{1C}$ site are not the same comes from data with yohimbine, which was 3000 times more active in the functional test than in the binding assay, and from ($-$) and ($+$) propranolol, which showed appreciable stereoselective displacement from the 5-HT$_{1C}$ recognition site but were only weakly stereoselective in the functional test. Our data add further weight to the prevailing view (see Cohen and Fludzinski 1987) that the fundus does not contain the 5-HT$_{1C}$ site.

Similarly, our data do not support the possibility that the fundus 5-HT receptor and the 5-HT$_{1D}$ site are the same. Whilst the potency and stereoselectivity of the interaction of yohimbine with the two sites is consistent with the suggestion, the majority of the data obtained with both agonists and antagonists do not support such a conclusion.

Recently, Sumner and Humphrey (1989) demonstrated that the high affinity [^{3}H]-5-HT binding sites labelled under "5-HT$_{1D}$ conditions" (i.e. using pig caudate tissue and 5-HT$_{1A}$ and 5-HT$_{1C}$ receptor masks),

158

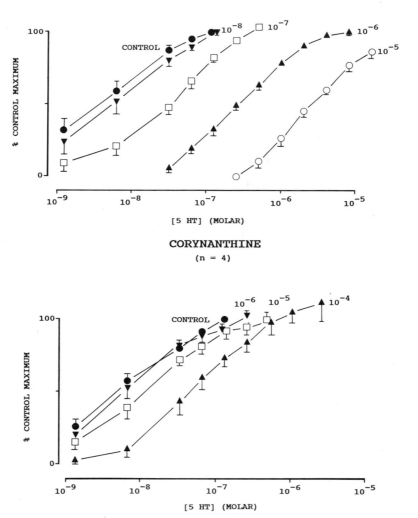

Figure 2. Blockade of responses to 5-HT on rat isolated stomach fundus by yohimbine and corynanthine. The molar concentrations of antagonists are given adjacent to each curve.

may not be homogeneous. In fact, the limited data available are consistent with the notion that the low affinity binding site identified by Sumner and Humphrey may be similar to the rat fundus 5-HT receptor; thus, 5-CT was about 10-fold less potent that 5-HT, whereas sumatriptan was a poor displacer at this particular site. Other workers have

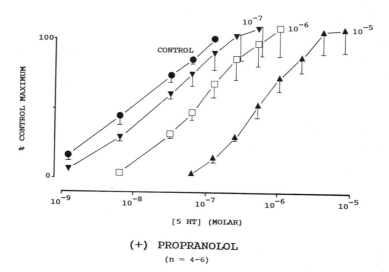

Figure 3. Blockade of responses to 5-HT on rat isolated stomach fundus by ($-$) and ($+$)
propranolol. The molar concentrations of antagonists are given adjacent to each curve.

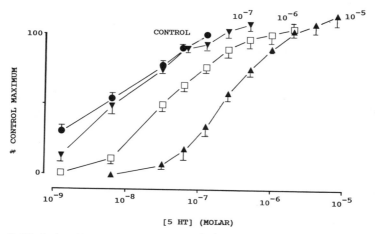

similarly drawn attention to the likely heterogeneity of 5-HT$_{1D}$ receptors (Leonhardt et al. 1989, Xiong and Nelson 1989). It may, therefore,
be premature to dismiss categorically a link between the fundus 5-HT
receptor and the 5-HT$_{1D}$ recognition site as currently defined.

In only a few instances has silent competitive inhibition of 5-HT-in-
duced contraction of the rat fundus been described. The potent compet-

itive antagonism by Org GC 94 is therefore particularly noteworthy and especially since the compound is a close derivative of mianserin, which was found to be either a weak partial agonist (Frankhuyzen and Bonta 1974) or a relatively poor antagonist (Cohen and Fludzinski 1987). Although none of the compounds used in this study can be considered selective for the 5-HT receptor of the fundus, our data do further define the site pharmacologically and hence contribute to the characterisation of this presently "unclassifiable" receptor.

Acknowledgements

We thank Dr D. Hoyer for permission to include his unpublished binding data and Mrs M. L. Part for expert technical assistance.

References

Arunlakshana, O., and Schild, H. O. (1959). Some quantitative uses of drug antagonists. Br. J. Pharmacol. Chemother. 14: 45–58.

Buchheit, K. H., Engel, G., Hagenbach, A., Hoyer, D., Kalkman, H. O., Seiler, M. P. (1986). The rat isolated stomach fundus strip, a model for 5-HT$_{1C}$ receptors. Br. J. Pharmacol. 88: 367P.

Bradley, P. B., Engel, G., Feniuk, W., Fozard, J. R., Humphrey, P. P. A., Middlemiss, D. N., Mylecharane, E. J., Richardson, B. P., Saxena, P. R. (1986). Proposals for the classification and nomenclature of functional receptors for 5-hydroxytryptamine. Neuropharmacol. 25: 563–576.

Clineschmidt, B. V., Reiss, D. R., Pettibone, D. J., Robinson, J. L. (1985). Characterization of 5-hydroxytryptaime receptors in rat stomach fundus. J. Pharm. Exp. Ther. 235: 696–708.

Cohen, M. L., and Wittenauer, L. A. (1986). Further evidence that the serotonin receptor in the rat stomach fundus is not 5-HT$_{1A}$ or 5-HT$_{1B}$. Life Sci. 38: 1–5.

Cohen, M. L., and Fludzinski, L. A. (1987). Contractile serotonergic receptor in rat stomach fundus. J. Pharm. Exp. Ther. 243: 264–269.

Frankhuyzen, A. L., and Bonta, I. L. (1974). Effect of mianserin, a potent anti-serotonin agent on the isolated rat stomach fundus preparation. Eur. J. Pharmacol. 25: 40–50.

Hoyer, D. (1989). 5-Hydroxytryptamine receptors and effector coupling mechanisms in peripheral tissues. In: Fozard, J. R. (ed.), Peripheral actions of 5-hydroxytryptamine. Oxford: Oxford University Press, pp. 72–99.

Leonhardt, S., Herrick-Davis, K., and Titeler, M. (1989). Detection of a novel serotonin receptor subtype (5-HT$_{1E}$) in human brain: interaction with a GTP-binding protein. J. Neurochem. 53: 465–471.

Sumner, M. J., and Humphrey, P. P. A. (1989). 5-HT$_{1D}$ binding sites in porcine brain can be sub-divided by GR43175. Br. J. Pharmacol. 98: 29–31.

Xiong, W., and Nelson, D. L. (1989). Characterization of a [^{3}H]-5-hydroxytryptamine binding site in rabbitcaudate nucleus that differs from the 5-HT$_{1A}$, 5-HT$_{1B}$, 5-HT$_{1C}$ and 5-HT$_{1D}$ subtypes. Life Sci. 45: 1433–1442.

Serotonin: Molecular Biology, Receptors and Functional Effects
ed. by J. R. Fozard/P. R. Saxena
© 1991 Birkhäuser Verlag Basel/Switzerland

Temperature Dependence of Agonist and Antagonist Affinity Constants at 5-HT$_1$-like and 5-HT$_2$ Receptors

D. J. Prentice, V. J. Barrett, S. J. MacLennan, and G. R. Martin

Analytical Pharmacology Group, Biochemical Sciences, Wellcome Research Laboratories, Beckenham, Kent BR3 3BS, UK

Summary. The studies described here show that in isolated vascular rings contractile responses to 5-HT mediated via both 5-HT$_1$-like and 5-HT$_2$ receptors were potentiated by cooling from 37°C to 25°C. This potentiation could be accounted for mainly by an increase in agonist affinity. Antagonist affinity at the two receptor types was unaltered by cooling. These differences in ligand-receptor thermodynamics have clear cut implications for pharmacological receptor classification schemes based upon agonists as well as antagonists. In addition, the results demonstrate that 5-HT$_1$-like as well as 5-HT$_2$ receptors participate in hypothermia-induced vascular supersensitivity to 5-HT, implying a possible role for either receptor type in cold-induced vasospasm.

Introduction

Hypothermia-induced supersensitivity to 5-hydroxtryptamine (5-HT) has been described in isolated intact tissues, suggesting that temperature has a fundamental influence on the agonist behaviour of this hormone (Van Neuten et al. 1984, Arneklo-Nobin 1987). In principle, these changes might result from alterations in affinity and/or occupancy-effect coupling, but no attempt has been made to establish which of these is affected. In either case, the possible implications for (i) receptor classification and (ii) the development of novel potential therapies for cold-induced vasospasm (e.g. Raynaud's Phenomenon) make desirable a better understanding of the mechanisms underlying such temperature effects. To this end we have used operational model fitting methods (Black and Leff 1983; Leff et al. 1990) to examine the actions of 5-HT and 5-carboxamidotryptamine (5-CT) at the 5-HT$_1$-like receptor mediating contraction of the rabbit saphenous vein (RbSV), and the actions of 5-HT at the 5-HT$_2$ receptor mediating contraction in the rabbit aorta (RbA). We have also examined the effect of temperature upon the affinity estimates obtained for antagonists in these two systems and estimated the thermodynamic parameters associated with agonist- and antagonist-receptor interactions.

Methodology and Protocol

Methods

Right and left lateral saphenous veins were obtained from male New Zealand White rabbits (2.5–3.0 kgs) killed by intravenous administration of pentobarbitone sodium (Sagatal, 120 mg/kg). Vessels were cannulated *in situ* using polypropylene tubing (o.d. = 1.0 mm), cleared of any adhering connective tissue and cut into rings 3–4 mm in length. These were carefully transferred from the cannula onto two fine tungsten wire hooks (diameter 250 μm) using the method of Hooker et al. (1977). The thoracic aorta was removed from the same rabbits. The vessel was cleared of adhering connective tissue after mounting on a polypropylene cannula (o.d. = 2.5 mm). Rings of approximately 3 mm were prepared according to the method of Stollak and Furchgott (1983), preserving the plane of the circular muscle.

Changes in tissue isometric force were recorded from vascular rings suspended in 20 ml organ baths containing Krebs solution of the following composition (μmol/l): NaCl 118.41, NaHCO$_3$ 25.00, KCl 4.75, KH$_2$PO$_4$ 1.19, MgSO$_4$ 1.19, glucose 11.10 and CaCl$_2$ 2.50. This was maintained at either 37°C or 25°C and continually gassed with 95%O$_2$:5%CO$_2$. Decreasing the temperature to 25°C produced a small decrease in pH (less than 0.2 pH unit). Data, presented as the mean $\pm$ s.e.m., are the averages obtained from 4–18 preparations taken from a minimum of 3 animals.

Experimental Protocol

At the beginning of each experiment a force was applied to each tissue and subsequently re-applied twice at intervals of 15 min (2 g for RbSV and 3 g for RbA). During this 30 min period, tissues were exposed to pargyline (500 μmol/l) which irreversibly inhibited MAO. At the same time α-adrenoceptors were inactivated using either phenoxybenzamine (0.3 μmol/l:RbSV) or benextramine tetrahydrochloride (10 μmol/l: RbA). In the RbSV PBZ treatment also irreversibly blocked the intraneuronal uptake and accumulation of 5-HT. After washout of excess inhibitors tissues were challenged with 80 μmol/l KCl and subsequently with 5-HT (RbSV, 1.0 μmol/l:RbA, 10 μmol/l) in order to establish viability and to provide reference contractions by which subsequent E/[A] curves could be normalised.

Studies with Agonists

After recovery from the initial challenges with KCl and 5-HT a cumulative E/[A] curve to 5-HT or 5-CT was constructed at 37°C. Following washout of the agonist, the organ bath temperature was either maintained at 37°C or reduced to 25°C over a period of approximately 30 min and a second curve constructed. Subsequently, tissues were again washed free of agonist and then incubated for 30 min with an irreversible ligand or vehicle: in the RbSV, benextramine tetrahydrochloride (1.0 μmol/l at 37°C and 8.0 μmol/l at 25°C) was used and in the RbA, phenoxybenzamine (PBZ: 1.0 μmol/l at both temperatures). At the end of this period unreacted ligand was removed by five exchanges of the organ bath Krebs solution over a period of 30 min. Finally a third cumulative curve was constructed. Responses were expressed as a percentage of the KCl contraction. It was necessary to increase the concentration of BHC 8-fold at 25°C in the RbSV to achieve a similar degree of receptor inactivation to that observed at 37°C. Presumably this resulted from a slowing of the kinetics of the drug-receptor interaction and/or access of the drug to the biophase at the lower temperature. Interestingly, this was not the case for PBZ in the RbA.

Studies with Antagonists

Tissues were exposed to antagonists for a period of two hours prior to construction of 5-HT $E/[A]$ curves. In each tissue, responses were normalised by scaling them to the response obtained with the initial 5-HT challenge.

Data Analysis

Logistic Curve Fitting

Two types of experiment were performed; those in which only a single $E/[A]$ curve was obtained in each preparation and those in which three successive curves were obtained in each preparation before and after specific interventions. In both cases $E/[A]$ curve data were fitted to a logistic function to provide estimates of α, $[A_{50}]$ and m, the asymptote, midpoint location and slope parameters respectively. Location parameters were actually estimated as negative logarithms ($p[A_{50}]$). Parameter estimates obtained for successive curves were used to calculate a midpoint concentration ratio ($\Delta p[A_{50}] =$ control $p[A_{50}]$ minus test $p[A_{50}]$) and maximum ratio (max ratio = test maximum divided by control maximum). For each type of intervention control curves were compared in this way and when significant differences were apparent the treatment curves from parallel experiments were adjusted accordingly. Paired control curve max ratios were within the range 0.94–1.13 and ΔpA_{50} values within the range -0.05–0.19.

Estimation of Antagonist Dissociation Constant (K_B)

When antagonists were studied a one way analysis of variance compared computed estimates of curve slope and maximum within and between treatment groups. If the antagonism observed was surmountable and produced parallel curve displacements, suggestive of competitive antagonism, computed $p[A_{50}]$ values were fitted to a form of the Schild equation in order to estimate the antagonist dissociation constant (K_B) and the slope of the Schild regression (n). If n was not significantly different from unity it was constrained to a value of one in order to estimate K_B (see Trist & Leff 1985; Leff et al. 1986).

Estimation of Agonist Dissociation Constant (K_A)

The $E/[A]$ curve data obtained at 37°C and 25°C in the absence and presence of an irreversible alkylating agent were fitted to the operational model of agonism (Black & Leff 1983; Leff et al. 1990);

$$E = \frac{E_m \tau^n [A]^n}{(K_A + [A])^n + \tau^n [A]^n} \tag{1}$$

in which K_A is the agonist dissociation constant, τ is the efficacy of the agonist in a particular tissue, E_m is the maximum possible effect in the receptor system and n defines the slope of the occupancy-effect relation. Regression analysis adjusting for differences between animals was used to compare parameters estimated at 37°C and 25°C.

Thermodynamic Analysis

Equilibrium thermodynamic parameters of binding were determined using the classical thermodynamic equations. The standard Gibbs free energy change ($\Delta G°$) of association was calculated from the equation $\Delta G° = -RT \ln K_D$ where R is the gas constant (1.99 cal/mol-deg), T is the temperature in degrees Kelvin and K_D is the equilibrium association constant ($1/K_A$). Enthalpy change ($\Delta H°$) was calculated from the Van't Hoff plot of dependence of K_D on temperature between 25°C and 37°C (298 & 310°K). The slope of the plot ($\ln K_D$ versus $1/T$) equals $-\Delta H°/R$. Entropy change ($\Delta S°$) was calculated from the values of $\Delta G°$ and $\Delta H°$ using the following equation: $\Delta G° = \Delta H° - T \Delta S°$ (parameters tabulated are for 37°C).

Drugs

The following drugs were used (source in parentheses): pargyline hydrochloride (Sigma, St. Louis, MO, USA); phenoxybenzamine hydrochloride (Smith, Kline and French, Welwyn Garden City, UK); (+)p-fluoro-N-[2-[4-[2-(hydroxymethyl)-1,4-benzodioxan-5-yl]-1-piperazinyl]ethyl]benzamide hydrochloride (flesinoxan: A gift from Duphar, Weesp, The Netherlands); benextramine tetrahydrochloride (Sigma); 5-hydroxytryptamine creatinine sulphate (Sigma); spiperone (Janssen Pharmaceutica, Beerse, Belgium); yohimbine hydrochloride (Sigma); 5-carboxamidotryptamine maleate (5-CT, synthesised by Dr. A. Robertson, Medicinal Chemistry Department, Wellcome Research Laboratories, Beckenham, Kent, UK). Phenoxybenzamine was dissolved and diluted in absolute ethanol. All other drugs were dissolved and diluted in distilled water except for spiperone which was dissolved in DMSO at 2 mM and subsequently diluted 10-fold in a 50:50 v/v mixture of distilled water and DMSO; any further dilutions were made up in distilled water. The final concentration of vehicles in the organ bath ($<0.01\%$ v/v) did not affect tissue responsiveness.

Results

Agonist effects at the 5-HT$_1$-like Receptor in RbSV

Figure 1 (panel b) illustrates the effect of reducing the organ bath temperature from 37°C to 25°C on responses to 5-HT. There was a significant leftward shift of 5-HT $E/[A]$ curves constructed at 25°C

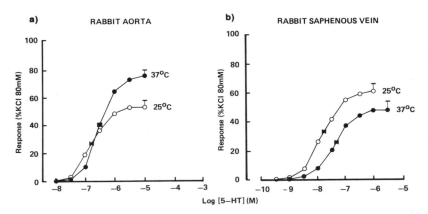

Figure 1 Illustrates paired concentration effect curves to 5-HT at 37°C (●) and at 25°C (○) in RbA (panel a) and in RbSV (panel b). Data are the averages of 17 and 18 $E/[A]$ curves respectively.

compared with those constructed at 37°C ($\Delta p[A_{50}] = -0.43 \pm 0.05$; $p < 0.001$) and a small but significant increase in maximum (max ratio = 1.19 ± 0.05; $p < 0.05$). Partial irreversible receptor occulsion using BHC at 37°C and 25°C produced a significant rightward shift with concomitant depression of the 5-HT curve maximum at both temperatures (at 37°C $\Delta p[A_{50}] = 0.33 \pm 0.03$, max ratio = 0.65 ± 0.04; $p < 0.001$; at 25°C $\Delta p[A_{50}] = 0.32 \pm 0.05$, max ratio = 0.56 ± 0.03; $p < 0.001$). Estimates of agonist affinity and efficacy obtained using these data indicated that the sensitisation could be accounted for primarily by a change in affinity of the agonist for the receptor. As shown in Table 1, lowering the temperature to 25°C increased affinity approximately two-fold (pK_A at 37°C = 7.12, pK_A at 25°C = 7.48; $p < 0.05$). There was no evidence for any significant difference in efficacy or, indeed, in any of the other parameters estimated at the two temperatures.

Similar effects of temperature upon 5-CT $E/[A]$ curves were observed. Again, lowering the temperature from 37°C to 25°C resulted in a significant leftward shift of the 5-CT $E/[A]$ curves ($\Delta p[A_{50}] = -0.29 \pm 0.05$; $p < 0.05$) and a small but significant increase in the maximum (max ratio = 1.19 ± 0.07; $p < 0.05$). As with 5-HT, there were significant rightward shifts with concomitant depression of maxima between $E/[A]$ curves constructed prior to and after incubation with BHC at 37°C and at 25°C (at 37°C $\Delta p[A_{50}] = 0.46 \pm 0.05$, max ratio = 0.34 ± 0.06; $p < 0.001$ and at 25°C $\Delta p[A_{50}] = 0.31 \pm 0.04$, max ratio = 0.54 ± 0.05; $p < 0.001$). Again the sensitisation to the effects of 5-CT was due principally to an increase in affinity (pK_A at 37°C = 7.70, pK_A at 25°C = 8.11; $p < 0.05$). However, in this case there were also small decreases in E_m and n at 25°C as compared to 37°C which just reached significance at the 5% level.

Table 1. Model fitting parameters for 5-HT and 5-CT in the rabbit saphenous vein (RbSV) and 5-HT in the rabbit aorta (RbA).

Agonist	Temperature	Parameter estimates $\pm$ s.e.m. (95% confidence limits)			
		pK_A	$\tau_{control}$	n	E_m
RbSV					
5-HT	37°C	7.12 ± 0.08	$2.81 (2.18-3.63)$	1.30 ± 0.05	93.2 ± 6.2
5-HT	25°C	7.48 ± 0.06	$1.84 (1.18-2.94)$	1.38 ± 0.16	82.1 ± 7.8
5-CT	37°C	7.70 ± 0.09	$1.45 (0.82-2.40)$	1.55 ± 0.22	134.9 ± 17.8
5-CT	25°C	8.11 ± 0.12	$1.88 (0.64-5.40)$	2.47 ± 0.43	121.1 ± 9.2
RbA					
5-HT	37°C	6.67 ± 0.04	$1.99 (1.71-2.32)$	2.76 ± 0.14	99.6 ± 4.3
5-HT	25°C	7.09 ± 0.06	$1.21 (1.09-1.33)$	3.68 ± 0.26	81.5 ± 6.4

Antagonist Effects at the 5-HT₁ like Receptor in the RbSV

Figure 2 (panel a) illustrates E/[A] curves to 5-HT in the absence and presence of flesinoxan at 37°C and 25°C. There were no significant differences between slopes and maxima of the E/[A] curves in the absence and presence of flesinoxan at either temperature ($p > 0.05$), and Schild plot slopes not significantly different from unity were obtained in

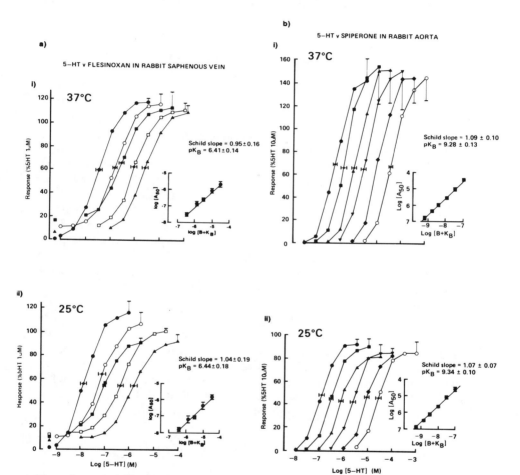

Figure 2 panel a) Illustrates 5-HT concentration effect curves constructed in RbSV in the absence (●) and presence of flesinoxan at 1 μM (○), 3 μM (■), 10 μM (□) & 30 μM (▲) after a 2 hour incubation period at i) 37°C (data are the averages of 4-7 E/[A] curves) and ii) 25°C (data are the averages of 3-6 E/[A] curves). These data are also displayed in the form of Clark plots (inset). Figure 2b) Illustrates 5-HT concentration effect curves constructed in RbA in the absence (●) and presence of spiperone at 1 nM (■), 3 nM (▲), 10 nM (▼), 30 nM (◆) and 0.1 μM (○) at i) 37°C (data are the averages of 4 E/[A] curves) and ii) 25°C (data are the averages of 4 E/[A] curves). These data are also displayed in the form of Clark plots (inset).

both cases ($p > 0.05$, slope $= 0.95 \pm 0.16$ at 37°C and 1.04 ± 0.19 at 25°C). pK_B values for flesinoxan of 6.41 ± 0.14 and 6.44 ± 0.03 were obtained at 37°C and 25°C respectively. Similar results were obtained when the effects of yohimbine were studied at 37°C and 25°C. There were no significant differences between slopes or maxima of the $E/[A]$ curves at either temperature and Schild plot slopes not significantly different from unity were obtained (slope $= 0.94 \pm 0.06$ at 37°C and 1.01 ± 0.11 at 25°C; $p > 0.05$). pK_B values for yohimbine of 6.79 ± 0.06 and 6.70 ± 0.15 were obtained at 37°C and 25°C respectively.

Agonist Effects at the 5-HT$_2$ Receptor in the RbA

Only effects of 5-HT were studied in this tissue. Figure 1 (panel a) illustrates paired $E/[A]$ curves to 5-HT at 37°C and 25°C. In contrast to the experiments performed using RbSV, decreasing the temperature from 37°C to 25°C elicited a profound relaxation of RbA ($-19.7 \pm 1.1\%$ cf KCl 80 mM). Nevertheless, there were significant leftward shifts of 5-HT $E/[A]$ curves ($\Delta p[A_{50}] = -0.25 \pm 0.02$; $p < 0.05$), accompanied, in this tissue, by a decrease in $E/[A]$ curve maxima (max ratio $= 0.70 \pm 0.03$; $p < 0.05$). There were significant rightward shifts with concomitant depression of maxima between $E/[A]$ curves constructed prior to and after incubation with PBZ at 37°C and at 25°C (at 37°C $\Delta p[A_{50}] = 0.52 \pm 0.04$, max ratio $= 0.19 \pm 0.02$; $p < 0.05$, and at 25°C $\Delta p[A_{50}] = 0.24 \pm 0.03$, max ratio $= 0.24 \pm 0.015$; $p < 0.05$). Estimates of agonist affinity and efficacy obtained using these data indicated that the sensitisation could be accounted for primarily by a change in affinity of the agonist for the receptor. As shown in table 1, lowering the temperature to 25°C increased affinity approximately 3-fold (pK_A at 37°C $= 6.67$, pK_A at 25°C $= 7.05$; $p < 0.05$). Regression analysis indicated that there were significant differences between all of the parameters estimated for 5-HT at 25°C as compared with 37°C. However, whereas the differences between the values of E_m and n only just reached significance at the 5% level, there were clear differences between ($p < 0.001$) pK_A and $\tau_{control}$ estimates at the two temperatures.

Antagonist Effects at the 5-HT$_2$ Receptor in the RbA

Figure 2 (panel b) illustrates $E/[A]$ curves to 5-HT in the absence and presence of spiperone. There were no significant differences between slopes and maxima of $E/[A]$ curves in the absence and presence of spiperone at 37°C and 25°C and Schild plot slopes not significantly different from unity were obtained (slopes $= 1.09 \pm 0.10$ and 1.07 ± 0.07 at 37°C and 25°C respectively). pK_B estimates of 9.28 ± 0.13 and 9.34 ± 0.10 were obtained for spiperone at 37°C and 25°C respectively.

Table 2. Equilibrium thermodynamic parameters for agonist and antagonist interactions at 5-HT$_1$-like and 5-HT$_2$ receptors in the rabbit saphenous vein (RbSV) and rabbit aorta (RbA) respectively. Values quoted are those for 37°C

| Ligand | Parameters ± s.e.m. | | |
	$\Delta G°$	$\Delta H°$	$\Delta S°$
RbSV			
5-HT	$-10.11 \pm 0.12(n = 12)$	$-10.41 \pm 3.15(n = 3)$	$-0.7 \pm 10.0(n = 3)$
5-CT	$-10.94 \pm 0.12(n = 10)$	$-13.06 \pm 5.17(n = 3)$	$-6.9 \pm 16.9(n = 3)$
flesinoxan	$-9.15 \pm 0.04(n = 4)$	$0.78 \pm 1.84(n = 4)$	$32.0 \pm 6.0(n = 4)$
yohimbine	$-9.76 \pm 0.05(n = 4)$	$4.07 \pm 1.86(n = 4)$	$44.3 \pm 6.3(n = 4)$
RbA			
5-HT	$-9.48 \pm 0.05(n = 4)$	$-9.78 \pm 2.73(n = 4)$	$-0.5 \pm 9.0(n = 4)$
spiperone	$-13.12 \pm 0.03(n = 4)$	$-2.89 \pm 0.56(n = 4)$	$33.0 \pm 1.8(n = 4)$

Thermodynamics

The results of the thermodynamic analyses are given in Table 2. These data indicate that agonist-receptor interactions were associated with decreases in enthalpy ($\Delta H < 0$) whilst antagonist-receptor interactions were associated with increases in entropy ($\Delta S > 0$).

Discussion

In this study we have examined the effects of hypothermia upon agonist and antagonist interactions at 5-HT$_2$ and 5-HT$_1$-like receptors mediating contraction of the rabbit aorta and rabbit saphenous vein respectively. Preliminary studies showed that contractile responses to agonists acting at both of these receptor types are potentiated by cooling therefore we sought to determine the basis for this sensitisation.

Direct model fitting of $E/[A]$ curves obtained at 37°C and 25°C before and after exposure to BHC showed that hypothermia-induced supersensitivity to 5-HT and to 5-CT at the 5-HT$_1$-like receptor apparently arose from a 2-3-fold increase in affinity of these agonists for the receptor, rather than an increase in the efficiency of occupancy-effect coupling. In agreement with this, lowering the temperature also failed to alter consistently the other tissue dependent variables, namely E_m and n. Although there were small decreases in both of these parameters at 25°C for 5-CT (which just reached significance at the 5% level), these differences were not evident when 5-HT was the agonist. In our view these discrepancies most probably derive from experimental variation, particularly since an increase in E_m rather than a decrease might be predicted if hypothermia induced supersensitivity to 5-HT agonists was elicited in part by a simple amplification. $E/[A]$ curves to 5-HT mediated via 5-HT$_2$ receptors in rabbit aorta were likewise shifted leftwards on cooling but in this tissue there was a concomitant depression of the curve maxima. Analysis of the curves obtained at 37°C and 25°C, before and after exposure to PBZ, indicated that again, this profile arose from an increase in affinity, but with a concomitant decrease in efficacy. Whereas, in RbSV cooling per se had little or no effect upon basal smooth muscle tone, in the RbA cooling from 37°C to 25°C elicited a quite profound relaxation and it is possible that this decrease in basal tone functionally antagonised contractile responses to 5-HT with the result that at 25°C the efficacy was significantly lower than that measured at 37°C. In principle, this apparent loss in efficacy could act to oppose the sensitisation caused by an increase in affinity. If this self-cancelling phenomenon is typical of 5-HT$_2$ receptor systems rather than a particular property of RbA, it might be argued that 5-HT$_1$-like receptors become relatively more important in mediating hypothermia-induced supersensitivity to 5-HT. Obviously, further studies are necessary to establish whether or not these findings reflect tissue or receptor system differences. Interestingly, the increases in affinity observed in the 5-HT receptor systems studied here concur with the observation of Janssens and Vanhoutte (1987) who found that hypothermia-induced supersensitivity to noradrenaline could be accounted for by an increased affinity of this hormone for its receptor.

Both flesinoxan and yohimbine, over the concentration ranges used, behaved as simple competitive antagonists at the 5-HT$_1$-like receptor in the rabbit saphenous vein at 25°C and at 37°C, indicating that these compounds had reached equilibrium within the incubation period. There was no difference between affinity estimates for these two antagonists at the two temperatures. Similarly, the affinity of the simple competitive 5-HT$_2$ receptor antagonist spiperone was unaltered by decreasing the temperature. A greater temperature dependence of agonist affinity than antagonist affinity has previously been noted for other systems eg beta-adrenoceptors (Weiland et al.. 1980) and dopamine D$_2$ receptors (Kilpatrick et al.. 1986). Calculation of the equilibrium thermodynamic parameters for the agonist- and antagonist-receptor interactions studied here indicated that the former were enthalpy driven ($\Delta H° < 0$) and the latter entropy driven ($\Delta S° > 0$). Since the affinity estimates were made at only two temperatures, the values of ΔH and ΔS given will be associated with relatively large errors. However, the nature of the energy changes observed here accord with data observed by other workers in various binding assays: Weiland et al. (1980) and Bree et al. (1986) found that antagonist binding to beta-adrenoceptors was largely entropy driven, whereas agonist binding was associated with a decrease in enthalpy although there was also a decrease in entropy. The only previously reported study carried out in isolated tissues examined the interaction between noradrenaline and the α_1-adrenoceptor in the rabbit aorta and found it to be enthalpy driven (Raffa et al. 1985).

Weiland et al. (1980) interpreted the large decreases in enthalpy and associated decreases in entropy observed for agonist-receptor interactions as a reflection of an agonist-induced conformational change in the receptor and demonstrated a good correlation between agonist efficacy and the enthalpy and entropy changes observed. These authors also suggested that antagonist-receptor interactions might be dependent upon hydrophobic bonding, the increases in entropy being associated with displacement of ordered water molecules around the binding site. However, Testa et al. (1987) have reported a good correlation between lipophilicity and observed entropy change for both agonists and antagonists at dopamine D_2 receptors. Similarly Bree et al. (1986) found a relationship between lipophilicity and entropy change for binding of agonists and antagonists at beta-adrenoceptors on rat lung membranes. Both of these groups suggested that the widespread observation that agonism is enthalpy driven and antagonism is entropy driven may be explained by the fact that most agonists tend to be hydrophilic whereas antagonists tend to be lipophilic. Clearly thermodynamic analysis gives an insight into the type of interaction occurring between ligand and receptor, however, there is controversy over whether thermodynamic parameters associated with the formation of drug-receptor complexes correlate with efficacy or whether they are simply a measure of lipophilicity. Since in this study the antagonists are more lipophilic than the agonists, it would be unwise to draw any definitive conclusions from the thermodynamic analyses carried out here until a number of other compounds with varying lipophilicities and efficacies have been studied. However, it is clear that the temperature dependence of agonist but not antagonist affinity estimates has implications in receptor classification schemes based upon such information.

The hypothermia-induced alterations in affinity of 5-HT for both 5-HT$_2$ and 5-HT$_1$-like receptors may also be significant in cold induced vasospasm, particularly since responses to 5-HT mediated via both of these receptor types synergise with those elicited by other platelet-derived vasospastic mediators such as thromboxone A$_2$ (MacLennan, unpublished data; Young et al. 1986). Secondary Raynaud's phenomenon (resulting from systemic sclerosis) is associated with platelet hyper-reactivity (Friedhoff et al. 1985; Biondi & Marasini 1989) and elevated plasma levels of both 5-HT and thromboxane A$_2$ (Reilly et al. 1986; Marasini et al. 1988). An increase in availability of the two hormones coupled with an increase in affinity of 5-HT for 5-HT$_1$-like and 5-HT$_2$ receptors mediating contraction suggests an attractive potential mechanistic basis for cold-induced vasospasm. In this regard it is interesting that clinical studies with ketanserin show that blockade of 5-HT$_2$ receptors affords only modest relief in secondary Raynaud's phenomenon, the benefit being not different in patients with idiopathic (primary) Raynaud's phenomenon in which platelet hyper-reactivity is less evident (see Coffman et al. 1990). In view of the long-standing implied role for 5-HT in this and associated conditions, a role for 5-HT$_1$-like receptors, acting perhaps in concert with 5-HT$_2$ and/or thromboxane A$_2$ receptors cannot be ruled out.

References

Arneklo-Nobin, B., Bodelsson, M., Nobin, A., Owman, C., and Tornebrandt, K. (1987). In Nobin, A., Owman, C., and Arneklo-Nobin, B. (eds.), Neuronal messengers in vascular function. Elsevier Science Publishers (Biomedical Division), pp. 91–103.

Biondi, M. L., and Marasini, B. (1989). Abnormal platelet aggregation in patients with Raynauds phenomenon. J. Clin. Pathol. 42: 716–718.

Black, J. W., and Leff, P. (1983). Operational models of pharmacological agonism. Proc. R. Soc. Lond. B. 220: 141–162.

Bree, F., El Tayar, N., Van De Waterbeemd, H., Testa, B., and Tillement, J-P. (1986). The binding of agonists and antagonists to rat lung β-adrenergic receptors as investigated by thermodynamic and structure-activity relationships. J. Receptor Res. 6: 381–409.

Coffman, J. D., Dormandy, J. A., Murray, G. D., Janssens, M. (1990). International study of ketanserin in Raynaud's phenomenon. In Paoletti, R., Vanhoutte, P. M., Brunello, N., and Maggi, F. M. (eds), Serotonin: from cell biology to pharmacology and therapeutics. Kluwer Academic Publishers, pp 429–433.

Francis, J. L., Roath, O. S., Challenor, V. F., and Waller, D. G. (1988). The effect of nisoldipine on whole blood platelet aggregation in patients with Raynauds phenomenon. Br. J. Clin. Pharmacol. 25: 751–754.

Friedhoff, L. T., Seibold, J. R., Kim, H. C., and Simester, K. S. (1984). Serotonin induced platelet aggregation in systemic sclerosis. Clin. Exp. Rheumatol. 2: 119–123.

Hooker, C. W., Calkins, P. J., and Fleisch, J. H. (1977). On the measurement of vascular and respiratory smooth muscle responses *in vitro*. Blood Vessels 14: 1–11.

Janssens, W. J., and Vanhoutte, P. M. (1978). Instantaneous changes of alpha-adrenoceptor affinity caused by moderate cooling in canine cutaneous veins. Am. J. Physiol. 234: H330–337.

Kilpatrick, G. J., El Tayar, N., Van De Waterbeemed, H., Jenner, P., Testa, B., and Marsden, C. D. (1986). The thermodynamics of agonist and antagonist binding to dopamine D-2 receptors. Mol. Pharmac. 30: 226–234.

Leff, P., Prentice, D. J., Giles, H., Martin, G. R., and Wood, J. (1990). Estimation of agonist affinity and efficacy by direct, operational model-fitting. J. Pharm. Meths. 23: 225–237.

Marasini, B., Biondi, M. L., Bianchi, E., Dell'Orto, P., and Agostoni, A. (1988). Ketanserin treatment and serotonin in patients with primary and secondary Raynaud's phenomenon. Eur. J. Clin. Pharmac. 35: 419–421.

Raffa, R. B., Aceto, J. F., and Tallarida, R. J. (1985). Measurement of thermodynamic parameters for norepinephrine contraction of isolated rabbit thoracic aorta. J. Pharmac. Exp. Ther. 235: 596–600.

Reilly, I. A. G., Roy, L., and Fitzgerald, G. A. (1986). Biosynthesis of thromboxane in patients with systemic sclerosis and Raynaud's phenomenon. B. M. J. 292: 1037–1039.

Stollak, J. S., and Furchgott, R. F. (1983). Use of selective antagonists for determining the type of receptors mediating the action of 5-HT and tryptamine in the isolated rabbit aorta. J. Pharmac. Exp. Ther. 224: 215–221.

Testa, B., Jenner, P., Kilpatrick, G. J., El Tayar. N., Van De Waterbeemd, H., and Marsden, C. D. (1987). Do thermodynamic studies provide information on both the binding to and the activation of dopaminergic and other receptors? Biochem. Pharmac. 36: 4041–4046.

Van Neuten, J. M., DE Ridder, W., and Vanhoutte, P. M. (1984). Ketanserin and vascular contractions in response to cooling. Eur. J. Pharmac. 99: 329–332.

Weiland, G. A., Minneman, K. P., and Molinoff, P. B. (1980). Thermodynamics of agonist and antagonist interactions with mammalian α-adrenergic receptors. Mol. Pharmac. 18: 341–347.

Young, M. S., Iwanov, V., and Moulds, R. F. W. (1986). Interaction between platelet-released serotonin and thromboxane A_2 on human digital arteries. Clin. Fxp. Pharm. Physiol. 13: 143–152.

Serotonin: Molecular Biology, Receptors and Functional Effects
ed. by J. R. Fozard/P. R. Saxena

Characterization of 5-HT$_3$-like Receptors in the Rat Cortex: Electrophysiological and Biochemical Studies

R. Y. Wang, C. R. Ashby, Jr., and E. Edwards

Department of Psychiatry and Behavioral Science, SUNY at Stony Brook, Stony Brook, NY 11794–8790, USA

Summary. Microiontophoresis of the 5-HT$_3$ receptor agonists, 2-methyl-5-HT and phenyl-biguanide (PBG), similar to the action of 5-HT, dose-dependently suppressed the firing of medial prefrontal cortical (mPFc) cells. There was little or no desensitization or tachyphylaxis to 5-HT or to the 5-HT$_3$ receptor agonists. The continuous iontophoresis of 1 M magnesium chloride markedly attenuated or blocked the suppressant effect produced by electrical stimulation of the ascending 5-HT pathway but did not alter the action of 2-methyl-5-HT, suggesting that the action of 2-methyl-5-HT is direct. The suppressant action of 2-methyl-5-HT on mPFc cells was blocked by selective 5-HT$_3$ receptor antagonists but not by other receptor antagonists. Furthermore, the S-enantiomer of zacopride was significantly more effective than the R-enantiomer in antagonising 2-methyl-5-HT indicating the steroselective antagonist action of zacopride.

In parallel to the electrophysiological studies, we have shown that both 2-methyl-5-HT and PBG mimicked the action of 5-HT in producing a significant increase in phosphoinositide (PI) turnover in fronto-cingulate and entorhinal cortical slices. The action of 2-methyl-5-HT and PBG on PI turnover was not altered by tetrodotoxin, suggesting that the action of these agonists is not dependent upon the impulse flow and that they do not alter PI turnover via interneurons. The stimulatory action of PBG and 2-methyl-5-HT was completely blocked by 5-HT$_3$ receptor antagonists but not by other receptor antagonists. Moreover, S-zacopride was more potent than R-zacopride in blocking 2-methyl-5-HT.

These results suggest that 5-HT$_3$ receptor agonists suppress the firing of mPFc neurons and stimulate PI turnover by a direct interaction with 5-HT$_3$-like receptors. Furthermore, our data suggest that the biochemical and electrophysiological properties of 5-HT$_3$-like receptors in the mPFc are different from those in the periphery.

Introduction

The 5-HT$_3$ receptor has been well defined in the periphery as a result of the development of selective 5-HT$_3$ receptor antagonists (Richardson et al. 1985). These receptors mediate the excitatory response of 5-HT on enteric and sympathetic neurons, neuroblastoma cell lines and primary cultures of central and peripheral neurons (Derkach et al. 1989; Ireland and Tyers 1987; Lambert et al. 1989; Neijt et al. 1988; Round and Wallis 1986; Yakel and Jackson 1988). 5-HT and 5-HT$_3$ receptor agonists induce a fast, transient, dose-dependent depolarization which is due to an increase in cation conductance and which desensitizes rapidly and can be blocked by 5-HT$_3$ receptor antagonists (see Andrade and Chaput 1991 for review).

Radioligand binding studies have shown that 5-HT$_3$ binding sites are present in a relatively high density in the mesocorticolimbic areas such as the nucleus accumbens, amygdala, entorhinal, cingulate and frontal cortices (Kilpatrick et al. 1987; Gehlert et al. 1990). Although the functional role of 5-HT$_3$ binding sites in the CNS is not entirely clear, numerous studies indicate that 5-HT$_3$ receptor antagonists might be effective in the treatment of emesis, migraine, anxiety, psychosis and symptoms of withdrawal from drug addition (see Tricklebank 1989 for review).

We have begun to identify and characterize 5-HT$_3$-like receptors in the rat brain, using combinations of electrophysiological and biochemical approaches. Our results suggest that 5-HT$_3$-like receptors in the mPFc are different from those reported in the periphery and in clonal cell lines. The neuronal responses mediated by 5-HT$_3$-like receptors in the mPFc have a slow onset and a relatively long duration of action, show little or no desensitization and might be coupled to PI hydrolysis (Ashby et al. 1991; Edwards et al. 1991).

Materials and Methods

Electrophysiological Studies

Male albino Sprague-Dawley rats (Taconic, Germantown, NY) were anesthetized with chloral hydrate (400 mg/kg, i.p.) and placed in a stereotaxic instrument. Standard extracellular single-unit recording and microiontophoretic techniques were used for recording from mPFc neurons (Ashby et al. 1990). The recording sites were primarily in the mPFc layers I–III, which contains a relatively high density of 5-HT$_3$ binding sites (Gehlert et al. 1990). A concentric electrode (NE-100, David Kopf Instruments) was implanted in the caudal linear raphe nucleus (CLi, anterior 2.2 mm to Lambada, lateral 0.0 mm to the midline and ventral 6.8 mm to the dura), a brain region through which the majority of the ascending 5-HT fibers pass (Azmitia and Segal 1978; Moore and Halais 1978), so that the effect produced by electrical stimulation of CLi on the mPFc cells could be studied systematically. Pulse trains were delivered at a rate of 15 Hz for 10 s with a 1 ms pulse width and an intensity of 0.1–1.0 mA. At the end of each experiment, the final site of the recording micropipette and the stimulation electrode was marked by ejection of Fast green dye and by passing a small DC current, respectively, and verified histologically. The procedures for assessment of drug-induced effect, construction dose-response curves for 5-HT and 5-HT$_3$ receptor agonists, and statistical analyses have been described in detail (Ashby et al. 1990; Hu and Wang 1988).

176

Biochemical Studies

Male Sprague-Dawley rats were sacrificed by decapitation and the fronto-cingulate and entorhinal cortices were rapidly dissected out bilaterally from brain slices and then pooled as described previously (Bannon et al. 1981; Edwards et al. 1991). Agonist-stimulated formation of ^{3}H-inositol phosphates was measured according to the methods of Berridge et al. (1982) with some modifications. The detailed procedures we used have been described (Edwards et al. 1991).

Results

Electrophysiological Studies

The 5-HT$_3$ receptor agonists, 2-methyl-5-HT and PBG, mimicked the action of 5-HT and produced a current (dose)-dependent suppression of the firing rate of spontaneously active mPFc neurons (Figure 1C). In addition, 2-methyl-5-HT was also effective in suppressing the firing of glutamate (GLU)-activated quiescent mPFc cells (Figure 1A). Note that the suppressant action of 2-methyl-5-HT has a relatively slow onset and a long duration with little or no desensitization; mPFc cells often

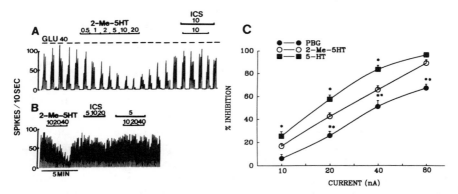

Figure 1. 2-Methyl-5-HT and PBG mimicked the inhibitory action of 5-HT and current (dose)-dependently suppressed the firing rate of mPFc cells. A and B are representative rate histograms showing that 2-methyl-5-HT dose-dependently suppresses the firing rate of both glutamate-activated (A) and spontaneously active (B) mPFc cells. Moreover, the inhibitory effect of 2-methyl-5-HT was blocked by the 5-HT$_3$ receptor antagonist ICS 205-930. The lines and numbers above each trace represent the duration of microiontophoretic application of drugs and currents in nanoamperes, respectively. Current (dose)-response curves for 2-methyl-5-HT, 5-HT and PBG to suppress the firing rate of mPFc cells is shown in C. The vertical bars represent SEM. *Values are significantly greater than those of 2-methyl-5-HT and PBG, $p < 0.01$; **Values are significantly less than those of 5-HT and 2-methyl-5-HT, $p < 0.01$, ANOVA plus Student Newman-Keuls (adapted from Ashby et al. 1991).

did not return to baseline firing until 2–3 min after the 2-methyl-5-HT or PBG application.

To determine whether the action of 2-methyl-5-HT on mPFc cell firing is direct, we examined the effect of iontophoretically applied magnesium chloride (Mg^{++}) on the suppressant action of 2-methyl-5-HT in the mPFc. The continuous iontophoresis (10–20 min) of 1M Mg^{++} markedly attenuated or blocked the suppressant effect produced by electrical stimulation of the ascending 5-HT pathway but did not alter the action of 2-methyl-5-HT (see Ashby et al. 1991). This result suggests that the action of 2-methyl-5-HT on mPFc cells is direct rather than indirectly mediated by releasing other neurotransmitters.

The suppressant action of 2-methyl-5-HT on mPFc cells was blocked effectively by a number of structurally diverse and seletive 5-HT$_3$ receptor antagonists, with a rank order of effectiveness as follows: ICS 205-930 = (±)zacopride > ganisetron = ondansetron = LY 278584 > MDL 72222 (Figure 2). (±)-Zacopride also antagonized the suppressant action of PBG, with an effectiveness similar to that observed against 2-methyl-5-HT. Furthermore, intravenous administration of (±)-zaco-pride ($0.95 \pm 0.17\ \mu g/kg$, $n = 7$) antagonized the action of 2-methyl-5-HT and PBG on mPFc cells.

To determine whether the action of 2-methyl-5-HT could be blocked by the enantiomers of zacopride in a stereospecific manner, we compare the action of microiontophoretically applied R- and S-enantiomers of

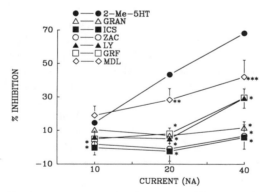

Figure 2. Current (dose)-response curves for the antagonist effects of various 5-HT$_3$ receptor antagonists against 2-methyl-5-HT (2-Me-5-HT). These curves were constructed by comparing the effects of 2-methyl-5-HT on mPFc cell firing in the presence of 10 nA of the 5-HT$_3$ receptor antagonists. GRAN, granisetron; ICS, ICS 205-930; ZAC, (±) zacopride; LY, LY 278584; GRF, ondansetron; MDL, MDL 72222. Standard error bars less than 6% are not shown. *Values are significantly different from that of 2-methyl-5-HT, $p < 0.01$. **The value is significantly different from that of 2-methyl-5-HT and those of all other antagonists, $p < 0.01$. ***The value is significantly different from that of 2-methyl-5-HT and those of GRAN, ICS 205-930 and ZAC, $p < 0.01$, ANOVA and Student-Newman-Keuls (from Ashby et al. 1991).

zacopride on the suppressant action of 2-methyl-5-HT. S-Zacopride appeared to be about 10 times more effective than the R-enantiomer in blocking 2-methyl-5-HT (Ashby et al. 1991), which is consistent with the results obtained from radioligand binding (Pinkus et al. 1990) and the PI turnover study (see below).

To test specificity of action of 2-methyl-5-HT, we investigated the effects of a number of other receptor antagonists to block the effects of 2-methyl-5-HT in mPFc cells. The microiontophoretic administration of metergoline, ($\pm$)-pindolol, SCH 23390, 1-sulpiride or SR 95103 all failed to block the action of 2-methyl-5-HT (see Ashby et al. 1991). The iontophoresis of metergoline also failed to block PBG.

Biochemical Studies

We have investigated the possible coupling of 5-HT_3 receptors to PI turnover in the rat fronto-cingulate and entorhinal cortices, two brain regions with relatively high density of this receptor subtype. First, we confirmed the previous finding (Conn and Sanders-Bush 1985) that 5-HT dose-dependently increases PI turnover, with an EC_{50} value of 0.5 μM and 0.3 μM for fronto-cingulate and entorhinal cortices, respectively. This effect could be blocked by granisetron, ondansetron and ICS 205-930 (Edwards et al. 1991), suggesting that 5-HT_3-like receptors in the fronto-cingulate and entorhinal cortices play an important role in mediating the action of 5-HT.

The 5-HT_3 receptor agonists, 2-methyl-5-HT and PBG, mimicked the action of 5-HT and dose-dependently increased PI turnover (46–76% of the response to 5-HT) in the rat fronto-cingulate and entorhinal cortices. Both 2-methyl-5-HT and PBG were without effect in the cerebellum, an area of the rat brain identified as having few or no 5-HT_3 binding sites (Kilpatrick et al. 1987). This indicates that 5-HT_3 receptor agonist-stimulated PI turnover is regionally specific. The stimulatory action of 2-methyl-5-HT was completely blocked by granisetron, ondansetron and ICS 205-930 but not by other receptor antagonists (Figure 3) such as ($\pm$)-pindolol, methysergide, ritanserin (see Figure 3), SR 95103, scopolamine (mACh), ($-$)-eticlopride (D2), SCH 23390 and parazosin (α_1) (see Edwards et al. 1991).

The effect of the zacopride enantiomers (Pinkus et al. 1990) on the PI stimulation produced by 1 μM 2-methyl-5-HT was also evaluated. (S)-Zacopride dose-dependently (1–100 nM) inhibited 2-methyl-5-HT-stimulated PI turnover whereas (R)-zacopride in the same concentrations was inactive (Edwards et al. 1991).

To determine whether the action of 5-HT and 5-HT_3 receptor agonists is dependent upon the propagation of action potentials along axons, the sodium channel blocker tetrodotoxin (1 μM) was applied. Tetrodotoxin

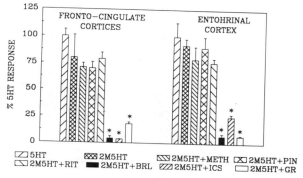

Figure 3. Representative bar graphs showing that 2-methyl-5-HT (2M5HT, 100 μM) stimulated phosphoinositide hydrolysis in the fronto-cingulate and entorhinal cortices was antagonized by 5-HT$_3$ receptor antagonists (10 μM) granisetron (BRL), ICS 205-930 (ICS) and ondansetron (GR) but not by other receptor antagonists (10 μM) such as methysergide (METH), pindolol (PIN) and ritanserin (RIT). The average increase of IP formation by 5-HT was $58 \pm 6.2\%$ above the baseline levels and is represented as 100% of 5-HT response. Data are from four separate experiments done in triplicate. Vertical bars represent S.E.M. *Values are significantly smaller than that of 2-methyl-5-HT ($p < 0.01$, Dunnett's t-test).

did not block the phosphoinositide response produced by 100 μM 5-HT, 2-Me-5HT and PBG, Nor did tetrodotoxin have any effect upon phosphoinositide turnover when tested alone (Edwards et al. 1991). The concentration of tetrodotoxin used in our experiments has been demonstrated to completely inhibit the phosphoinositide response induced by maximal concentration of veratrine (Conn and Sanders-Bush 1986). These results suggest that 5-HT, 2-methyl-5-HT and PBG-stimulated PI turnover is unlikely to be mediated by interneurons.

The effects of 5-HT, 2-methyl-5-HT and ($\pm$)-DOI [1-(2,5-dimethoxy-4-iodephenyl)-2-aminopropane, an agonist for 5-HT$_{1C}$ and 5-HT$_2$ receptors] on PI hydrolysis were compared because in rat cerebral cortical slices, 5-HT stimulated PI hydrolysis is thought to be mediated primarily through 5-HT$_2$ binding sites (Conn and Sanders-Bush 1985, 1986; Sanders-Bush and Conn 1986). DOI stimulated PI turnover in the fronto-cingulate and entorhinal cortices at 10 μM as previously reported for cerebral cortex (Pierce and Peroutka 1988). At 100 μM, DOI's effect was actually decreased. The response to DOI was inhibited by ketanserin and ritanserin (Edwards et al. 1991). 2-methyl-5-HT was approximately twice as effective as DOI in stimulating PI hydrolysis (Figure 4). When cortical tissue was incubated with 2-methyl-5-HT (100 μM) and DOI (10 μM), which respectively produced a maximal response, they produced an additive effect on PI turnover. The summation of the effect produced by 2-methyl-5-HT plus DOI on PI turnover is equivalent to the PI response produced by 5-HT (100 μM) itself. Interestingly, when 5-HT$_{1C}$ and 5-HT$_2$ receptors were blocked by ritanserin

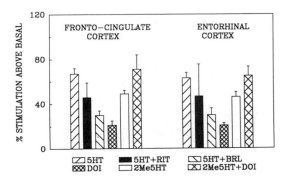

Figure 4. Comparison of the effects of 5-HT (100 μM), ritanserin (RIT, 10 μM) plus 5-HT, granisetron (BRL, 10 μM) plus 5-HT, DOI (10 μM), 2-methyl-5-HT (2Me5HT, 10 μM) and DOI plus 2-methyl-5-HT on PI turnover in the fronto-cingulate and entorhinal cortices. Data are generated from three separate experiments done in triplicate and are presented as per cent stimulation above baseline (from Edwards et al. 1991).

(10 μM), the effect of 5-HT on the PI turnover appears to be the same as the effect produced by 2-methyl-5-HT. Moreover, the effect of 5-HT on PI turnover was approximately equivalent to that of DOI when the 5-HT$_3$ receptors were blocked by ganisetron (10 μM).

Discussion

Our results show that the microiontophoresis of 5-HT, 2-methyl-5-HT and PBG produce a current (dose)-dependent suppression of the firing rate of mPFc cells in anesthetized rats. In contrast to the "fast and transient action" mediated by ligand gated ion channels as shown in the periphery (see Andrade and Chaput 1991 for review), the onset of the suppressant action of 2-methyl-5-HT and PBG in the mPFc was relatively slow. In addition, 2-methyl-5-HT and PBG produced a relatively prolonged suppression of baseline firing after cessation of microiontophoresis, and showed little or no desensitization or tachyphylaxis.

The suppressant action of 2-methyl-5-HT and PBG is blocked by various potent and selective 5-HT$_3$ receptor antagonists, but not by other neurotransmitter receptor antagonists. Moreover, the action of 2-methyl-5-HT on mPFc cells could be blocked in a stereospecific manner by the enantiomers of zacopride. These results suggest that the suppressant action of 2-methyl-5-HT and PBG in the mPFc is a result of their interaction at brain 5-HT$_3$ sites.

It has been shown that 2-methyl-5-HT (Blandina et al. 1989; Jiang et al. 1990) and PBG (Schmidt and Black 1989) release dopamine in rat brain. Thus, it is possible that both 2-methyl-5-HT and PBG could produce an inhibition of cell firing by releasing dopamine in the mPFc.

However, this appears unlikely since it has been reported that the iontophoresis of dopamine at high currents only produces a minimal suppression of neuronal firing in layers I–III of the mPFc (Bunney and Aghajanian 1976). Furthermore, we have demonstrated that the microiontophoresis of the relatively selective dopamine D1 receptor antagonist, SCH 23390, and the selective D2 receptor antagonist, 1-sulpiride, which have been shown previously to be effective in antagonizing the suppressant action of the D1 receptor agonist, SKF 38393, and the D2 receptor agonist, quinpirole, in the nucleus accumbens and caudate-putamen (Hu and Wang 1988; White and Wang 1986), fail to block the action of 2-methyl-5-HT. In addition, we have shown that the continuous microiontophoresis of 1 M Mg^{++}, which significantly attenuates neurotransmitter release as determined from the electrical stimulation of the ascending 5-HT pathway, does not alter the suppressant effect of 2-methyl-5-HT. Taken together, these results strongly suggest that the action of 2-methyl-5-HT is direct.

Although 2-methyl-5-HT has a relatively low affinity for cortical $5-HT_{1C}$ sites (Ismaiel et al. 1990), a result from a recent study suggests that 2-methyl-5-HT may interact with cortical $5-HT_{1C}$ sites (Barnes et al. 1989), which could potentially explain the observed suppressant action of 2-methyl-5-HT. However, we have shown that although microiontophoresis of the $5-HT_{1C}$, $5-HT_2$ receptor agonist DOI also suppresses the firing of mPFc cells, this effect can not be blocked by the $5-HT_3$ receptor antagonist granisetron (Ashby et al. 1990). Moreover, the microiontophoresis of metergoline, a potent $5-HT_{1C}$, $5-HT_2$ receptor antagonist, blocked the effect of DOI but not that of 2-methyl-5-HT (Ashby et al. 1990, 1991). These results indicate that 2-methyl-5-HT is probably not interacting with $5-HT_{1C}$ sites.

Complementary to the results of electrophysiological studies, we have shown that $5-HT_3$-like receptors in the fronto-cingulate and entorhinal cortices might be coupled to PI hydrolysis. Thus, both 2-methyl-5-HT and PBG mimicked the action of 5-HT and produced a dose-dependent increase in PI turnover. The effects of 2-methyl-5-HT and PBG on PI turnover were potently blocked by the selective $5-HT_3$ receptor antagonists granisetron, ICS 205-930 ondansetron and zacopride. Interestingly, as shown in our electrophysiological studies, the effects of 2-methyl-5-HT were inhibited stereoselectively by zacopride. Thus, only the active enantiomer S-zacopride (Pincus et al. 1990), but not the less active R-zacopride, blocked 2-methyl-5-HT-induced PI turnover. By contrast, other receptor antagonists such as methysergide (a $5-HT_1$, $5-HT_2$ receptor antagonist), pindolol (a $5-HT_{1A}$, $5-HT_{1B}$, β adrenoceptor antagonist), eticlopride (a dopamine D2 receptor antagonist), prazosin (an α_1 adrenoceptor antagonist), SCH 23390 (a $5-HT_{1C}$, $5-HT_2$, D1 receptor antagonist) scopolamine (a muscarinic ACh receptor antagonist) and SR 95103 (a $GABA_A$ receptor antagonist) did not block the action of 2-methyl-5-HT at a concentration of 10 μM in either the fronto-cingulate or the entorhinal cortical slices. These results indicate that the effects produced by 2-methyl-5-HT and PBG are most likely mediated by 5-HT$_3$-like receptors. It should be mentioned, however, that there was an apparent 20–100 fold difference between the potency of 5-HT$_3$ receptor antagonists at blocking 2-methyl-5-HT stimulated phosphoinositide hydrolysis and competing for 5-HT$_3$ binding sites (see Edwards et al. 1991). Further experiments (Schild plots) are needed to ascrtain the absolute potency of various 5-HT$_3$ antagonists on 2-methyl-5-HT induced PI turnover.

It has been repeatedly shown that 5-HT stimulated PI turnover is linked to the stimulation of 5-HT$_2$ receptor subtype in the rat cerebral cortex (Conn and Sanders-Bush 1985, 1986; Pierce and Peroutka 1988; Sanders-Bush and Conn 1986). In the present study, we have demonstrated that phosphoinositide hydrolysis is probably linked to 5-HT$_{1C}$, 5-HT$_2$ and 5-HT$_3$ receptors in the fronto-cingulate and entorhinal cortices. Thus, agonists for 5-HT$_{1C}$, 5-HT$_2$ and 5-HT$_3$ receptors increase PI turnover. However, 2-methyl-5-HT and PBG, two 5-HT$_3$ receptor agonists, are more efficacious than the 5-HT$_{1C}$, 5-HT$_2$ receptor agonist DOI in stimulating phosphoinositide hydrolysis. Moreover, when cortical tissue was incubated with maximally effective concentrations of 2-methyl-5-HT and DOI, they produced an additive effect on PI turnover. The summation of the effect produced by 2-methyl-5-HT plus DOI on PI turnover is similar to the PI response produced by 5-HT itself. These results suggest that in the rat fronto-cingulate and entorhinal cortices, 2-methyl-5-HT and DOI exert their actions through different 5-HT receptor subtypes and that the effect of 5-HT is mediated by 5-HT$_{1C}$, 5-HT$_2$ and 5-HT$_3$ receptors. Consistent with this view, following blockade of 5-HT$_{1C}$, 5-HT$_2$ receptors with ritanserin, the effects of 5-HT on the stimulation of PI turnover appears to be the same as the effect produced by 2-methyl-5-HT. The latter finding also suggests that 2-methyl-5-HT is a full 5-HT$_3$ receptor agonist at the phosphoinositide hydrolysis pathway.

It should be pointed out, however, that activation of a given receptor resulting in stimulation of phosphoinositide hydrolysis does not always mean that the receptor is directly linked to phosphoinositide hydrolysis because the latter could be the result of stimulated arachidonic acid metabolism, the release of another neurotransmitter (Conn and Sanders-Bush 1985, 1986) or the activation of phospholipase C due to an increase in calcium influx (Kendall and Nahorski 1984). In a series of experiments, Conn and Sanders-Bush (1986) have systematically ruled out these possibilities for 5-HT-stimulated phosphoinositide hydrolyis and concluded that phosphoinositide hydrolysis is the transducing mechanism of the 5-HT$_2$ receptor. We have examined the possibility that the 5-HT and 5-HT$_3$ receptor agonist stimulated phosphoinositide hydrolysis might be mediated indirectly by activating interneurons and a subsequent neurotransmitter release. Our results do not support this view. In our experiments, the sodium channel blocker tetrodotoxin which abolishes the propagation of action potentials along axons had no effect of 5-HT, 2-methyl-5-HT and PBG stimulation of phosphoinositide response. Although further studies need to be done, our present results tentatively suggest that phosphoinositide hydrolysis is also linked to the 5-HT_3 receptor and that 5-HT stimulated phosphoinositide hydrolysis is the result of its action on both 5-HT_2 and 5-HT_3 receptors.

In conclusion, our electrophysiological and biochemical studies provide several lines of evidence indicating that the pharmacological and physiological characteristics of 5-HT_3-like receptors in the fronto-cingulate cortices of the rat are different from those in the periphery. Thus, our results indicate that 2-methyl-5-HT and PBG produce a suppressant action on the firing rate of mPFc cells with slow onset, prolonged action and little or no desensitization, whereas 5-HT and 5-HT_3 receptor agonists elicit a depolarizing action on peripheral neurons with a fast onset and rapid desensitization. Furthermore, we have shown that 5-HT_3-like receptors in the fronto-cingulate and entorhinal cortices might be linked to PI hydrolysis, whereas in the periphery and neuronal cell lines, it is thought that 5-HT_3 receptors are ligand-gated ion channels. Finally, 5-HT and 2-methyl-5-HT increase acetylcholine release from neurons in the myenteric plexus, whereas 2-methyl-5-HT decrease acetylcholine release from rat cortical slices. In view of these distinct differences, we feel that the central 5-HT_3 sites should be referred to as "5-HT_3-like". Alternatively, the site we have characterized in the mPFc may represent a "new 5-HT receptor subtype". Obviously, additional studies are needed to further characterize the functional role of this binding site.

Acknowledgements

We thank Joyce Shepherd for typing this manuscript and Shu-Xin Tian for the photography. We also thank Beecham (granisetron), Eli Lilly (LY 278584), FarmItalia (metergoline), Glaxo (ondansetron), Merrell Dow (MDL 72222), Ravizza (1-sulpiride), Sanofi (SR 95103), Sandoz (ICS 205–930) and Schering (SCH 23390) and Wyeth-Ayerst (zacopride) for their donation of drugs. This research was supported by USPHS grants MH-41440 and MH-00378 to R.Y.W. and MH-09791 to C.R.A., Jr.

References

Andrade, R., and Chaput, Y. (1991). The electrophysiology of serotonin receptor subtypes. In Peroutka, S. J. (ed), Serotonin Receptor Subtypes: Basic and Clinical Aspects. New York: Wiley-Liss, pp. 103–124.

184

Ashby, Jr., C. R., Jiang, L. H., Kasser, R. J., and Wang, R. Y. (1990). Electrophysiological characterization of 5-hydroxytryptamine$_2$ receptors in the rat medial prefrontal cortex. J. Pharmacol. Exp. Ther. 251: 171–178.

Ashby, Jr., C. R., Minabe, Y., Edwards, E., and Wang, R. Y. (1991). 5-HT$_3$-like receptors in the rat medial prefrontal cortex: an electrophysiological study. Brain Res. (in press).

Azmitia, E. C., and Segal, M. (1978). An autoradiographic analysis of the differential ascending projections of the dorsal and median raphe nuclei in the rat. J. Comp. Neurol. 179: 641–668.

Bannon, M. J., Michaud, R. L., and Roth, R. H. (1981). Mesocortical dopamine neurons lack of autoreceptors modulating dopamine synthesis. Mol. Pharmac. 19: 270–275.

Barnes, J. M., Barnes, N. M., Costall, B., Naylor, R. J., and Tyers, M. B. (1989). 5-HT$_3$ receptors mediate the inhibition of acetylcholine release in cortical tissue. Nature 338: 498–499.

Berridge, M. J., Downes, C., and Hanley, M. R. (1982). Lithium amplifies agonists-dependent phosphatidylinositol responses in brain and salivary glands. Biochem. J. 206: 587–595.

Blandina, P., Goldfarb, J., Craddock-Royal, B., and Green, J. P. (1989). Release of endogenous dopamine by stimulation of 5-hydroxytryptamine$_3$ receptors in rat striatum. J. Pharmacol. Exp. Ther. 251: 803–809.

Bunney, B. S., and Aghajanian, G. K. (1976). Dopamine and norepinephrine innervated cells in the rat prefrontal cortex: pharmacological differentiation using microiontophoretic techniques. Life Sci. 19: 1783–1792.

Conn, P. J., and Sanders-Bush, E. (1985). Serotonin-stimulated phosphoinositide turnover: mediated by the S2 binding site in rat cerebral cortex in subcortical regions. J. Pharmacol. Exp. Ther. 234: 195–203.

Conn, P. J., and Sanders-Bush, E. (1986). Biochemical characterization of serotonin stimulated phosphoinositide turnover. Life Sci. 38: 663–669.

Derkach, V., Surprenant, A., and North, R. A. (1989). 5-HT$_3$ receptors and membrane ion channels. Nature 339: 706–709.

Edwards, E., Harkins, K., Ashby, Jr., C. R., and Wang, R. Y. (1991). The effect of 5-HT$_3$ receptor agonists on phosphoinositide hydrolysis in the rat fronto-cingulate and entorhinal cortices. J. Pharmacol. Exp. Ther. 256: 1025–1032.

Gehlert, D. R., Gackenheimer, S. L., Wong, D. T., and Robertson, D. W. (1990). Autoradiographic localization of 5-HT$_3$ receptors in the rat brain using ^{3}H-LY 278584. Neurosci. Abstr. 16: 1300.

Hu, X. T., and Wang, R. Y. (1988). Comparison of the effects of D-1 and D-2 dopaminergic receptor agonists on the neruons in the rat caudate-putamen: an electrophysiological study. J. Neurosci. 8: 4340–4348.

Ireland, S. J., and Tyers, M. B. (1987). Pharmacological characterization of 5-hydroxytryptamine-induced depolarization of the rat isolated vagus nerve. Br. J. Pharmacol. 90: 229–238.

Ismaiel, A. M., Titeler, M., Miller, K. J., Smith, T. S., and Glennon, R. A. (1990). 5-HT$_1$ and 5-HT$_2$ binding profile of the serotonergic agents α-methylserotonin and 2-methylserotonin. J. Med. Chem. 755–758.

Jiang, L. H., Ashby, Jr., C. R., Kasser, R. J., and Wang, R. Y. (1990). The effect of intraventricular administration of the 5-HT$_3$ receptor agonist 2-methylserotonin on the release of dopamine in the nucleus accumbens: an *in vivo* chronocoulometric study. Brain Res. 513: 156–160.

Kendall, D. A., and Nahorski, S. R. (1984). Inositol phospholipid hydrolysis in rat cerebral cortical slices: II. Calcium requirement. J. Neurochem. 42: 1388–1394.

Kilpatrick, G. J., Jones, B. J., and Tyers, M. B. (1987). Identification and distribution of 5-HT$_3$ receptors in rat brain using radioligand binding. Nature 330: 746–748.

Lambert, J. J., Peters, J. A., Habes, T. G., and Dempster, J. (1989). The properties of 5-HT$_3$ receptors in clonal cell lines studied by patch-clamp techniques. Br. J. Pharmacol. 97: 27–40.

Moore, R. Y., Halaris, A. E., and Jones, B. E. (1978). Serotonin neurons of the midbrain raphe: ascending projections. J. Comp. Neurol. 180: 417–438.

Neijt, H. C., TeDuits, I. J., and Vijuerberg, H. P. M. (1988). Pharmacological characterization of serotonin 5-HT$_3$-receptor mediated electrical response in cultured mouse neuroblastoma cells. Neuropharmacol. 27: 301–307.

Pierce, P. A., and Peroutka, S. J. (1988). Antagonism of 5-hydroxytryptamine$_2$ receptor mediated phosphatidyl inositol turnover by d-lysergic acid diethylamide. J. Pharm. Exp. Ther. 247: 918–925.

Pinkus, L. M., Sarbin, N. S., Gordon, J. C., and Munson, Jr., H. R. (1990). Antagonism of ^{3}H-zacopride binding to 5-HT$_3$ recognition sites by R- and S-enantiomers. Eur. J. Pharmacol. 179: 231–235.

Richardson, B. P., Engel, G., Donatsch, P., and Stadler, P. A. (1985). Identification of serotonin M-receptor subtypes and their specific blockade by a new class of drugs. Nature 316: 126–131.

Round, A., and Wallis, D. I. (1986). The depolarizing action of 5-hydroxytryptamine on rabbit vagal afferent and sympathetic neurons *in vitro* and its selective blockade by ICS 205-930. Br. J. Pharmacol. 88: 485–494.

Sanders-Bush, E., and Conn, P. J. (1986). Effector systems coupled to serotonin receptors in brain: Serotonin-stimulated phosphoinositide hydrolysis. Psychopharmacol. Bull. 22: 829–836.

Schmidt, C. J., and Black, C. K. (1989). The putative 5-HT$_3$ agonist phenylbiguanide induces carrier-mediated release of [^{3}H]-dopamine. Eur. J. Pharmacol. 167: 309–310.

Trickleband, M. D. (1989). Interactions between dopamine and 5-HT$_3$ receptors suggest new treatments for psychosis and drug addiction. TIPS 10: 127–129.

White, F. J., and Wang, R. Y. (1986). Electrophysiological evidence for the existence of D$_1$ and D$_2$ receptors in the rat nucleus accumbens. J. Neurosci. 6: 274–280.

Yakel, J. L., and Jackson, M. B. (1988). 5-HT$_3$ receptors mediate rapid responses in cultured hippocampus and a clonal cell line. Neuron. 1: 615–621.

Serotonin: Molecular Biology, Receptors and Functional Effects
ed. by J. R. Fozard/P. R. Saxena
© 1991 Birkhäuser Verlag Basel/Switzerland

Binding Characteristics of a Quaternary Amine Analog of Serotonin: 5-HTQ

R. A. Glennon[1], S. J. Peroutka[2], and M. Dukat[1]

[1]Department of Medicinal Chemistry, School of Pharmacy, Medical College of Virginia, Virginia Commonwealth University, Richmond, VA 23298-0540, USA; [2]Department of Neurology, Stanford University Medical Centre, Stanford, CA 94305-5235, USA

Summary. Because tryptamine derivatives typically bind at multiple populations of 5-HT receptors, they are not normally considered as templates for the design of site-selective serotonergic agents. Taking advantage of the proposal that 5-HT_3 receptors are ligand-gated ion channel receptors, and that such receptors usually accept positively charged ligands, we synthesized and evaluated the N,N,N-trimethyl quaternary amine analog of serotonin (5-HTQ). 5-HTQ binds at 5-HT_3 receptors with ten times the affinity of 5-HT and displays considerable selectivity relative to the other populations of 5-HT receptors. Although 5-HTQ may not readily penetrate the blood-brain barrier, limiting its application for *in vivo* investigations, it should be useful for *in vitro* studies.

Introduction

The two most widely used 5-HT_3 receptor agonists are serotonin (5-HT) and 2-methyl 5-HT. Obviously, 5-HT is a non-selective agonist. Although 2-methyl 5-HT is more selective than 5-HT, it possesses a lower affinity than 5-HT for 5-HT_3 receptors and is somewhat less potent than 5-HT as a 5-HT_3 receptor agonist (see Fozard 1990; Kilpatrick et al. 1990 for recent reviews). The goal of the present investigation was to develop a novel 5-HT_3 receptor selective agonist.

Results and Discussion

Conformational Considerations

In general, many ergolines bind with high affinity but with little selectivity at various populations of 5-HT receptors. However, ergolines typically display little to no affinity for 5-HT_3 receptors. For example, (+)-LSD and related ergolines bind at 5-HT_3 receptors with an affinity (i.e., Ki value) of $>10,000$ nM (Barnes et al. 1988; Peroutka and Hamik 1988). Although both 5-HT and LSD are tryptamine derivatives, LSD represents a conformationally restricted tryptamine and some insight might be gained from examining the relative conformations of these molecules. The distance between the aromatic (benzene) ring and the basic amine nitrogen atom in LSD is 5.2 Å. Because 5-HT is a conformationally flexible molecule, the corresponding distance in 5-HT can range from 5.3 to 6.4 Å (Glennon et al. 1990a). Thus, although 5-HT can mimic the ring-to-amine distance of the ergolines, unlike LSD it would also be capable of interacting at a receptor that requires a longer ring-to-amine distance. It might be concluded then that 5-HT interacts at 5-HT$_3$ receptors in a conformation that does not exactly mimic that defined by the ergolines. There are also several other possible explanations to account for the low affinity of the erogolines. For example, the 5-HT$_3$ receptor may not accommodate the added bulk of the ergolines that is not present in the simpler tryptamines such as 5-HT. This seems unlikely however because 8-hydroxy, N,N-dipropyl-amino tetralin (8-OH-DPAT) (an aminotetralin which is structurally simpler, yet conformationally similar, to LSD) also binds at 5-HT$_3$ receptors with low affinity (Peroutka and Hamik 1988). Nevertheless, as a working hypothesis, it was assumed that 5-HT interacts at 5-HT$_3$ receptors in a non-ergoline conformation. Because the exact conformation for binding is unknown, it was decided to restrict our initial investigations to conformationally flexible molecules.

Tryptamines as Templates

Tryptamines generally display little selectivity for the various types of 5-HT binding sites and are not typically considered to be good templates for the design of selective serotonergic agents. 5-HT, for example, binds at 5-HT$_{1A}$, 5-HT$_{1B}$, 5-HT$_{1C}$, 5-HT$_{1D}$, and [^{3}H]DOB-labeled 5-HT$_2$ receptors with high affinity (Ki = 2–10 nM). Nevertheless, with an understanding of the structure-affinity relationships (SAFIR) and structural requirements for each of the different types of 5-HT receptors, it should eventually be possible to design tryptamine analogs that display some degree of selectivity. Indeed, we have previously used this approach to develop tryptamine analogs that display selectivity for 5-HT$_{1A}$ versus 5-HT$_{1B}$ receptors (Glennon et al. 1988). More recently, we have designed and synthesized the 5-HT$_2$-selective tryptamine derivative, 5-methoxy-1-n-propyl-α-methyltryptamine (Glennon et al. 1990b). Thus, this type of approach should be successful if there is sufficient SAFIR information on each of the different populations of serotonin receptors.

"Extended" Tryptamine Analogs

Because the possibility exists that 5-HT$_3$ receptors may accommodate ligands with a greater ring-to-amine distance than that found in 5-HT, we examined two "extended" tryptamine analogs, or tryptamine analogs where an additional methylene group has been inserted in the alkyl side chain: homotryptamine [i.e., 3-(3-amino-n-propyl)indole] and N,N-dimethylhomotryptamine. Homotryptamine binds at 5-HT$_3$ receptors with low affinity (Ki > 2,000 nM). N,N-Dimethylhomotryptamine also binds with low affinity (Ki = 730 nM) but reveals that the two terminal amine methyl groups contribute to binding. Additional studies with these types of agents are in progress.

Quaternary Tryptamine Analogs

5-HT$_3$ receptors appear to be uniquely different from the other types of 5-HT receptors in that they are believed to represent ligand-gated ion channels. Such receptors often accommodate charged species. To date, essentially nothing is known regarding whether or not the 5-HT$_1$ and 5-HT$_2$ populations of 5-HT receptors will accommodate permanently charged (i.e., quaternary amine) molecules. What little is known, however, suggests that quaternary amines bind with low affinity. For example, we have previously shown that QDOB (i.e., the N,N,N-trimethyl quaternary amine analog of the 5-HT$_2$ receptor agonist DOB) binds with a significantly lower affinity (Ki = 8,250 nM) at 5-HT$_2$ receptors than does DOB itself (Ki = 0.8 nM) (Glennon et al. 1987). Because the mono- and di-methyl analogs of DOB bind with high to moderate affinity (Ki = 7.7 and 94 nM, respectively), it would seem that it is the positive charge of the quaternary analog that decreases the affinity of QDOB. More recently, we have also found that the quaternary amine methiodide derivative (Ki = 6,500 nM) of the 5-HT$_{1A}$ ligand 8-methoxy-2-(N-methyl-N-n-propylamino)tetralin (Ki = 5 nM) binds with very low affinity at 5-HT$_{1A}$ receptors (unpublished data). In the case of aminotetralins, small alkyl substituents actually increase affinity for 5-HT$_{1A}$ receptors (Naiman et al. 1989); thus, here too, the lower affinity of the quaternary amine analog is probably associated with the positive charge and not simply to the presence of alkyl substituents. In contrast, it has been demonstrated that 5-HT$_3$ receptors can accommodate quaternary amines. For example, the quaternary amine analogs of the 5-HT$_3$ antagonist ICS 205-930 binds at 5-HT$_3$ receptors with high affinity (Watling et al. 1988). Theoretically then, a quaternary amine analog of 5-HT should also bind at 5-HT$_3$ receptors. Furthermore, whereas introduction of small alkyl groups at the terminal amine (i.e., tertiary amine analogs) decreases the affinity of certain tryptamines for 5-HT$_{1B}$, 5-HT$_{1D}$ and 5-HT$_2$ receptors, such substituents may not be detrimental, and indeed may even enhance affinity, at 5-HT$_3$ receptors (for example, see the above discussions). Thus, our goal was to synthesize and evaluate a simple, conformationally flexible, quaternary amine

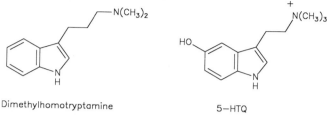

Figure 1. Structures of dimethylhomotryptamine and 5-HTQ.

tryptamine analog. The simplest such derivative of 5-HT is N,N,N-trimethyl 5-HT iodide or 5-HTQ (see Figure 1 for structure). This agent had been previously reported (Wieland et al. 1934) and was resynthesized from bufotenine by direct alkylation with methyl iodide.

5-HTQ binds at [^{3}H]LY 278584-labeled rat cortex 5-HT$_3$ receptors (Ki = 17 nM) with higher affinity than does 5-HT (Ki = 200 nM). Further, because N,N-dimethyl 5-HT (bufotenine; Ki = 80 nM) also binds with higher affinity than 5-HT, it appears that both the alkyl groups as well as the quaternary nature of 5-HTQ contribute to binding. A binding profile for 5-HTQ (Table 1) reveals that it is quite selective for 5-HT$_3$ receptors.

Table 1. Binding profile of 5-HTQ compared to 5-HT

Receptor	Tritiated ligand	Tissue	Ki (nM) value ($\pm$SEM) 5-HTQ		5-HT[a]
5-HT$_{1A}$	8-OH DPAT	Rat Cortex	200	($\pm$60)	2
5-HT$_{1B}$	5-HT	Rat Cortex	4,300	($\pm$200)	5
5-HT$_{1C}$	Mesulergine	Pig Cortex	>10,000[b]		10
5-HT$_{1D}$	5-HT	GP[c] Caudate	4,100	($\pm$1000)	3
5-HT$_2$	Ketanserin	Rat Cortex	5,300	($\pm$1000)	950
5-HT$_3$	LY 278584	Rat Cortex	17	($\pm$7)	200
α_1-Adrenergic	Raulwoscine	Rat Cortex	>10,000		—
α_2-Adrenergic	WB 4101	Rat Cortex	>10,000		—
β-Adrenergic	DHA	Rat Cortex	>10,000		—
Benzodiazepine	Flunitrazepam	Rat Cortex	>10,000		—

[a] Ki values for 5-HT, except at 5-HT$_3$ receptors, are from Ismaiel et al. (1990). Ki value for 5-HT at 5-HT$_3$ receptors was determined as for 5-HTQ.
[b] SEM not determined where Ki > 10,000.
[c] GP = guinea pig.

Conclusions

Taking advantage of the fact that ligand-gated ion channel receptors can accommodate quaternary amines, we prepared and evaluated an example of a quaternary amine analog of serotonin: 5-HTQ. 5-HTQ binds at 5-HT$_3$ receptors with about four times the affinity of its corresponding tertiary amine counterpart (i.e., bufotenine) and ten times the affinity of 5-HT. Unlike 5-HT and bufotenine, 5-HTQ is quite selective for the 5-HT$_3$ population of 5-HT receptors (Table 1). It should be noted however, that Wallis and Nash (1980) have previously reported that the quaternary amine analog of 5-HT can act at nicotinic receptors on rabbit cervical ganglion.

A literature search on 5-HTQ had been conducted prior to undertaking this project and little of recent relevance was discovered. Coincidentally, we subsequently discovered a paper by Richardson et al. (1985) in which N,N,N-trimethyl 5-HT is included in a table. Although there is no discussion of this agent in the body of the text and although no binding data are provided, it is shown that this agent is a 5-HT$_3$ receptor agonist with a potency comparable to that of 5-HT.

In conclusion, it appears (a) that 5-HT$_{1A}$, 5-HT$_{1B}$, 5-HT$_{1C}$, 5-HT$_{1D}$ and 5-HT$_2$ receptors do not readily accommodate quaternary amine analogs, (b) that tryptamines can be structurally modified so as to achieve some selectivity, (c) that terminal amine substitution of 5-HT enhances its affinity for 5-HT$_3$ receptors, and (d) that 5-HTQ, a quaternary amine analog of 5-HT, binds with high affinity and selectivity at 5-HT$_3$ receptors. It should be possible to utilize this approach to increase the 5-HT$_3$ affinity and selectivity of other non-selective 5-HT agonists.

Acknowledgements

This work was supported in part by PHS grant NS 23520 (RAG).

References

Barnes, N. M., Costall, B., and Naylor, R. J. (1988). [^{3}H]Zacopride: Ligand for the identification of 5-HT$_3$ recognition sites. J. Pharm. Pharmacol. 40: 548–551.

Fozard, J. R. (1990). 5-HT$_3$ Receptors. In: Paoletti, R., Vanhoutte, P. M., Brunello, N., and Maggi, F. M. (eds.), Serotonin: From Cell Biology to Pharmacology and Therapeutics. Dordrecht: Kluwer, pp. 331–338.

Glennon, R. A., Titeler, M., Seggel, M. R., and Lyon, R. A. (1987). N-Methyl derivatives of the 5-HT$_2$ agonist 1-(4-bromo-2,5-dimethoxyphenyl)-2-aminopropane. J. Med. Chem. 30: 930–932.

Glennon, R. A., Titeler, M., Lyon, R. A., and Slusher, M. (1988). N,N-di-*n*-propylserotonin: Binding at serotonin binding sites and a comparison with 8-hydroxy-2-(di-*n*-propylamino)tetralin. J. Med. Chem. 31: 867–870.

Glennon, R. A., Westkaemper, R. B., and Bartyzel, P. (1990a). Medicinal chemistry of serotonergic agents. In: Peroutka, S. J. (ed.), Serotonin Receptor Subtypes. New York: Wiley-Liss, in press.

Glennon, R. A., Chaurasia, C., Titeler, M., and Weisberg, E. L. (1990b). Development of a high-affinity tryptamine analog for 5-HT$_2$ sites. Second IUPHAR Satellite Meeting on Serotonin, Basel, Switzerland. July 11–13, 1990, Abs p. 50.

Ismaiel, A. M., Titeler, M., Miller, K. J., Smith, T. S., and Glennon, R. A. (1990). 5-HT1 and 5-HT2 Binding profiles of the serotonergic agents α-methylserotonin and 2-methylserotonin. J. Med. Chem. 33: 755–758.

Kilpatrick, G. J., Jones, B. J., and Tyres, M. B. (1990). Brain 5-HT$_3$ Receptors. In: Paoletti, R., Vanhoutte, P. M., Brunello, N., and Maggi, F. M. (eds.), Serotonin: From Cell Biology to Pharmacology and Therapeutics. Dordrecht: Kluwer, pp. 339–345.

Naiman, N. A., Lyon, R. A., Bullock, A. E., Rydelek, L. Y., Titeler, M., and Glennon, R. A. (1989). 2-Alkylaminotetralin derivatives: Interaction with 5-HT$_{1A}$ binding sites. J. Med. Chem. 32: 253–256.

Peroutka, S. J., and Hamik, A. (1988). [^{3}H]Quipazine labels 5-HT$_3$ recognition sites in rat cortical membranes. Eur. J. Pharmacol. 148: 297–299.

Richardson, B. P., Engel, G., Donatsch, P., and Stadler, P. A. (1985). Identification of 5-hydroxytryptamine M-receptor subtypes and their specific blockade by a new class of drugs. Nature 316: 126–131.

Wallis, D. I., and Nash, H. L. (1980). The action of methylated derivatives of 5-hydroxy-tryptamine at ganglionic receptors. Neuropharmacology 19: 465–472.

Watling, K. J., Aspley, S., Swain, C. J., and Saunders, J. (1988). [^{3}H]-Quaternized ICS 205–930 labels 5-HT$_3$ receptor binding sites in rat brain. Eur. J. Pharmacol. 149: 397–398.

Wieland, H., Konz, W., and Mittasch, H. (1934). Die Konstitution von Bufotenin und Bufotenidin. Justus Liebigs Ann. Chem. 513: 1–25.

Serotonin: Molecular Biology, Receptors and Functional Effects
ed. by J. R. Fozard/P. R. Saxena
© 1991 Birkhäuser Verlag Basel/Switzerland

5-HT$_1$-Like Receptors Unrelated to the Known Binding Sites?

D. van Heuven-Nolsen, C. M. Villalón*, M. O. den Boer, and
P. R. Saxena

*Department of Pharmacology, Faculty of Medicine and Health Sciences,
Erasmus University Rotterdam, Post Box 1738, 3000 DR Rotterdam, The Netherlands*

Summary. It is well recognized that 5-HT$_1$-like receptors are heterogeneous and since in the radioligand studies at least four different 5-HT$_1$ binding site subtypes (5-HT$_{1A}$, 5-HT$_{1B}$, 5-HT$_{1C}$ and 5-HT$_{1D}$) have been discriminated, four such receptor sites may well exist. Despite this multiplicity of 5-HT receptor subtypes, it appears that some 5-HT$_1$-like receptor-mediated functional responses, elicited by agonists such as 5-carboxamidotryptamine and blocked by methiothepin but not ketanserin, may not correspond to the above subtypes. These effects seem to include, among others, tachycardia in the cat, constriction of arteriovenous anastomoses in the pig and contraction of the rabbit isolated saphenous vein. In the cat heart the potency orders established for both the agonists (5-Carboxamidotryptamine > 5-HT $\gg$ BEA 1654 $\gg$ 8-OH-DPAT = RU 24969 = sumatriptan) and antagonists (methiothepin > methysergide > mesulergine $\gg$ metergoline $\gg$ mianserin > ketanserin > ($\pm$)propranolol) do not correlate with the order of affinity constants for any of the 5-HT$_1$ binding site subtypes. Similarly, there are differences between the 5-HT$_1$-like receptors involved in the constriction of the porcine arteriovenous anastomoses and the rabbit saphenous vein and the 5-HT$_{1A}$, 5-HT$_{1B}$, 5-HT$_{1C}$ and 5-HT$_{1D}$ binding sites.

Introduction

A functional response mediated by 5-hydroxytryptamine$_1$-like (5-HT$_1$-like) receptors is defined on the basis of the following criteria (Bradley et al. 1986): (i) susceptability to antagonism by non-selective 5-HT receptor antagonists, like methiothepin or methysergide, (ii) resistance to blockade by selective 5-HT$_2$ and 5-HT$_3$ receptor antagonists, such as ketanserin and 1αH,3α,5αH-tropan-3yl-3,5 dichlorobenzoate (MDL 72222), respectively and (iii) mimicry by selective agonists at 5-HT$_1$-like receptors, e.g. 5-carboxamidotryptamine. It is well recognized that 5-HT$_1$-like receptors are heterogeneous and, at present, four distinct subtypes of 5-HT$_1$ binding sites, namely 5-HT$_{1A}$, 5-HT$_{1B}$, 5-HT$_{1C}$ and 5-HT$_{1D}$, have been defined (Hoyer 1989). The affinity values of a number of drugs for 5-HT$_1$ binding subtypes in mammalian brain membranes are summarized in Table 1.

Pedigo et al. (1981) were the first to point out that 5-HT$_1$ binding sites can be subdivided into 5-HT$_{1A}$ and 5-HT$_{1B}$ subtypes according to

*Supported by a Science and Technology grant from the EEC.

a high and low affinity, respectively, for spiperone. 8-OH-DPAT was shown to display a high and selective affinity for the 5-HT_{1A} subtype (Middlemiss and Fozard 1983). Subsequently, 5-HT_{1C} binding sites, found in particularly high concentrations in the choroid plexus of rat, pig and man, were proposed on the basis of the displacement of [³H]5-HT by low concentration of mesulergine (Pazos et al. 1984; Hoyer et al. 1985, 1986). More recently, the 5-HT_{1D} sites were described in the bovine, porcine and human brain (Heuring and Peroutka 1987; Hoyer et al. 1988; Waeber et al. 1988a; b). Despite the multiplicity of 5-HT_1 binding sites, it appears that some functional responses mediated by 5-HT_1-like receptors may not correlate with the above subtypes of the 5-HT_1 binding site. This will be illustrated in this publication for the tachycardiac response in the cat, the constriction of the arteriovenous anastomoses in the pig and the contraction of the rabbit isolated saphenous vein by 5-HT.

5-HT_1-Like Receptors Mediating Tachycardia in the Cat

In 1960 Trendelenburg suggested that the 5-HT-induced cardiostimulation in the cat isolated atria, being unaffected by reserpine, dichloroisoprenaline or cocaine but blocked by lysergide, is mediated by the 5-HT "D" receptor, as classified by Gaddum and Picarelli (1957).

In the intact cat, 5-HT usually decreases heart rate via a chemoreceptor reflex (von Bezold-Jarisch reflex) mediated by 5-HT_3 receptors present on cardiac vagal afferents (Saxena et al. 1985). The cardioaccelerator response can be revealed in vivo after vagotomy (plus spinal section) or treatment with MDL 72222 (Saxena et al. 1985). This response is not altered by bilateral adrenalectomy, guanethidine, propranolol or burimamide, again suggesting that 5-HT acts directly on the atria (Saxena et al. 1985; Connor et al. 1986). The tachycardiac response to 5-HT in spinal cats is little affected by MDL 72222 or the 5-HT_2 receptor antagonists, ketanserin, ritanserin or cyproheptadine, although pizotifen and mianserin, two other 5-HT_2 receptor antagonists, showed some antagonism. Remarkably, methysergide and methiothepin, which have high affinity for both 5-HT_1 and 5-HT_2 binding sites, potentially antagonized the 5-HT-induced tachycardia, indicating mediation by 5-HT_1-like receptors. Moreover, selective agonists at 5-HT_1-like receptors, namely 5-carboxamidotryptamine and BEA 1654, mimicked 5-HT to support the notion that the tachycardiac response to 5-HT in spinal cats is mediated by 5-HT_1-like receptors (Saxena 1988; Saxena et al. 1985; Connor et al. 1986).

The activity of a number of 5-HT receptor agonists and antagonists on the cat heart have been arranged according to their affinities at the 5-HT_{1A}, 5-HT_{1B}, 5-HT_{1C} and 5-HT_{1D} binding sites (Figure 1). Admit-

194

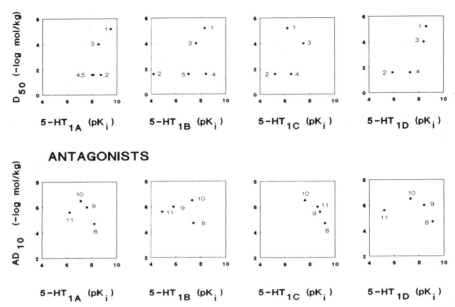

Figure 1. Comparison of the effects of agonists (D_{50}) and antagonists (AD_{10}) on the tachycardiac response to 5-HT in the cat with the binding affinities (pK_i) of the compounds to the different 5-HT_1 binding sites. The data are from Hoyer (1989), Saxena (1988) and Schoeffter and Hoyer (1989). The numbers adjacent to the individual data points correspond with the numbers of the different compounds in Table 1. D_{50}, Dose of the agonist required to increase heart rate by 50 beats/min; AD_{10}, Dose of antagonists required to raise D_{50} of 5-HT by a factor of 10.

tedly, one has to recognize the interference by pharmacokinetic factors in the intact organism; it is nevertheless quite clear that the function potencies of the different compounds do not correlate with the binding affinities at these sites. 8-OH-DPAT, with selective activity for the 5-HT_{1A} binding site, and RU 24969, with preferential affinity for both 5-HT_{1A} and 5-HT_{1B} sites (Table 1), had little activity. BEA 1654 was far less active than 5-HT in causing tachycardia while the affinity at the 5-HT_{1A} and 5-HT_{1B} is only 0.5 log unit less. Propranolol, which shows high affinity for the 5-HT_{1B} site and to a lesser extent the 5-HT_{1A} site (Table 1), was almost ineffective in antagonizing the 5-HT-induced tachycardia in the cat. These observations exclude the involvement of 5-HT_{1A} and 5-HT_{1B} receptors. Since mesulergine, which has a relatively high affinity for the 5-HT_{1C} site (Table 1), was relatively less active than methiothepin or methysergide as an antagonist and 5-carboxami-dotryptamine, which has a low affinity for the 5-HT_{1C} site (Table 1), was highly active as an agonist, mediation of tachycardia by 5-HT_{1C}

Table 1. Affinity values (pK_i) of some drugs for 5-HT_1 binding sites in mammalian brain membranes

	5-HT_{1A}	5-HT_{1B}	5-HT_{1C}	5-HT_{1D}
Agonists				
1. 5-Carboxamidotryptamine	9.5	8.3	6.2	8.6
2. 8-OH-DPAT	8.7	4.2	5.2	5.9
3. 5-HT	8.5	7.6	7.5	8.4
4. RU 24969	8.1	8.4	6.5	7.3
5. BEA 1654	8.0	7.0	ND	ND
6. Indorenate	7.8	5.4	6.5	6.7
7. Sumatriptan	6.1	6.4	4.1	7.5
Antagonists				
8. Metergoline	8.1	7.4	9.2	9.1
9. Methysergide	7.6	5.8	8.6	8.4
10. Methiothepin	7.1	7.3	7.6	7.3
11. Mesulergine	6.2	4.9	8.8	5.2
12. Propranolol	6.8	7.3	6.8	5.5
13. Ketanserin	5.9	5.7	7.0	6.0

BEA 1654, N-(3-acetylaminophenyl)piperazine; 8-OH-DPAT, 8-Hydroxy-2-(di-n-propy-lamino)tetralin; RU 24969,5-methoxy-3-(1,2,3,6-tetrahydro-4-pyridinyl)-1H-indole succinate. Based on data from Hoyer (1989 and personal communication) and Schoeffter and Hoyer (1989). ND = no data.

receptors can also be excluded. The contribution of 5-HT_{1D} receptors seems similarly highly unlikely. Despite having affinity for the 5-HT_{1D} site, 8-OH-DPAT, RU 24969 and, particularly, sumatriptan proved inactive or much less active (unpublished observation) than could be expected from their pK_i value. Lastly, it is noteworthy that metergoline, which has a higher affinity than methiothepin for all four 5-HT_1 binding site subtypes (Hoyer 1989), was only weakly active in antagonizing the tachycardiac response to 5-HT or 5-carboxamidotryptamine (Saxena 1988). Thus, the functional 5-HT_1-like receptors that mediate tachycardia in the cat seem to be unrelated to the four 5-HT_1 binding site subtypes so far delineated in brain membranes.

5-HT_1-Like Receptors Mediating Constriction of the Rabbit Isolated Saphenous Vein

5-HT contracts ring preparations of the rabbit saphenous vein (Martin and McLennan 1990; Van Heuven-Nolsen et al. 1990). This effect is resistant to blockade by the selective 5-HT_3 receptor antagonist MDL 72222. Although the response is blocked by ketanserin and spiperone, the antagonism is weaker than would be expected from their affinities for the 5-HT_2 binding site. Results with agonists suggest mediation by a 5-HT_1-like receptor; the effects of 5-HT were mimicked with high potency by 5-carboxamidotryptamine and other 5-HT_1-like agonists,

196

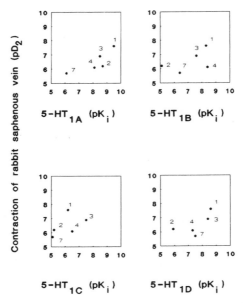

Figure 2. Comparison of the effects of 5-HT$_1$-like receptor agonists on the contraction of the rabbit isolated saphenous vein (pD$_2$) with the binding affinities (pK$_i$) of the agonists to the different 5-HT$_1$ binding sites. The data are from Hoyer (1989), Schoeffter and Hoyer (1989) and Van Heuven-Nolsen et al. (1991). The numbers adjacent to the individual data points correspond with the numbers of the different compounds in Table 1.

like 8-OH-DPAT, RU 24969 and sumatriptan. 5-HT$_1$-like receptor activation is further supported by the fact that the contractile effect of 5-HT is potently antagonized by methiothepin.

Since all the agonist drugs that elicit contraction (Figure 2) have affinity for the 5-HT$_{1A}$ binding site (Table 1), the involvement of 5-HT$_{1A}$ receptors may be argued. However, this seems unlikely since the response to 5-HT is not blocked by cyanopindolol and propranolol, and is only little affected by spiperone; all these antagonists display high affinity for the 5-HT$_{1A}$ binding site and the first two also for the 5-HT$_{1B}$ site (Table 1). Since, in addition, 5-carboxamidotryptamine, 8-OH-DPAT and RU 24969 do not have high affinity for the 5-HT$_{1C}$ binding site, the 5-HT$_1$-like receptor mediating contraction in the rabbit isolated saphenous vein does not appear to be related to the 5-HT$_{1A}$, 5-HT$_{1B}$ or 5-HT$_{1C}$ subtype. The contribution of the 5-HT$_{1D}$ receptors cannot be completely ruled out for 8-OH-DPAT, RU 24969 and, particularly, sumatriptan have affinities for this subtype (Table 1). However, as pointed out below, these compounds do have effects on receptors that do not seem to correlate with 5-HT$_1$ binding subtypes.

5-HT₁-Like Receptors Mediating Constriction of Arteriovenous Anastomoses in the Pig

It has been demonstrated that intracarotid administration of 5-HT in the anesthetized pig decreases common carotid blood flow by reducing the arteriovenous anastomotic blood flow, whereas the arteriolar blood flow increases, especially to the skin and ears (Saxena and Verdouw 1982). As shown in Table 2, the response to 5-HT in the arteriovenous anastomoses is not antagonized by the 5-HT_3 receptor antagonist MDL 72222 nor by the 5-HT_2 receptor antagonists, cyproheptadine, ketanserin and WAL 1307 (Verdouw et al. 1984; Saxena et al. 1986). The constriction of the arteriovenous anastomoses seems to involve a 5-HT_1-like receptor since the effect is mimicked by 5-carboxamidotryptamine (Saxena and Verdouw 1985), BEA 1654 (Verdouw et al. 1985), methysergide (Saxena and Verdouw 1984), 8-OH-DPAT (Bom et al. 1989a), RU 24969 (Bom et al. 1989b), indorenate (Villalón et al. 1990) and sumatriptan (Den Boer et al. 1991), all of which are partial or full agonists at 5-HT_1-like receptors. Methiothepin almost completely abolishes the reduction of arteriovenous anastomotic blood flow by these drugs. However, it needs to be mentioned that, unlike 5-HT and 5-carboxamidotryptamine, the other agonists only slightly increase the arteriolar blood flow. This indicates that even within the carotid circulation the 5-HT_1-like receptors appear to be heterogeneous.

The high affinity of 5-carboxamidotryptamine, 8-OH-DPAT, RU 24969, BEA 1654, indorenate and the moderate affinity of sumatriptan

Table 2. 5-HT_1-like receptors and drug effects on arterioles and arteriovenous anastomoses in the porcine carotid artery bed

Drug type	Arterioles	AVAs	Antagonism by	Resistance to
5-HT	− − − −	+ + + +	Methiothepin	Cyproheptadine, Ketanserin, WAL 1307, MDL 72222
5-CT	− − − −	+ + + +	—	Cyproheptadine
Methysergide	−	+ + +	—	—
BEA 1654	− −	+ + +	—	Ketanserin
8-OH-DPAT	−	+ + + +	Methiothepin	Ketanserin
Ipsapirone	0	0	—	—
Indorenate	−	+ + + +	Methiothepin	Ketanserin, Metergoline
RU 24969	−	+ + + +	Methiothepin	Ketanserin, (±)Pindolol
Sumatriptan	0/−	+ + +	Methiothepin	Ketanserin

AVAs, arteriovenous anastomoses; −, dilatation; +, contraction; 0, no effect. The number of − and + indicate the magnitude of effect. Based on Saxena and Villalón (1990).

for the 5-HT_{1A} site appears to suggest that the 5-HT_{1A} receptor may mediate the reduction of the arteriovenous anastomotic blood flow by 5-HT. However, the selective 5-HT_{1A} receptor agonist ipsapirone (Peroutka 1986) is inactive in the pig carotid circulation (Bom et al. 1988) and the reduction in arteriovenous anastomotic blood flow by RU 24969, a 5-HT_{1A} and 5-HT_{1B} receptor agonist (Table 1), cannot be blocked by ($\pm$)-pindolol (Bom et al. 1989b), an antagonist at both these receptors (Hoyer 1989). Therefore, neither 5-HT_{1A} nor 5-HT_{1B} receptors seem to be involved. The involvement of 5-HT_{1C} receptors is also unlikely because of the very high potency of 5-carboxamidotryptamine (Saxena and Verdouw 1985). Compared to the above 5-HT_1-like receptor subtypes, the 5-HT_{1D} receptor would seem a more likely candidate, since sumatriptan, methysergide, RU 24969, 8-OH-DPAT and indorenate all bind with a reasonable affinity to the 5-HT_{1D} receptor (see Hoyer 1989). On the other hand, the sumatriptan-induced contraction of the dog saphenous vein (Humphrey et al. 1988) and the indorenate-induced reduction in the porcine carotid arteriovenous anastomotic blood flow (Villalón et al. 1990) are both resistant to metergoline, which has a high affinity for 5-HT_{1A}, 5-HT_{1B}, 5-HT_{1C} and 5-HT_{1D} receptors (Hoyer 1989) and has been shown to antagonize a putative 5-HT_{1D} receptor-mediated effect, namely the endothelium-dependent relaxation of the pig coronary artery by 5-HT (Schoeffter and Hoyer 1990). Though in another putative functional model for 5-HT_{1D} receptors, inhibition of adenylate cyclase production in calf substantia nigra, metergoline behaved almost as a full agonist (Schoeffter et al. 1988), no agonist effect of metergoline was observed in the cranial arteriovenous anastomoses (Villalón et al. 1990). Taken together, the evidence to-date indicates that the 5-HT_1-like receptor mediating the reduction of arteriovenous anastomotic blood flow in the pig does not fully correspond to any of the brain subtypes characterized so far. Rather the receptor would seem to be the same as that characterized in the dog saphenous vein and intracerebral vessels in a variety of species (Perren et al. 1991).

Concluding Remarks

The pharmacology of the 5-HT_1-like receptors is still a research area with many enigmas to be resolved. Differences between species, brain binding sites and peripheral functional receptors complicate the discussion. In the present communication we argue that at least the receptor(s) mediating reduction in arteriovenous anastomotic blood flow in the pig and the tachycardia in the cat do not seem to be related to the 5-HT_{1A}, 5-HT_{1B}, 5-HT_{1C} or 5-HT_{1D} subtype.

References

Bom, A. H., Saxena, P. R., and Verdouw, P. D. (1988). Further characterization of the 5-HT$_1$-like receptors in the carotid circulation of the pig. Br. J. Pharmacol. 94: 327P.

Bom, A. H., Verdouw, P. D., and Saxena, P. R. (1989a). Carotid hemodynamics in pigs during infusions of 8-OH-DPAT: reduction in arteriovenous shunting is mediated by 5-HT$_1$-like receptors. Br. J. Pharmacol. 96: 125–132.

Bom, A. H., Villalón, C. M., Verdouw, P. D., and Saxena, P. R. (1989b). The 5-HT$_1$-like receptor mediating reduction in porcine carotid arteriovenous shunting by RU 24969 is not related to either 5-HT$_{1A}$ or the 5-HT$_{1B}$ subtype. Eur. J. Pharmacol. 171: 87–96.

Bradley, P. B., Engel, G., Feniuk, W., Fozard, J. R., Humphrey, P. P. A., Middlemiss, D. N., Mylecharane, E. J., Richardson, B. P., and Saxena, P. R. (1986). Proposals for the classification and nomenclature of functional receptors for 5-hydroxytryptamine. Neuropharmacol. 25: 563–576.

Connor, H. E., Feniuk, W., Humphrey, P. P. A., and Perren, M. J. (1986). 5-Carboxamidotryptamine is a selective agonist at 5-hydroxytryptamine receptors mediating vasodilation and tachycardia in anaesthetized cats. Br. J. Pharmacol. 87: 417–426.

Den Boer, M. O., Villalón, C. M., Heiligers, J. P. C., Humphrey, P. P. A., and Saxena, P. R. (1991). Role of 5-HT$_1$-like receptors in the reduction of porcine cranial arteriovenous anastomotic shunting by sumatriptan. Br. J. Pharmacol. 102: 323–330.

Gaddum, J. H., and Picarelli, Z. P. (1957). Two kinds of tryptamine receptors. Br. J. Pharmacol. 12: 323–328.

Heuring, R. E., and Peroutka, S. J. (1987). Characterization of a novel ^{3}H-5-hydroxytryptamine binding site subtype in bovine brain membranes. J. Neurosci. 7: 894–903.

Hoyer, D. (1989). 5-Hydroxytryptamine receptors and effector coupling mechanisms in peripheral tissues. In Fozard, J. R. (ed.), The peripheral actions of 5-hydroxytryptamine, Oxford: Oxford University Press, pp. 72–99.

Hoyer, D., Engel, G., and Kalkman, H. O. (1985). Molecular pharmacology of 5-HT$_1$ and 5-HT$_2$ recognition sites in rat and pig brain membranes: radioligand binding studies with ^{3}H 5-HT, ^{3}H 8-OH-DPAT, (-)^{125}I iodocyanopindolol, ^{3}H mesulergine and ^{3}H ketanserin. Eur. J. Pharmacol. 118: 13–23.

Hoyer, D., Pazos, A., Probst, A., and Palacios, J. M. (1986). Serotonin receptors in the human brain II. Characterization and autoradiographic localization of 5-HT$_{1C}$ and 5-HT$_2$ recognition sites. Brain Res. 376: 97–107.

Hoyer, D., Waeber, C., Pazos, A., Probst, A., and Palacios, J. M. (1988). Identification of a 5-HT$_1$ recognition site in human brain membranes different from 5-HT$_{1A}$, 5-HT$_{1B}$ and 5-HT$_{1C}$ sites. Neurosci. 85: 357–362.

Humphrey, P. P. A., Feniuk, W., Perren, M. J., Connor, H. E., Oxford, A. W., Coates, I. H., and Butina, D. (1988). GR43175, a selective agonist for the 5-HT$_1$-like receptor in dog isolated saphenous vein. Br. J. Pharmacol. 94: 1123–1132.

Martin, G. R., and MacLennan, S. J. (1990). Analysis of the 5-HT receptor in rabbit saphenous vein exemplifies the problems of using exclusion criteria for receptor classification. Naunyn-Schm. Arch. Pharmacol. 342: 111–119.

Middlemiss, D. N., and Fozard, J. R. (1983). 8-Hydroxy-2-(di-n-propylamino)-tetralin discriminates between subtypes of the 5-HT$_1$ recognition site. Eur. J. Pharmacol. 90: 151–153.

Pazos, A., Hoyer, D., and Palacios, J. M. (1984). The binding of serotonergic ligands at the porcine choroid plexus: characterization of a new type of serotonin recognition site. Eur. J. Pharmacol. 106: 539–546.

Pedigo, N. W., Yamamura, H. I., and Nelson, D. L. (1981). Discrimination of multiple ^{3}H-hydroxytryptamine-binding sites by the neuroleptic spiperone in rat brain. J. Neurochem. 36: 220–226.

Peroutka, S. J. (1986). Pharmacological differentiation and characterization of 5-HT$_{1A}$, 5-HT$_{1B}$ and 5-HT$_{1C}$ binding sites in rat frontal cortex. J. Neurochem. 47: 529–540.

Perren, M. J., Feniuk, W., and Humphrey, P. P. A. (1991). Vascular 5-HT$_1$-like receptors which mediate contraction of the dog is isolated saphenous vein and carotid arterial vasoconstriction in anaesthetised dogs are not of the 5-HT$_{1A}$ or 5-HT$_{1D}$ subtype. Br. J. Pharmacol. In press.

Saxena, P. R. (1988). Further characterization of 5-hydroxytryptamine$_1$-like receptors mediating tachycardia in the cat: no apparent relationship to known subtypes of the 5-hydroxy-tryptamine$_1$ binding site. Drug Dev. Res. 13: 245–258.

Saxena, P. R., and Verdouw, P. D. (1982). Redistribution by 5-hydroxytryptamine of carotid arterial blood at the expense of arteriovenous blood flow. J. Physiol. (Lond.) 332: 501–520.

Saxena, P. R., and Verdouw, P. D. (1984). Effects of methysergide and 5-hydroxytryptamine on carotid blood flow distribution in pigs: Further evidence for the presence of atypical 5-HT receptors. Br. J. Pharmacol. 82: 817–826.

Saxena, P. R., and Verdouw, P. D. (1985). 5-Carboxamide tryptamine, a compound with high affinity for 5-hydroxytryptamine$_1$-like binding sites, dilates arterioles and constricts arteri-ovenous anastomoses. Br. J. Pharmacol. 84: 533–544.

Saxena, P. R., and Villalón, C. M. (1990). Cardiovascular effects of serotonin agonists and antagonists. J. Cardiovasc. Pharmacol. 15 (suppl. 7): S17–S35.

Saxena, P. R., Duncker, D. J., Bom, A. H., Heiligers, J., and Verdouw, P. D. (1986). Effects of MDL 72222 and methiothepin on carotid vascular responses to 5-hydroxytryptamine in the pig: Evidence for the presence of "5-hydroxytryptamine$_1$-like" receptors. Naunyn-Schm. Arch. Pharmacol. 333: 198–204.

Saxena, P. R., Mylecharane, E. J., and Heiligers, J. (1985). Analysis of the heart rate effects of 5-hydroxytryptamine in the cat; mediation of tachycardia by 5-HT$_1$-like receptors. Naunyn-Schm. Arch. Pharmacol. 330: 121–129.

Schoeffter, P., and Hoyer, D. (1998). How selective is GR 43175? Interactions with functional 5-HT$_{1A}$, 5-HT$_{1B}$, 5-HT$_{1C}$ and 5-HT$_{1D}$ receptors. Naunyn-Schm. Arch. Pharmacol. 340: 135–138.

Schoeffter, P., and Hoyer, D. (1990). 5-Hydroxytryptamine (5-HT)-induced endothelium-de-pendent relaxation of pig coronary arteries is mediated by 5-HT receptors similar to the 5-HT$_{1D}$ receptor subtype. J. Pharmacol. Exp. Ther. 252: 387–395.

Schoeffter, P., Waeber, C., Palacios, J. M., and Hoyer, D. (1988). The 5-hydroxytryptamine 5-HT$_{1D}$ receptor subtype is negatively coupled to adenylate cyclase in calf substantia nigra. Naunyn-Schm. Arch. Pharmacol. 337: 602–608.

Trendelenburg, U. (1960). The action of histamine and 5-hydroxytryptamine on isolated mammalian atria. J. Pharmacol. Exp. Ther. 130: 450–460.

Van Heuven-Nolsen, D., Tysse Klasen, T. H. M., Luo, Q., and Saxena, P. R. (1990). 5-HT$_1$-like receptors mediate contractions in the rabbit saphenous vein. Eur. J. Pharmacol. 191: 375–382.

Verdouw, P. D., Jennewein, H. M., Heiligers, J., Duncker, D. J., and Saxena, P. R. (1984). Redistribution of carotid artery blood flow by 5-HT: effects of the 5-HT$_2$ receptor antagonists ketanserin and WAL 1307. Eur. J. Pharmacol. 102: 499–509.

Verdouw, P. D., Jennewein, H. M., Mierau, J., and Saxena, P. R. (1985). N-(3-acety-laminophenyl) piperazine hydrochloride (BEA 1654), a putative 5-HT$_1$ agonist, causes constriction of arteriovenous anastomoses and dilatation of arterioles. Eur. J. Pharmacol. 107: 337–346.

Villalón, C. M., Bom, A. H., Heiligers, J., Den Boer, M. O., and Saxena, P. R. (1990). Constriction of porcine arteriovenous anastomoses by indorenate is unrelated to 5-HT$_{1A}$, 5-HT$_{1B}$, 5-HT$_{1C}$ or 5-HT$_{1D}$ receptor subtype. Eur. J. Pharmacol. 190: 167–176.

Waeber, C., Dietl, M. M., Hoyer, D., Probst, A., and Palacios, J. M. (1988a). Visualization of a novel serotonin recognition site (5-HT$_{1D}$) in the human brain by autoradiography. Neurosci. Lett. 88: 11–16.

Waeber, C., Schoeffter, D., Palacios, J. M., and Hoyer, D. (1988b). Molecular pharmacology of 5-HT$_{1D}$ recognition sites: radioligand binding studies in human, pig and calf brain membranes. Naunyn-Schm. Arch Pharmacol. 337: 595–601.

Part II

Peripheral and Central Pharmacology

Serotonin: Molecular Biology, Receptors and Functional Effects
ed. by J. R. Fozard/P. R. Saxena
© 1991 Birkhäuser Verlag Basel/Switzerland

The Electrophysiology of 5-HT

D. I. Wallis and P. Elliott

Department of Physiology, University of Wales College of Cardiff, PO Box 902, Cardiff CF1 1SS, UK

Summary. A number of 5-HT actions where the mechanism is one of opening an ion channel, resulting in a fast depolarization, a slow depolarization or a hyperpolarization are compared with actions where the mechanism is a closure of ion channels, resulting in a slow depolarization, altered transmitter release or altered membrane accommodation. The relationship of these mechanisms to a particular receptor subtype and to a particular second messenger system is likely to become apparent in the near future. At present, the extent to which differentiation of an electrophysiological action of 5-HT in terms of mechanism is predictive of involvement of a particular 5-HT receptor is unclear. Evidence from the nervous system suggests that a particular mechanism is usually associated with a certain sub-type of 5-HT receptor and, thus, classification by function provides a useful initial framework provided it does not become a dogma.

Introduction

In this chapter, we have not attempted a comprehensive review of the effects of 5-HT observed with electrophysiological techniques, because this would have resulted in a book, and a large one at that. Rather we have selected certain observations which we think are important in understanding the neurobiology of 5-HT and taken the opportunity of placing some of our own work in the context of mechanisms rather than of receptor pharmacology.

One of the earliest reported electrophysiological effects of 5-HT was described by A. S. Paintal (1954). He showed that certain vagal sensory nerve fibres could be provoked to discharge by 5-HT or phenyl biguanide, a selective 5-HT$_3$ receptor agonist. The substances were injected via an aortic cannula, whose tip lay just upstream of the coeliac artery, in the anaesthetized cat. The vagus nerve was teased out so that recordings could be made from single units excited by an identified physiological stimulus. Gastric stretch receptors were noted to be particularly sensitive to phenyl biguanide, whereas other visceral receptors were not excited. However, other types of vagal visceral receptor, apart from stretch receptors, were also shown to be sensitive to 5-HT or phenyl biguanide.

Possibly the first intracellular observations of postsynaptic responses to 5-HT were those of Gerschenfeld and Paupardin-Tritsch (1974a; b).

In two papers in the Journal of Physiology, they showed that neurones in the buccal ganglion of *Aplysia* responded to 5-HT in a complex manner; these neurones receive a serotoninergic input from two giant cerebral neurones. Six different responses to 5-HT applied by iontophoresis were observed in *Aplysia* and *Helix* buccal neurones. These six responses not only displayed differences in pharmacology, but were generated by differing ionic mechanisms. The latter involved opening of sodium, potassium or chloride channels; alternatively, closure of potassium or sodium/potassium-selective channels. Fast depolarizations generated by 5-HT and opening of sodium channels desensitized rapidly and were blocked by (+)tubocurarine, LSD, 7-methyltryptamine and bufotenine. Bufotenine, but not the other three agents, also blocked slow depolarizations generated by a more slowly activated sodium conductance increase. No antagonist was found for the depolarization generated by a closure of potassium channels. As will become apparent below, it is scarcely possible at present to relate these molluscan neurone responses, at least pharmacologically, to responses recorded from mammalian neurones. Possibly molecular biology will allow us to bridge this divide.

A current difficulty in understanding the actions of 5-HT in the nervous system is relating electrophysiological responses to a 5-HT receptor classification scheme. Receptor classification, although based to some extent on functional studies of non-nervous tissues, is grounded in binding data. A binding site may not have a functional correlate, of course, or the functional correlate may be a biochemical event not associated with any electrical event at the cell membrane. In neither case will anything be recordable by the electrophysiologist.

Channel Opening and Channel Closing

One useful way of considering the action of a neurotransmitter like 5-HT is to separate potential changes generated by the opening of ion channels – the classical mode of action (Figure 1, upper panel) – from potential changes resulting from channel closure and the reduction of a current normally flowing across the cell membrane (lower panel). Either mechanism may generate a depolarization, but whereas channel opening does so by permitting Na ion flux into the cell, channel closure usually elicits a depolarization by closure of K channels and, thus, removing an existing outward current. Channel opening may be the consequence of a direct interaction of the 5-HT molecule and the channel protein to evoke a conformation change. Channel closure is usually thought to be an indirect action, the G protein system being involved in the change in channel state.

CHANNEL OPENING

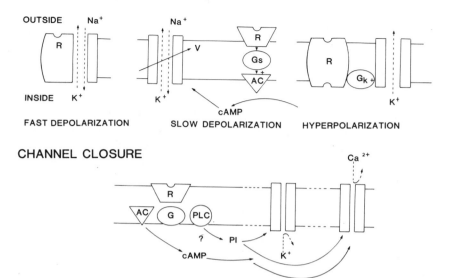

Figure 1. Schematic diagram illustrating the ionic basis of 5-HT actions. Association of 5-HT with a receptor opening an ion channel may lead to a rapid depolarization, a slow depolarization or a hyperpolarization, depending on the permeability characteristics of the channel. The slow depolarization is generated indirectly via cyclic AMP which increases a non-selective cationic conductance termed I_h. Association of 5-HT with a receptor closing an ion channel is shown as indirect action, mediated via a G protein and/or generation of phosphoinositide (PI) or cyclic AMP. Generation of the intracellular second messenger may result in closure of K channels or voltage-sensitive Ca channels.

Channel Opening

5-HT₃ Receptor-Mediated Population Responses

An interest over many years in the 5-HT receptors of mammalian sympathetic ganglia makes these tissues a convenient initial example of channel opening mechanisms. Wallis and Woodward (1974) observed that the facilitation of transmission through the rabbit superior cervical ganglion, induced by a conditioning stimulus to the preganglionic nerve, was much enhanced by 5-HT. They recorded the postganglionic compound action potential from a ganglion in a simple *in vitro* superfusion system. The facilitation of this response was less than 10% under control conditions, i.e. the endogenous facilitation was less than 10%. Facilitation was increased to about 40% by 5-HT ($1-30\ \mu$M) in a concentration-related manner. It was realised later that some of these effects were due to a direct depolarizing action of 5-HT on the ganglion cells. The depolarization arising in a population of cells was recorded using a technique known as the sucrose-gap method. This technique had been developed with Hans Kosterlitz and Gordon Lees in Aberdeen (Wallis et al. 1975).

In some of our first studies of 5-HT depolarizations, it emerged that injections of a bolus of 5-HT (0.2 μmol in saline) evoked a depolarization dependent upon Na^+ (Wallis and Woodward 1975). The restoration of ionic equilibrium resulted in expulsion by the Na pump of Na^+ ions, and the creation of a hyperpolarization following the depolarization. This hyperpolarization was blocked by ouabain, as would be expected for a potential generated by activity of the pump. Thus, secondary responses which are hyperpolarizing can be generated by 5-HT. In our version of the sucrose-gap apparatus, the ganglion was suspended across a series of chambers perfused with different solutions. The recording of the surface potential difference is facilitated by reducing the extracellular shunt of potential between two regions. This is achieved by superfusing an intervening region with a solution of sucrose in de-ionized water and, thus, of high electrical resistance. The result is that depolarizations evoked by 5-HT in one region of the tissue are short-circuited to a much smaller degree over the remainder of the tissue. A large signal is recorded. The chambers in which the ganglion is suspended are slender perspex units held together in a clamp. The sucrose chamber has a rubber membrane around it in which holes of the appropriate size have been punched to accommodate the post-ganglionic nerve from the ganglion. These holes provide an effective seal from the neighbouring chambers containing saline.

The nodose ganglion of the vagus contains the cell bodies of a large population of visceral afferents. This can be used in the same apparatus (Wallis et al. 1982), although the punched holes need to be slightly larger. Dose-related depolarizations to 5-HT can be evoked from both the rabbit superior cervical and rabbit nodose ganglia. Repeated dose-response curves can be determined and the effect of antagonists examined quantitatively. Surmountable antagonism is displayed by a number of selective 5-HT$_3$ receptor antagonists at least at low concentrations. Equilibrium of an antagonist, such as MDL 72222, with the tissue is apparently slow. The apparent pA_2 values obtained for MDL 72222, metoclopramide, quipazine and ICS 205-930, ranging from 7.1 to 10.4, on sympathetic neurones and on vagal primary afferents were found to be similar in the two tissues (Round and Wallis 1987).

Population responses recorded from axons of the preganglionic cervical sympathetic nerve (Elliott and Wallis 1988) and from primary afferent axons of the vagus nerve (Elliott and Wallis 1990a; Elliott et al. 1990) have also proved useful in studying the quantitative pharmacology of the 5-HT$_3$ receptors present on their surfaces. In these studies, we used the grease-gap technique described by Marsh et al. (1987) to provide the high resistance between test and reference areas on the tissue surface. The potencies of the selective 5-HT$_3$ receptor antagonists, MDL 72222, ICS 205-930 and SDZ 206-830 were determined on the sympathetic axons (Elliott and Wallis 1988), while the antagonist effects of BRL 43694 and metoclopramide on the vagus nerve were compared (Elliott et al. 1990).

5-HT$_3$ Receptor-Mediated Responses in the Single Cell

The depolarizations recorded extracellularly of population responses typically extend over many minutes. Intracellular recording is required for insight into what is happening at the level of the single cell. The depolarizations evoked by 5-HT in a single superior cervical ganglion cell of the rabbit (Wallis and North 1978) showed many of the characteristics typical of 5-HT$_3$ responses. 5-HT was applied iontophoretically and evoked a rapid, phasic depolarization. Brief injections of hyperpolarizing current through the intracellular electrode evoked electrotonic potentials whose amplitude was proportional to the resistance of the cell membrane. The diminution in these potentials during the 5-HT depolarization indicated that channels had opened, which had decreased the resistance of the membrane. The 5-HT responses were reduced by superfusion with 5-HT itself at a concentration of 1–50 μM (Wallis and North 1978), thus exemplifying the characteristic tachyphylaxis displayed by 5-HT$_3$ receptor mediated responses.

The channel opened by 5-HT was shown by ionic substitution studies to be one which permits passage of both Na and K ions. Consistent with this view, the reversal potential for the action of 5-HT on rabbit scg cells was -24 mV (Wallis and North 1978). A very similar value was obtained by Wallis and Dun (1987) for the action of 5-HT on guinea-pig sympathetic neurones in the coeliac ganglion. From these measurements one can calculate the relative permeability to Na and K of the channels opened by 5-HT using the method of Takeuchi and Takeuchi (1960; see also Higashi and Nishi 1982). The Na/K permeability ratio is estimated to be close to 1.0. Some recent data from other types of peripheral neurone, and also neuroblastoma cells, are shown in Table 1. Patch clamp techniques were used to estimate the conductance of the channel opened by 5-HT – the 5-HT$_3$ receptor channel. The Na/K permeability ratio has usually been found to be greater than 2, indicating that the channels are at least twice as permeable to Na as to K. Typically the Hill slopes for the interaction are nearer 2 than 1.0. The characteristics of the single channels opened by 5-HT are still being actively pursued by a number of laboratories; some of the principal groups are shown in Table 1. There are indications that there may be 3 different channels linked to the 5-HT$_3$ receptor or 3 different conductance states of the channel. Three different unitary conductances of 0.3,

Table 1. Characteristics of the 5-HT$_3$ receptor ion channel, as measured in voltage-clamp and patch-clamp (whole cell configuration or outside-out) studies

Tissue	Na/K permeability ratio	Hill slope	Unitary conductance (pS)
Guinea-pig submucous plexus	2.1–2.7[1]	2.3–2.4[1]	9 [1,2] 15 – 17
Rabbit nodose	2.3[3]	1.6–1.7[3]	17[4,7]
Neuroblastoma cells NCB – 20 hamster NIE – 115 mouse	2.4[5]	2.8[6]	0.3[7]

[1]Surprenant (1990); [2]Derkach et al. (1989); [3]Higashi and Nishi (1982); [4]Malone et al. (1990); [5]Neijt et al. (1989); [6]Neijt et al. (1988); [7]Peters, J. A. (personal communication)

9 and about 17 picoSiemens have been measured. These may occur in different tissues or species. Alan North and his colleagues have shown that 5-HT$_3$ channel activity can be evoked up to 5 hours after excision of an outside-out patch, suggesting that neither G proteins nor a cytoplasmic messenger are necessary for the action of 5-HT at 5-HT$_3$ receptors (Derkach et al. 1989; see also Lambert et al. 1989).

Functional 5-HT$_3$ Receptor-Mediated Responses

Functionally, 5-HT$_3$ receptor-mediated responses fall into 3 categories, all mediated via neurones. (a) Phasic depolarizations have been recorded from a variety of peripheral neurones, such as sympathetic ganglion cells, parasympathetic ganglion cells, vagal neurones and enteric neurones. 5-HT$_3$ receptor-mediated depolarizations do not as yet appear to have been recorded from CNS neurones *in situ*; however, such depolarizations dependent upon a fast inward current have been seen in about 5–10% of mouse cultured hippocampal or striatal neurones (Yakel et al. 1988). (b) Electrophysiological recording has also identified the excitation of different sorts of sensory endings by 5-HT. This action is selective; certain types of sensory ending are sensitive to 5-HT, while others are not. Thus, for example, carotid body chemoreceptors, heart epicardial/ventricular endings, pulmonary J (deflation) receptors and human skin nociceptors are excited by the amine (for references see McQueen 1990). (c) Whether the 5-HT$_3$-triggered release of transmitter from nerve endings, where this occurs, is also through membrane depolarization has still to be firmly established, but both central and peripheral terminals appear to bear receptors of this type. Noradrenaline release from sympathetic endings in the heart (Fozard and Mwaluko 1976), acetylcholine release from bronchi (Aas 1983) and dopamine release in the rat striatum (Blandina et al. 1990) evoked by

5-HT$_3$ receptor activation have been described. It has yet to be established whether the inhibition of acetylcholine release in rat entorhinal cortex mediated by 5-HT$_3$ receptors (Barnes et al. 1989) is a direct action or an indirect action via release of an inhibitory intermediary.

Augmentation of Rectifier Currents

Another mechanism of depolarization dependent upon channel opening (Figure 1) is seen in rat nucleus prepositus hypoglossal neurones. Bobker and Williams (1989) showed that 5-HT acts to augment a time- and voltage-dependent inward current, I$_h$, which is activated by hyperpolarization. I$_h$ has a reversal potential of about -32 mV. The ion channels involved admit Na and K ions and are activated at membrane potentials more negative than -50 mV. These 5-HT responses are slow, do not desensitize and may be mediated by generation of cAMP via a 5-HT-sensitive adenyl cyclase (Figure 1). The response to 5-HT is mimicked by 5-CT ($0.3-10\ \mu$M), but not by 8-OH-DPAT ($1\ \mu$M). The only antagonist that blocked the effect of 5-HT was spiperone ($1-5\ \mu$M), although caesium could block I$_h$. I$_h$ is found in many types of neurones and may regulate the spontaneous firing rate of the cell. Pape and McCormick (1989) have demonstrated a similar response in lateral geniculate (thalamic) neurones of guinea-pig and cat. Enhancement by 5-HT of the mixed Na$^+$/K$^+$ current reduced the ability of thalamic neurones to generate rhythmic bursts of firing and promoted excitability. This 5-HT response was blocked by caesium and also by methysergide, but not by ritanserin ($5\ \mu$M). 8-OH-DPAT did not mimic the response.

Although a similar inward rectifier current (I$_{IR}$) has been observed in spinal motoneurones of $2-3$ day old rats using the whole-cell patch clamp mode of recording, and this current was enhanced by 5-HT (Takahashi and Berger 1990), this does not appear to be the mechanism underlying the 5-HT depolarization of motoneurones seen at later stages of development (see below). The depolarizing response to 5-HT identified by Takahashi and Berger was blocked by spiperone ($5\ \mu$M) and mimicked by 8-OH-DPAT ($10\ \mu$M).

Opening of K Channels

A final example of a channel opening mechanism is where membrane hyperpolarization is generated (Figure 1). Expression of 5-HT$_{1A}$ receptor action in hippocampal pyramidal neurones is via opening of K channels, which are also voltage-sensitive as shown by Colino and Halliwell (1987). The response probably involves a pertussis-toxin sensitive G protein, which may directly couple the receptor to the K channel (Andrade et al. 1986; see Figure 1). The voltage responses and currents generated under voltage-clamp in their experiments were blocked by spiperone ($5 \mu M$), while responses to baclofen which activates GABA receptors coupled to the same channels were not blocked by spiperone. Since 5-HT$_{1A}$ receptors are involved in this response (Colino and Halliwell 1987), these papers provide important evidence that this 5-HT$_{1A}$ receptor mechanism, at least, does not involve mediation by a change in the level of cAMP, does not involve a rise in intracellular Ca^{2+} nor inositol phospholipid turnover, but does involve a G protein (Andrade et al. 1986).

Channel Closure

Depolarization of Motoneurones

5-HT responses generated by channel closure (Figure 1) are typically slow and display a substantial latency; they probably involve coupling via a G protein system. 5-HT is thought to excite several types of neurone by increasing the level of an intracellular second messenger responsible for closure of K channels. These kind of responses may be expected to be linked to either 5-HT$_1$ or 5-HT$_2$ receptors.

An example of this kind of response on which we have worked for a number of years is the action of 5-HT on spinal motoneurones. A section of hemisected lumbar spinal cord, isolated from a 7–12 day-old rat, is set up for rapid superfusion with oxygenated Krebs solution. Suction electrodes are used to stimulate a ventral root and the segmental dorsal root. Motoneurones (MNs) impaled with intracellular electrodes usually responded to 5-HT with a slow depolarization (Figures 2, 4). The amplitude of the depolarisation was related to the concentration of 5-HT superfused, but uptake of 5-HT, presumably neuronal uptake, considerably affected the sensitivity to the amine (Elliott and Wallis 1990b). When uptake was blocked with citalopram, there was an approximately 10-fold increase in sensitivity to 5-HT.

The depolarization evoked by 5-HT sometimes showed 2 phases, the secondary phase of MN depolarization being associated with intense action potential discharge by the cell (Figure 2). A regular finding was that the spontaneous synaptic activity impinging on a cell was suppressed by 5-HT through a second, unrelated 5-HT action. The 5-HT$_2$ receptor antagonist, ritanserin, had rather selective effects on the 5-HT actions on the cell illustrated in Figure 2. At a concentration of 0.1 μM ritanserin, the discharge at peak depolarization was decreased, although the depolarization itself was little affected. At a 10 times higher concentration, the depolarization was blocked. However, the suppression of

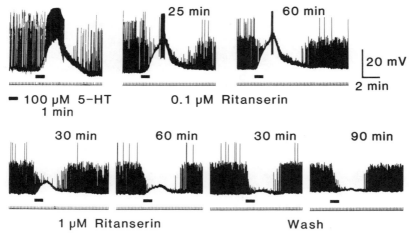

25 min   60 min

20 mV

2 min

■ 100 μM 5-HT   0.1 μM Ritanserin
1 min

30 min   60 min   30 min   90 min

1 μM Ritanserin   Wash

Figure 2. Intracellular records using a KCl-filled electrode from a lumbar motoneurone of neonate rat showing the intense background activity and the depolarization evoked by superfusion of the cord with 5-HT. Note that in this cell the depolarization evoked by 5-HT involved an initial phase and a second phase accompanied by intense firing; the background activity (spontaneous psps) was inhibited by 5-HT. Ritanserin (0.1 μM) reduced the firing evoked by 5-HT without affecting the depolarization. A higher concentration of ritanserin (1 μM) reduced the depolarization but did not prevent the inhibitory action of 5-HT on the background activity. The antagonist could not apparently be washed from the tissue within 90 minutes.

spontaneous activity by 5-HT was not affected by ritanserin (Figure 2). This suppression may be related to the inhibition by 5-HT of reflex responses evoked from a ventral root.

Figure 3 illustrates the complex actions of 5-HT on the responses generated in a particular lumbar motoneurone. The top row of traces shows the complex synaptic potential evoked by dorsal root stimulation with several spikes being discharged by the MN. These responses are probably elicited by monosynaptic and polysynaptic pathways. 5-HT greatly inhibited the synaptic potential. At the same time, 5-HT was able to increase the excitability of the motoneurone itself (Figure 3B). When the ventral root was stimulated under control conditions, no spike invaded the cell soma. In the presence of 5-HT, invasion of the soma occurred and a full soma-dendritic action potential was generated. This increase in excitability induced by 5-HT was also seen to current injection through the impaling electrode (Figure 3C). Passage of a current of 0.2 nAmps across the cell membrane elicited spike discharge in the presence of 5-HT compared to no spike under control conditions; a current of 1 nAmp evoked a single spike under control conditions, but repetitive spiking in the presence of 5-HT.

The nature of the ion channel affected by 5-HT is indicated by the change in membrane responses to hyperpolarizing current pulses (Figure 4). The upper traces show the experimental protocol and the slow

212

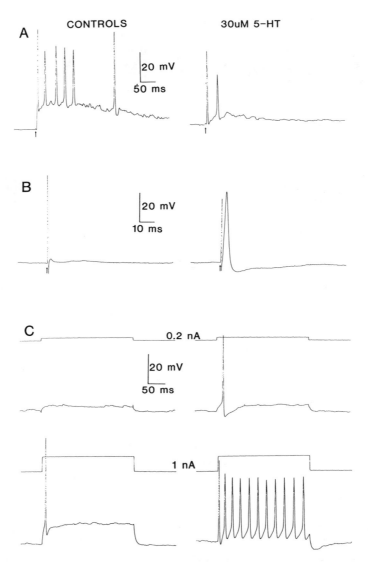

Figure 3. A) Motoneurone responses of an 11-day old rat to stimulation of dorsal root (3V, 0.1 ms) and inhibition of longer latency components of response by 30 μM 5-HT. 5-HT depolarized the cell by 16 mV.

B) In the same cell, ventral root stimulation (10 V, 0.1 ms) elicited an action potential blocked at the junction of axon and initial segment (M response). In the presence of 5-HT full invasion of the soma to give an SD spike occurred.

C) In the same cell, 5-HT caused an increase in excitability to depolarizing current. Current injection (0.2 and lnA) elicited a single spike at the higher current (left), but in the presence of 5-HT the higher current evoked repetitive spike discharge. In the presence of 5-HT, input resistance increased by 26% (not illustrated).

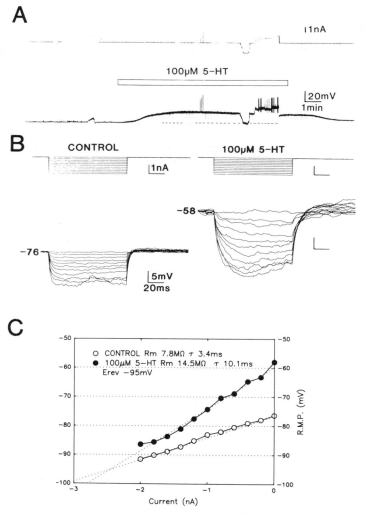

Figure 4. Membrane properties of a neonatal rat lumbar motoneurone under control conditions and on superfusion with 100 µM 5-HT.

A) Current injection protocol (upper trace) and membrane potential (lower trace). A slow depolarization of around 20 mV was elicited by 5-HT.

B) Upper traces show magnitude of hyperpolarizing current pulses injected across the membrane and lower traces the resulting electrotonic potentials. Note that traces are superimposed. In the presence of 5-HT, injected current elicited larger electrotonic potentials and the cell depolarized by 18 mV.

C) I/V relationship of peak electrotonic potential and injected current. Open circles are the values for control conditions and closed symbols are values in the presence of 5-HT. The fitted lines indicate that the slope resistance increased from 7.8 MΩ to 14.5 MΩ in the presence of 5-HT. The reversal potential for 5-HT action in this cell was −95 mV.

depolarization evoked by 5-HT. As the middle voltage traces show, larger hyperpolarizations were generated in the presence of 5-HT indicating that membrane resistance had increased due to closure of ion channels. On plotting these electrotonic potential amplitudes against injected current, an I/V relationship is generated under control conditions (Figure 4, open symbols) and on exposure to 5-HT, 100 μM (closed symbols). The fitted lines intersect at a point close to the equilibrium potential for K ions; i.e. 5-HT evoked no current flow at this potential because K ions are involved and they are at equilibrium. K channel closure has been proposed as a mechanism underlying 5-HT depolarization of facial MNs by Vandermaelen and Aghajanian (1980) and Larkman and Kelly (1988).

The precise identity of the K channel that is closed by 5-HT is still unclear. 5-HT depolarization of ventral hippocampal neurones has been ascribed to reduction of the leak conductance and the voltage-dependent M current by 5-HT (Halliwell and Colino 1990). The 5-HT receptor mediating the response, although uncharacterised, is not a 5-HT$_2$ receptor because neither ketanserin nor spiperone blocked it (Colino and Halliwell 1987). In rat nucleus accumbens neurones, 5-HT depolarizes principally by reducing an inward rectifier conductance (K current, G_{ir}) that was sensitive to barium (North and Uchimura 1989). The receptor involved is a 5-HT$_2$ receptor blocked by nanomolar concentrations of ketanserin and spiperone, which may be coupled to activation of protein kinase C. The reduction of a K conductance may be manifested also as a reduction in the accommodation of the cell membrane, i.e. a maintenance of excitability. This was noted as a distinct action in hippocampal neurones (Halliwell and Colino 1990) and attributed to suppression of a slow, voltage-insensitive, Ca-activated K-conductance.

Pharmacological characterization of the 5-HT receptor on spinal MNs is easier to achieve recording population responses from a ventral root. A brief summary of this pharmacology (Table 2) suggests that the receptor is broadly classifiable as a 5-HT$_2$ receptor; however, the receptor may not be identical to the 5-HT$_2$ binding site in the CNS.

Thus, 5-HT, αMe5-HT, 5-CT and 5-methoxytryptamine (5-MOT) had similar depolarizing actions, with 5-HT the most potent agonist providing neuronal uptake of 5-HT was blocked. The order of potency is shown in Table 2. 8-OH-DPAT, RU 24969 and 2-methyl-5-HT did not depolarize MNs. A number of antagonists have been found to antagonize the 5-HT depolarization, including the non-selective 5-HT receptor ligands, methysergide, metergoline, mesulergine and cyproheptadine. In these studies, noradrenaline which also depolarizes MNs was used as a control agonist. These antagonists characteristically caused a progressive rightward shift of the concentration-response curves but, at higher concentrations of antagonist, the curve was flattened. Spiperone

Table 2. Pharmacology of 5-HT responses in neonate rat spinal motoneurones. The two responses are direct depolarization of the motoneurone membrane and depression of the short latency synaptic response evoked by stimulating the ipsilateral dorsal root of the same segment

	Agonists	Antagonists	Ineffective
Depolariz- ation	5HT > alphaMe5HT > 5-CT > 5MOT ≫ T	Methysergide ketanserin ritanserin spiperone mesulergine cyproheptadine metergoline DOI	Methiothepin ICS 205-930 MDL 72222 Cyanopindolol 8-OH-DPAT RU 24969 2-Me-5HT quipazine
Inhibition of synaptic input	5-CT 8-OH-DPAT RU 24969 TFMPP methysergide	—	Ketanserin spiperone ritanserin methiothepin

had some antagonist action, but there may be a resistant component of the response insensitive to this agent. The 5-HT$_2$ receptor agonist, dimethoxy-iodo-phenyl-isopropylamine (DOI) had an unselective antagonist action, blocking responses to both 5-HT and noradrenaline. Ketanserin and ritanserin caused an unsurmountable antagonism in the concentration range 10^{-7}–10^{-6} M. Thus, the receptor has the characteristics of a 5-HT$_2$ receptor, but neither ketanserin nor spiperone have the potency they appear to have against 5-HT$_2$ responses in rat nucleus accumbens neurones (North and Uchimura 1989) and 5-CT is a weak full agonist, while at the nucleus accumbens 5-HT$_2$ receptor it is reported as being inactive.

5-HT Receptors on Spinal Reflex Pathways

The depression of the reflex responses evoked from MNs is mediated by a different 5-HT receptor (Table 2). Whether this receptor is located presynaptically has not been determined with certainty. Both the mono- and polysynaptic components of the reflex (Figure 5) are depressed by 5-HT in a concentration-dependent manner (Crick and Wallis 1990). The inhibition of neuronal 5-HT uptake with citalopram (10^{-7}–10^{-6} M) results in the sensitivity to 5-HT increasing several hundred fold. The depressant action of 5-HT on the monosynaptic reflex is mimicked by 5-CT, 8-OH-DPAT, RU 24969, trifluoro-methyl-phenylpiperazine (TFMPP) and also methysergide. Apart from 5-HT and 5-CT these agents have a much weaker depressant action on the polysynaptic reflex. As yet we have not identified an effective antagonist

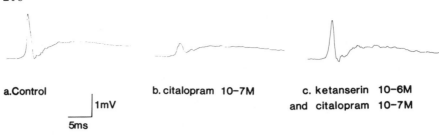

a.Control

$|$ 1mV

5ms

b. citalopram 10-7M

c. ketanserin 10-6M
and citalopram 10-7M

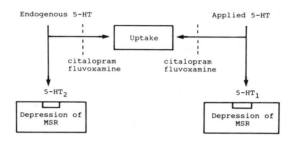

Figure 5. Upper traces show reflex responses recorded from a ventral root on stimulation of the segmental dorsal root in the hemisected lumbar spinal cord of the neonate rat. The initial rapid potential change is the monosynaptic component to the response (MSR) and the delayed elongated potential is the polysynaptic component. In the presence of citalopram both responses were depressed, but this depression was relieved after superfusion with ketanserin for 30–60 min. The diagram illustrates our conclusions that endogenous 5-HT accumulates in the cord after blockade of neuronal uptake by citalopram and depresses reflex responses, especially the monosynaptic reflex, via a 5-HT$_2$ receptor. Exogenous 5-HT applied by superfusion acts mainly via ketanserin-insensitive 5-HT$_1$-like receptors to depress the MSR.

of this action of 5-HT in depressing the monosynaptic reflex. Ketanserin, ritanserin, spiperone and methiothepin were not effective.

Intriguingly, the depressant action of 5-HT was also mimicked by 5-HT uptake blockers such as citalopram (Figure 5) and fluvoxamine. We assume this comes about because of the accumulation of endogenous 5-HT within the cord after uptake blockade. But, if so, the depressant action of this endogenous 5-HT is mediated via a 5-HT receptor which is blocked by ketanserin – as indicated in the schematic diagram in Figure 5. Thus, it appears that endogenous and applied 5-HT may act at different receptors to inhibit transmission in the spinal cord.

Block of Ca Channels: Presynaptic Inhibition

Transmission block via a presynaptic receptor can be used as a second example of a 5-HT action to close channels (Figure 1). In this case, the amine appears to be acting to prevent voltage-sensitive Ca channels from opening. A direct action on membrane Ca current, which was mimicked by 8-OH-DPAT, has been demonstrated in acutely dissociated dorsal raphe neurones by Kelly and Penington (1989). The 5-HT action on transmission is exemplified by depression of orthodromic responses recorded from a single superior cervical ganglionic neurone (Wallis and Elliott 1990). The depression of transmission is via a 5-HT$_1$-like action. 5-HT, RU 24969 and 5-CT all produced the same effect. In these experiments the stimulus intensity to the preganglionic nerve was adjusted so as to be just supra-threshold for eliciting an EPSP and spike in the postganglionic cell. 5-HT, 5-CT and RU 24969 reduced the EPSP sufficiently so that a spike was only elicited occasionally – sometimes not at all, sometimes after a longer latency. 5-CT and RU 24969 produced no discernible change in postsynaptic membrane properties, so that it could be concluded that a presynaptic action was involved. Dun and Karczmar (1981) showed that 5-HT reduced the number of transmitter quanta released at the ganglion cell synapse. Whether this action of 5-HT is G-protein-mediated and dependent on intracellular second messengers has still to be established. Thus, the arrows in Figure 1 are merely to suggest that, like some other 5-HT$_1$-receptor-mediated membrane events, this will prove also to be G-protein-coupled.

Summary and Conclusion

We have considered *firstly*, a number of 5-HT actions where the mechanism is one of opening some species of ion channel, resulting in a fast depolarization, a slow depolarization or a hyperpolarization, and *secondly*, actions where the mechanism is a closure of ion channels, resulting in a slow depolarization, altered transmitter release or altered membrane accommodation. Understanding these mechanisms in terms of the linkage to receptors and to second messenger systems is likely to become clearer in the near future as more information becomes available. It is not clear to what extent differentiation of the electrophysiological actions of 5-HT in terms of mechanisms is likely to be predictive of different 5-HT receptors linked to each action. The evidence to date from the nervous system does suggest that a particular mechanism is usually associated with a certain sub-type of 5-HT receptor and, thus, classification by function can provide a useful initial framework. However, it must not become a straitjacket because exceptional linkages of mechanism with receptor sub-type may quite possibly be found. One example may be the 5-HT$_{1A}$ receptor in the hippocampus; another unsubstantiated one is the linkage of 5-HT$_3$ receptors with inhibition of transmitter release in cerebral cortex (Barnes et al. 1989).

218

Acknowledgements

Work by the authors included in this chapter was supported by the Wellcome Trust and the Welsh Scheme for the Development of Health and Social Research.

References

Aas, P. (1983). Serotonin-induced release of acetylcholine from neurones in the bronchial smooth muscle of the rat. Acta Physiological Scandinavica 117: 477–480.

Andrade, R., Malenka, R. C., and Nicoll, R. A. (1986). A G protein couples serotonin and $GABA_B$ receptors to the same channels in hippocampus. Science 234: 1261–1265.

Barnes, J. M., Barnes, N. M., Costall, B., Naylor, R. J., and Tyers, M. B. (1989). $5-HT_3$ receptors mediate inhibition of acetylcholine release in cortical tissue. Nature 338: 762–763.

Blandina, P., Goldfarb, J. and Green, J. P., (1990). The continuing story of 5-hydroxy-tryptamine receptors: A $5-HT_3$ receptor modulates dopamine release from rat striatal slice. In: Saxena, P. R., Wallis, D. I., Wouters, W., and Bevan, P. (eds.), Cardiovascular pharmacology of 5-hydroxytryptamine. Dordrecht, Kluwer, pp. 117–126.

Bobker, D. H., and Williams, J. T. (1989). Serotonin augments the cationic current I_h in central neurons. Neuron. 2: 1535–1540.

Crick, H., and Wallis, D. I. (1990). Effect of 5-HT on reflex responses of neonate rat spinal motoneurones. Br. J. Pharmac. 100: 308P.

Colino, A., and Halliwell, J. A. (1987). Differential modulation of three separate K-conduc-tances in hippocampal CA1 neurones by serotonin. Nature 328: 73–77.

Derkach, V., Surprenant, A., and North, R. A. (1989). $5-HT_3$ receptors are membrane ion channels. Nature 339: 706–709.

Dun, N. J., and Karczmar, A. G. (1981). Evidence for a presynaptic inhibitory action of 5-hydroxytryptamine in a mammalian sympathetic ganglion. J. Pharmacol. Exp. Ther. 217: 714–718.

Elliott, P., and Wallis, D. I. (1988). The depolarizing action of 5-hydroxytryptamine on rabbit isolated, preganglionic cervical sympathetic nerves. Naunyn-Schmeideberg's Arch. Pharma-col. 338: 608–615.

Elliott, P., and Wallis, D. I. (1990a). Analysis of the actions of 5-hydroxytryptamine on the rabbit isolated vagus nerve. Naunyn-Schmeideberg's Arch. Pharmacol. 341: 494–502.

Elliott, P., and Wallis, D. I. (1990b). The action of 5-hydroxytryptamine on lumbar motoneu-rones in neonatal rat spinal cord in vitro. J. Physiol. 426: 54P.

Elliott, P., Seemungal, B. M., and Wallis, D. I. (1990). Antagonism of the effects of 5-HT on the rabbit isolated vagus nerve by BRL 43694 and metoclopramide. Naunyn-Schmeide-berg's Arch. Pharmacol. 341: 503–509.

Fozard, J. R., and Mwaluko, G. M. P. (1976). Mechanism of the indirect sympathomimetic effect of 5-hydroxytryptamine on the isolated heart of the rabbit. Brit. J. Pharmacol. 57: 115–125.

Gerschenfeld, H. M., and Paupardin-Tritsch, D. (1974a). Ionic mechanisms and receptor properties underlying the responses of molluscan neurones to 5-hydroxytryptamine. J. Physiol. 243: 427–456.

Gerschenfeld, H. M., and Paupardin-Tritsch, D. (1974b). On the transmitter function of 5-hydroxytryptamine at excitatory and inhibitory monosynaptic junctions. J. Physiol. 243: 457–481.

Halliwell, J. A., and Colino, A. (1990). Control of hippocampal pyramidal cell excitability by 5-HT. Neurosci. Letts. 38: S116.

Higashi, S., and Nishi, S. (1982). 5-hydroxytryptamine receptors of visceral primary afferent neurones in rabbit nodose ganglia. J. Physiol. 323: 543–567.

Kelly, J. S., and Penington, N. J. (1989). 5-hydroxytryptamine inhibits the voltage-dependent calcium current of acutely dissociated central neurones from the adult rat dorsal raphe nucleus. J. Physiol. 418: 35P.

Lambert, J. J., Peters, J. A., Hales, T. G., and Dempster, J. (1989). The properties of $5-HT_3$ receptors in clonal cell lines studied by patch-clamp techniques. Br. J. Pharmacol. 97: 27–40.

Larkman, P. M., and Kelly, J. S. (1988). The effects of serotonin (5-HT) and antagonists on rat facial motoneurones in the *in vitro* brainstem slice. J. Neurosci. Methods 24: 199.

Malone, H. M., Callachan, H., Lambert, J. J., and Peters, J. A. (1990) The effects of 5-hydroxytryptamine on adult rabbit nodose ganglion neurones in cell culture. Neurosci. Letts. 38: S116.

Marsh, S. J., Stansfeld, C. E., Brown, D. A., Davey, R., and McCarthy, D. (1987). The mechanism of action of capsaicin on sensory C-type neurones and their axons in vitro. Neuroscience 23: 257–289.

McQueen, D. S. (1990). Cardiovascular reflexes and 5-hydroxytryptamine. In: Saxena, P. R., Wallis, D. I., Wouters, W., and Bevan, P. (eds.), Cardiovascular pharmacology of 5-hydroxytryptamine. Dordrecht, Kluwer, pp. 233–245.

Neijt, H. C., Te Duits, I. J., and Vijverberg, H. P. M. (1988). Pharmacological characterization of serotonin 5-HT_3 receptor-mediated electrical response in cultured mouse neuroblastoma cells. Neuropharmacology 27: 301–307.

Neijt, H. C., Plomp, J. J., and Vijverberg, H. P. M. (1989). Kinetics of the membrane current mediated by serotonin 5-HT_3 receptors in cultured mouse neuroblastoma cells. J. Physiol. 411: 257–269.

North, R. A., and Uchimura, N. (1989). 5-hydroxytryptamine acts at 5-HT_2 receptors to decrease potassium conductance in rat nucleus accumbens neurones. J. Physiol. 417: 1–12.

Paintal, A. S. (1954). The response of gastric stretch receptors and certain other abdominal and thoracic vagal receptors to some drugs. J. Physiol. 126: 271–285.

Pape, H. C., and McCormick, D. A. (1989). Noradrenaline and serotonin selectivity modulate thalamic burst firing by enhancing a hyperpolarization-activated cation current. Nature 340: 715–718.

Round, A. A., and Wallis, D. I. (1987). Further studies on the blockade of 5-HT depolarizations of rabbit vagal afferent and sympathetic ganglion cells by MDL 72222 and other antagonists. Neuropharmacology 26: 39–48.

Surprenant, A. (1990). Whole cells and single channel current produced by activation of 5-HT_3 receptors in enteric neurones. Neurosci. Letts. 38: S115.

Takahashi, T., and Berger, A. J. (1990). Direct excitation of rat spinal motoneurones by serotonin. J. Physiol. 423: 63–76.

Takeuchi, A., and Takeuchi, N. (1960). On the permeability of end-plate membrane during the action of transmitter. J. Physiol. 154: 52–67.

Vandermaelen, C. P., and Aghajanian, G. K. (1980). Intracellular studies showing modulation of facial motoneurone excitability by serotonin. Nature 287: 346–347.

Wallis, D. I., and Dun, N. J. (1987). Fast and slow depolarizing responses of guinea-pig coeliac ganglion cells to 5-hydroxytryptamine. J. Aut. Nerv. System. 21: 185–194.

Wallis, D. I., and Elliott, P. (1990). 5-HT and related drugs and autonomic ganglia. In: Saxena, P. R., Wallis, D. I., Wouters, W., and Bevan, P. (eds). Cardiovascular pharmacology of 5-HT: prospective therapeutic applications. Dordrecht, Kluwer, pp. 177–190.

Wallis, D. I., and North, R. A. (1978). The action of 5-hydroxytryptamine on single neurones of the rabbit superior cervical ganglion. Neuropharmacology 17: 1023–1028.

Wallis, D. I., Lees, G. M., and Kosterlitz, H. W. (1975). Recording resting and action potentials by the sucrose-gap method. Comp. Biochem. Physiol. 50C: 199–216.

Wallis, D. I., Stansfeld, C. E., and Nash, H. L. (1982). Depolarizing responses recorded from nodose ganglion cells of the rabbit evoked by 5-hydroxytryptamine and other substances. Neuropharmacology 21: 31–40.

Wallis, D. I., and Woodward, B. (1974). The facilitatory actions of 5-hydroxytryptamine and bradykinin in the superior cervical ganglion of the rabbit. Br. J. Pharmac. 51: 521–532.

Wallis, D. I., and Woodward, B. (1975). Membrane potential changes induced by 5-hydroxytryptamine in the rabbit superior cervical ganglion. Br. J. Pharmac. 55: 199–212.

Yakel, J. L., Trussell, L. O., and Jackson, M. B. (1988). Three serotonin responses in cultured mouse hippocampal and striatal neurons. J. Neurosci. 8: 1273–1285.

Serotonin: Molecular Biology, Receptors and Functional Effects
ed. by J. R. Fozard/P. R. Saxena
© 1991 Birkhäuser Verlag Basel/Switzerland

Pharmacological Characterization of Brain 5-HT$_4$ Receptors: Relationship between the Effects of Indole, Benzamide and Azabicycloalkybenzimidazolone Derivatives

J. Bockaert, L. Fagni, M. Sebben, and A. Dumuis

Centre CNRS-INSERM de Pharmacologie-Endocrinologie, Rue de la Cardonille, 34094 Montpellier Cédex 5, France

Summary. 5-HT$_4$ receptors have recently been discovered in brain (foetal mice neurons, adult guinea pig hippocampus) and also in guinea pig ileum and ascending colon, rat oesophagus, pig and human heart. These 5-HT$_4$ receptors have been shown to stimulate cAMP production in brain and heart and to inhibit voltage-dependent K^+ channels in colliculi neurons.

The pharmacological characteristics of 5-HT$_4$ receptors are very particular. Indole derivatives, including 5 substituted tryptamine agonists, are generally active (especially 5-methoxytryptamine, 5-carboxamidotryptamine, RU 28253). All the 2-methoxy-4-amino-5-chloro substituted benzamide derivatives tested so far, are agonists as are the azabicycloalkyl-benzimidazolone derivatives. Classical pharmacological criteria indicate that indole, benzamides and azabicycloalkylbenzimidazolone derivatives act on the same receptor molecule entity. However, [^{3}H]-renzapride (benzamide) binding sites on colliculi membranes are not displaced by indole derivatives including 5-HT, 5-methoxytryptamine and RU 28253. This indicates that either [^{3}H]-renzapride binding sites are unrelated to 5-HT$_4$ receptors or that benzamides and indole derivatives interact at different subsites of the 5-HT$_4$ receptor. The only 5-HT$_4$ receptor antagonist described to date is ICS 205 930 (a competitive antagonist). The affinity of this drug is far lower (μM) than its affinity for 5-HT$_3$ receptors (nM). Finally, another particularity of the 5-HT$_4$ rceptor is its rapid desensitization.

Introduction

The discovery of a new type of 5-HT receptor in the brain that we designated as the 5-HT$_4$ receptor (Dumuis et al. 1988, 1989a; b; Bockaert et al. 1990), has brought excitement to the community of 5-HT receptor specialists (Clarke et al. 1989).

Over the past three years, several biological preparations have been shown to be regulated by 5-HT receptors neither belonging to the 5-HT$_1$, 5-HT$_2$ nor 5-HT$_3$ receptor families, but with a pharmacology very close to that of the 5-HT$_4$ receptors that we described in foetal mice colliculi neurons (Dumuis et al. 1988, 1989a; b) and in adult guinea-pig hippocampus (Bockaert et al. 1990). These preparations included the guinea-pig ileum (Sanger 1987; Buchheit et al. 1985; Craig and Clarke 1990; Craig et al. 1990), the ascending colon (Elswood et al. 1990), the rat oesophagus (Baxter and Clarke 1990), the pig and human heart

(Villallòn et al. 1990; Kaumann et al. 1990a; b) and the pyramidal neurons of the CA_1 region of rat hippocampus (Chaput et al. 1990).

In this report we have summarized all the information presented to date on brain 5-HT$_4$ receptors paying particular attention to the relationship between the effects of indole, benzamide and azabicycloalkybenzimidazolone derivatives.

Agonists at 5-HT$_4$ Receptors

The 5-HT$_4$ receptors were first discovered in primary cultures of foetal mice colliculi neurons. These types of neuronal cultures are grown in serum-free medium and contain a very low percentage of glial cells. They constitute a good model to study receptor coupling to second messenger systems (Bockaert et al. 1986; Weiss et al. 1986).

5-HT$_4$ receptors coupled to adenylate cyclase are able to stimulate 3 to 4 fold basal cAMP production measured in the presence of IBMX (0.75 mM) and forskolin (0.1 μM) which potentiates the response, but alone, has no effect (Weiss et al. 1985).

Indole Derivatives: As shown in Table 1, indolalkylamines including tryptamine and 5-substituted tryptamines, such as 5-hydroxytryptamine, 5-methoxytryptamine, 5-carboxamidotryptamine, 5-hydroxydimethyl-tryptamine (bufotenine), 5-methoxy-N-N-dimethyltryptamine are agonists with different potencies (Dumuis et al. 1988). It is worth noting that analogs of 5-HT substituted at position 5 of the indole ring, especially 5-methoxytryptamine, are good 5-HT$_4$ receptor agonists having no significant effect on 5-HT$_3$ receptors (Dumuis et al. 1988; Craig and Clarke 1990; Craig et al. 1990; Fozard 1990). In contrast, 2-methyl-5-HT, a selective 5-HT$_3$ receptor agonist has no effect on 5-HT$_4$ receptors. When the indole group is substituted at both positions 5, with a methoxy group, and 3, with a tetrahydropyridine, this leads to compounds which are either active [RU 28 253: 5-methoxy-3-(1, 2, 5, 6-tetrahydropyridine-4-yl) indole] or inactive [RU 24 969:5-methoxy-3-(1, 2, 3, 6-tetrahydro-pyridine-4-yl) indole] (Table 1).

Benzamide Derivatives: 2-methoxy-4-amino-5-chloro substituted benzamide derivatives including metoclopramide, BRL 24 924 (renzapride), BRL 20 627, cisapride and zacopride are agonists on our model (Dumuis et al. 1988, 1989a; b; Bockaert et al. 1990) (Table 1). We have now extended this series to several cisapride analogs (R 76186, R 66621, R 48895) (Table 1), as well as to other 2-methoxy-4-amino-5-chloro substituted benzamides like clebopride which is a dopaminolytic drug. All the benzamides bearing the 2-methoxy-4-amino-5-chloro substitution shown in Figure 1, which have been tested so far, have been found to be agonists. In contrast, other benzamides with different substitution groups on the benzamide ring, such as sulpiride (2-methoxy-5-aminosul-

Table 1. Affinities and efficacies of three series of compounds: the indole, the benzamide and the azabicycloalkyl benzimidalolone derivatives which act as agonists on 5-HT receptors in colliculi neurons

Compounds	Affinity EC50 (nM)	Efficacy % 5-HT response
Indole derivatives		
5-MeOT	100 ± 12	100 ± 5
5-HT	109 ± 17	100
RU 28253	560 ± 120	85 ± 8
Bufotenine	1580 ± 350	64 ± 7
5-CT	3160 ± 630	100 ± 11
5-MeO-N,N-DMT	17780 ± 3800	100 ± 12
Tryptamine	>10000	50 ± 4
2-CH3-5-HT	>10000	<50
RU 24 969	inactive	0
Benzamide derivatives		
R 76186	27 ± 9	134 ± 14
R 66621	72 ± 12	113 ± 20
Cisapride	72 ± 22	142 ± 17
Renzapride	116 ± 14	133 ± 15
Clebopride	136 ± 18	122 ± 14
R 48895	182 ± 23	104 ± 16
Zacopride	1122 ± 140	144 ± 22
BRL 2067	3450 ± 320	60 ± 8
Metoclopramide	4570 ± 580	44 ± 5
R 60 918	inactive	0
Sulpiride	inactive	0
Tiapride	inactive	0
Azabicycloalkyl benzimidazolone derivatives		
BIMU 8	72 ± 18	124 ± 8
BIMU 1	360 ± 120	16 ± 4
DAU 6215	2600 ± 540	16 ± 4

The effect of compounds on adenylate cyclase activity were determined as previously described (Dumuis et al. 1988). Affinities are expressed as EC_{50}, means ± standard errors of at least four separate experiments.

The efficacies (Emax) of the agonists are expressed as a percentage of stimulation relative to the maximal stimulatory effect of 5-HT taken as 100%.

fonyl substituted benzamide), tiapride (2-methoxy-5-methyl-sulfonyl substituted benzamide) or the cisapride analog: R 60 918 (2-methoxy-4-acetamido substituted benzamide), lack agonistic activity (Figure 1, Table 1). It would be interesting to systematically analyse the chemical requirements at positions 2, 4 and 5 of the benzamide ring. The nature of the substitution (R_4) determines the relative potency and efficacy of the drug but not the agonistic nature of the compound.

Benzamides like cisapride, renzapride and (RS) zacopride appeared to be full agonists at 5-HT_4 receptors in foetal mice colliculi neurons (Dumuis et al. 1989b) and at 5-HT_4 receptors involved in the potentiation of electrically-evoked contraction in guinea pig ileum (Craig and

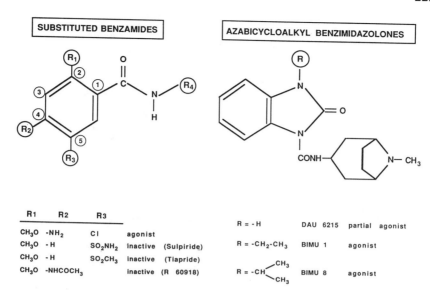

R1	R2	R3		
CH₃O	-NH₂	Cl	agonist	
CH₃O	-H	SO₂NH₂	inactive	(Sulpiride)
CH₃O	-H	SO₂CH₃	inactive	(Tiapride)
CH₃O	-NHCOCH₃		inactive	(R 60918)

R = -H	DAU 6215	partial agonist
R = -CH₂-CH₃	BIMU 1	agonist
R = -CH(CH₃)CH₃	BIMU 8	agonist

Figure 1. General structures of non-indole 5-HT₄ receptor agonists. Substituted benzamides: Whatever the nature of the substitution at R4 position, the 2-methoxy-4-amino-5-chloro benzamide derivatives were agonists. Some of them were partial agonists (metoclopramide, BRL 20 627, see Table 1).
Azabicycloalkyl benzimidazolones.

Clarke 1990). However, these drugs are only partial agonists at 5-HT₄ receptors in guinea-pig hippocampal membranes (Bockaert et al. 1990), in guinea-pig ascending colon and in the contraction of resting guinea-pig ileum (Elswood et al. 1990). It is possible that benzamides are even antagonists in other systems such as the pyramidal cells of the CA₁ region of rat hippocampus (Chaput et al. 1990). Whether or not these differences correspond to a change in the coupling at 5-HT₄ receptors with their cellular effectors or to a slight difference in receptor structure, remains to be determined. In all these systems, metoclopramide is a weak partial agonist. In foetal mice colliculi neurons (Dumuis et al. 1989a; b) and in rat oesophagus (Baxter and Clarke 1990), the rank order and relative agonist potencies of the classical benzamides is cisapride > renzapride > (RS) zacopride > metoclopramide. In guinea-pig ileum, Craig and Clarke found a slightly different rank order of agonist potencies in guinea-pig ileum: renzapride > (RS)zacopride > cisapride (Craig and Clarke 1990). In contrast, we have recently found that a good correlation (r = 0.9, Figure 2) exists between the potencies of 9 drugs (5-HT and 8 benzamides) at 5-HT₄ receptors in foetal mice colliculi neurons and 5-HT receptors in guinea-pig ileum contraction (Dumuis et al. 1990a).

Azabicycloalkyl Benzimidazolone Derivatives: In an attempt to find new drugs which act on 5-HT₄ receptors, we focused our attention on

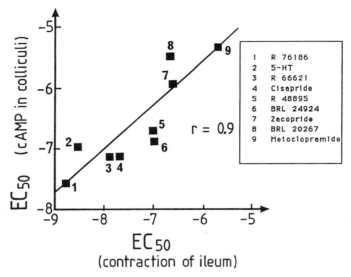

Figure 2. Correlation between the EC_{50}'s of several 5-HT$_4$ agonists to stimulate cAMP production in colliculi neurons and electrically-evoked guinea-pig ileum contractions. Intact ileum segments and longitudinal muscle myenteric plexus were used to test the effects of benzamides and 5-HT, respectively (for more details, see methods section in Dumuis et al. 1990a). Cylcic AMP production in colliculi neurons and determination of the EC_{50} of drugs on this system were performed as previously described (Dumuis et al. 1988). The significance probability for the correlation coefficient (r) using Student's t-test is $p < 0.05$.

gastrointestinal prokinetic compounds. A clear mechanism of action remains, however to be established. Among these compounds is a series of azabicycloalkyl benzimidazolne derivatives (Figure 1) developed by Boehringer Ingelheim (Turconi et al. 1990; Rizzi et al. 1990): BIMU 1: (N-(endo-8-methyl-8-azabicyclo(3.2.1.)oct-3-yl)-2,3-dihydro-3-ethyl-2-oxo-1H-benzimidazol-1-carboxamide)-hydrochloride. BIMU 8: (N-(endo-8-methyl-8-azabicyclo(3.2.1.)oct-3-yl)-2,3-dihydro-3-propyl-2-oxo-1H-benzimidazol-1-carboxamide)-hydrochloride and DAU 6215 (N-(endo-8-methyl-8-azabicyclo(3.2.1.)-oct-3-yl)-2,3-dihydro-2-oxo-1H-benzimidazol-1-carboxamide)-hydrochloride. BIMU 1 and BIMU 8 were agonists at 5-HT$_4$ receptors in foetal mice colliculi neurons, whereas DAU 6215 was a poor partial agonist (Dumuis et al. 1990b; 1991) (Table 1).

ICS 205-930, the Only Antagonist Available for 5-HT$_4$ Receptors

Classical antagonists at 5-HT$_1$ receptors (metergoline, methysergide, methiothepin, spiperone, mesulergine and (−)-pindolol), 5-HT$_2$ receptors (spiperone, ketanserin and methiothepin) or 5-HT$_3$ receptors

(MDL 72222, ondansetron, BRL 43684 and cocaïne) had no significant effect on 5-HT-induced cAMP production in foetal mice colliculi neurons nor on adult guinea-pig hippocampal membranes (Dumuis et al. 1988; Bockaert et al. 1990). The only drug which competitively blocked 5-HT and the benzamide stimulation of cAMP, was ICS 205-930 (Dumuis et al. 1988, 1990a, 1991; Craig and Clarke 1990). Its affinity for 5-HT$_4$ is however relatively low (μM), and far lower than its affinity for 5-HT$_3$ receptors (nM) (Richardson et al. 1985). In all preparations where 5-HT$_4$ receptors have been described (guinea-pig ileum, colon and hippocampus, rat oesophagus, pyramidal cells from the CA$_1$ region of rat hippocampus, pig and human heart), ICS 205-930 has been found to be an antagonist with μmolar affinity (Baxter and Clarke 1990; Bockaert et al. 1990; Chaput et al. 1990; Craig and Clarke 1990; Elswood, 1990; Kaumann et al. 1990a; b).

Pharmacological Evidence for a Common Site of Action of 5-HT, Benzamides and Azabicycloalkylbenzimidazolone Derivatives on 5-HT$_4$ Receptors

We have now obtained most of the pharmacological criteria necessary to conclude that 5-HT, benzamides and azabicycloakylbenzimidazolones act on the same receptor molecule entity. These criteria are:

1) ICS 205-930 competitively blocks the effect of these agonists with almost the same affinity (Dumuis et al. 1988; 1989a; b; 1991),

2) In colliculi neurons and guinea-pig hippocampal membranes, the effects of 5-HT on cAMP production are not additive with those of benzamides or azabicycloalkylbenzimidazolones (Dumuis et al. 1989b; 1991; Bockaert et al. 1990)

3) Partial agonists such as DAU 6215 or metoclopramide reduce 5-HT activity (Dumuis et al. 1991)

4) 5-HT, benzamides and azabicycloalkylbenzimidazolones cross-desensitize 5-HT$_4$-mediated cAMP production (Dumuis et al. 1989b; 1991).

Rapid desensitization of 5-HT$_4$ receptors has been described both in colliculi neurons and guinea-pig ileum (Dumuis et al. 1989b; Craig et al. 1990; Fozard 1990). This charasteristic can now be used to identify this receptor and also to eliminate a 5-HT$_4$ receptor mediated response in a tissue containing several types of 5-HT receptors. Indeed, Craig et al. (1990) eliminated the 5-HT$_4$ response in guinea-pig ileum by preincubating with 5-methoxytryptamine, a 5-HT$_4$ receptor agonist having no affinity for 5-HT$_3$ receptors. In contrast, they eliminated the 5-HT$_3$

response in the same organ by preincubating with 2-methyl-5-HT, a 5-HT$_3$ receptor agonist having no affinity for 5-HT$_4$ receptors (table 1).

[^{3}H]-Renzapride Binding Sites in Foetal Mice Colliculi Membranes. Binding Sites with or without Connection with 5-HT$_4$ Receptors?

The discovery of a specific radioligand would rapidly extend our knowledge on 5-HT$_4$ receptors and especially on brain 5-HT$_4$ receptors. In particular, its regional distribution in animal as well as in human brain would greatly benefit from this discovery. For this purpose, we used the more potent 5-HT$_4$ interacting drugs i.e. [^{3}H]-cisapride and [^{3}H]-renzapride, for binding experiments on membranes of colliculi neurons, a preparation in which we are certain that 5-HT$_4$ receptors are present. Using the protocol described in Figure 3, we were unable to detect any displaceable [^{3}H]-cisapride binding sites. In contrast, we measured displaceable [^{3}H]-renzapride binding using 300 μM renzapride to define non-specific binding. At 60 nM [^{3}H]-renzapride, specific binding represented 36% of total binding. Higher concentrations of [^{3}H]-renzapride could not be used because the non-specific binding was too high and impaired the accuracy of the determinations. Using the specific binding obtained with [^{3}H]-renzapride concentrations below 60 nM, a linear Scatchard plot was obtained yielding a Kd of 280 ± 125 nM and a Bmax of 360 ± 90 fmoles/mg protein (n = 8) (data not shown).

Unlabeled renzapride displaced [^{3}H]-renzapride binding (20 nM) with an EC$_{50}$ of 1 μM. R76186, a cisapride derivative, was also able to completely displace the binding, whereas BRL 20 627 and metoclopramide were very weak displacers (EC$_{50} > 100$ μM) (Figure 3A). This order of potency is similar, although not identical to the order described for cAMP production in intact colliculi neurons (Table 1). In particular, R76186 was at least as potent as renzapride in stimulating cAMP production (Table 1) but was not as active as renzapride in displacing [^{3}H]-renzapride binding (Figure 3A).

BIMU 8, an azabicycloalkyl benzimidazolone derivative, having 5-HT$_4$ receptor agonistic properties was only able to displace part of [^{3}H]-renzapride binding (Figure 3B). This was surprising but indicated that [^{3}H]-renzapride binding was heterogeneous. ICS 205-930, the only 5-HT$_4$ receptor antagonist, was able to completely block [^{3}H]-renzapride binding (Figure 3C) with an EC$_{50}$ of 10 μM. Since the chemical structure of ICS 205-930 and renzapride are quite different, this result can be taken as an argument favoring a relationship between 5-HT$_4$ receptors and [^{3}H]-renzapride binding sites.

However, another surprising result was that indole derivatives including 5-HT, 5-methoxytryptamine and RU 28253, were unable to displace [^{3}H]-renzapride binding (Figure 3D). Other compounds having

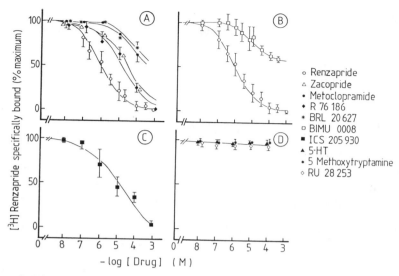

Figure 3. Displacement of [³H]-renzapride binding by various agonists and antagonists in colliculi neuronal membranes. Membrane preparation: Neuronal cultures were prepared as described by Bockaert et al. (1986). At 11 days, harvested cells were centrifuged at 900 x g at 4°C in NaCl 9% during 5 min. The cell pellet was homogenized in Tris-HCl buffer 50 mM pH 7.4 containing 2 mM MgCl₂ and 1% BSA (Buffer A), centrifuged for 20 min at 4°C at 50 000 x g. Then the pellet was used for binding in a final protein concentration of 100 μg/tube in buffer A.

Binding experiments: The binding assay of [³H] renzapride (16.2 Ci/m mole) (generously donated by Beecham – Pharmaceuticals), was performed in triplicate at 37°C for 30 min. Incubations in a final volume of 100 μl contained buffer A, 100 μg membranes, 20 nM [³H] renzapride and drugs (dissolved all in buffer A). Incubations were initiated by addition of membranes and stopped by dilution with 1 ml buffer A, followed by a rapid filtration and washing 3 times with 4 ml ice-cold Tris-HCl pH 7.4 over a Whatman GF/C filter.

The results are expressed as a percent of specific [³H]-renzapride binding. The specific binding is defined as the difference between the total [³H]-renzapride binding and the remaining [³H]-renzapride binding in the presence of 300 μM renzapride.
A) 5-HT₄ agonists
B) Comparison of displacement by renzapride and BIMU 8.
C) Displacement by ICS 205 930.
D) Displacement by indole derivatives.

no displaceable activity included 5-HT₃ receptor antagonists, such as ondansetron and BRL 43694, and 5-HT₁ and 5-HT₂ receptor antagonists such as methysergide and ketanserin.

Two hypotheses can be proposed to explain these results: 1) [³H]-renzapride binding sites are unrelated to 5-HT₄ receptors. 2) 5-HT₄ receptors contain two subsites, one for benzamides, the other for 5-HT and indoles. Occupation of only one of these subsites triggers the coupling of 5-HT₄ receptors to the Gs protein, leading to adenylate cyclase activation. It is possible that binding of benzamides or 5-HT on their specific subsites are mutually exclusive. The binding of ICS 205-930

228

would overlap the benzamide and the indole subsites and would therefore competitively block both the benzamide and 5-HT stimulation of adenylate cyclase (Dumuis et al. 1988; 1989a; b; Bockaert et al. 1990). Only the use of better radioligands would decide which one of these hypotheses is correct.

Possible Role of 5-HT$_4$ Receptors in Brain

In order to find what could be the cellular physiological consequences of 5-HT$_4$ receptor activation in colliculi neurons, two guidelines were retained: 1) if in colliculi neurons, 5-HT$_4$ receptors trigger the same cellular functions as they are expected to do so in enteric nerves terminals, we should look for a cellular response leading to neuronal depolarization and, as a possible consequence, an increase in neurotransmitter release. 2) in Aplysis and snail neurons, 5-HT receptors have been described which, like 5-HT$_4$ receptors, are able to stimulate cAMP (Deterre et al. 1982; Kandel and Schwartz 1982). In these invertebrate systems, the increase in cAMP is followed by a closure of voltage-

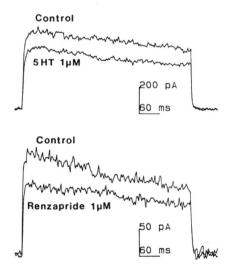

Figure 4. Effects of 5-HT and renzapride on K$^+$ channels of colliculi neurons. Recordings were performed at room temperature on 7 day *in vitro* neurons, using the whole-cell configuration of the patch-clamp technique, described by Hamill et al. (1981). External solution contained (mM): NaCl (130), KCl (3), MgCl$_2$(2), CaCl$_2$(2), Hepes (10), D-glucose (10), IBMX (1), TTX (3.3 μM), pH 7.4.

Patch pipettes (3–5Ω) were filled with an internal medium containing (mM): KCl (130), MgCl$_2$ (2), CaCl$_2$ (0.5), EGTA (5), Hepes (10), ATPNa$_2$ (4), GTPNa$_3$ (0.5), pH 7.2.

Drugs were applied using the slightly modified perfusion technique described by Johnson and Ascher (1987).

dependent K$^+$ channels (Deterre et al. 1982; Kandel and Schwartz 1982) and in Aplysia by an increase in transmitter release (Kandel and Schwartz 1982).

We therefore followed the possibility that 5-HT$_4$ receptors in colliculi neurons also block K$^+$ channels. Forskolin (50 μM), well-known to stimulate cAMP whatever the cell considered, blocks voltage-dependent K$^+$ channels in foetal mice colliculi neurons (data not shown). Preliminary results indicate that 5-HT, as well as renzapride, were also able to block K$^+$ channels (Figure 4). Although a full characterization of the receptor involved is needed, the 5-HT and renzapride effects are likely to be mediated through 5-HT$_4$ receptors. Therefore, the cellular mechanisms of action of 5-HT receptors in Aplysia neurons and those of 5-HT$_4$ receptors in vertebrate brain appear to be similar. This is particularly interesting since the 5-HT receptors in Aplysia neurons are implicated both in short and long-term memory of these animals, as demonstrated by Kandel and colleagues (Kandel and Schwartz 1982).

Conclusion

In conclusion, the existence of a 5-HT receptor family differing from the 5-HT$_1$, 5-HT$_2$, and 5-HT$_3$ receptor families now seems established. The pharmacology of this new receptor is so different from 5-HT$_1$, 5-HT$_2$ and 5-HT$_3$ receptors that it has been called 5-HT$_4$. Whether this new family has one or several members or not, remains to be demonstrated. As summarized in the present report, we already possess pharmacological tools to study 5-HT$_4$ receptors. However, even better compounds are required to fully characterize these receptors.

Acknowledgements

We would like to thank Miss M. Aliaga and Mrs M. Passama for their help in preparing this manuscript. Support was provided by the Centre National de la Recherche Scientifique (CNRS), the Institut Nationale de la Santé et de la Recherche Médicale (INSERM), Boehringer – Ingelheim (De Angeli Institute, Milan) and Bayer – Troponwerke (France – FRG).

We would also like to thank Dr G. J. Sanger and Beechan Pharmaceuticals (Harlow, UK) for generously providing us with [^{3}H]-renzapride.

References

Baxter, G. S., and Clarke, D. E. (1990). Putative 5-HT$_4$ receptors mediate relaxation of rat oesophagus. Abstract of the second IUPHAR satellite meeting on Serotonin 11–13 July, Basel. p. 85.

Bockaert, J., Gabrion, J., Sladeczek, F., Pin, J. P., Récasens, M., Sebben, M., Kemp, D. E., Dumuis, A., and Weiss, S. (1986). Primary culture of striatal neurons: a model of choice for pharmacological and biochemical studies of neurotransmitter receptors. J. Physiol. 81: 219–227.

Bockaert, J., Sebben, M., and Dumuis, A. (1990). Pharmacological characterization of 5-hydroxytryptamine$_4$ (5-HT$_4$) receptors positively coupled to adenylate cyclase in adult guinea pig hippocampal membranes: Effect of substituted benzamide derivatives. Mol. Pharmacol. 37: 408–411.

Buchheit, K. H., Engel, G., Muschler, E., and Richardson, B. P. (1985). Study of the contractile effect of 5-hydroxytryptamine (5-HT) in the isolated longitudinal muscle strip from guinea pig ileum. Evidence for two distinct release mechanisms. Naunyn-Schmiedeberg's Arch. Pharmacol. 329: 36–41.

Chaput, Y., Araneda, R. C., and Andrade, R. (1990). Pharmacological and functional analysis of a novel serotonin receptor in rat hippocampus. Eur. J. Pharmacol. 182: 441–456.

Clarke, D. E., Craig, D. A., and Fozard, J. R. (1989). The 5-HT$_4$ receptor: naughty but nice. Trends Pharmacol. Sci., 10: 385–386.

Craig, D. A., and Clarke, D. E. (1990). Pharmacological characterisation of neuronal receptor for 5-hydroxytryptamine in guinea pig ileum with properties similar to the 5-HT$_4$ receptor. J. Pharmacol. Exp. Ther. 252: 1378–1386.

Craig, D. A., Eglen, R. M., Walsh, L. K. M., Perkins, L. A., Whiting, R. L., and Clarke, D. E. (1990). 5-Methoxytryptamine and 2-methyl-5-hydroxytryptamine induced desensitization as a discriminative tool for the 5-HT$_3$ and putative 5-HT$_4$ receptors in guinea pig ileum. Naunyn-Schmiedeberg's Arch. Pharmacol. 342: 9–16.

Deterre, P., Paupardin-Tritsch, D., Bockaert, J., and Gerschenfeld, H. M. (1982). cAMP-mediated decrease in K$^+$ conductance evoked by serotonin and dopamine in the same neuron: A biochemical and physiological single cell study. Proc. Natl. Acad. Sci. USA 79: 7934–7938.

Dumuis, A., Bouhelal, R., Sebben, M., Cory, R., and Bockaert, J. (1988). A non-classical 5-hydroxytryptamine receptor positively coupled with adenylate cyclase in the central nervous system. Mol. Pharmacol. 34: 880–887.

Dumuis, A., Sebben, M., and Bockaert, J. (1989a). BRL 24924: a potent agonist at a non classical 5-HT receptor positively coupled with adenylate cyclase in colliculi neurons. Eur. J. Pharmacol. 162: 381–384.

Dumuis, A., Sebben, M., and Bockaert, J. (1989b). The gastrointestinal prokinetic benzamine derivatives are agonists at the non-classical 5-HT receptor (5-HT$_4$) positively coupled to adenylate cyclase in neurons. Naunyn Schmiedeberg's Arch. Pharmacol. 340: 403–410.

Dumuis, A., Schuurkes, J., Ghoos, E., Sebben, M., and Bockaert, J. (1990a) Comparison between the potencies of a series of benzamide derivatives which stimulate 5-HT$_4$ receptors of mouse colliculi neurons and guinea-pig ileum (submitted for publication).

Dumuis, A., Sebben, M., Monferini, E., Nicola, M., Ladinski, H., and Bockaert, J. (1990b). Benzimidazolone derivatives as a novel class of potent agonists at the 5-HT$_4$ receptor positively coupled to adenylate cyclase in brain. Abstract of second IUPHAR satellite meeting on Serotonin 14–13 July, Basel. p. 42.

Dumuis, A., Sebben, M., Monferini, E., Nicola, M., Turconi, M., Ladinski, H., and Bockaert, J. (1991). Azabicycloalkylbenzimidazolone derivatives as a novel class of potent agonists at the 5-HT$_4$ receptor coupled to adenylate cyclase in brain. Naunyn Schmiedeberg's Arch. Pharmacol. 343: 245–251.

Elswood, C. J., Bunce, K. T., and Humphrey, P. P. A. (1990). Identification of 5-HT$_4$ receptors in guinea-pig ascending colon. Abstract of the second IUPHAR satellite meeting on Serotonin 11–13 July, Basel p. 86.

Fozard, J. R. (1990). Agonists and antagonists of 5-HT$_3$ receptors. In: Saxena, P. R., Wallis, D. I., Wouters, W., and Bevan, P. (eds). Cardiovascular Pharmacology of 5-hydroxytryptamine. Kluwer Academic Publishers, Dordrecht, pp. 101–115.

Hamill, P. O., Marty, A., Neher, E., Sackmann, B., and Sigworth, F. K. (1981). Improved patch-clamp techniques for high resolution current recording from cells and cell-free membrane patches. Pflügers Arch. 391: 85–100.

Johnson, J. W., and Ascher, P. (1987). Glycine potentiates the NMDA response in cultured mouse brain neurons. Nature. 325: 529–531.

Kandel, E. R., and Schwartz, J. H. (1982). Molecular biology of learning: modulation of transmitter release. Science. 218: 433–443.

Kaumann, A. J., Sanders, L., Brown, A. M., Murray, K. J., and Brown, M. J. (1990a). A 5-hydroxytryptamine receptor in human atrium. Br. J. Pharmacol. 100: 879–885.

Kaumann, A. J., Sanders, L., Brown, A. M., Murray, K. J., and Brown, M. J. (1990b). 5-HT and the human heart. Abstract of the Second IUPHAR satellite meeting on Serotonin 11–13 July, Basel. p. 141.

Richardson, B. P., Engel, G., Donatsch, P., and Stadler, P. A. (1985). Identification of serotonin M-receptor subtypes and their specific blockade by a new class of drugs. Nature 316: 126–131.

Rizzi, C. A., Onori, L., Sanguini, A. M., Coccini, T., and Tonini, M. (1990). Facilitation of cholinergic transmission and peristalsis induced by a novel class of benzimidazolone derivatives in the guinea-pig small intestine. Abstract of the second IUPHAR satellite meeting on serotonin 11–13 July, Basel p. 152.

Sanger, G. J. (1987). Increased gut cholinergic activity and antagonism of 5-hydroxytryptamine M-receptors by BRL 24924: potential clinical importance of BRL 24924. Br. J. Pharmacol. 91: 77–87.

Turconi, M., Nicola, M., Quintero, M. G., Maiocchi, L., Micheletti, R., Giraldo, E., and Donetti, A. (1990). Synthesis of a new class of 2,3-dihydro-2-oxo-1H-benzimidazolone-1-carboxylic acid derivatives as highly potent 5-HT$_3$ receptor antagonists. J. Med. Chem. 33: 2101–2108.

Villalòn, C. M., den Boer, M. O., Heiligers, J. P. C., and Saxena, P. R. (1990). Mediation of 5-hydroxytryptamine-induced tachycardia in the pig by the putative 5-HT$_4$. Br. J. Pharmacol. 100: 665–667.

Weiss, S., Sebben, M., Garcia-Sainz, A. J., and Bockaert, J. (1985). D2-dopamine receptor-mediated inhibition of cyclic AMP formation in striatal neurons in primary culture. Mol. Pharmacol. 27: 595–599.

Weiss, S., Pin, J.-P., Sebben, M., Kemp, D. E., Sladeczeck, F., Gabrion, J., and Bockaert, J. (1986). Synaptogenesis of cultured striatal neurons in serum-free medium: A morphological and biochemical study. Proc. Natl. Acad. Sci. USA 83: 2238–2242.

Serotonin: Molecular Biology, Receptors and Functional Effects
ed. by J. R. Fozard/P. R. Saxena
© 1991 Birkhäuser Verlag Basel/Switzerland

Pharmacological Properties of the Putative 5-HT$_4$ Receptor in Guinea-Pig Ileum and Rat Oesophagus: Role in Peristalsis

D. E. Clarke[1], G. S. Baxter[1], H. Young[2], and D. A. Craig[1]

[1]Institute of Pharmacology, Syntex Research, 3401 Hillview Ave., Palo Alto, CA 94303, USA;
[2]Division of Gastroenterology, Stanford University School of Medicine, Stanford, CA 94305, USA

Summary. The occurrence and pharmacological profile of the putative 5-HT$_4$ receptor in peripheral tissues is discussed, and evidence is presented that the putative receptor mediates relaxation of the rat oesophagus and functions to facilitate the peristaltic reflex in guinea-pig ileum.

Introduction

In September of 1988, Craig and Clarke submitted an abstract to the British Pharmacological Society describing a novel receptor for 5-hydroxytryptamine (5-HT) in guinea-pig ileum (Craig and Clarke 1989). The pharmacology of the receptor turned out to be remarkably similar to a receptor site for 5-HT described by Bockaert and colleagues in the CNS which they published in December of the same year (Dumuis et al. 1988). The unique pharmacological characteristics of the receptor (see also Shenker et al. 1987) prompted Dumuis et al. (1988) to classify the site as a 5-HT subtype distinct from those delineated by Bradley et al. (1986). Thus, the 5-HT$_4$ appellation entered the scientific literature, and there now seems little doubt that the receptor to which it refers will be the focus of much future basic and applied research.

At the time of writing, positive identification of the 5-HT$_4$ receptor has been made in six peripheral tissues (see Table 1), and the list can be expected to increase at a rapid rate. Thus, in the periphery the 5-HT$_4$ receptor is associated with the myocardium and structures of the alimentary canal, and as with the 5-HT$_4$ receptor in the CNS, the peripheral 5-HT$_4$ receptor is found across species.

The present article deals with the pharmacological characterization of the 5-HT$_4$ receptor in guinea-pig ileum and rat oesophagus and discusses the role of the 5-HT$_4$ receptor in the physiology and pharmacology of the peristaltic reflex.

Table 1. Putative 5-HT$_4$ receptors in peripheral tissues[a]

Tissue	Reference
1. Guinea-Pig ileum. (contraction)	Craig and Clarke (1989, 1990) Clarke et al. (1989) Eglen et al. (1990)
2. Guinea-Pig colon. (contraction)	Elswood et al. (1990)
3. Guinea-Pig atrium. (positive chronotropism)	Eglen (unpublished)
4. Pig heart (*in vivo*). (positive chronotropism)	Villalón et al. (1990)
5. Human atrial appendage. (positive inotropism)	Kaumann et al. (1990a)
6. Rat oesophagus. (relaxation)	Reeves et al. (1989) Baxter and Clarke (1990)

[a]The 5-HT$_4$ receptor was named by Dumuis et al. (1988) and is not as yet officially recognized by the 5-HT Receptor Nomenclature Committee.

Materials and Methods

Pieces of guinea-pig ileum were set up for recording tension in longitudinal muscle as described by Craig and Clarke (1990) and Craig et al. (1990).

The tunica muscularis mucosae of rat oesophagus was set up as described by Bieger and Triggle (1985) and was contracted to steady-state tension with carbachol (3 μM). Cumulative inhibition curves to 5-HT and other agonists were constructed. In order to achieve equilibrium conditions and to pharmacologically isolate the receptor of interest, monoamine oxidase was inhibited with pargyline (100 μM), and cocaine (30 μM), corticosterone (30 μM) and methysergide (1 μM) were added to the Tyrode solution. Initial experiments with ondansetron (5 μM), N-acetyl-5-hydroxytryptophyl-5-hydroxytryptophan amide (10 μM) and indomethacin (3 μM) ruled out the presence of 5-HT$_3$ and 5-HT$_{1P}$ receptors (Mawe et al. 1986) as well as 5-HT-induced release of prostanoids.

The peristaltic reflex was studied in 5 cm segments of distal guinea-pig ileum utilizing a modification (Craig and Clarke 1991) of the method described by Bulbring and Crema (1958). Intraluminal pressure was raised in increments of 0.1 mmHg every 1 min to determine the threshold pressure for peristalsis or to conduct experiments at sub-threshold pressures (about 50% threshold). Fluid was introduced into the oral end and expelled via the aboral end; both pressures were recorded.

234

Results

Figure 1 summarizes the effect of 5-HT on longitudinal smooth muscle tension in guinea-pig ileum and lists the key pharmacological probes for the 5-HT$_4$ receptor. The 5-HT$_4$ phase of the biphasic concentration-effect curve to 5-HT is cholinergically mediated and is abolished by atropine. It is also tetrodotoxin-sensitive. These results suggest that the 5-HT$_4$ receptor is located neuronally, either directly on cholinergic nerves or upstream.

The lack of selectivity of ICS 205-930 for the 5-HT$_4$ receptor (Figure 1) and its non-specific actions (Scholtysik 1987; Scholtysik et al. 1988) at or near concentrations required for 5-HT$_4$ antagonism prompted Craig et al. (1990) to investigate alternative strategies for 5-HT$_4$ receptor discrimination. In this regard, the susceptibility of 5-HT receptors to desensitization was exploited using the two key indole agonists given in Figure 1.

Experiments revealed that incubation of ilea with 5-methoxy-tryptamine (10 μM) desensitized the 5-HT$_4$ receptor while leaving the 5-HT$_3$ fully functional. Conversely, 2-methyl-5-HT (10 μM) desensitized the 5-HT$_3$ receptor without affecting the 5-HT$_4$ receptor. Other studies demonstrated that desensitization with 5-methoxytryptamine, but not

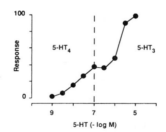

Key pharmacological probes for the 5-HT$_4$ receptor

---

1. Agonized potently by 5-HT and 5-methoxytryptamine, but not 2-methyl-5-HT.

2. Agonized by substituted benzamide derivatives: renzapride, zacopride, and cisapride. (These agents act as antagonists at the 5-HT$_3$ receptor.)

3. Inhibited by ICS 205-930 with an affinity of 3μM. (Not selective; exerts a higher affinity for 5-HT$_3$ receptors.)

Figure 1. Putative 5-HT$_4$ receptor in guinea-pig ileum as studied on longitudinal smooth muscle tension in non-stimulated segments. The figure is based on data from Clarke et al. (1989), Craig and Clarke (1990) and Craig et al. (1990).

2-methyl-5-HT, blocked fully but reversibly the agonistic action of renzapride.

The presence of the 5-HT$_4$ receptor in guinea-pig ileum and the established prokinetic activity of certain substituted benzamides (Figure 1; Schuurkes et al. 1987; King and Sanger 1988; Stanniforth and Pennick 1990) suggested that the fundamental reflex of the gut, the peristaltic reflex, might be modulated by the 5-HT$_4$ receptor. Indeed, Figure 2 shows that renzapride functioned to evoke peristalsis in segments of guinea-pig ileum and that this action was blocked by ICS 205-930 (3 μM), but not by selective inhibition of the 5-HT$_3$ receptor with ondansetron (5 μM). Peristalsis in response to renzapride developed slowly (about 1 min to onset) but continued intermittently over a subsequent 5 min observation period. As illustrated in Figure 2, in some experiments renzapride exhibited a tendency to produce a contracture of the ileum. Figure 3 shows that 5-HT (0.1 μM) can also function to evoke the peristaltic reflex. Again, peristalsis was slow in onset and was characterized by intermittent firing of the reflex, with periodic contractures of circular muscle in several experiments. As with renzapride, the response to 5-HT was resistant to ondansetron (5 μM) but was blocked by ICS 205-930 (3 μM).

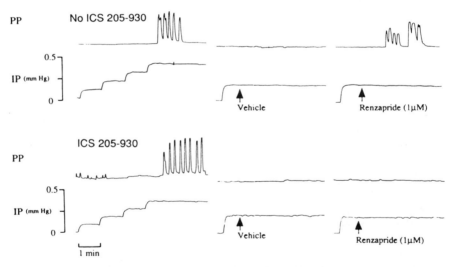

Figure 2. Recordings of intraluminal pressure (IP) and peristaltic pressure (PP) from paired segments of ileum from a single guinea-pig using Krebs' solution containing ondansetron (5 μM) to block 5-HT$_3$ receptors. Upper panels: Left, IP was raised at 1 min intervals to determine the threshold pressure for peristalsis. Center, IP was maintained at 1/2 threshold pressure and vehicle (H$_2$O, 50 μl), added to the bathing fluid, failed to elicit peristalsis. Right, IP was maintained at 1/2 threshold pressure and after a brief delay, renzapride (1 μM), added to the bathing fluid, produced peristalsis. Lower panels: The same as in the upper panels except the bathing fluid contained ICS 205-930 (3 μM) to block 5-HT$_4$ receptors. In the presence of ICS 205-930 (3 μM), renzapride failed to evoke peristalsis. The same result was obtained in 7 additional experiments from 7 different guinea-pigs (Craig and Clarke 1991, with permission).

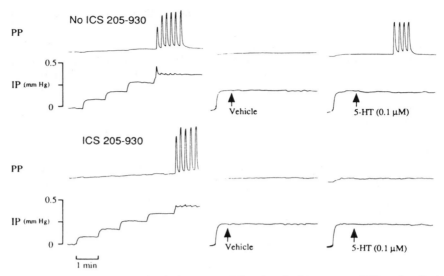

Figure 3. Recordings of intraluminal pressure (IP) and peristaltic pressure (PP) as described in Figure 3. Thus, the facilitatory effect of 5-HT on the peristaltic reflex is blocked by ICS 205-930 (3 μM). The same result was obtained in 4 additional experiments from 4 different guinea-pigs (Craig and Clarke 1991, with permission).

Two other interesting findings (not illustrated) have been made from the studies on peristalsis: (a) the absence of a significant effect of ICS 205-930, itself, on the threshold pressure for firing the reflex and (b) the ability of renzapride to lower the threshold for initiation of peristalsis even in the presence of 0.1 μM atropine (atropine-resistant peristalsis). This concentration of atropine is 100 times its equilibrium dissociation constant for muscarinic cholinoceptors (Clague et al. 1985). It should be noted that atropine, itself, raised the intraluminal pressure threshold for firing the reflex.

The absence of a reported high affinity probe for the 5-HT$_4$ receptor precludes ligand binding studies at the 5-HT$_4$ receptor and necessitates that drug discovery programs utilize functional assays systems for drug development. In this regard, the complexity of the guinea-pig ileum and its inherent variability (see Discussion) prompted Baxter and Clarke (1990) to identify a more simple and stable 5-HT$_4$ receptor assay. In this context, extensive previous work (Bieger and Triggle 1985; Triggle et al. 1988) has demonstrated that relaxation of the rat oesophagus by 5-HT is mediated via an "atypical" 5-HT receptor (see also abstracts by Reeves et al. (1989)).

Experiments conducted on the oesophagus revealed an inhibitory 5-HT$_4$ receptor at which the order and relative potency (equipotent concentration-ratios) of 5-HT and related compounds is as follows:

5-HT (pIC_{50} = 8.1; 1) > 5-methoxytryptamine (2) > (RS)-alpha-methyl-5-HT (10) > cisapride and (S)-zacopride (12) > renzapride (21) > (RS)-zacopride (86) > 5-carboxamidotryptamine (95) > (R)-zacopride (110) > metaclopramide (128) > tryptamine (906) > 2-methyl-5-HT (1540). Other studies with ICS 205-930 (3 μM) gave agonist independent pA_2 values in the region of 6.5, thus fingerprinting the 5-HT$_4$ receptor. ICS 205-930 did not affect the concentration-effect curves to isoproterenol or papaverine but yielded a pA_2 value of 5.3 versus carbachol. Finally, all indole agonists acted as full agonists relative to 5-HT, whereas the substituted benzamides exhibited about 80 to 90% of the maximal response to 5-HT. At high concentrations the substituted benzamides appeared to exert muscarinic cholinoceptor blocking activity which became manifest as a second phase to their concentration-effect curves. This phase occurred after the maximum response at the 5-HT$_4$ receptor and led to complete inhibition of carbachol-induced tone.

Discussion

It is now clear that the first phase of the biphasic concentration-effect curve to 5-HT in guinea-pig ileum emanates from a unique 5-HT receptor distinct from the 5-HT$_1$, 5-HT$_2$ and 5-HT$_3$ subtypes (Craig and Clarke 1990). Equally well, it is clear that the second phase of the biphasic concentration-effect curve is mediated by 5-HT$_3$ receptors which are identical to Gaddum's M receptor (Gaddum and Picarelli 1957; Clarke et al. 1989; Craig and Clarke 1990).

The pharmacology of the novel ileal receptor exhibits considerable overlap with that defining the 5-HT$_4$ receptor in the CNS (Dumuis et al. 1988) and is without doubt a peripheral example of such a receptor subgroup. Nevertheless, it is pertinent to note that the characterization and classification of the 5-HT$_4$ receptor is weighted toward its unique agonist profile, and this in turn depends critically upon validation by ICS 205-930. This antagonist, however, is not a selective probe for the 5-HT$_4$ receptor, as it exhibits at least a 30-fold higher affinity for 5-HT$_3$ receptors. Furthermore, ICS 205-930, at concentrations close to those utilized for 5-HT$_4$ receptor blockade, evokes muscarinic cholinoceptor blockade (this study) as well as effects upon potassium, sodium and calcium currents (Scholtysik 1987; Scholtysik et al. 1988). Thus, at the present time the 5-HT Receptor Nomenclature Committee is reticent to recognize the 5-HT$_4$ receptor as a new, major subdivision of 5-HT receptors.

The above situation underlines the importance of procuring a high affinity, selective, competitive antagonist for receptor definition. In this vein, the receptor desensitization studies of Craig et al. (1990) in guinea-pig ileum enhanced confidence that the 5-HT$_4$ receptor is an entity distinct from the 5-HT$_3$ receptor and reinforced the conclusion that the prokinetic agent renzapride functioned as a 5-HT$_4$ receptor agonist. In the ileum, 5-HT$_4$ receptor desensitization with 5-methoxytryptamine proved to be highly selective in that responses to potassium chloride, carbachol, 1,1-dimethyl-4-phenylpiperazinium, histamine, substance P as well as responses to 5-HT at the 5-HT$_3$ receptor were unaffected. Furthermore, the affinity of ICS 205-930 for the 5-HT$_3$ receptor was also unaffected by desensitization with 5-methoxy-tryptamine. Thus, desensitization proved to be a useful tool for receptor identification and characterization in the ileum. Indeed, this approach may extend in applicability to other cell types and systems as desensitization of 5-HT$_4$ receptors is also evident in the CNS (Dumuis et al. 1989) and rat oesophagus (Baxter and Clarke, unpublished observations). On a cautionary note, however, it must be recalled that 5-methoxytryptamine is a potent agonist at 5-HT$_1$ and 5-HT$_2$ receptors and cannot, therefore, be regarded as a specific probe for the 5-HT$_4$ receptor. In this regard, desensitization of the 5-HT$_4$ receptor with a substituted benzamide (Figure 1) would seem to be a useful additional strategy.

A problem with using the guinea-pig ileum for quantitative studies on the 5-HT$_4$ receptor is that the first phase of the biphasic concentration-effect curve to 5-HT is subject to variation, even between ileal segments from the same guinea-pig. Thus, the plateau region of the curve (see Figure 1) may vary from as low as 5% to as much as 30 to 40% of the maximum response to 5-HT. The magnitude of the first phase relative to the second can be increased to about 60 to 65% by evoking responses to 5-HT upon a background of submaximal electrical stimulation of enteric cholinergic nerves (Craig and Clarke 1990; Craig et al. 1990). However, this does not overcome the problem of variation, and the instability of the preparation is increased. Because of these factors, and the associated presence of the 5-HT$_3$ receptor, a search was made for a more simple and quantitative preparation expressing the 5-HT$_4$ receptor. In this regard, the finding of a 5-HT$_4$ receptor in rat oesophagus is of significance. Furthermore, under our defined experimental conditions, no evidence was obtained for other 5-HT receptors, and the preparation was found to respond in a stable and reproducible manner. In view of the muscarinic cholinoceptor blocking activity of ICS 205-930 it would have been preferable to contract the oesophagus with a non-cholinergic agonist but, unfortunately, no other agonist class was found to hold developed tension as effectively as carbachol.

Whereas the presently described pharmacological profile of the 5-HT$_4$ receptor in rat oesophagus is in general agreement with the work of others (Bieger and Triggle 1985; Reeves et al. 1989), there are important differences. In particular, 5-methoxytryptamine was found to be over

100-times less potent than 5-HT by Reeves et al. (1989), and several benzamide derivatives (renzapride, cisapride, metaclopramide) acted as unsurmountable antagonists toward 5-HT (Triggle et al. 1988; Reeves et al. 1989). These differences are not easy to explain, but a potent agonistic action of 5-methoxytryptamine of approximate equivalency to 5-HT is essential for definition of the $5-HT_4$ receptor (Dumuis et al. 1988; Clarke et al. 1989; Craig and Clarke 1990). Antagonism by the benzamides may reflect the resultant of a number of factors, such as: a lower intrinsic efficacy relative to 5-HT; low receptor density or coupling efficiency or both; receptor desensitization, etc. In addition, in high concentration the substituted benzamides, as with ICS 205-920, exert muscarinic cholinoceptor blocking activity which may further complicate the interpretation of experimental results.

Care was taken in the rat oesophagus to work under equilibrium conditions. Therefore, tissue uptake of 5-HT was blocked with cocaine and monoamine oxidase was irreversibly inactivated with pargyline to preserve lipophilic analogues of 5-HT which simply diffuse into cells. Thus, the high potency of 5-HT ($pIC_{50} = 8.1$) and 5-methoxytryptamine relative to that reported by some others at $5-HT_4$ receptor (Dumuis et al. 1988, 1989; Shenker et al. 1987; Reeves et al. 1989) may be attributed, at least in part, to this maneuver. Blockade of indole uptake and metabolism may also explain why 5-HT and 5-methoxytryptamine are 12 times more potent than cisapride or renzapride in rat oesophagus, whereas 5-HT and renzapride (Dumuis et al. 1988) and 5-HT and cisapride (Bockaert et al. 1990) have been reported to be equipotent in mouse colliculi neurons and guinea-pig hippocampal membranes respectively. Kaumann et al. (1990b) also suppressed uptake of 5-HT with cocaine in human atrial appendage and, likewise, found 5-HT to be more potent than renzapride (about 10 times).

An exciting result from the present study is the demonstration, for the first time, that $5-HT_4$ receptor agonism functions to facilitate the peristaltic reflex. Thus, renzapride and 5-HT, in concentrations fully compatible with $5-HT_4$ receptor agonism, initiated intermittent peristalsis at subthreshold intraluminal pressures. This action was blocked by ICS 205-930 but not ondansetron, clearly implicating the $5-HT_4$ receptor over an action at $5-HT_3$ receptors. Complete blockade by ICS 205-930 suggests that agonism by 5-HT and renzapride at ileal $5-HT_{1A}$, "$5-HT_1$-like", $5-HT_2$ and $5-HT_{1P}$ receptors is not an important factor. All of these 5-HT receptor sites are present in guinea-pig ileum, but all are resistant to antagonism by ICS 205-930 (Bradley et al. 1986; Mawe et al. 1986). Furthermore, blockade with ICS 205-930 can hardly be attributed to its affinity for muscarinic cholinoceptors, as renzapride was found to evoke "atropine-resistant" peristalsis and ICS 205-930, itself, did not parallel atropine in raising the threshold pressure for peristalsis. In addition, as judged from previous work (Craig and Clarke

1990) and from the present study, the concentration of ICS 205-930 employed is below its equilibrium dissociation constant for muscarinic cholinoceptors (see Results). Thus, 5-HT$_4$ receptor agonism by renzapride would seem to offer a singular, long awaited explanation (see Schuurkes et al. 1987; King and Sanger 1988) for its prokinetic activity, as firing of the fundamental reflex involved in gut transit is facilitated. Similarly, despite years of inconclusive study, a rational explanation is offered for the documented facilitatory role of 5-HT on the peristaltic reflex (Bulbring and Crema 1958). Although certain prokinetic benzamides such as metoclopramide (Okwvasaba and Hamilton 1976) and cisapride (Tonini et al. 1989) have been shown to enhance or promote peristalsis, the findings have been equivocal and have not resulted in a clear mechanistic base. The present findings suggest strongly that in guinea-pig ileum the 5-HT$_4$ receptor may be the target site. Facilitation of acetylcholine release to enhance the reflex would be fully consistent with previous studies (Sanger 1987; Craig and Clarke 1990). However, additional neurotransmitters may be involved, such as substance P. Others have shown that atropine-resistant peristalsis (Tonini et al. 1981) can be blocked by desensitization with substance P (Yokoyama and North 1983) or by a substance P antagonist (Bartho et al. 1982; Grider and Makhlouf 1986). Finally the failure of ICS 205-930, itself, to affect the peristaltic reflex suggests that the 5-HT$_4$ receptor may play a secondary or modulatory role in the reflex, rather than being an essential, key link.

Acknowledgement

This work was supported by NIH Grant NS 24871.

References

Bartho, L., Holzer, P., Donnerer, J., and Lembeck, F. (1982). Effects of substance P, cholecystokinin octapeptide, bombesin, and neurotensin on the peristalic reflex of the guinea-pig ileum in the absence and in the presence of atropine. Naunyn-Schmiedeberg's Arch. Pharmacol. 321: 321–328.

Baxter, G. S., and Clarke, D. E. (1990). Putative 5-HT$_4$ receptors mediate relaxation of rat oesophagus. The Second IUPHAR Satellite Meeting on Serotonin. Abstract P85: 78.

Bieger, D., and Triggle, C. R. (1985). Pharmacological properties of mechanical responses of the rat oesophageal muscularis mucosae to vagal and field stimulation. Br. J. Pharmacol. 84: 93–106.

Bockaert, J., Sebben, M., and Dumuis, A. (1990). Pharmacological characterization of 5-hydroxytryptamine$_4$ (5-HT$_4$) receptors positively coupled to adenylate cyclase in adult guinea-pig hippocampal membranes: effect of substituted benzamide derivatives. Mol. Pharmacol. 37: 408–411.

Bradley, P. B., Engel, G., Fozard, J. R., Humphrey, P. P. A., Middlemiss, D. N., Mylecharane, E. J., Richardson, B. P., and Saxena, P. R. (1986). Proposals for the classification and nomenclature of functional receptors for 5-hydroxytryptamine. Neuropharmacology. 25: 563–576.

Bulbring, E., and Crema A. (1958). Observations concerning the action of 5-hydroxy-tryptamine on the peristaltic reflex. Br. J. Pharmacol. 13: 444–457.

Clague, R. U., Eglen, R. M., Strachan, A. C., and Whiting, R. L. (1985). Action of agonists and antagonists at muscarinic receptors present on ileum and atria in vitro. Br. J. Pharmacol. 64: 293–300.

Clarke, D. E., Craig, D. A., and Fozard, J. R. (1989). The 5-HT$_4$ receptor: Naughty but nice. Trends Pharmacol. Sci. 10: 385–386.

Craig, D. A., and Clarke, D. E. (1989). 5-Hydroxytryptamine and cholinergic mechanisms in guinea-pig ileum. Br. J. Pharmacol. 96: 247P.

Craig, D. A., and Clarke, D. E. (1990). Pharmacological characterization of a neuronal receptor for 5-hydroxytryptamine in guinea-pig ileum with properties similar to the 5-hydroxytryptamine$_4$ receptor. J. Exp. Pharmacol. Ther. 252: 1378–1386.

Craig, D. A., Eglen, R. M., Walsh, L. K. M., Perkins, L. A., Whiting, R. L., and Clarke, D. E. (1990). 5-Methoxytryptamine and 2-methyl-5-hydroxytryptamine-induced desensitization as a discriminative tool for 5-HT$_3$ and putative 5-HT$_4$ receptors in guinea-pig ileum. Naunyn-Schmiedeberg's Arch. Pharmacol. 342: 9–16.

Craig, D. A., and Clarke, D. E. (1991). Peristalsis evoked by 5-HT and renzapride: evidence for putative 5-HT$_4$ receptor activation. Br. J. Pharmacol. 102: 563–564.

Dumuis, A., Bouhelal, R., Sebben, M., Cory, R., and Bockaert, J. (1988). A nonclassical 5-hydroxytryptamine receptor positively coupled with adenylate cyclase in the central nervous system. Mol. Pharmacol. 34: 880–887.

Dumuis, A., Sebben, M., and Bockaert, J. (1989). BRL 24924: A potent agonist at a nonclassical 5-HT receptor positively coupled with adenylate cyclase in colliculi neurons. Eur. J. Pharmacol. 162: 381–384.

Eglen, R. M., Swank, S. R., Dubuque, L. K., and Whiting R. L. (1990). Characterization of serotonin receptors mediating guinea-pig ileal contractions in vitro. Br. J. Pharmacol. 101: 513–520.

Elswood, C. J., Bunce, K. T., and Humphrey, P. P. A. (1990). Identification of 5-HT$_4$ receptors in guinea-pig ascending colon. The Second IUPHAR Satellite Meeting on Serotonin. Abstract P86: 78.

Gaddum, J. H., and Picarelli, Z. P. (1957). Two kinds of tryptamine receptor. Br. J. Pharmacol. 12: 323–328.

Grider, J. R., and Makhlouf, G. M. (1986). Colonic peristaltic reflex: identification of vasoactive peptide as mediator of descending relaxation. Am. J. Physiol. 251: G40–G45.

Kaumann, A. J., Sander, L., Brown, A. M., Murray, K. J., and Brown, M. J. (1990a). A 5-hydroxytryptamine receptor in human atrium. Br. J. Pharmacol. 100: 879–995.

Kaumann, A. J., Sanders, L., Brown, A. M., Murray, K. J., and Brown, M. J. (1990b). Human atrial 5-HT receptors: similarity to rodent neuronal 5-HT$_4$ receptors. Br. J. Pharmacol. 100: 319P.

King, F. D., and Sanger, G. J. (1988). Gastrointestinal motility enhancing agents. In: Allen, R. C. (ed.), Annual Reports in Medicinal Chemistry. San Diego: Academic Press, vol. 23, pp. 201–210.

Mawe, G. M., Branchek, T. A., and Gershon, M. D. (1986). Blockade of 5-HT-mediated enteric slow EPSPs by BRL 24924: gastrokinetic effects. Am. J. Physiol. 257: G386–G396.

Okwuasaba, F. K., and Hamilton, J. T. (1976). The effect of metoclopramide on intestinal muscle responses and the peristaltic reflex in vitro. Can. J. Physiol. Pharmacol. 54: 393–404.

Reeves, J. J., Bunce, K. T., Humphrey, P. P. A., and Gunning S. J. (1989). Further characterization of the 5-HT receptor mediating smooth muscle relaxation in rat oesophagus. Br. J. Pharmacol. 99: 800P.

Sanger, G. J. (1987). Increased gut cholinergic activity and antagonism of 5-hydroxytryptamine M receptor by BRL 24924: Potential clinical importance of BRL 24924 Br. J. Pharmacol. 91: 77–81.

Scholtysik, G. (1987). Evidence for inhibition by ICS 205-930 and stimulation by BRL 34915 of K$^+$ conductance in cardiac muscle. Naunyn-Schmiedeberg's Arch. Pharmacol. 335: 692–696.

Scholtysik, G., Imoto, Y., Yatani, A., and Brown A. M. (1988). 5-Hydroxytryptamine antagonist ICS 205-930 blocks cardiac potassium, sodium and clacium currents. J. Pharmacol. Exp. Ther. 245: 773–777.

Schuurkes, J. A. J., Megens, A. A. H. P., Niemegeers, C. J. E., Leysen, J. E., Van Nueten, J. M. (1987). A comparative study of the cholinergic vs the anti-dopaminergic properties of benzamides with gastrointestinal prokinetic activity. In: Szurszewski, J. H. (ed.), Cellular physiology and clinical studies of gastrointestinal prokinetic activity. New York: Elsevier, pp. 231–247.

Shenker A., Maayani, S., Weinstein, H., and Green J. P. (1987). Pharmacological characterization of two 5-hydroxytryptamine receptors coupled to adenylate cyclase in guinea-pig hippocampal membranes. Mol. Pharmacol. 31: 357–367.

Stanniforth, D. H., and Pennick, M. (1990). Human pharmacology of renzapride: a new gastrokinetic benzamide without dopamine antagonist properties. Eur. J. Clin. Pharmacol. 38: 161–164.

Tonini, M., Frigo, G., Lecchini, S., D'Angelo, L., and Crema, A. (1981). Hyoscine-resistant peristalsis in guinea-pig ileum. Eur. J. Pharmacol. 71: 375–381.

Tonini, M., Galligan, J. J., and North R. A. (1989). Effects of cisapride on cholinergic neurotransmission and propulsive motility in the guinea-pig ileum. Gastroenterology. 96: 1257–1264.

Triggle, C. R., Ohia, S. E., and Bieger, D. (1988). 5-Hydroxytryptamine-induced relaxation of rat and mouse oesophageal smooth muscle. Pharmacologist. 30: A126.

Villalón, C. M., den Boer, M. O., Heiligers, J. P. C., and Saxena, P. R. (1990). Mediation of 5-hydroxytryptamine-induced tachycardia in the pig by the putative 5-HT$_4$ receptor. Br. J. Pharmacol. 100: 665–667.

Yokoyama, S., and North, R. A. (1983). Electrical activity of longitudinal and circular muscle during peristalsis. Am. J. Physiol. 244: G83–G88.

Serotonin: Molecular Biology, Receptors and Functional Effects
ed. by J. R. Fozard/P. R. Saxena
© 1991 Birkhäuser Verlag Basel/Switzerland

5-HT$_{1D}$ and 5-HT$_4$ Receptor Agonists Stimulate the Peristaltic Reflex in the Isolated Guinea Pig Ileum

K. H. Buchheit and T. Buhl

Preclinical Research (386/543), Sandoz Pharma Ltd., CH-4002 Basel, Switzerland

Summary. The effect of serotonin (5-HT) and of more selective agonists on the peristaltic reflex evoked in the isolated guinea pig ileum was investigated. Using the Trendelenburg technique, peristaltic contractions were elicited by increasing intraluminal pressure and rhythmic contractions of the longitudinal and circular muscle were measured after serosal administration of the drugs. 5-HT potently stimulated contractions of the longitudinal muscle. Of the potent 5-HT$_{1A}$ receptor agonists, 8-OH-DPAT, 5-carboxamidotryptamine (5-CT) and dipropyl-5-CT (DP-5-CT), only 5-CT caused a substantial stimulation. Of the 5-HT$_{1C}$/5-HT$_2$ receptor agonists DOI and 5-methoxytryptamine (5-MeOT), DOI was inactive whereas 5-MeOT potently stimulated longitudinal muscle contractions. Compounds with agonist activity at 5-HT$_{1D}$ receptors (5-CT, 5-MeOT and sumatriptan) had a stimulatory effect. The 5-HT$_3$ receptor agonist 2-methyl-5-HT and 5-HT$_3$ receptor antagonists did not influence the peristaltic reflex. Various 5-HT$_4$ receptor agonists stimulated longitudinal muscle activity. 6-OH-Indalpine, a 5-HT$_{1P}$ agonist, was without any effect. This data suggest that 5-HT stimulates the peristaltic reflex in the isolated guinea pig ileum by activation of 5-HT$_4$ and 5-HT$_{1D}$ receptors; other 5-HT receptor subtypes appear not to play a significant role in the modulation of this reflex.

Introduction

In mammalian small intestine, 5-HT is stored in enteric neurons (Gershon et al. 1977) and in mucosal enterochromaffin cells (Erspamer and Asero 1952), and is secreted into the lumen and into the submucosal tissue (Forsberg and Miller 1982). 5-HT has profound effects on the motility of the small intestine. In most species, e.g. in dogs (Haverback et al. 1957), cats (Erspamer 1966) and guinea pigs (Holzer and Lembeck 1979), 5-HT stimulates motility. In man, however, a stimulatory effect precedes an inhibitory one (Schmid and Kinzlmeier 1959). The same kind of response is found in the isolated guinea pig ileum if 5-HT is administered to the serosal side, whereas mucosal administration produces only stimulation (Bülbring and Crema 1958). Not very much is known about the character of the 5-HT receptors involved. We therefore re-investigated the effects of 5-HT in isolated guinea pig ileum with the objective of identifying the receptor subtype which mediates the effects on small intestinal motility. This question was approached by investigating the effect of various 5-HT receptor agonists and antagonists on the peristaltic reflex in the isolated guinea pig ileum (Trendelenburg 1917). To determine whether the effect of a 5-HT compound in the Trendelenburg assay might be explained by its affinity for a particular 5-HT receptor subtype, the potency to modify the peristaltic reflex in the guinea pig ileum was compared to values from other functional assays in the literature.

Materials and Methods

Female guinea pigs, 200–400 g were stunned and bled. Segments of terminal ileum, 4–5 cm long, were removed. The tissue was suspended in an organ bath under an initial load of 1g; it was bathed with Tyrode solution (NaCl 136.8; $CaCl_2$ 1.8; KCl 2.7; NaH_2PO_4 0.42; $MgCl_2$ 1.05; $NaHCO_3$ 11.9; glucose 5.6 mM) at pH 7.4 maintained at $37°C$ and bubbled with 5% CO_2 in oxygen. Peristalsis was elicited for 30 s by increasing intraluminal pressure from 0 to 1 cm H_2O in 6 min intervals. Longitudinal muscle responses were recorded using an isotonic force displacement transducer. A pressure transducer measured intraluminal pressure as an indication of circular muscle activity. Longitudinal and circular muscle activity were measured by determining the area under the curve of peristaltic contractions.

Concentration response curves were established by plotting the mean area under the curve from 4–6 experiments expressed as percentage of the control preparations. Drugs were applied to the serosal side in a cumulative way at intervals of at least 12 min. Each concentration was left in contact with the tissue for 10 min. Drugs were characterized by the concentration yielding half-maximal stimulation of the peristaltic concentrations, expressed as pEC_{50} values. These values were graphically determined from the concentration response curves.

5-HT (Fluka, Buchs, Switzerland), (±)-8-hydroxydipropylaminotetralin (8-OH-DPAT), 5-carboxamidotryptamine (5-CT), (±)-1-(2,5-dimethoxy-4-iodophenyl)-2-aminopropane (DOI) and metoclopramide (RBI, Natick, USA) were purchased, dipropyl-5-carboxamidotryptamine (DP-5-CT), 5-methoxytryptamine (5-MeOT), sumatriptan, ICS 205-930 ([3α-tropanyl)-]-1H-indole-3-carboxylic acid ester), granisetron, 2-methyl-5-HT (2-Me-5-HT), renzapride, zacopride and 6-OH-indalpine were synthesized at Sandoz Pharma Ltd., Basel. Cisapride and BRL 20627 ([2α, 6β, 9aα)-(±)-4-amino-5-chloro-2-methoxy-N-(octahydro-6-methyl-2H-quinolizin-2-yl)benzamide] were gifts by Janssen (Beerse, Belgium) and Beecham (Harlow, UK), respectively.

Results

Elevating the level of the buffer in the reservoir 1 cm over that in the organ bath, fills the lumen of the ileum. This causes a radial distension

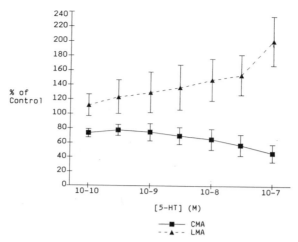

Figure 1. Effect of 5-HT on peristaltic activity in the isolated guinea pig ileum. Shown are effects on longitudinal (LMA) and circular (CMA) muscle activity elicited by a pressure stimulus of 1 cm of water. All values represent means ± S.D.; n = 5; activity of control = 100%.

of the muscle and elicits a peristaltic reflex consisting of coordinated contractions and relaxations of the longitudinal and the circular muscle layer. Under control conditions, these responses are constant for several hours, without signs of fatigue. After serosal administration, 5-HT sensitized the ileum against pressure stimuli and thus augmented the rhythmic peristaltic contractions elicited by the pressure increase of 1 cm H_2O. As a consequence, longitudinal contractions were potently increased, whereas contractions of the circular muscle layer were decreased (Figure 1). These reductions of circular muscle contractions parallel the stimulatory effect on longitudinal muscle and are not a specific response to 5-HT but can be observed after stimuli of various origins, such as carbachol, CCK or histamine (Buchheit, unpublished). (For this reason, only effects on longitudinal muscle contractions will be considered in the following sections.) The maximal increase obtained with 10^{-7} mol/l 5-HT was 200% of the control value, and a pEC_{50} value of 8.1 was determined (Table 1).

For the $5-HT_{1A}$ receptor agonists 8-OH-DPAT, 5-CT and DP-5-CT, no consistent reaction could be obtained (Table 1). Only 5-CT caused a potent and marked increase in peristaltic movements with a pEC_{50} value of 8.2 and an E_{max} of 180% (Figure 2). 8-OH-DPAT was completely inactive at concentrations up to 10^{-6} mol/l. Higher concentrations caused a slight reduction of longitudinal muscle contractions. DP-5-CT stimulated the peristaltic reflex with a pEC_{50} value of 6.8. A maximal increase of 220% was obtained at 10^{-6} mol/l. Higher concentrations caused a diminished effect.

Table 1. Stimulatory effect of various selective 5-HT compounds on longitudinal muscle activity in the Trendelenburg preparation

Compound	Selectivity	pEC_{50}	E_{max} (%)
5-HT	no	8.1	201
8-OH-DPAT	1A	<5	—
5-CT	1A/1D	8.2	180
DP-5-CT	1A	6.8	220
5-MeOT	1C/1D/4	8.0	220
DOI	1C/2	<6	—
Sumatriptan	1D	7.1	270
ICS 205-930	3	<7	—
Granisetron	3	<7	—
2-Me-5-HT	3	<5	—
Metoclopramide	3/4	5.7	340
Cisapride	3/4	7.7	220
Renzapride	3/4	7.3	166
Zacopride	3/4	5.6	170
BRL 20627	3/4	6.6	200
6-OH-Indalpine	1P	<5	—

Maximal effect (E_{max}) expressed as percentage of control; selectivity for 5-HT receptor subtype(s) is indicated.

5-MeOT and DOI were investigated as non-selective but potent 5-HT$_{1C/2}$ receptor agonists (Table 1). Only 5-MeOT caused a potent stimulation with a pEC_{50} value of 8.0 ($E_{max} = 220\%$). Concentrations of 5-MeOT higher than 10^{-7} mol/l were found to have inhibitory effects on longitudinal contractions, and thus the concentration response curve of 5-MeOT has a bell-shaped appearance. DOI on the other hand did not induce significant changes in the peristaltic contractile response of the longitudinal muscle in the concentration range between 1 nmol/l and 1 μmol/l.

The 5-HT$_{1D}$ receptor agonist sumatriptan stimulated peristaltic contractions of the longitudinal muscle very effectively (Table 1, Figure 2) with a pEC_{50} value of 7.1. At a concentration of 1 μmol/l, the increase of contractile activity was 270%. The response to sumatriptan was qualitatively not different to that of 5-HT or 5-CT.

ICS 205-930, granisetron and 2-Me-5-HT were tested as representatives for 5-HT$_3$ receptor antagonists and agonists respectively. ICS 205-930 had no effects at concentrations between 10^{-9} and 10^{-7} mol/l. Higher concentrations reduced contractions of the longitudinal and circular muscle, with the circular muscle being more sensitive to ICS 205-930. Effects in the concentration range above 10^{-7} mol/l showed great variability. Similar results were obtained for granisetron which was ineffective at concentrations between 10^{-9} and 10^{-6} mol/l and reduced contractions of both muscle layers at 10^{-5} mol/l. The 5-HT$_3$ receptor agonist 2-Me-5-HT had no influence on the peristaltic reflex at concentrations below 10^{-6} mol/l. The effect of 5-HT could not be

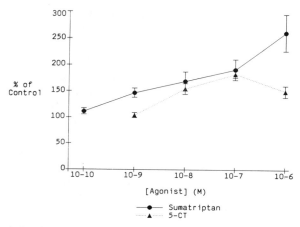

Figure 2. Stimulation by sumatriptan and 5-carboyamidotriptamine (5-CT) of the peristaltic activity in the isolated guinea pig ileum. Shown are effects on longitudinal muscle activity elicited by a pressure stimulus of 1 cm of water. All values represent means ± S.D.; n = 5; activity of control = 100%.

antagonized by ICS 205-930 at a concentration of 10^{-7} mol/l which is 10 times above the K_D concentration of ICS 205-930 for 5-HT_3 receptors in the guinea pig ileum.

Several substituted benzamides, such as metoclopramide, BRL 20627, renzapride, zacopride and cisapride, which have been characterized as 5-HT_4 receptor agonists (Dumuis et al. 1989) were also investigated. All these compounds stimulated the contractile activity of the longitudinal muscle layer with varying potencies (Table 1). All 5 compounds caused a qualitatively identical response. Metoclopramide and zacopride were only weakly active (pEC_{50} values: 5.7 and 5.6 respectively). BRL 20627 was about 10 times more potent ($pEC_{50} = 6.6$). Renzapride and cisapride stimulated contractions with a pEC_{50} value of 7.7 and 7.3, respectively. Metoclopramide was the most efficacious stimulator of longitudinal muscle activity with an increase of contractile activity of 340%. The E_{max} values for all other benzamides were in the range between 170 and 220%. The 5-HT_{1P} receptor agonist 6-OH-indalpine, as characterized by Gershon and colleagues (Mawe et al. 1986), was tested in the concentration range between 10^{-9} and 10^{-5} mol/l and did not alter the peristaltic response to pressure stimuli.

Discussion

The peristaltic reflex is a complex biological response to increased intraluminal pressure in the gastrointestinal tract. Due to the presence of 5-HT in enterochromaffin cells (Erspamer 1966) within the mucosa of the gut, and due to the presence of various 5-HT receptor subtypes both on neuronal and on smooth muscle tissue, 5-HT has therefore the potential to regulate and modify the peristaltic reflex. The Trendelenburg preparation, using a piece of isolated guinea pig ileum, offers a simple method for the investigation of such effects. Using this preparation, problems arise from the quantification of the peristaltic responses, due to the great variability of contraction height and frequency even within one preparation. This problem can be overcome by integration of the area under the contraction time curves as a measurement for contractile activity. Using these integrals, it can be demonstrated that contractions of the circular and longitudinal muscle layer, evoked by increases of intraluminal pressure, remain stable for some hours. Thus, this technique can be used to quantify drug effects on the peristaltic reflex.

In contrast to a previous report (Bülbring and Crema 1958), in our hands 5-HT had only stimulatory effects on the peristaltic reflex. An inhibition could be seen at no stage of the experiment. No explanation can be given for this difference. Only the longitudinal muscle coat was stimulated by 5-HT, whereas contractility of the circular muscle layer was reduced. This latter event, however, is not related to the effect of 5-HT directly, but rather seems to be a consequence of the stimulatory effect on the longitudinal muscle.

In order to elucidate the type of 5-HT receptor mediating the stimulatory effect on the peristaltic reflex, various 5-HT receptor agonists and antagonists which are more selective for a particular receptor subtype than 5-HT itself, were investigated. The following compounds failed to induce a specific effect on the peristaltic reflex (Table 1): the 5-HT_{1A} receptor agonist 8-OH-DPAT, the $5\text{-HT}_{1C/2}$ receptor agonist DOI, the 5-HT_3 receptor agonist 2-Me-5-HT, as well as ICS 205-930 and granisetron which are antagonists at the 5-HT_3 receptor subtype, and finally 6-OH-indalpine, a 5-HT_{1P} receptor agonist.

Among the compounds which mimic the effect of 5-HT and thus stimulate the peristaltic reflex, two groups can be discriminated, the substituted benzamides metoclopramide, BRL 20627, cisapride, renzapride and zacopride, and the indole compounds 5-HT, 5-CT, 5-MeOT and sumatriptan (Table 1).

What conclusions can be drawn from these results? In order to sort out whether the stimulatory effect of a compound on the peristaltic reflex is due to its effect at a particular 5-HT receptor subtype, the potency to stimulate the contractile activity, expressed as pEC_{50} value, was compared to potencies from functional assays in the literature, as compiled in Table 2.

Tables 1 and 2 reveal that the stimulatory effect of the indole derivatives cannot be explained by 5-HT_{1A} receptor agonism, as 5-MeOT and sumatriptan are weak 5-HT_{1A} receptor agonists but potent stimulators

Table 2. Potency of 5-HT receptor ligands to stimulate or inhibit 5-HT receptor subtypes in functional *in vitro* assays. Data are expressed as pEC_{50} or pA_2 values. pK_D values from binding experiments are given when no functional data were available

Compound	5-HT_{1A}	5-HT_{1C}	5-HT_{1D}	5-HT_2	5-HT_3	5-HT_4
5-HT	6.9(1)	6.5(5)	7.6(6)	6.2(7)	6.4(8)	7.0(10)
8-OH-DPAT	7.0(1)	<4(5)	5.8(6)	<4(7)	<5(8)	<5(11)
5-CT	7.2(1)	5.7(5)	8.1(6)	<5(7)	5.0(8)A	5.5(11)
DP-5-CT	7.4(1)	4.5(4)	6.5(6)	5.3(4)	<5(8)	n.d.
5-MeOT	6.6(1)	6.2(5)	7.3(6)	5.7(7)	<5(8)	7.0(11)
DOI	5.6(2)	7.0(5)	5.7(2)	7.8(2)	n.d.	n.d.
Sumatriptan	5.6(3)	4.1(3)	6.3(3)	3.7(3)+	n.d.	n.d.
ICS 205-930	<5(1)	<5(4)	n.d.	<5(4)	7.9(9)A	6.4(10)
Granisetron	<5(2)	<5(2)	n.d.	<5(2)	7.8(8)A	<5(10)
2-Me-5-HT	4.4(1)	5.8(4)	6.4(4)	5.0(4)	5.1(8)	<5(11)
Metoclopramide	5.2(4)	5.4(4)	n.d.	5.9(4)	5.2(8)A	5.3(10)
Cisapride	5.7(2)	6.3(2)	n.d.	n.d.	7.2(8)A	7.1(10)
Renzapride	4.9(2)	5.4(2)	n.d.	4.7(2)	7.0(8)A	6.9(10)
Zacopride	<4(2)	5.9(2)	n.d.	n.d.	8.0(8)A	5.9(10)
BRL 20627	n.d.	n.d.	n.d.	n.d.	6.4(8)A	5.5(10)
6-OH-Indalpine	4.4(2)	4.4(2)	n.d.	4.7(2)	<5(8)	n.d.

n.d. = no data; A = antagonist; + = binding data; (1) Markstein et al. 1986; (2) unpublished binding data, D. Hoyer; (3) Schoeffter and Hoyer 1989; (4) binding data, Hoyer et al. 1989; (5) Hoyer et al. 1989; (6) Schoeffter et al. 1988; (7) Cory et al. 1987; (8) Buchheit, unpublished; (9) Buchheit et al. 1985; (10) Dumuis et al. 1989; (11) Dumuis et al. 1988.

of the peristaltic reflex and as the potent 5-HT_{1A} receptor agonists 8-OH-DPAT and DP-5-CT are weak stimulators of contractile activity. At a first glance, the effect of 5-MeOT in the isolated guinea pig ileum might advocate a role for 5-HT_{1C}-/5-HT_2 receptors. However, the 5-HT_{1C}-/5-HT_2 receptor agonist DOI is inactive in the Trendelenburg assay. As at 5-HT_{1C} receptors in pig choroid plexus DOI is even more potent than 5-MeOT, and the pK_d value of DOI for 5-HT_2 receptors is higher than that of 5-MeOT (Table 2), it is unlikely that 5-HT_{1C}- or 5-HT_2 receptors mediate the effects of 5-HT or 5-MeOT.

Three compounds with substantial affinity for 5-HT_{1D} receptors were investigated: 5-CT, 5-MeOT and sumatriptan (Table 2). All 3 compounds stimulated the peristaltic reflex potently and massively (Table 1). The resulting pEC_{50} values correspond quite well with functional pD_2 values for inhibition of adenylate cyclase in calf substantia nigra. These results favor the assumption that 5-HT_{1D} receptor agonism is responsible for the stimulatory effect of the four active indole derivatives.

Another indole derivative, 6-OH-indalpine, which had been characterized as a 5-HT_{1P} receptor agonist inducing slow depolarizations in neurons of the myenteric plexus (Mawe et al. 1986, did not modify the peristaltic reflex in nanomolar to micromolar concentrations. This does not argue for the involvement of 5-HT_{1P} receptors in the regulation of peristalsis.

Excitatory 5-HT_3 receptors have been identified on neurons of the myenteric plexus in the guinea pig ileum and thus were good candidates for mediators of the stimulatory effect of 5-HT. The inactivity of the 5-HT_3 receptor agonist 2-Me-5-HT argues against such a role. In addition, the other stimulators of the peristaltic reflex, 5-CT, 5-MeOT and sumatriptan are devoid of affinity for 5-HT_3 receptors. Furthermore, the effect of 5-HT on the peristaltic reflex could not be antagonized by the 5-HT_3 receptor antagonists ICS 205-930 and granisetron, both failing to modify the peristaltic reflex when tested at concentrations meaningful for 5-HT_3 receptors. This also argues against the involvement of 5-HT_3 receptors.

How do the benzamides stimulate the peristaltic reflex? All benzamides used are antagonists at 5-HT_3 receptors (Table 2). But as shown above, this effect is unrelated to the stimulatory effect on peristalsis. In addition to their antagonism at 5-HT_3 receptors, these compounds have been shown to be agonists at the so called 5-HT_4 receptor (Dumuis et al. 1989). Their potency at this receptor, as measured by the stimulation of adenylate cyclase in mouse colliculi neurons, is shown in Table 2. A comparison of the pEC_{50} values for the stimulation of the peristaltic reflex (Table 1) with the pEC_{50} values for the stimulation of adenylate cyclase in mouse colliculi neurons (Table 2) reveals a significant correlation ($r = 0.92$; $p = 0.011$) between the two data sets. This suggests that the benzamides stimulate the peristaltic reflex in vitro by activation of 5-HT_4 receptors.

Taken together, the results obtained with various 5-HT receptor agonists and antagonists, provided no evidence for an involvement of 5-HT_{1A}-, 5-HT_{1C}-, 5-HT_2, 5-HT_3- or 5-HT_{1P} receptors in the regulation of the peristaltic reflex in the isolated guinea pig ileum. There is evidence, however, that in this preparation peristaltic contractions can be stimulated independently by activation of 5-HT_{1D} and 5-HT_4 receptors. Activation of these receptors leads to sensitization of the tissue against intraluminal pressure stimuli and as a consequence, the longitudinal muscle layer responds with increased contractions. Whether these mechanisms can also be activiated in man to achieve a prokinetic effect on the small intestine remains to be determined. Substituted benzamides such as metoclopramide, renzapride and cisapride clearly have such an effect, it is unclear however, whether this effect is elicited by facilitation of the peristaltic reflex. So far, no data about the effect of compounds with substantial agonism at 5-HT_{1D} receptors on human gastrointestinal motility are available, but with such compounds being in clinical development, this question will be answered in the near future.

Acknowledgements

The authors are indebted to Dr. D. Hoyer for the permission to use unpublished data. The skilfull technical assistance of A. Bertholet, J. L. Runser and P. B. Wingrove is gratefully acknowledged.

References

Buchheit, K. H., Engel, G., Mutschler, E., and Richardson, P. (1985). Study of the contractile effect of 5-hydroxytryptamine (5-HT) in the isolated longitudinal muscle strip from guinea pig ileum. Naunyn-Schmiedeberg's Arch. Pharmacol. 329: 36–41.

Bülbring, E., and Crema, A. (1958). Observations concerning the action of 5-hydroxy-tryptamine on the peristaltic reflex. Br. J. Pharmacol. 13: 444–457.

Cory, R. N., Rouot, B., Guillon, G., Sladeczek, F., Balestre, M. N., and Bockaert, J. (1987). The 5-hydroxytryptamine (5-HT$_2$) receptor stimulates inositol phosphate formation in intact and broken WRK1 cells: Determination of occupancy-response relationships for 5-HT agonists. J. Exp. Pharmacol. Ther. 241: 258–267.

Dumuis, A., Bouhelal, R., Sebben, M., Cory, R., and Bockaert, J. (1988). A non-classical 5-hydroxytryptamine receptor positively coupled with adenylate cyclase in the central nervous system. Mol. Pharmacol. 34: 880–887.

Dumuis, A., Sebben, M., and Bockaert, J. (1989). The gastrointestinal prokinetic benza-mide derivatives are agonists at the non-classical 5-HT receptor (5-HT$_4$) positively cou-pled to adenylate cyclase in neurons. Naunyn-Schmiedeberg's Arch. Pharmacol. 340: 403–410.

Erspamer, V. (1966). Occurrence of indolealkylamines in nature. In Erspamer, V. (ed.), Handbook of experimental pharmacology. Vol. 19, 5-Hydroxytryptamine and related indolealkylamines. New York: Springer, pp. 132–182.

Erspamer, V., and Asero, B. (1952). Identification of enteramine, the specific hormone of the enterochromaffin cell system as 5-hydroxy-tryptamine. Nature 169: 800–801.

Forsberg, E. J., and Miller, R. J. (1982). Cholinergic agonists induce vectorial release of serotonin from duodenal enterochromaffin cells. Science 217: 355–356.

Gershon, M. D., Dreyfus, C. F., Pickel, V. M., Joh, T. H., and Reis, D. J. (1977). Serotoninergic neurons in the peripheral nervous system: Identification in gut by immuno-histochemical localization of tryptophan hydroxylase. Proc. Natl. Acad. Sci. USA 74: 3086–3089.

Haverback, B. J., Hogben, C. A. M., Moran, N. C., and Terry, L. L. (1957). Effect of serotonin (5-hydroxytryptamine) and related compounds on gastric secretion and intesti-nal motility in the dog. Gastroenterology 32: 1058–1065.

Holzer, P., and Lembeck, F. (1979). Effect of neuropeptides on the efficiency of the peristaltic reflex. Naunyn-Schmiedeberg's Arch. Pharmacol. 307: 257–264.

Hoyer, D. (1989). 5-Hydroxytryptamine receptors and effector coupling mechanisms in peripheral tissues. In Fozard, J. R. (ed.), The peripheral actions of 5-hydroxytryptamine. Oxford: Oxford University Press, pp. 72–99.

Hoyer, D., Waeber, C., Schoeffter, P., Palacios, J. M., and Dravid, A. (1989). 5-HT$_{1C}$ receptor mediated stimulation of inositol phosphate production in pig choroid plexus. Naunyn-Schmiedeberg's Arch. Pharmacol. 339: 252–258.

Markstein, R., Hoyer, D., and Engel, G. (1986). 5-HT$_{1A}$ receptors mediate stimulation of adenylate cyclase in rat hippocampus. Naunyn-Schmiedeberg's Arch. Pharmacol. 333: 335–341.

Mawe, G. M., Branchek, T. A., and Gershon, M. D. (1986). Peripheral neural serotonin receptors: Identification and characterization with specific antagonists and agonists. Proc. Natl. Acad. Sci. USA 83: 9799–9803.

Schmid, E., and Kinzlmeier, H. (1959). Das Verhalten der Magenacidität und der Motilität im Verdauungstrakt der Menschen bei Infusion von Serotonin. Naunyn-Schmiedeberg's Arch. Pharmacol. 236: 51–54.

Schoeffter, P., and Hoyer, D. (1989). How selective is GR 43175? Interactions with functional

5-HT$_{1A}$, 5-HT$_{1B}$, 5-HT$_{1C}$ and 5-HT$_{1D}$ receptors. Naunyn-Schmiedeberg's Arch. Pharmacol. 340: 135–138.

Schoeffter, P., Waeber, C., Palacios, J. M., and Hoyer, D. (1988). The 5-HT$_{1D}$ receptor subtype is negatively coupled to adenylate cyclase in calf substantia nigra. Naunyn-Schmiedeberg's Arch. Pharmacol. 337: 602–608.

Trendelenburg, P. (1917). Physiologische und pharmakologische Versuche über die Dünndarmperistaltik. Arch. Exp. Path. Pharmacol. 81: 55–129.

Serotonin: Molecular Biology, Receptors and Functional Effects
ed. by J. R. Fozard/P. R. Saxena
© 1991 Birkhäuser Verlag Basel/Switzerland

The 5-HT$_4$ Receptor Mediating Tachycardia in the Pig

C. M. Villalón*, M. O. den Boer, J. P. C. Heiligers, and
P. R. Saxena

*Department of Pharmacology, Faculty of Medicine and Health Sciences,
Erasmus University Rotterdam, Post Box 1738, 3000 DR Rotterdam, The Netherlands*

Summary. The tachycardic response to 5-HT in the anaesthetized pig was mimicked by the indole derivatives, 5-methoxytryptamine and α-methyl-5-HT, and, to a lesser extent, by the benzamide derivatives (in order of potency) zacopride, renzapride, cisapride, metoclopramide and dazopride; the latter derivatives behaved as partial agonists. ICS 205-930 in high doses (1 and 3 mg/kg), but not other 5-HT receptor antagonists (methiothepin, metergoline, methysergide, mesulergine, ketanserin, cyproheptadine, mianserin, pizotifen, granisetron and MDL 72222) or dopamine antagonists (haloperidol and domperidone), antagonized this response. These results definitely exclude the involvement of 5-HT$_1$-like, 5-HT$_2$, 5-HT$_3$ or dopamine receptors and strongly suggest that the putative 5-HT$_4$ receptor mediates the positive chronotropic action of 5-HT in the anaesthetized pig. This receptor displays a pharmacological profile which indicates similarity with the 5-HT$_4$ receptor present on the neurons of the mouse embryo colliculi and guinea-pig ileum, as well as to that in the human heart. Since the pig heart is devoid of other 5-HT receptors, the heart rate responses to 5-HT in the pig can be utilized as a convenient experimental model for discovering drugs selective at the 5-HT$_4$ receptor. Lastly, the involvement of this novel 5-HT receptor in the above functional responses argues in favour of the extension of the 5-HT receptor classification to include the 5-HT$_4$ receptor.

Introduction

It is well known that 5-hydroxytryptamine (5-HT) elicits complex cardiovascular effects comprising bradycardia or tachycardia, hypotension or hypertension and vasodilatation or vasoconstriction. The eventual response depends – amongst other factors – upon the species, basal vascular tone, dose employed, vascular segment under study and, most importantly, the nature of the 5-HT receptors involved. With respect to the latter, there is little doubt that the receptor pharmacology of 5-HT has gone through a major transition over the last few years (Bradley et al. 1986; Fozard 1987; Saxena and Villalón 1990). This chapter outlines some of the most recent advances in the pharmacology of 5-HT receptors, with special emphasis on the analysis of the mechanism involved in the tachycardiac responses induced by indole- and benzamide derivatives in the pig.

*Supported by a Science and Technology grant from the EEC.

Cardiac Responses to 5-HT and Related Drugs

5-HT can exert multiple cardiac effects either directly or indirectly (see Saxena and Villalón 1990). Such effects encompass positive inotropic effects as well as decreases and increases in heart rate. Relatively little is known about the nature of the myocardial receptors involved in 5-HT-induced positive inotropic effects. For example, both 5-HT and lysergide are able to increase cardiac contractility in molluscs (Greenberg 1960; Wright et al. 1962), presumably through an increase in adenyl cyclase activity (Sawada et al. 1984). Furthermore, in the feline right ventricle (papillary muscle) and human atrial myocardium 5-HT induces a positive inotropic effect, which is mediated by an atypical 5-HT receptor unrelated to either 5-HT_1-like 5-HT_2 or 5-HT_3 receptors (Kaumann 1985; Kaumann et al. 1990a; b). On the other hand, bradycardia induced by 5-HT is mediated by 5-HT_3 receptors in most species, via the activation of the von Bezold Jarisch reflex. In marked contrast, 5-HT-induced tachycardiac response is notoriously species-dependent and is mediated, directly or indirectly, either by 5-HT_1-like (cat), 5-HT_2 (rat, dog), 5-HT_3 (rabbit, dog) or 'novel' 5-HT (pig) receptors, or by tyramine-like (guinea-pig) or unidentified mechanisms (see Saxena 1986; Saxena and Villalón 1990).

Evidence for a Novel Receptor Mediating 5-HT-Induced Tachycardia in the Pig

In the anaesthetized pig, intravenous administration of 5-HT increases heart rate (Duncker et al. 1985; Bom et al. 1988). Interestingly, this tachycardiac response was not antagonized by drugs affecting autonomic receptors (phentolamine, propranolol, hexamethonium and atropine), histamine H_1 (mepyramine) and H_2 (cimetidine) receptors, dopamine receptors (haloperidol), calcium channels (verapamil), 5-HT_1-like (methiothepin, methysergide, metergoline and mesulergine), 5-HT_2 (ketanserin, cyproheptadine, pizotifen, mianserin and those mentioned above as 5-HT_1-like receptor antagonists), or 5-HT_3 (MDL 72222 and ICS 205-930) receptors (see Table 1). Furthermore, the selective inhibitors of 5-HT-uptake, namely indalpine and fluvoxamine, potentiated the tachycardiac response to 5-HT (see Table 1), while a number of selective agonists at 5-HT_1-like (5-CT, 8-OH-DPAT, RU 24969 or BEA 1654) or 5-HT_3 (2-methyl-5-HT) receptors failed to induce tachycardia (Duncker et al. 1985; Bom et al. 1988). Therefore, the 5-HT-induced tachycardia in the pig does not involve endogeneous catecholamines, dopamine or histamine, Ca^{2+} transport into the sino-atrial node cells, or the 5-HT_1-like, 5-HT_2 or 5-HT_3 receptors, but seems to be mediated by a new type of 5-HT receptor.

Table 1. Effect of various antagonists on 5-HT-induced tachycardia in the pig

Antagonist	Dose mg/kg	n	Increase in heart rate (beats/min)					
			5-HT (3 μg/kg)		5-HT (10 μg/kg)		5-HT (30 μg/kg)	
			Before	After	Before	After	Before	After
Non-5-HT antagonists								
Phentolamine	1.0	5	26 ± 4	24 ± 6	42 ± 4	38 ± 6	58 ± 5	51 ± 12
Propranolol	0.5	5	24 ± 4	27 ± 4	43 ± 3	43 ± 4	60 ± 4	60 ± 3
Atropine	0.5							
+propranolol	0.5	3	37 ± 4	35 ± 4	50 ± 5	51 ± 1	55 ± 6	58 ± 2
Atropine	1.0							
+hexamethonium	10.0	4	9 ± 2	10 ± 2	21 ± 11	37 ± 7	32 ± 7	53 ± 8
Cimetidine	1.0	4	38 ± 5	33 ± 3	48 ± 5	47 ± 4	58 ± 5	56 ± 6
Mepyramine	1.0	4	31 ± 5	35 ± 3	45 ± 7	47 ± 5	55 ± 7	53 ± 5
Haloperidol	1.0	3	31 ± 4	35 ± 2	45 ± 5	46 ± 3	51 ± 6	55 ± 4
Verapamil	0.1	3	23 ± 4	23 ± 2	44 ± 2	34 ± 7	58 ± 2	45 ± 5
5-HT antagonists								
Methiothepin	0.5	6	22 ± 3	21 ± 4	40 ± 5	39 ± 4	53 ± 5	53 ± 3
Metergoline	0.5	5	29 ± 2	22 ± 3	49 ± 3	46 ± 6	63 ± 4	63 ± 6
Methysergide	0.5	6	26 ± 3	28 ± 3	50 ± 4	49 ± 3	63 ± 5	63 ± 6
Mesulergine	0.5	3	25 ± 1	26 ± 1	40 ± 2	45 ± 2	52 ± 4	56 ± 4
Ketanserin	0.5	4	20 ± 4	18 ± 4	42 ± 8	41 ± 7	58 ± 7	57 ± 7
Cyproheptadine	0.5	5	17 ± 3	23 ± 3*	40 ± 3	45 ± 4*	54 ± 3	60 ± 4*
Pizotifen	0.5	5	29 ± 2	26 ± 2	49 ± 3	45 ± 2	61 ± 5	56 ± 4
Mianserin	0.5	5	24 ± 2	23 ± 2	45 ± 1	44 ± 3	59 ± 3	54 ± 4
MDL 72222	0.3	4	28 ± 4	22 ± 5*	45 ± 4	45 ± 5	60 ± 5	58 ± 6
ICS 205-930	0.3	3	29 ± 2	21 ± 3	43 ± 4	41 ± 2	54 ± 6	56 ± 1
5-HT-uptake blockers								
Indalpine	1.0	3	20 ± 3	48 ± 3*	30 ± 1	58 ± 3*	45 ± 3	54 ± 7*
Fluvoxamine	1.0	4	11 ± 2	26 ± 2*	21 ± 2	40 ± 8*	30 ± 3	48 ± 9*

*, $P < 0.05$, after vs before. Data from Bom et al. (1988).

Tachycardic Action of Some Indole- and Benzamide Derivatives in the Pig

Firstly we investigated the effects of šome indole- (5-HT, 5-methoxy-tryptamine, α-methyl-5-HT, 5-CT, sumatriptan, indorenate and RU 24969) and benzamide (renzapride, zacopride, cisapride, metoclo-pramide and dazopride) derivatives on porcine heart rate. With respect to the indole derivatives, only 5-methoxytryptamine and α-methyl-5-HT mimicked 5-HT, and 5-CT displayed only weak (if any) agonist activity; the 5-HT$_1$-like receptor agonists sumatriptan (Humphrey et al. 1989), indorenate (Villalón et al. 1991) and RU 24969 (Bom et al. 1989) were inactive at the doses tested (Figure 1; Villalón et al. 1990b; 1991). With respect to the benzamide derivatives, only renzapride, zacopride and cisapride displayed substantial agonist effects which, however, were less marked than those of 5-HT, 5-methoxytryptamine or α-methyl-5-HT; metoclopramide and dazopride behaved as very weak agonists in this

256

INDOLE DERIVATIVES

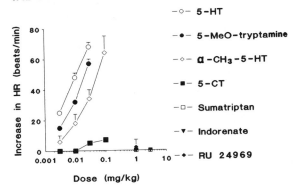

BENZAMIDE DERIVATIVES

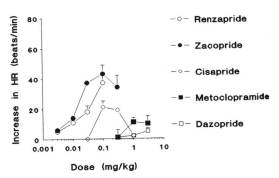

Figure 1. Increases in heart rate (HR) by indole- and benzamide derivatives in the anaes-thetized pig. 5-HT (n = 35), 5-methoxytryptamine (5-MeO-tryptamine; n = 30), α-methyl-5-HT (α-CH₃-5-HT; n = 3), sumatriptan (n = 4); indorenate (n = 4); renzapride (n = 8); zacopride (n = 6), cisapride (n = 5), metoclopramide (n = 8); and dazopride (n = 4). From Villalón et al. 1990b; 1991.

experimental model (Figure 1). At the doses used, the duration of action of the benzamide derivatives was much longer than that of the indole derivatives (Villalón et al. 1990b; 1991).

Antagonism by ICS 205-930 of the Tachycardic Responses to Indole- and Benzamide Derivatives by ICS 205-930

The tachycardia produced by 5-HT, 5-methoxytryptamine and isopre-naline remained essentially unchanged after physiological saline (Figure 2; Villalón et al., 1990b). The increases in heart rate induced by 5-HT

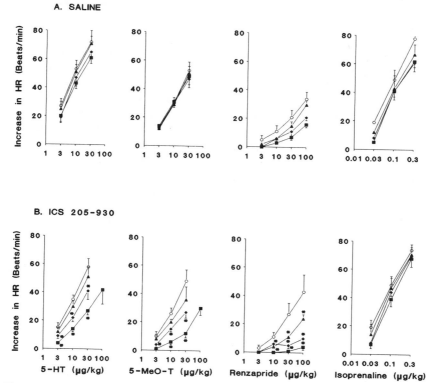

A. SALINE

B. ICS 205-930

Figure 2. The effects of ICS 205-930 (O——O), control; ▲——▲, 0.3 mg/kg; ◆——◆, 1.0 mg/kg and ■——■, 3.0 mg/kg) or the corresponding volumes of saline on the tachycardic responses to 5-HT (n = 5, saline; n = 7, ICS 205-930), 5-methoxytryptamine (5-MeO-T; n = 5, saline; n = 5, ICS 205-930), renzapride (n = 5, saline; n = 3, ICS 205-930) and isoprenaline (n = 3, saline; n = 5, ICS 205-930). HR, Heart rate; *, significant change (P < 0.05) in the control response to the agonist drug by ICS 205-930 as compared to the corresponding injection of saline. From Villalón et al. 1990b.

and 5-methoxytryptamine were clearly antagonized by ICS 205-930 (1 and 3 mg/kg) in a dose dependent manner. The responses to repeated doses of renzapride in the control animals progressively decreased, but these reductions were more marked after ICS 205-930 (Figure 2; Villalón et al. 1990b). Similarly, 3 mg/kg of ICS 205-930 inhibited the tachycardiac responses to α-methyl-5-HT, zacopride, cisapride, metoclopramide and dazopride (Villalón et al., 1991). In marked contrast, ICS 205-930 (up to 3 mg/kg) did not affect the isoprenaline-induced tachycardia (Figure 2), precluding antagonism of cardiac β-adrenoceptors by ICS 205-930. Therefore, ICS 205-930 behaves as a low affinity antagonist at the receptor mediating tachycardia by the indole- and benzamide derivatives in the pig.

258

Benzamide Derivatives as Partial Agonists at the Pig Heart 5-HT Receptor

As mentioned above, tachycardia induced by zacopride, cisapride, metoclopramide and dazopride was much less marked than that seen with 5-HT or 5-methoxytryptamine (see Figure 1), thus raising the possibility that the benzamides may behave as partial agonists. Indeed, the tachycardic responses to both 5-HT and 5-methoxytryptamine in the pig were antagonized by the benzamide derivatives in a dose-dependent manner (Figure 3; Villalón et al. 1991). Renzapride also exhibited similar effects (Villalón et al. 1990b). It has to be emphasized that the tachycardic effects of 5-HT and 5-methoxytryptamine were not 'masked' by the increase in heart rate induced by the benzamide derivatives, as the responses to 5-HT and 5-methoxytryptamine were elicited when the tachycardiac effect of the benzamides had worn off (Villalón et al. 1990b; 1991). These results suggest that the indole- and benzamide derivatives act on a common receptor site.

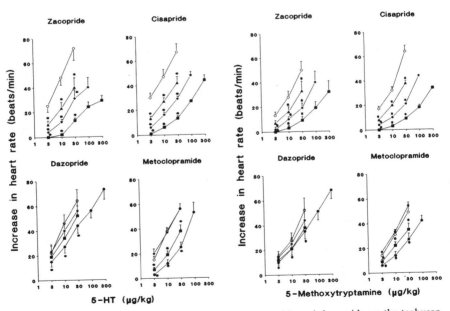

Figure 3. The effects of zacopride, cisapride, metoclopramide and dazopride on the tachycardiac responses to 5-HT and 5-methoxytryptamine. The doses of the antagonists were: ○——○, 0 mg/kg (control); ▲——▲, 0.1 mg/kg; ◆——◆, 0.3 mg/kg; ■——■, 1.0 mg/kg, and ☉——☉, 3.0 mg/kg). n = 4–8. *, Significantly different from the corresponding control response to 5-HT or 5-methoxytryptamine (P < 0.05). From Villalón et al. 1991.

Dissimilarity Between the Pig Heart 5-HT Receptor and 5-HT$_3$ Receptor

ICS 205-930 and the benzamide derivatives are potent 5-HT$_3$ receptor antagonists (for references see Fozard 1990). However, the following points suggest that the novel 5-HT receptor in the pig heart is not of the 5-HT$_3$ type: (i) both α-methyl-5-HT and 5-methoxytryptamine are inactive at the 5-HT$_3$ receptor (Richardson et al. 1985; Richardson and Engel 1986; Fozard 1990); (ii) the selective 5-HT$_3$ receptor agonists 2-methyl-5-HT and 1-phenyl-biguanide (Fozard 1990) were essentially inactive in eliciting tachycardia (Bom et al. 1988; Villalón et al. 1991); (iii) ICS 205-930 in a dose (0.3 mg/kg) sufficient to block the 5-HT$_3$ receptor (Bradley et al. 1986) did not modify 5-HT or 5-methoxytryptamine-induced tachycardic responses (Table 1: Figure 1; Villalón et al. 1990b); and (iv) after high doses (3 mg/kg) of the selective 5-HT$_3$ receptor antagonists MDL 72222 (Fozard 1984; 1990) and granisetron (Sanger and Nelson 1989; Fozard 1990), the responses to both 5-HT and 5-methoxytryptamine remained unchanged (Table 1; Villalón et al. 1990b; 1991).

Dissimilarity Between the Pig Heart 5-HT Receptor and 5-HT$_2$ Receptor

Although both 5-methoxytryptamine and α-methyl-5-HT can interact with 5-HT$_2$ (and 5-HT$_1$-like) receptors (Richardson and Engel 1986; Martin et al. 1987; Hoyer, 1988; Ismaiel et al. 1990), the tachycardic action of 5-methoxytryptamine and α-methyl-5-HT (Villalón et al. 1991), as well as that of 5-HT (Table 1; Bom et al. 1988), was not modified after ketanserin (0.5 mg kg^{-1}), a dose which effectively blocks the 5-HT$_2$ receptor (Van Nueten et al. 1981; Saxena and Lawang 1985). Therefore, the 5-HT$_2$ receptor is not involved in the tachycardic effects of the indole- and benzamide-derivatives.

Dissimilarity Between the Pig Heart 5-HT Receptor and Dopamine Receptors

A number of benzamide derivatives investigated by us have gastrokinetic actions (Alphin et al. 1986; Copper et al. 1986; van Daele et al. 1986; Sanger 1987; Schuurkes et al. 1985). Although most of these drugs are devoid of significant dopamine blocking activity, metoclopramide displays high affinity for the dopamine DA$_2$ receptor (Cooper et al. 1986). Moreover, ICS 205-903, zacopride and other 5-HT$_3$ receptor antagonists are able to inhibit the release of dopamine by 5-HT and

2-methyl-5-HT in the central nervous system (Blandina et al. 1988; Tricklebank 1989). It is for these reasons that it was considered important to rule out whether domperidone, a DA_2 receptor antagonist with gastrokinetic action, either antagonizes 5-HT-induced tachycardia or itself causes tachycardia in the pig. As with haloperidol (Table 1; Bom et al. 1988), domperidone (3 mg/kg) did not modify the tachycardic responses to either 5-HT or 5-methoxytryptamine. Moreover, domperidone was also devoid of any effect on the pig heart rate *per se* (Villalón et al. 1991). Therefore, the positive chronotropic effect induced by both tryptamine and benzamide derivatives in the pig heart is unrelated to a possible interaction with dopaminergic pathways and/or receptors. Moreover, domperidone has no agonist activity at the cardiac 5-HT receptor in the pig.

Similarities Between the Pig Heart 5-HT Receptor and the 5-HT$_4$ Receptor

The 5-HT receptor mediating increases in cAMP in the mouse embryo colliculi neurons has been named 5-HT$_4$ on the basis of the agonist action of 5-methoxytryptamine, 5-carboxamidotryptamine (low affinity) and certain benzamide derivatives (renzapride, metoclopramide, cisapride) and the antagonist action of ICS 205-930 (in high concentrations). Moreover, α-methyl-5-HT and 2-methyl-5-HT are inactive as agonists, and MDL 72222, granisetron and ondansetron do not act as antagonists (Dumuis et al. 1988, 1989; Bouhelal et al. 1988; Clarke et al. 1989). The pharmacological characteristics of the porcine heart 5-HT receptor, though exhibiting several similarities, differed in some important respects: (i) 5-carboxamidotryptamine, apparently because of the low affinity, did not show activity in the pig heart (Duncker et al. 1985; Bom et al. 1988) in doses which are highly active at the cat heart (Saxena et al. 1985); (ii) α-methyl-5-HT, which has little activity in the mouse brain (Dumuis et al. 1988, 1989), was highly active in the porcine heart; (iii) the potency order reported by Dumuis et al. (1989) using mouse embryo colliculi (cisapride > renzapride > zacopride > 5-HT > metoclopramide) was at variance from that found at the pig heart (5-HT > 5-methoxytryptamine > α-methyl-5-HT > zacopride ≥ renzapride > cisapride > metoclopramide > dazopride) (Villalón et al. 1990b; 1991); and (iv) the benzamide derivatives cisapride and renzapride, which are full agonists at the mouse brain receptor, behaved as partial agonists at the pig heart receptor. Several possible explanations for these differences in agonist potencies include: use of "second messenger" (cAMP) and functional (tachycardia) responses; tissue-dependent factors such as the number of receptors and coupling efficiency; and/or drug-dependent factors such as the affinity of 5-HT and related agonists for each of these novel receptors.

The 5-HT$_4$ receptor may also mediate the 5-HT-induced enhancement of cholinergic activity in the guinea pig isolated ileum (Sanger 1987; Craig and Clarke 1990) and ascending colon (Elswood et al. 1990), as well as relaxation of the rat oesophagus (Baxter and Clarke 1990). As in the case of the pig heart, the tryptamine derivatives, 5-methoxytryptamine and α-methyl-5-HT, and some benzamides mimic, and ICS 203-930 antagonizes 5-HT at the 5-HT$_4$ receptor in the guinea pig gastrointestinal tract (Craig and Clarke 1990; Elswood et al. 1990) and the rat oesphagus (Baxter and Clarke 1990). Moreover, the order of potency at these cholinergic neurons (5-HT > 5-methoxytryptamine > renzapride > α-methyl-5-HT > zacopride = cisapride; Craig and Clarke, 1990) is practically identical to that found in the porcine heart.

The 5-HT$_4$ receptor is apparently also involved in the inotropic action of 5-HT, mediated via cAMP increase, in the human atria; the response to 5-HT is not modified by ketanserin, methysergide, lysergide, methiothepin, yohimbine, ($\pm$)propranolol, ($-$)pindolol or MDL 72222, but is blocked by a high concentration (2 μM) of ICS 205-930 (Kaumann et al. 1990a; b). The precise role of these receptors in cardiac function and cardiovascular pathologies remains to be determined.

Acknowledgement

We are grateful to the British Journal of Pharmacology for the use of figures.

References

Alphin, R. S., Smith, W. L., Jackson, C. B., Droppleman, D. A., and Sancillo, L. F. (1986). Zacopride (AHR-11190 B): A unique and potent gastrointestinal prokinetic and antiemetic agent in laboratory animals. Dig. Dis. Sci. 31: 4825.

Baxter, G. S., and Clarke, D. E. (1990). Putative 5-HT$_4$ receptors mediate relaxation of rat oesphagus. Proceedings of the 2nd IUPHAR Satellite Meeting on Scrotonin, Basel, July 11-13, 1990. Abs P85.

Blandina, P., Goldfarb, J., and Green, P. J. (1988). Activation of a 5-HT$_3$ receptor releases dopamine from rat striatal slice. Eur. J. Pharmacol. 155: 349–350.

Bom, A. H., Duncker, D. J., Saxena, P. R., and Verdouw P. D. (1988). 5-Hydroxytryptamine-induced tachycardia in the pig: possible involvement of a new type of 5-hydroxytryptamine receptor. Br. J. Pharmacol. 93: 663–671.

Bom, A. H., Villalón, C. M., Verdouw, P. D., and Saxena, P. R. (1989). The 5-HT$_1$-like receptor mediating reduction of porcine carotid arteriovenous shunting by RU 24969 is not related to either the 5-HT$_{1A}$ or the 5-HT$_{1B}$ subtype. Eur. J. Pharmacol. 171: 87–96.

Bouhelal, R., Sebben, M., and Bockaert, J. (1988). A 5-HT receptor in the central nervous system, positively coupled with adenyl cyclase, is antagonized by ICS 205-930. Eur. J. Pharmacol. 146: 187–188.

Bradley, P. B., Engel, G., Feniuk, W., Fozard, J. R., Humphrey, P. P. A., Middlemiss, D. N., Mylecharane, E. J., Richardson, B. P., and Saxena, P. R. (1986). Proposals for the classification and nomenclature of functional receptors for 5-hydroxytryptamine. Neuropharmacology 25: 563–576.

Clarke, D. E., Craig, D. A., and Fozard, J. R. (1989). The 5-HT$_4$ receptor: naughty but nice. Trends Pharmacol. Sci. 10: 385–386.

Cooper, S. M., and McClelland, M., McRitchie, B., and Turner, D. H. (1986). BRL 24924: A new and potent gastric motility stimulant. Br. J. Pharmacol. 88: 383P.

Craig, D. A., and Clarke, D. E. (1990). Pharmacological characterization of a neuronal receptor for 5-hydroxtryptamine in guinea pig ileum with properties similar to the 5-hydroxytryptamine$_4$ receptor. J. Pharmacol. Exp. Ther. 252: 1378–1386.

Dumuis, A., Bouhelal, R., Sebben, M., Cory, R., and Bockaert, J. (1988). A nonclassical 5-hydroxytryptamine receptor positively coupled with adenylate cyclase in the central nervous system. Mol. Pharmacol. 34: 880–887.

Dumuis, A., Sebben, M., and Bockaert, J. (1989). The gastrointestinal prokinetic benzamide derivatives are agonists at the non-classical 5-HT receptor (5-HT$_4$) positively coupled to adenylate cyclase in neurones. Naunyn-Schmied. Arch. Pharmacol. 340: 403–410.

Duncker, D. J., Saxena, P. R., and Verdouw, P. D. (1985). 5-Hydroxytryptamine causes tachycardia in pigs by acting on receptors unrelated to 5-HT$_1$, 5-HT$_2$ or M type. Br. J. Pharmacol. 86: 596P.

Elswood, C. J., Bunce, K. T., and Humphrey, P. P. A. (1990). Identification of 5-HT$_4$ receptors in guinea-pig ascending colon. Proceedings of the 2nd IUPHAR Satellite Meeting on Serotonin, Basel, July 11–13, 1990. Abs. P86.

Fozard, J. R. (1984). MDL 72222: a potent and highly selective antagonist at neuronal 5-hydroxytryptamine receptors. Naunyn-Schmied. Arch. Pharmacol. 326: 36–44.

Fozard, J. R. (1987). 5-HT: The enigma variations. Trends Pharmacol. Sci. 8: 501–506.

Fozard, J. R. (1990). Agonists and antagonists at 5-HT$_3$ receptors. In Saxena, P. R., Wallis, D. I., Wouters, W., and Bevan, P. (eds.), The cardiovascular pharmacology of 5-hydroxytryptamine: prospective therapeutic applications. Dordrecht: Kluwer, pp. 101–115.

Greenberg, M. J. (1960). Structure-activity relationship of tryptamine analogues on the heart of venus mercenaria. Brit. J. Pharmacol. 15: 375–388.

Hoyer, D. (1988). Functional correlates of serotonin 5-HT$_1$ recognition sites. J. Receptor Res. 8: 59–81.

Humphrey, P. P. A., Perren, M. J., Feniuk, W., Connor, H. E., and Oxford, A. W. (1989). The pharmacology of a novel 5-HT$_1$-like receptor agonist, GR43175. Cephalalgia 9 (Suppl. 9): 23–33.

Ismaiel, A. M., Titler, M., Miller, K. J., Smith, T. S., and Glennon, R. A. (1990). 5-HT$_1$ and 5-HT$_2$ binding profiles of the serotonergic agents α-methylserotonin and 2-methylserotonin. J. Med. Chem. 33: 755–758.

Kaumann, A. J. (1985). Two classes of myocardial 5-hydroxytryptamine receptors that are neither 5-HT$_1$ nor 5-HT$_2$. J. Cardiovasc. Pharmacol. 7 (Suppl. 7): S76–S78.

Kaumann, A. J., Murray, K. J., Brown, A. M., Sanders, L., and Brown, M. L. (1990a). A 5-hydroxytryptamine receptor in human atrium. Br. J. Pharmacol. 100: 879–885.

Kaumann, A. J., Murray, K. J., Brown, A. M., Frampton, J. E., Sanders, L., and Brown, M. J. (1990b). Heart 5-HT receptors. A novel 5-HT receptor in human atrium. In Paoletti, R., Vanhoutte, P. M., Brunello, N., and Maggi, F. M. (eds.), Serotonin: From cell biology to Pharmacology and Therapeutics. Dordrecht: Kluwer, 347–354.

Martin, G. R., Leff, P., Cambridge, D., and Barret, V. J. (1987). Comparative analysis of two types of 5-hydroxytryptamine receptor mediating vasorelaxation: differential classification using tryptamines. Naunyn-Schmied. Arch. Pharmacol. 336: 365–373.

Richardson, B. P., and Engel, G. (1986). The pharmacology and function of 5-HT$_3$ receptors. Trends Neurosci. 9: 424–428, 1986.

Richardson, B. P., Engel, G., Donatsch, P., and Stadler, P. A. (1985). Identification of serotonin M-receptor subtypes and their blockade by a new class of drugs. Nature 316: 126–131.

Sanger, G. J. (1987). Increased gut cholinergic activity and antagonism of 5-hydroxytryptamine M-receptors by BRL 24924: potential clinical importance of BRL 24924. Br. J. Pharmacol. 91: 77–87.

Sanger, G. J., and Nelson, D. R. (1989). Selective and functional 5-hydroxytryptamine$_3$ receptor antagonism by BRL 43694 (granisetron). Eur. J. Pharmacol. 159: 113–124.

Saxena, P. R. (1986). Nature of the 5-hydroxytryptamine receptors in mammalian heart. Prog. Pharmacol. 6: 173–185.

Saxena, P. R., and Lawang, A. (1985). A comparison of cardiovascular and smooth muscle effects of 5-hydroxytryptamine and 5-carboxamidotryptamine, a selective agonist at 5-HT$_1$ receptors. Arch. Int. Pharmacodyn. 277: 235–252.

Saxena, P. R., and Villalón, C. M. (1990). Cardiovascular effects of serotonin agonists and antagonists. J. Cardiovasc. Pharmacol. 15 (Suppl. 7): S17–S34.

Saxena, P. R., Mylecharane, E. J., and Heiligers, J. (1985). Analysis of the heart rate effects of 5-hydroxytryptamine in the cat; mediation of tachycardia by 5-HT$_1$-like receptors. Naunyn-Schmied. Arch. Pharmacol. 330: 121–129.

Sawada, M., Ichinose, M., Ito, I., Maeno, T., and McAdoo, D. J. (1984). Effects of 5-hydroxytryptamine on membrane potential, contractility, accumulation of cyclic AMP, and Ca^{2+} movements in anterior aorta and ventricle of aplysia. J. Neurophysiol. 51: 361–374.

Schuurkes, J. A. J., Van neuten, J. M., Van Daele, P. G. H., Reyntjes, A. J., and Janssen, P. A. J. (1985). Motor-stimulating properties of cisapride on isolated gastrointestinal preparations of the guinea pig. J. Pharmacol. Exp. Ther. 234: 775–783.

Tricklebank, M. D. (1989). Interactions between dopamine and 5-HT$_3$ receptors suggest new treatments for psychosis and drug addiction. Trends Pharmacol. Sci. 10: 127–129.

Van Daele, G. H. P., De Bruyn, M. F. L., Sommen, F. M., Janssen, M., Van Neuten, J. M., Schuurkes, J. A. J., Niemegeers, C. J. E., and Leysen, J. E. (1986). Synthesis of cisapride, a gastrointestinal stimulant derived from cis-4-amino-3-methoxypiperidine. Drug Dev. Res. 8: 225–232.

Van Nueten, J. M., Janssen, P. A. J., Van Beek, J., Xhonneux, R., Verbeuren, T. J., and Vanhoutte, P. M. (1981). Vascular effects of ketanserin (R 41 468), a novel antagonist at 5-HT$_2$ serotonergic receptors. J. Pharmacol. Exp. Ther. 218: 217–230.

Villalón, C. M., Den Boer, M. O., Heiligers, J. P. C., and Saxena, P. R. (1990a). Porcine arteriovenous anastomotic constriction by indorenate is unrelated to 5-HT$_{1A}$, 5-HT$_{1B}$, 5-HT$_{1C}$ or 5-HT$_{1D}$ receptor subtype. Eur. J. Pharmacol. 190: 167–176.

Villalón, C. M., Den Boer, M. O., Heiligers, J. P. C., and Saxena, P. R. (1990b). Mediation of 5-hydroxytryptamine-induced tachycardia in the pig by the putative 5-HT$_4$ receptor. Br. J. Pharmacol. 100: 665–667.

Villalón, C. M., Den Boer, M. O., Heiligers, J. P. C., and Saxena, P. R. (1991). Further characterization, using tryptamine and benzamide derivatives, of the putative 5-HT$_4$ receptor mediating tachycardia in the pig. Brit. J. Pharmacol. 102: 107–112.

Wright, A. M., Moorhead, M. A., and Welsh, J. H. (1962). Actions of derivatives of lysergic acid on the heart of venus mercenaria. Brit. J. Pharmacol. 18: 440–450.

Serotonin: Molecular Biology, Receptors and Functional Effects
ed. by J. R. Fozard/P. R. Saxena
© 1991 Birkhäuser Verlag Basel/Switzerland

Contractile 5-HT$_{1D}$ Receptors in Human Brain Vessels

E. Hamel and D. Bouchard

Laboratory of Cerebrovascular Research, Montreal Neurological Institute, 3801 University St., Montréal, Canada H3A 2B4

Summary. The receptor which mediates the contraction elicited by 5-hydroxytryptamine (5-HT, serotonin) in human pial arterioles has been characterized. The overall rank order of agonist potency was 5-carboxamidotryptamine (5-CT) > 5-HT > RU 24969 = α-CH$_3$-5-HT = methysergide ≫ MDL 72832 = 2-CH$_3$-5-HT ≫ 8-OH-DPAT, with maximal contractile responses being comparable for most agonists. An excellent correlation was obtained for the vascular potencies of agonists and their published affinities for rat cortical 5-HT$_{1B}$ (r = 0.86; $p < 0.01$) and human caudate 5-HT$_{1D}$ (r = 0.98; $p < 0.005$) binding sites. The 5-HT-induced contraction was not blocked by 5-HT$_{1B}$ (propranolol), 5-HT$_{1C}$ (mesulergine, mianserin), 5-HT$_2$ (ketanserin, mianserin, MDL 72832) or 5-HT$_3$ (MDL 72222) antagonists. In contrast, methiothepin (non-selective 5-HT$_1$-like/5-HT$_2$) and metergoline (non-selective 5-HT$_{1D}$) were potent inhibitors of this vasomotor effect with respective pA$_2$ values on the order of 8.51 and 7.06. A correlation analysis between the overall agonist/antagonist vascular potencies and their reported potencies for 5-HT-mediated second messenger effects; i.e. adenylate cyclase activity (5-HT$_{1A}$, 5-HT$_{1B}$ or 5-HT$_{1D}$-coupled effect) or inositol phosphate metabolism (5-HT$_{1C}$-mediated response), showed a positive correlation for 5-HT$_{1B}$ (r = 0.89; $p < 0.001$) and 5-HT$_{1D}$ (r = 0.90; $p < 0.001$) functional receptors. Our study suggests that a 5-HT$_1$-like receptor, most likely of the 5-HT$_{1D}$ subtype, mediates the vasoconstriction to 5-HT in human brain vessels.

Introduction

In all mammalian species studied including man, cerebral arteries are highly sensitive to 5-HT which elicits potent vasoconstriction of the cerebrovascular smooth muscle (for review, see Young et al. 1987; Lee 1989). Among other cerebrovascular disorders, 5-HT has been intimately associated to the etiology of migraine (see Fozard 1987; 1988; 1989). Although its precise role remains unclear, it is undeniable that a common characteristic of a variety of drugs used in the prophylaxis of migraine is their ability to antagonize diverse receptor-mediated actions of 5-HT (Titus et al. 1986; Fozard 1988, 1989). Recently, a 5-HT$_1$-like agonist (GR 43175, sumatriptan) has reached Phase III clinical trials and was found to be beneficial in the treatment of acute migraine headache (Doenicke et al. 1988). Sumatriptan as well as some ergot derivatives and other drugs used in the treatment of acute migraine attacks have been proposed to share a common 5-HT$_1$-like prejunctional mechanism involved in blocking the trigemino-vascular neurogenic inflammation (Saito et al. 1988; Buzzi and Moskowitz, 1990; Buzzi et al. 1990), a probable cause of pain associated with migraine (Moskowitz 1989). Alternatively, the anti-migraine action of sumatriptan has been ascribed by others to its ability to selectively constrict the cranial vasculature via specific 5-HT$_1$-like receptors (Saxena and Ferrari 1989; Humphrey et al. 1990).

Considering the multiple subtypes of receptors with which 5-HT can interact and the well documented species-related variations in 5-HT cerebrovascular receptors (Young et al. 1987; Lee 1989), it appears that a better understanding of the possible mechanism(s) of some anti-migraine agents could only be achieved by a clear identification of human 5-HT cerebrovascular receptors. Original studies on human cerebral blood vessels, performed before the availability of selective 5-HT agonists and antagonists, failed to clearly identify the receptor(s) involved in the 5-HT vasocontractile response (Hardebo et al. 1978; Edvinsson et al. 1978; Foster and Whalley 1982; Müller-Schweinitzer 1983). Slightly thereafter, Peroutka and Kuhar (1984) visualized by radioautography the presence of 5-HT$_1$, but not 5-HT$_2$, muscular binding sites in human basilar artery. A more detailed pharmacological investigation of these receptors has recently shown that they appear identical to the contractile 5-HT$_1$-like receptors of the dog sapheneous vein and cerebral blood vessels from dog and primate (Parsons et al. 1989).

In the present study, we have pharmacologically characterized the 5-HT contractile receptors in human pial arterioles, a vascular segment of primary importance in the regulation of blood supply to the brain. We determined the vascular potencies of several selective 5-HT agonists and antagonists and compared these to their published affinities at various subtypes of 5-HT binding sites or functional 5-HT receptor-coupled second messenger effects.

Materials and Methods

Human pial arterioles used (0.8 mm outside diameter) were temporal ramifications of the middle cerebral artery and were obtained post-mortem (delay 4–12 hours – Montreal Douglas Hospital Brain Bank) from 11 subjects (6 males, 5 females, aged 61–88 years) who had died from various causes. The vessels were cleaned of surrounding tissue under a dissecting microscope and placed in a Krebs' Ringer solution (for details see Hamel et al. 1989 and Hamel and Bouchard 1991). Small vascular segments were mounted between two L-shaped metal prongs in temperature-controlled (37°C) tissue baths (5 ml) for recording of the isometric tension (Hogestätt et al. 1983) on a Grass Polygraph (Model 7E to which force displacement transducers GRASS FT 103D are connected).

Agonist potency: Arteriolar segments were given a mechanical tension of 0.4 g and allowed to stabilize at this level for 90 min. Then, log concentration-response curves were generated for each agonist. The maximal contractile capacity of each arteriolar segment was determined by exposure to a depolarizing solution of K^+. Relative agonist potency was determined by their pD_2 values ($-\log EC_{50}$) calculated according to Van den Brink (1977) with the following equation:

$$pD_2 = -\log[A] - \log\left(\frac{E_{A\,max}}{E_A} - 1\right)$$

The maximal response ($E_{A\,max}$) of each agonist was compared with the $E_{A\,max}$ obtained for 5-HT in the same preparations, and expressed as percentage of 5-HT $E_{A\,max}$. For reference, the $E_{A\,max}$ elicited by 10 μM 5-HT corresponded to 1.64 ± 0.09 g or $71 \pm 5\%$ of the maximal contractile capacity of the vascular segments determined with K^+.

Antagonist potency: Concentration-response curves to 5-HT were obtained in the absence (control) and presence of a fixed (1 μM) or graded (10 nM $-$ 1 μM) concentrations of antagonist. The antagonist was in contact with the vessels for a 20 min period before reconstruction of the 5-HT dose-response curve. The antagonist potency was evaluated by the use of pA_2 values for dual antagonist (competitive/metactoid) as described by Van den Brink (1977) and calculated as:

$$pA_2 = -\log[B] + \log\left(\frac{[A_2]}{[A_1]} - 1\right)$$

In addition, the pA_2 values were estimated from the regression lines obtained in the Schild plot analysis (Arunlakshana and Schild 1959) which was used to assess the competitive nature of the antagonism by determination of the slope of the regression line.

Statistics: All results are presented as mean $\pm$ S.E.M. Differences in agonist potency (pD_2 and $E_{A\,max}$ values) were evaluated by one way analysis of variance (ANOVA) for unequal sample size and Student-Newman-Keuls multiple comparison test. Linear regression lines and correlation coefficients were calculated in order to detect and quantify correlations between vascular and binding site, or second messenger-coupled functional receptor potencies. A $P \leqslant 0.05$ was considered significant.

Drugs: The following compounds were kindly donated: ($\pm$) α-CH$_3$-5-HT, ($\pm$) 2-CH$_3$-5-HT, methysergide and mesulergine (Sandoz, Basel, Switzerland); 5-methoxy-3 (1,2,3,6-tetrahydro 4-pyridinyl) 1H indole succinate (RU 24969, Roussel UCLAF, Paris, France); α-phenyl-1-(2-phenylethyl)-4-piperidine methanol (MDL 11939); 1αH, 3α, 5αH-tropan-3yl-3,5-dichlorobenzoate (MDL 72222) and 8-[4-(1,4-benzodioxan-2-yl methylamino) butyl-]-8-azaspiro [4, 5] decane-7,9-dione (MDL 72832) were from Le Centre de Recherche Merrell Dow

International, Strasbourg, France; methiothepin (Hoffman-Laroche, Basel, Switzerland) and metergoline (Dr. R. Quirion, Douglas Hospital Research Centre, Verdun, Canada). 5-HT and (±) propranolol were purchased from Sigma (U.S.A.) and ketanserin, mianserin and 8-OH-DPAT ((±)-2-dipropylamino-8-hydroxy-1,2,3,4-tetrahydro-naphthalene) hydrobomide from R.B.I. (U.S.A.).

Results

Except for methysergide which elicited a contractile response consider-ably smaller than that induced by 5-HT (53% of 5-HT $E_{A\,max}$, $p < 0.05$, Table 1), all other agonists resulted in a full constriction of human pial arterioles in a dose-dependent manner (Figure 1). The most potent agonist was 5-CT, followed by 5-HT, RU 24969, α-CH$_3$-5-HT and methysergide while the others (5-HT$_{1A}$ and 5-HT$_3$ agonists) were 100 to 1000 fold less potent than 5-HT itself, as evaluated from their EC$_{50}$ ratios (Table 1). A correlation analysis performed between the agonist vascular potencies (pD_2 values from Table 1) and their published

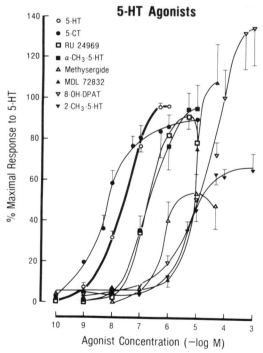

Figure 1. Concentration-response curves to 5-HT and 5-HT agonists on human pial arterioles under resting tension. Vertical bars show S.E.M. of n = 4 to 33 (see Table 1 for more details).

Table 1. 5-HT agonists potency for inducing contraction on human pial arterioles.

Agonists	n	$E_{A\,max}$ (% of 5-HT $E_{A\,max}$)	pD$_2$ ($-\log$ EC$_{50}$)	$\dfrac{\text{EC}_{50}\text{ Agonist}}{\text{EC}_{50}\text{ 5-HT}}$
5-HT	33	100	7.61 ± 0.08^b	1
5-CT	21	90 ± 5	8.18 ± 0.09^b	0.25
RU 24969	8	86 ± 7	6.85 ± 0.15^c	5
α-CH$_3$-5-HT	10	97 ± 9	6.57 ± 0.16^c	12
Methysergide	4	53 ± 14^a	6.30 ± 0.06^c	13
MDL 72832	5	103 ± 19	5.45 ± 0.08^d	96
2-CH$_3$-5-HT	18	68 ± 5	5.20 ± 0.13^d	450
8-OH-DPAT	10	137 ± 19^b	4.67 ± 0.21^b	1160

Values are mean $\pm$ s.e.m. from number of (n) individual segments as indicated. Those used for 5-HT represent pooled data obtained with all agonist experiments.
a; $P < 0.05$ when compared with all agonists except 2-CH$_3$-5-HT; b; $P < 0.05$ with respect to all other agonists; c; $P < 0.05$ with all agonists except α-CH$_3$-5-HT, methysergide or RU 2469, respectively; d; $P < 0.05$ with all compounds but 2-CH$_3$-5-HT and MDL 72832, respectively. ANOVA and student Newman-Keuls multiple test.

affinities at various subtypes of 5-HT binding sites (taken as pK$_D$, pK$_i$ or pIC$_{50}$ values from Engel et al. 1986; Glennon 1987; Mir et al. 1988; Hoyer 1988; Waeber et al. 1988 and Herrick-Davis et al. 1988) showed a significant correlation with 5-HT$_{1B}$ sites in rat cortex (r = 0.86; $p < 0.01$) and 5-HT$_{1D}$ sites in human caudate membranes (r = 0.98; $p < 0.005$). No significant correlation could be detected at 5-HT$_{1A}$, 5-HT$_{1C}$ or 5-HT$_2$ binding sites, with respective correlation coefficients (r) of 0.38; 0.41 or 0.005. Most antagonists tested failed to inhibit the 5-HT-induced constriction of human pial arterioles. Such was the case for 5-HT$_{1B}$ (propranolol), 5-HT$_{1C}$ (mesulergine, mianserin), 5-HT$_2$ (ketanserin, mianserin, MDL 11939) and 5-HT$_3$ (MDL 72222) antagonists. All these compounds were devoid of antagonistic activity up to 1 μM as

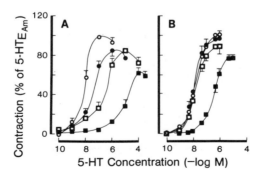

Figure 2. Concentration-response curves to 5-HT on human pial arterioles in the absence ($\bigcirc$, control) and presence of 10 nM ($\bullet$), 100 nM ($\square$) or 1000 nM ($\blacksquare$) methiothepin (A, n = 6) or metergoline (B, n = 4). Vertical bars shown S.E.M.

5-HT potency (pD_2) and maximal response ($E_{A\,max}$) were not significantly modified (data not shown). In some instances, the 5-HT response was slightly enhanced though not significantly (for ketanserin and mianserin). In contrast, in the presence of various concentrations of methiothepin and metergoline, an apparent parallel rightward shift in the 5-HT dose-response curves was obtained, with a slight decrease in the maximal effect at higher antagonist concentrations (Figure 2). The pA_2 values calculated according to Van den Brink for mixed antagonists were 8.46 ± 0.24 and 7.24 ± 0.09, for methiothepin and metergoline, respectively. However, in the Schild plot analysis, the slopes were significantly different from unity (1.41 ± 0.09 and 1.72 ± 0.23, respectively) and the pA_2 estimated from the regression lines were 8.55 ± 0.16 (methiothepin) and 6.88 ± 0.05 (metergoline). Attempts were made to correlate the overall agonist/antagonist cerebrovascular potencies (pD_2 or pA_2 values) with their corresponding pEC_{50} or pK_B measured in well established models for functional 5-HT_{1A}, 5-HT_{1B} or 5-HT_{1D} receptors (inhibition of adenylate cyclase activity in calf hippocampus, rat or calf substantia nigra, respectively; Schoeffter and Hoyer, personal communication) or 5-HT_{1C} receptors (production of inositol phosphates in pig

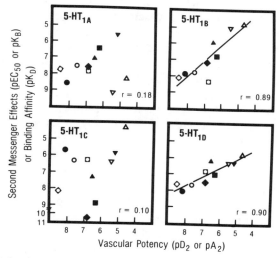

Figure 3. Correlation between agonist or antagonist vascular potencies (pD_2 or pA_2 values) in human pial arterioles and inhibition of adenylate cyclase in calf hippocampus (5-HT_{1A}), rat (5-HT_{1B}) or calf(5-HT_{1D}) substantia nigra, or stimulation of inositol phosphate metabolism in pig choroid plexus (5-HT_{1C}). Vascular potencies are from this study, those for the functional 5-HT receptor-coupled second messengers are from Hoyer, D., and Schoeffter, P. (personal communication) and represent pEC_{50} (agonists) or pK_B (antagonists) values. For $\alpha\text{-CH}_3\text{-5-HT}$, $2\text{-CH}_3\text{-5-HT}$ and MDL 72832, pK_D values at corresponding binding sites were used. ○: 5-HT; ●: 5-CT; □: RU 24969; ■: methysergide; △: 8-OH-DPAT; ▲: $\alpha\text{-CH}_3\text{-5-}$HT; ▽: MDL 72832; ▼: $2\text{-CH}_3\text{-5-HT}$; ◇: methiothepin; ◆: metergoline.

choroid plexus, Schoeffter and Hoyer personal communication). A significant correlation was found only with 5-HT$_{1B}$ (r = 0.98; $p < 0.001$) and 5-HT$_{1D}$ (r = 0.90; $p < 0.001$) functional receptors (Figure 3).

Discussion

The present pharmacological study suggests that a 5-HT$_1$-like receptor, which correlates with human caudate 5-HT$_{1D}$ binding sites or functional 5-HT$_{1D}$ receptor-coupled adenylate cyclase, is responsible for the 5-HT-induced vasoconstriction in human pial arterioles. This statement is based on our agonist and antagonist data as well as on the pharmacological properties of the various subtypes of 5-HT receptors.

The agonist and antagonist vascular effects quite clearly showed that 5-HT$_{1A,1C}$, 5-HT$_2$ and 5-HT$_3$ receptors were not involved in the vasomotor effect of 5-HT in this cerebrovascular bed. Firstly, the very low potency of two selective 5-HT$_{1A}$ agonists, 8-OH-DPAT (Gozlan et al. 1983) and MDL 72832 (Mir et al. 1988), to induce vasoconstriction (their vascular potencies being more than a 1000 fold less than their affinities at 5-HT$_{1A}$ sites (Hover et al. 1988; 1989), clearly suggested that these agents most probably interacted with other 5-HT receptor subtypes or α-adrenoceptors (Timmermans et al. 1984; Engel et al. 1986; Mir et al. 1988) on the cerebrovascular smooth muscle. Secondly, the finding that mesulergine and mianserin, two potent 5-HT$_{1C}$ antagonists, showed no inhibitory effect on this vascular response indicates that 5-HT$_{1C}$ receptors cannot be primarily responsible of the 5-HT-mediated effect. This is further supported by the lack of correlation between the agonist vascular potencies and their affinities at either 5-HT$_{1C}$ binding sites or functional 5-HT$_{1C}$ receptor-coupled inositol phosphate. Thirdly, the demonstration that selective 5-HT$_2$ antagonists such as ketanserin, mianserin and MDL 11939 (Engel et al. 1986; Hoyer 1988, 1989; Dudley et al. 1988) failed to block the 5-HT-induced contraction in human pial arterioles (this study, see also Hamel and Bouchard 1991), as reported also in human basilar artery (Parsons et al. 1989), rules out the involvement of vasocontractile 5-HT$_2$ receptors. This conclusion contrasts with our previous report on feline 5-HT cerebrovascular receptors (Hamel et al. 1989). Such discrepancies partly emphasize the species-related variations in the identity of contractile 5-HT cerebrovascular receptors (see Young et al. 1987; Lee 1989). Finally, the weak agonist activity of 2-CH$_3$-5-HT and the lack of antagonism of MDL 72222 in human pial vasculature, two compounds with high affinity for 5-HT$_3$ receptors (Fozard 1984; Richardson and Engel 1986) would exclude participation of this receptor subtype.

Several pieces of evidence strongly indicate, however, that a 5-HT$_1$-like receptor is being stimulated by 5-HT to induce contraction of human pial arterioles: (1) the higher potency of 5-CT as compared to 5-HT (Bradley et al. 1986); (2) the agonist potency of methysergide, as reported for other 5-HT$_1$-like vascular receptors (Engel et al. 1983); (3) the antagonistic activity of methiothepin (a non-selective 5-HT$_1$-like/5-HT$_2$ antagonist) together with the complete lack of effect of the most selective 5-HT$_2$ antagonists. According to our correlation study between the vascular agonist potencies and their binding site affinities, two subtypes of 5-HT$_1$-like receptors are likely to be involved, namely the 5-HT$_{1B}$ and 5-HT$_{1D}$ receptors. However, as reported earlier with cyanopindolol in human basilar artery (Parsons et al. 1989), we found that the 5-HT$_{1B}$ antagonist propranolol was inactive in blocking the 5-HT-induced contraction of human pial arterioles. This finding, together with the ability of metergoline to inhibit the 5-HT vasomotor effect with relatively high affinity and the reported absence of 5-HT$_{1B}$ sites in human brain (Hoyer et al. 1986) strongly indicates that the cerebrovascular profile of antagonist potencies is similar to that reported for 5-HT$_{1D}$ receptors (Waeber et al. 1988; Herrick-Davis et al. 1988). However, the fact that methiothepin was much more potent in inhibiting the 5-HT-induced vasoconstriction in human pial arteries than predicted from its affinity at human 5-HT$_{1D}$ binding sites (pK$_D$ of 6.77, Waeber et al. 1988) remains unexplained. The differences in cerebrovascular and neuronal potencies for metergoline and methiothepin might suggest that the contractile receptor is not identical to the caudate 5-HT$_{1D}$ binding site. Nevertheless, the presence of a cerebrovascular 5-HT$_{1D}$ contractile receptor is further supported by the good correlation obtained between the overall agonist/antagonist cerebrovascular potencies and functional 5-HT$_{1D}$ receptor-mediated inhibition of adenylate cyclase activity.

In summary, the present study provides strong evidence that the 5-HT$_1$-like receptor which mediates vasoconstriction in human pial arterioles is similar to the 5-HT$_{1D}$ receptor subtype. It cannot be excluded that we have identified a receptor slightly different from the brain caudate nucleus 5-HT$_{1D}$ binding site but the current status of 5-HT pharmacology, binding site affinities, second messenger-mediated effects and species distribution emphasize its similarity to the 5-HT$_{1D}$ subtype. Interestingly, in addition to 5-HT$_2$ receptors, such contractile 5-HT$_1$ receptors had been implicated in feline cerebrovascular bed (Hamel et al. 1989) and, more recently, a 5-HT$_{1D}$ receptor has been associated with endothelium-dependent relaxation of pig coronary arteries (Schoeffter and Hoyer 1990). From our results, it appears possible that the anti-migraine drug sumatriptan, which has high affinity at 5-HT$_{1D}$ binding sites (Peroutka and McCarthy 1989; Schoeffter and Hoyer 1989) could interact with the human cerebrovascular bed in order to induce vasoconstriction, a mechanism postulated to be beneficial in counteracting migraine-associated cerebrovascular dysfunctions

(Saxena and Ferrari 1989; Humphrey et al. 1990). However, this effect might not be exclusive and further work is needed to evaluate the presence of such functional receptors at other neurovascular levels at which sumatriptan and other anti-migraine drugs have been shown to be active (Saito et al. 1988; Buzzi and Moskowitz 1990; Buzzi et al. 1990).

Acknowledgements

This work was supported by a grant from the "Fondation des Maladies du Coeur du Québec" as well as the Medical Research Council of Canada and the Fonds de la Recherche en Santé du Québec. The authors are most grateful to Drs. Daniel Hoyer and Philippe Schoeffter (Preclinical Research, Sandoz Pharma Ltd, Basel, Switzerland) for their generous assistance in providing us with the potency values of various compounds at 5-HT_{1A}, 5-HT_{1B}, 5-HT_{1C} and 5-HT_{1D} functional receptors. We want to thank the various drug companies or individuals listed in Methods for providing us with compounds not commercially available. We also are most grateful to the Douglas Hospital Brain Bank for tissues, and to Ms. Carolyn Elliot for the preparation of this manuscript.

References

Arunlakshana, O., and Schild, H. O. (1959). Some quantitative uses of drug antagonists. Br. J. Pharmacol. 14: 48–58.

Bradley, P. B., Engel, G., Feniuk, W., Fozard, J. R., Humphrey, P. P. A., Middlemiss, D. N., Mylecharane, E. J., Richardson, B. P., and Saxena, P. R. (1986). Proposals for the classification and nomenclature of functional receptors for 5-hydroxytryptamine. Neuropharmacology 25: 563–576.

Buzzi, M. G., and Moskowitz, M. A. (1990). The antimigraine drug, sumatriptan (GR 43175), selectively blocks neurogenic plasma extravasation from blood vessels in dura mater. Br. J. Pharmacol. 99: 202–206.

Buzzi, M. G., Dimitriadou, V., Theoharides, T. C., and Moskowitz, M. A. (1990). Neurogenic inflammatory response in rat dura mater is blocked by drugs that share 5-HT_1-like mechanisms. Abstract 131, The Second IUPHAR Satellite Meeting on Serotonin, p. 101.

Doenicke, A., Brand, J., and Perrin, V. L. (1988). Possible benefit of GR 43175, a novel 5-HT_1 like receptor agonist, for the acute treatment of severe migraine. Lancet i: 1309–1311.

Dudley, M. W., Wiech, N. L., Miller, F. P., Carr, A. A., Cheng, H. C., Roebel, L. E., Doherty, N. S., Yamamura, H. I., Ursillo, R. C., and Palfreyman, M. G. (1988). Pharmacological effects of MDL 11,939: A selective, centrally acting antagonist of 5-HT_2 receptors. Drug Develop. Res. 13: 29–43.

Edvinsson, L., Hardebo, J. E., and Owman, C. (1978). Pharmacological analysis of 5-hydroxytryptamine receptors in isolated intracranial and extracranial vessels of cat and man. Circ. Res. 42: 143–151.

Engel, G., Göthert, M., Müller-Schweinitzer, E., Schlicker, E., Sistonen, L., and Stadler, P. A. (1983) Evidence for common pharmacological properties of [³H]5-hydroxytryptamine binding sites, presynaptic 5-hydroxytryptamine autoreceptors in CNS and inhibitory presynaptic 5-hydroxytryptamine receptors on sympathetic nerves. Naunyn Schmiedeberg's Arch. Pharmacol. 324: 116–124.

Engel, G., Göthert, M., Hoyer, D., Schlicker, E., and Hillenbrand, K. (1986). Identity of inhibitory presynaptic 5-hydroxytryptamine (5-HT) autoreceptor in the rat brain cortex with 5-HT_{1B} binding sites. Naunyn-Schmiedeberg's Arch Pharmacol. 332: 1–7.

Forster, C., and Whalley, E. T. (1982). Analysis of the 5-hydroxytryptamine induced contraction of the human basilar arterial strip compared with the rat aortic strip in vitro. Naunyn-Schmiedeberg's Arch. Pharmacol. 319: 12–17.

Fozard, J. R. (1984). MDL 72222: a potent and highly selective antagonist at neuronal 5-hydroxytryptamine receptors. Naunyn-Schmiedeberg's Arch. Pharmacol. 326: 36–44.

Fozard, J. R. (1987). The pharmacological basis of migraine treatment. In: Migraine: Clinical and Research Aspects. Blau, J. N. (ed.), The Johns Hopkins University Press, Baltimore, pp. 165–184.

Fozard, J. R. (1988). A critique of migraine therapy. In: The Management of Headache, In: Rose, C. F. (ed.), Raven Press, New York, pp. 97–114.

Fozard, J. R. (1989). 5-HT in migraine: evidence from 5-HT receptor antagonists for a neuronal aetiology. In: Migraine: A Spectrum of Ideas. Sandler, M. and Collins, G. M., (eds.) Oxford University Press, pp. 128–141.

Glennon, R. A. (1987). Central serotonin receptors as targets for drug research. J. Med. Chem. 30: 1–12.

Gozlan, H., EL Mestikawy, S., Pichat, L., Glowinski, J., and Hamon, M. (1983). Identification of presynaptic serotonin autoreceptors using a new ligand: ^{3}H-PAT. Nature 305: 140–142.

Hamel, E., Robert, J.-P., Young, A. R., and MacKenzie, E. T. (1989). Pharmacological properties of the receptor(s) involved in the 5-hydroxytryptamine-induced contraction of the feline middle cerebral artery. J. Pharmacol. Exp. Therap. 249(3): 879–889.

Hamel, E., and Bouchard, D. (1991). Contractile 5-HT$_1$ receptors in human pial arterioles: correlation with 5-HT$_{1D}$ binding sites. Br. J. Pharmacol. 102: 227–233.

Hardebo, J. E., Edvinsson, L., Owman, C., and Svendgaard, N.-A. (1978). Potentiation and antagonism of serotonin effects on intracranial and extracranial vessels. Neurology 28: 64–70.

Herrick-Davis, K., Titeler, M., Leonhardt, S., Struble, R., and Price, D. (1988). Serotonin 5-HT$_{1D}$ receptors in human prefrontal cortex and caudate: interaction with a GTP binding protein. J. Neurochem. 51: 1906–1912.

Högestätt, E. D., Andersson, K.-E., and Edvinsson, L. (1983). Mechanical properties of rat cerebral arteries as studied by a sensitive device for recording of mechanical activity in isolated small blood vessels. Acta. Physiol. Scand. 117: 49–61.

Hoyer, D., Pazos, A., Probst, A., and Palacios, J. M. (1986). Serotonin receptors in the human brain. I. Characterization and autoradiographic localization of 5-HT$_{1A}$ recognition sites. Apparent absence of 5-HT$_{1B}$ recognition sites. Brain Res. 376: 85–96.

Hoyer, D. (1988). Functional correlates of serotonin 5-HT$_1$ recognition sites. J. Recept. Res. 8(1–4): 59–81.

Hoyer, D. (1989). 5-Hydroxytryptamine receptors and effector coupling mechanisms in peripheral tissues. In: Fozard, J. R. (ed.), The Peripheral Actions of 5-Hydroxytryptamine. Oxford University Press, Oxford, New York, Tokyo, pp. 72–99.

Humphrey, P. P. A., Feniuk, W., Perren, M. J., Motevalian, M., and Whalley, E. T. (1990). The mechanism of anti-migrane action of sumatriptan. Abstract 132, The Second IUPHAR Satellite Meeting on Serotonin, p. 102.

Lee, T. J.-F. (1989). Recent studies on serotonin-containing fibers in cerebral circulation. In: Neurotransmission and Cerebrovascular Function II, Seylaz, J. & Sercombe, R. (eds.), Elsevier Science Publishers B. V. (Biomedical Division), Amsterdam, pp. 133–149.

Mir, A. K., Hibert, M., Tricklebank, M. D., Middlemiss, D. N., Kidd, E. J., and Fozard, J. R. (1988). MDL 72832: a potent and stereoselective ligand at central and peripheral 5-HT$_{1A}$ receptors. Eur. J. Pharmacol. 149: 107–120.

Moskowitz, M. A., Buzzi, M. G., Sakas, D. E., and Linnik, M. D. (1989). Pain mechanisms underlying vascular headaches. Rev. Neurol. 145; 181–193.

Müller-Schweinitzer, E. (1983). Vascular effects of ergot alkaloids: a study on human basilar arteries. Gen. Pharmac. 14: 95–102.

Parsons, A. A., Whalley, E. T., Feniuk, W., Connor, H. E., and Humphrey, P. P. A. (1989). 5-HT$_1$-like receptors mediate 5-hydroxytryptamine-induced contraction of human isolated basilar artery. Br. J. Pharmacol. 96: 434–440.

Peroutka, S. J., and Kuhar, M. J. (1984). Autoradiographic localization of 5-HT$_1$ receptors to human and canine basilar arteries. Brain Res. 310: 193–196.

Peroutka, S. J., and McCarthy, B. G. (1989). Sumatriptan (GR 43175) interacts selectively with 5-HT$_{1B}$ and 5-HT$_{1D}$ binding sites. Eur. J. Pharmacol. 163: 133–136.

Richardson, B. P., and Engel, G. (1986). The pharmacology and function of 5-HT$_3$ receptors. Trends Neurol. Sci., 424–428.

Saito, K., Markowitz, S., Moskowitz, M. A. (1988). Ergot alkaloids block neurogenic extravasation in dura matter: proposed action in vascular headaches. Ann. Neurol. 24: 732–737.

Saxena, P. R., and Ferrari, M. D. (1989). 5-HT$_1$-like receptor agonists and the pathophysiology of migraine. Trends Pharmacol. Sci. 10: 200–204.

Schoeffter, P., and Hoyer, D. (1989). How selective is GR 43175? Interactions with functional 5-HT$_{1A}$, 5-HT$_{1B}$, 5-HT$_{1C}$ and 5-HT$_{1D}$ receptors. Naunyn-Schmiedeberg's Arch. Pharmacol. 340: 135–138.

Schoeffter, P., and Hoyer, D. (1989). 5-Hydroxytryptamine (5-HT)-induced endothelium-dependent relaxation of pig coronary ateries is mediated by 5-HT receptors similar to the 5-HT$_{1D}$ receptor subtype. J. Pharmacol. Exp. Ther. 252(1): 387–395.

Timmermans, P. B. M. W. M., Mathy, M.-J., Wilffert, B., Kalkman, H. O., Smit, G., Dijkstra, D., Horn, A., and Van Zwietin, P. A. (1984). α_1/α_2-Adrenoceptor agonists selectivity of mono- and dihydroxy-2-N,N-Di-n-propylaminotetralins. Eur. J. Pharmacol. 97: 55–65.

Van Den Brink, F. G. (1963). General theory of drug-receptor interactions. Drug-receptor interaction models. Calculation of drug parameters. In: Kinetics of Drug Action, Van Rossum, J. M. (ed.), Springer-Verlag, Berlin pp. 169–254.

Waeber, C., Schoeffter, P., Palacios, J. M., and Hoyer, D. (1988). Molecular pharmacology of 5-HT$_{1D}$ recognition sites: radioligand binding studies in human, pig and calf brain membranes. Naunyn-Schmiedeberg's Arch. Pharmacol. 337: 595–601.

Young, A. R., Hamel, E., MacKenzie, E. T., Seylaz, J., and Verrecchia, C. (1987). The multiple actions of 5-hydroxytryptamine on cerebrovascular smooth muscle: In: Neuronal Messengers in Vascular Function, Nobin, A., Owman, C., and Arneklo-Nobin, B. (eds.), Elsevier Science Publishers, Amsterdam, pp. 57–74.

Serotonin: Molecular Biology, Receptors and Functional Effects
ed. by J. R. Fozard/P. R. Saxena

Human, Monkey and Dog Coronary Artery Responses to Serotonin and 5-Carboxamidotryptamine

N. Toda

Department of Pharmacology, Shiga University of Medical Sciences, Seta, Ohtsu 520-21, Japan

Summary. In beagle coronary arteries, serotonin-induced contractions are related directly to advancing age from 30 days to 12 years. This change is not associated with age-dependent endothelial disfunction. Contractile responses to serotonin of dog and monkey coronary arteries are evidently greater in the proximal portion than in the distal. Serotonin-induced contractions are in the order of human > monkey > dog arteries. On the other hand, 5-carboxamidotryptamine (5-CT), a $5\text{-}HT_1$-like receptor agonist, relaxes dog coronary arteries, but contracts the human and monkey arteries. Relaxations of the dog arteries are dependent on endothelium, and are suppressed by methysergide but not altered by ketanserin. Human and monkey artery contractions by 5-CT and serotonin are not potentiated by endothelium denudation, and are attenuated by methysergide and ketanserin. It appears that endothelial serotonergic receptors do not contribute to relaxations in primate coronary arteries, whereas endothelial $5\text{-}HT_1$-like receptor activation participates in the release of EDRF in dog coronary arteries. Contractions caused by 5-CT in human and monkey coronary arteries may be associated with predominant activation of $5\text{-}HT_1$-like receptor subtypes over $5\text{-}HT_2$ receptors, whereas those by serotonin are due mainly to $5\text{-}HT_2$ receptor activation.

Introduction

Despite abundant information concerning vascular actions of serotonin in experimental animals (Van Nueten et al. 1985), physiological and pathopohysiological roles in the cardiovascular system have not been clarified. Serotonin is one of the potent, endogenous vasoconstrictors in human coronary arteries (Kalsner and Richards 1984; Ginsburg et al. 1984; Toda et al. 1985), and ergonovine provokes coronary vasospasm possibly by stimulation of serotonergic receptors (Brasenor and Angus 1981); therefore, the amine is considered to participate in the genesis of coronary vasospasm. The data on coronary arteries from experimental mammals, such as dogs, rabbits and pigs, have been accumulated. However, information concerning primate coronary arteries would be more useful for the analysis of coronary circulatory disorders in patients, in relation to serotonin. This chapter describes the comparison of the responses to serotonin and 5-carboxamidotryptamine (5-CT), a $5\text{-}HT_1$-like receptor agonist (Feniuk et al. 1981; Engel et al. 1983), in helical strips of coronary arteries from dogs, Japanese monkeys and

276

humans. The human materials were obtained during autopsy within three hours after death (Toda and Okamura, 1989).

Serotonin Actions and Aging

We have demonstrated that the contractile response of coronary arteries from beagles is related directly to increasing age from 30 days to 12 years (Figure 1) (Toda et al. 1987). The contractions are suppressed by treatment with 10^{-6} M ketanserin, suggesting a major involvement of 5-HT_2 receptor subtype. In the arteries obtained from immature beagles, the serotonin-induced contraction is not potentiated by endothelium denudation. In addition, relaxations caused by acetylcholine, mediated by EDRF, are not greater in the immature beagle arteries than in the arteries from adult and senile beagles. Therefore, different magnitudes of contraction are not due to age-dependent decreases in the endothelial function. Function of 5-HT_2 receptors responsible for smooth muscle contraction is expected to increase in beagle coronary arteries in an age-dependent manner.

On the other hand, monkey coronary artery responsiveness is not altered so evidently with age as that seen in the beagle arteries. Baby monkey (3 weeks) coronary arteries respond to serotonin with a moder-

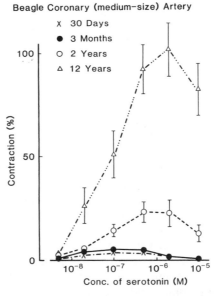

Figure 1. Age-dependent change in the respose to serotonin of medium-size coronary arteries from beagles of different age (from Toda et al. 1987; with permission). Contractions caused by 30 mM K^+ were taken as 100%.

ate contraction, which is markedly greater than that in the baby beagle arteries shown in Figure 1. Contractions caused by serotonin, normalized by K^+ (30 mM)-induced contractions, are less in the baby monkey arteries than in those from the older monkeys; however, the amine-induced contractions do not differ in the juvenile, adult and senile monkey arteries.

Serotonin Actions in Proximal and Distal Portions

Contractile responses to serotonin are dependent on the portions in dog and monkey coronary arteries (Miwa et al. 1984; Toda et al. 1987). Proximal coronary arteries respond to the amine with significantly greater contractions than the distal arteries (Figure 2). Such a regional difference is not associated with the endothelial function responsible for the amine-induced relaxation mediated via EDRF observed in pig and dog coronary arteries (Cocks and Angus 1983).

Actions of Serotonin and 5-CT in Different Mammals

Contractions induced by serotonin and 5-CT of coronary arteries of medium size from adult dogs, monkeys and humans are compared in Figure 3. The magnitude of serotonin-induced contraction relative to that caused by 30 mM K^+ is in the order of human > monkey > dog

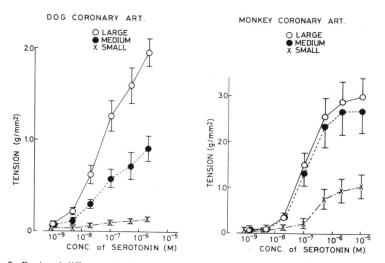

Figure 2. Regional difference in the response of serotonin of coronary arteries from dogs and monkeys (from Miwa and Toda 1984; with permission). The ordinate represents the amine-induced contraction/cross sectional area of the artery strips.

278

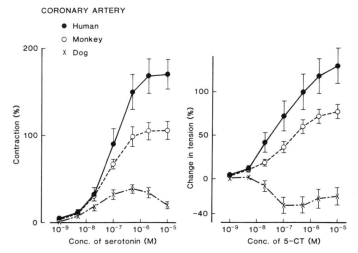

Figure 3. Species difference in the response to serotonin and 5-carboxamidotryptamine (5-CT) of coronary artery strips (Toda and Okamura 1990; with permission). Contractions induced by 30 mM K^+ were taken as 100% contraction, and relaxations by 10^{-4} M papaverine were taken as 100% relaxation.

arteries. In the dog arteries, 5-CT produces concentration-related relaxations. On the other hand, the human and monkey arteries respond to these amines only with contractions (Toda and Okamura 1990). The magnitude of contraction by 5-CT tended to be less than that with serotonin.

Endothelium denudation significantly potentiates the serotonin-induced contraction in dog coronary arteries, as already shown by other investigators in the cat, dog and pig coronary arteries (Griffith et al. 1984; Houston and Vanhoutte 1988; Shimokawa et al. 1989). 5-CT-induced relaxations in the dog arteries are suppressed by treatment with 10^{-6} M methysergide, a non-selective serotonergic receptor antagonist, but not altered by 10^{-6} M ketanserin, a $5-HT_2$ receptor antagonist (Van Nueten et al. 1981). The relaxation is reversed to a slight contraction by removal of endothelium. In the endothelium-denuded dog arteries, the 5-CT-induced contraction is abolished only by methysergide. These findings suggest that the induced relaxation is associated with the release of EDRF by stimulation of $5-HT_1$-like receptor subtypes in endothelium. 5-CT appears to produce slight contractions by acting also on $5-HT_1$-like receptors in smooth muscle.

In contrast to the dog arteries, responses to serotonin and 5-CT of monkey and human coronary arteries are not significantly influenced by removal of endothelium. In monkey coronary artery strips treated with 10^{-6} M ketanserin and partially contracted with prostaglandin $F_{2\alpha}$, serotonin and 5-CT produce slight but significant contractions

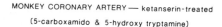

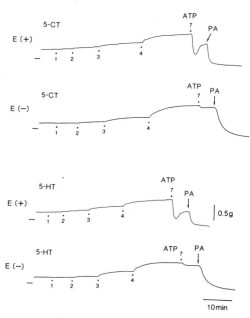

Figure 4. Responses to serotonin (5-HT) and 5-CT of monkey coronary artery strips with (E +) and without endothelium (E −). The strips were treated with 10^{-6} M ketanserin and partially contracted with prostaglandin $F_{2\alpha}$ from the level shown as horizontal lines just left of each tracing. Concentrations of 5-HT and 5-CT from 1 to 4 = 10^{-9}, 5×10^{-9}, 2×10^{-8} and 10^{-7} M, respectively. Endothelial integrity was determined by the addition of 10^{-7} M ATP. PA = 10^{-4} M papaverine to attain the maximal relaxation.

even in the presence of endothelium (Figure 4). The contractions are not potentiated by endothelium denudation. Such a result is also obtained in the arteries not treated with ketanserin, although the amine-induced contractions are markedly greater than those under treatment with the 5-HT₂ antagonist. In primate coronary arteries, serotonin does not appear to release significant amounts of relaxing factor from endothelium.

Methysergide contracts monkey coronary arteries and almost abolishes the contraction caused by 5-CT (10^{-9}–10^{-5} M), whereas ketanserin (10^{-8}–10^{-6} M) does not alter the artery tone but attenuates the 5-CT-induced contraction dose-dependently. The maximal attenuation by ketanserin is appreciably less than that by methysergide. Similar results are also obtained in human coronary arteries stimulated by 5-CT. Therefore, 5-CT-induced contractions are associated mainly with stimulation of 5-HT₁-like receptor subtypes and also with 5-HT₂ receptor stimulation. Treatment with 5-CT suppresses the serotonin-induced contraction in dog coronary arteries but does not alter the contraction in the monkey arteries. 5-CT may interfere with binding of serotonin to 5-HT$_1$-like receptors in the dog arteries. Ketanserin markedly inhibits the serotonin-induced contraction in the monkey and human arteries. It may be concluded that the primate coronary arteries contract in response to serotonin, possibly by predominant activation of 5-HT$_2$ over 5-HT$_1$-like receptors.

Ergonovine provokes coronary vasospasm in patients with variant angina pectoris (Heupler et al. 1978). This substance produces intense contractions of human coronary arteries, possibly by stimulation of 5-HT$_2$ receptors, since the contraction is suppressed by ketanserin. Serotonin markedly contracts human and monkey coronary arteries of proximal portion; however, the response is not potentiated by endothelium denudation. Serotonin derived from platelets may play a role in the genesis of coronary vasospasm. According to literatures (De Caterina et al. 1984; Noble and Drake-Holland 1988), effectiveness of ketanserin in relieving the coronary vasospasm is controversial. Recent studies have demonstrated possible cardio-protective actions of ketanserin in patients with cardiac ischemic episodes (reviewed by Van Zwieten 1990). Atherosclerotic lesions appear to participate in aggregating platelets and releasing serotonin and other vasoactive substances but not to potentiate the serotonin-induced contraction by eliminating the EDRF actions in human coronary arteries.

References

Brazenor, R. M., and Angus, J. A. (1981). Ergometrine contracts isolated canine coronary arteries by a serotonergic mechanism: no role for alpha adrenoceptors. J. Pharmacol. Exp. Ther. 218: 530–536.

Cocks, T. M., and Angus, J. A. (1983). Endothelium-dependent relaxation of coronary arteries by noradrenaline and serotonin. Nature 305: 627–630.

De Caterina, R., Carpeggiani, C., and L'Abbate, A. (1984). A double-blind, placebo-controlled study of ketanserin in patients with Prinzmetal's angina. Evidence against role for serotonin in the genesis of coronary vasospasm. Circulation 69: 889–894.

Engel, G., Göthert, M., Müller-Schweinitzer, E., Schliker, E., Sistonen, L., and Stadler, P. A. (1983). Evidence for common pharmacological properties of [^{3}H] 5-hydroxytryptamine binding sites. Presynaptic 5-hydroxytryptamine autoreceptors in CNS and inhibitory presynaptic 5-hydroxytryptamine receptors on sympathetic nerves. Naunyn-Schmiedeberg's Arch. Pharmacol. 324: 116–124.

Feniuk, W., Humphrey, P. P. A., and Watts, A. D. (1981). Further characterization of pre- and post-junctional receptors for 5-hydroxytryptamine in isolated vasculature. Br. J. Pharmacol. 73: 191P.

Ginsburg, R., Bristow, M. R., Davis, K., Dibiase, A., and Billingham, M. E. (1984). Quantitative pharmacologic responses of normal and atherosclerotic isolated human epicardial coronary arteries. Circulation 69: 430–440.

Griffith, T. M., Henderson, A. H., Edwards, D. H., and Lewis, M. J. (1984). Isolated perfused rabbit coronary artery and aortic strip preparations: the role of endothelium-derived relaxant factor. J. Physiol. 351: 13–24.

Heupler, F. A., Proudfit, W. L., Pazavi, M., Shirkey, E. K., Greenstreet, R., and Sheldon, W. C. (1978). Ergonovine maleate provocative test for coronary arterial spasm. Am. J. Cardiol. 41: 631–640.

Houston, D. S., and Vanhoutte, P. M. (1988). Comparison of serotonergic receptor subtypes on the smooth muscle and endothelium of the canine coronary artery. J. Pharmacol. Exp. Ther. 244: 1–10.

Kalsner, S., and Richards, R. (1984). Coronary arteries of cardiac patients are hyperreactive and contain stores of amines: a mechanism for coronary spasm. Science 223: 1435–1437.

Miwa, K., and Toda, N. (1984). Regional differences in the response to vasoconstrictor agents of dog and monkey isolated coronary arteries. Br. J. Pharmacol. 82: 295–301.

Noble, M. I. M., and Drake-Holland, A. J. (1988). Serotonin antagonism prevents myocardial infarction and death from coronary artery disease. Br. Heart J. 59: 100–101.

Shimokawa, H., Aarhus, L. L., and Vanhoutte, P. M. (1987). Porcine coronary arteries with regenerated endothelium have a reduced endothelium-dependent responsiveness to aggregating platelets and serotonin. Circ. Res. 61: 256–270.

Toda, N., Bian, K., and Inoue, S. (1987). Age-related changes in the response to vasoconstrictor and dilator agents in isolated beagle coronary arteries. Naunyn-Schmiedeberg's Arch. Pharmacol. 336: 359–364.

Toda, N., and Okamura, T. (1989). Endothelium-dependent and -independent responses to vasoactive substances of isolated human coronary arteries. Am. J. Physiol. 257: H988–H995.

Toda, N., and Okamura, T. (1990). Comparison of the response to 5-carboxamidotryptamine and serotonin in isolated human, monkey and dog coronary arteries. J. Pharmacol. Exp. Ther. 253: 676–682.

Toda, N., Okamura, T., Shimizu, I., and Tatsuno, Y. (1985). Postmortem functional changes in coronary and cerebral arteries from humans and monkeys. Cardiovasc. Res. 19: 707–713.

Van Nueten, J. M., Janssen, P. A. J., Van Beek, J., Xhonneux, R., Verbeuren, T. J., and Vanhoutte, P. M. (1981). Vascular effects of ketanserin (R41468), a novel antagonist of 5-HT$_2$ serotonergic receptors. J. Pharmacol. Exp. Ther. 218: 217–230.

Van Nueten, J. M., Janssens, W. J., and Vanhoutte, P. M. (1985). Serotonin and vascular smooth muscle. In Vanhoutte, P. M. (ed.), Serotonin and Vascular Smooth Muscle. New York: Raven Press, pp. 95–103.

Van Zwieten, P. A., Blauw, G. J., and Van Brummelen, P. (1990). Pathophysiological relevance of serotonin in cardiovascular diseases. Prog. Pharmacol. Clin. Pharmacol. 7: 63–75.

Serotonin: Molecular Biology, Receptors and Functional Effects
ed. by J. R. Fozard/P. R. Saxena
© 1991 Birkhäuser Verlag Basel/Switzerland

5-Carboxamidotryptamine Induced Renal Vasoconstriction in the Dog

D. Cambridge, M. V. Whiting, and L. J. Butterfield

Department of Pharmacology, Wellcome Research Laboratories, Beckenham, Kent BR3 3BS UK

Summary. Systemic administration of 5-Carboxamidotryptamine (5-CT) causes dose-related renal vasoconstrictor (ED_{50}, 10 ± 3 ng/kg) responses in the dog, in addition to its well-documented vasodilator effects. These renal vasoconstrictor responses are characterised by their rapid onset but transient profile, and cannot be ascribed to either renal sympathetic nerve activation, or endogenous 5-hydroxytryptamine (5-HT) release via an uptake mechanism. In addition, they are not attenuated by the presence of spiperone (antagonist at $5\text{-}HT_1$-like receptors mediating vasodilation), ketanserin (antagonist at $5\text{-}HT_2$ receptors), or MDL 72222 (antagonist at $5\text{-}HT_3$ receptors). However, they are attenuated by methiothepin (antagonist at $5\text{-}HT_1$-like receptors mediating vasoconstriction), and mimicked by the preferential $5\text{-}HT_1$-like receptor agonist, sumatriptan (ED_{50}, 290 ± 60 ng/kg). This suggests that the dog kidney contains a population of $5\text{-}HT_1$-like receptors mediating vasoconstriction, which appear to be pharmacologically similar to those identified in the cerebral vasculature. However, the exact role and location of these receptors in the kidney remains unclear.

Introduction

5-Carboxamidotryptamine (5-CT) is well-known for its marked systemic vasodilator actions *in vivo*, which are mediated through $5\text{-}HT_1$-like receptors (Connor et al. 1986; Martin et al. 1987; Cambridge and Whiting 1989; Whiting and Cambridge 1990). However, Wright and Angus (1989) reported that 5-CT also caused renal artery spasm in the rabbit, which could be attenuated by the $5\text{-}HT_2$ receptor antagonist, ketanserin, and the 5-HT uptake inhibitor, fluoxetine. They concluded that 5-CT was able to stimulate the release of endogenous 5-HT from platelets, via an uptake process, and that this endogenous 5-HT then acted upon $5\text{-}HT_2$ receptors in the renal artery, to cause renal artery spasm. In the course of some *in vivo* studies in dogs, designed to investigate the regional haemodynamic responses to tryptamines, we observed that 5-CT, a potent vasodilator of the carotid and coronary vasculature in these animals, was also a potent vasoconstrictor of the renal vasculature. However, the doses of 5-CT needed to produce these renal vascular responses were approximately one order of magnitude less than those required for its vasodilator actions, and from initial experiments did not appear to utilise the same mechanism of action as that described for the rabbit (Wright and Angus 1989). We therefore undertook the following studies to investigate the mechanism of this unexpected response to 5-CT.

Materials and Methods

Beagle dogs of either sex (body weight 12–18 kg) were used in these studies. Animals were anaesthetised with intravenous pentobarbitone (30–40 mg/kg), tracheotomised and artificially respired with room air via a Palmer large animal ventilator. Cannulae were placed into the upper abdominal aorta via a femoral artery, to measure aortic blood pressure, and into a femoral vein, for drug and anaesthetic administration. An ultrasound flowprobe was placed upon the left carotid artery to measure carotid blood flow. A left thoracotomy was performed at ribs 4 and 5 to expose the heart, and a pericardial cradle constructed to allow access to the left circumflex coronary artery in order to measure coronary artery blood flow using an ultrasound flow probe, and for the insertion of a cannula into the left atrium for the systemic administration of 5-CT in order to avoid the pulmonary vasculature. An electromagnetic flow probe was then placed on the left renal artery via a left flank incision in order to measure renal blood flow.

In initial experiments, the renal vasoconstrictor responses to 5-CT were evaluated after renal sympathectomy (sectioning the renal nerves), after ganglion blockade (mecamylamine 3 mg/kg), and after 5-HT-uptake blockade (citalopram 1 mg/kg). Dose-response curves to 5-CT were also constructed in the presence and absence of $5-HT_2$ (ketanserin 0.1 mg/kg) and $5-HT_3$ (MDL 72222 0.5 mg/kg) blockade to eliminate the possibility of the involvement of these receptors. These studies were then extended to evaluate the effects of additional blockade of the $5-HT_1$-like receptors that mediate vasodilation (spiperone 0.3 and 1 mg/kg) and vasoconstriction (methiothepin 0.1 and 1 mg/kg).

In these experiments, groups of 3–5 animals were used for each experimental protocol. Regional haemodynamic changes were measured as conductance changes (calculated from mean blood flow/mean aortic blood pressure). The mean ED_{50} (dose of 5-CT to produce 50% of its maximum response) was calculated from individual dose-response curves.

Results

5-CT, when administered via the intra-left atrial route (0.001–3 μg/kg) into 4 anaesthetised dogs, caused dose-related falls in blood pressure (ED_{50}, 171 $\pm$ 94 ng/kg) associated with small increases in heart rate, and dose-related increases in both carotid and coronary blood flow and

284

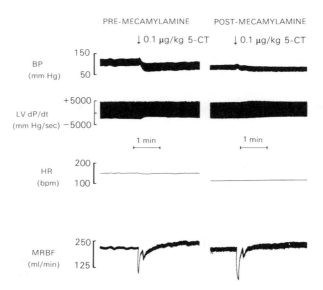

Figure 1. The effects of a single intra-left atrial bolus dose of 5-Carboxamidotryptamine (5-CT) in the anaesthetised dog, before and after ganglion-blockade with mecamylamine.

conductance (ED_{50}, 120 ± 30 and 160 ± 70 ng/kg, respectively). However, 5-CT also caused dose-related, transient reductions in renal blood flow (see Figure 1) and conductance (ED_{50}, 10 ± 3 ng/kg). Comparing these haemodynamic effects of 5-CT by examination of the ED_{50} doses, showed that the renal vasoconstrictor responses required approximately 0.06–0.08 times the dose required for the expected vasodilator responses.

The renal vasoconstrictor responses to 5-CT were reproducible on repeated dosing and were unaffected by renal sympathectomy. In addition, these responses were unaltered by administration of the ganglion blocker, mecamylamine or by the 5-HT uptake inhibitor, citalopram. The effects of ganglion-blockade on these renal responses are shown in Figure 1.

The renal vasoconstrictor responses to 5-CT were also unaffected by the presence of ketanserin and MDL 72222 (antagonists of the 5-HT_2 and 5-HT_3 receptors, respectively) and spiperone (antagonist of the 5-HT_1-like receptors mediating vasodilation: see Figure 2), although spiperone did significantly attenuate the carotid and coronary vasodilator responses to 5-CT. However, the renal vasoconstrictor responses were significantly attenuated in a dose-related manner by methiothepin (antagonist of the 5-HT_1-like receptors mediating vasoconstriction: see Figure 3). In these preparations, methiothepin also significantly attenuated the carotid and coronary vasodilator responses to 5-CT. Thus unlike spiperone, methiothepin was unable to distinguish the 5-HT_1-like receptors mediating vasoconstriction and vasodilation.

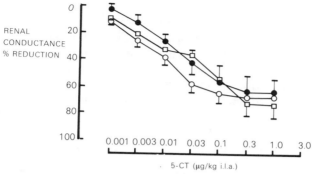

Figure 2. The effects of spiperone on 5-Carboxamidotryptamine (5-CT) induced changes in renal conductance in anaesthetised dogs in the presence of 5-HT$_2$/5-HT$_3$-blockade. control (○: $n = 4$); spiperone 0.3 mg/kg (●: $n = 4$); spiperone 1.0 mg/kg (□: $n = 4$).

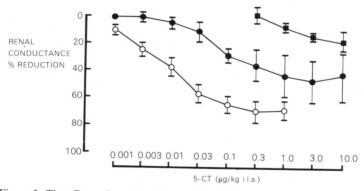

Figure 3. The effects of methiothepin on 5-Carboxamidotryptamine (5-CT) induced changes in renal conductance in anaesthetised dogs in the presence of 5-HT$_2$/5-HT$_3$-blockade. control (○: $n = 4$); methiothepin 0.1 mg/kg (●: $n = 4$); methiothepin 1.0 mg/kg (■: $n = 4$).

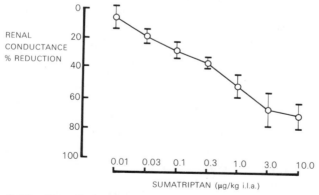

Figure 4. The effects of sumatriptan on renal conductance in anaesthetised dogs ($n = 4$) in the presence of 5-HT$_2$/5-HT$_3$-blockade.

In a further series of anaesthetised dogs ($n = 4$), the preferential 5-HT_1-like vasoconstrictor agonist, sumatriptan, administered by the same intra-left atrial route, was also found to cause dose-related reductions in renal blood flow and conductance (ED_{50}, 290 ± 60 ng/kg), (see Figure 4) which had a similar transient profile to those observed with 5-CT. The only other discernible haemodynamic effect of sumatriptan in these animals was its, now well-documented, dose-related reduction in carotid blood flow and conductance (ED_{50}, 9000 ± 6400 ng/kg). As with 5-CT the renal and carotid vasoconstrictor responses to sumatriptan were also attenuated by methiothepin.

Discussion

Systemic administration of 5-CT causes a dose-related reduction in renal blood flow in the dog which has a rapid onset but a short duration of action. This renal vasoconstrictor effect contrasts markedly with the systemic vasodilator responses, mediated by 5-HT_1-like receptors, which are more usually associated with this tryptamine analogue (Saxena and Lawang 1985; Saxena and Verdouw 1985; Connor et al. 1986; Martin et al. 1987; Cambridge and Whiting 1989; Whiting and Cambridge 1990).

It is possible that the mode of administration of 5-CT, directly into the left atrium, and its systemic vasodilator effects, could stimulate some reflex activation of the renal sympathetic nerves. However, neither renal sympathectomy (sectioning the renal nerves) nor the administration of a ganglion blocker, were able to influence the renal vasoconstrictor response to 5-CT. Thus in the dog there appears to be no role for the renal sympathetic innervation in the renal vasoconstrictor response to 5-CT.

Wright and Angus (1989) recently reported that 5-CT also causes renal artery spasm in the rabbit, and they were able to demonstrate that this response appears to be a consequence of the release of endogenous 5-HT from circulating platelets and subsequent stimulation of renal arterial 5-HT_2 receptors. However, the renal vasoconstrictor response to 5-CT in the dog was unaffected by prior treatment with citalopram, a specific 5-HT uptake inhibitor, nor was it affected by the presence of ketanserin and MDL 72222, 5-HT_2 and 5-HT_3 receptor antagonists, respectively. The mechanism of the 5-CT renal vasoconstrictor response in the dog therefore, appears to be quite different to that seen in the rabbit.

Recent studies have reported the existence of 5-HT_1-like receptors in the dog saphenous vein (Humphrey et al. 1988) and in the dog cerebral vasculature (Feniuk et al. 1989) which also mediate vascular contraction responses. This suggested that in the dog renal vasculature 5-CT might stimulate a similar population of 5-HT_1-like receptors. In the present study pretreatment with the 5-HT_1-like antagonist spiperone was ineffective at inhibiting the renal vasoconstrictor responses to 5-CT in the dog. However, as has been reported previously (Martin et al. 1987; Whiting and Cambridge 1990) spiperone was an effective antagonist of the vasodilator responses to 5-CT. Thus spiperone appears to have a preferential antagonist action at the vascular 5-HT$_1$-like receptor mediating vasodilation. By contrast, pretreatment with the 5-HT$_1$-like antagonist methiothepin, which has been previously shown to inhibit the 5-HT$_1$-like receptor mediated vasoconstrictor responses (Saxena et al. 1986; Humphrey et al. 1988; Feniuk et al. 1989), significantly attenuated the renal vasoconstrictor responses to 5-CT in the dog. Although methiothepin pretreatment also inhibited the vasodilator responses to 5-CT, these results are consistent with the stimulation, by 5-CT, of a population of 5-HT$_1$-like vasoconstrictor receptors in the renal vasculature of the dog. The existence of such receptors in the dog renal vasculature is further reinforced by our observation that the selective 5-HT$_1$-like agonist, sumatriptan (Humphrey et al. 1988; Feniuk et al. 1989) produces renal vasoconstrictor responses in the dog which are similar in profile to those seen with 5-CT and are also inhibited by methiothepin.

In conclusion, we have demonstrated the presence of a population of 5-HT$_1$-like receptors in the renal vasculature of the dog, which mediate renal vasoconstrictor responses. The exact location of these receptors within the renal vasculature has not yet been identified, although it is probable that these receptors are associated with resistance-type vessels. However, the haemodynamic profile of the renal vasoconstriction responses, produced by activation of these receptors, indicates that there are opposing intra-renal mechanisms which prevent any sustained reduction in renal blood flow. It therefore remains unclear as to the role of these receptors within the dog kidney.

References

Cambridge, D., and Whiting, M. V. (1989). Characterising the 5-HT receptor mediating coronary vasodilation in the dog using tryptamine analogues. Br. J. Pharmacol. 98 (Suppl.), 750P.

Connor, H. E., Feniuk, W., Humphrey, P. P. A., and Perren M. J. (1986). 5-Carboxamidotryptamine is a selective agonist at 5-hydroxytryptamine receptors mediating vasodilation and tachycardia in anaesthetised cats. Br. J. Pharmacol. 87: 417–426.

Feniuk, W., Humphrey, P. P. A., and Perren, M. J. (1989). The selective carotid arterial vasoconstrictor action of GR43175 in anaesthetised dogs. Br. J. Pharmacol. 96: 83–90.

Humphrey, P. P. A., Feniuk, W., Perren, M. J., Connor, H. E., Oxford, W. A., Coates, I. H., and Butina, D. (1988). GR43175, a selective agonist for the 5-HT$_1$-like receptor in dog isolated saphenous vein. Br. J. Pharmacol. 94: 1123–1132.

Martin, G. R., Leff, P., Cambridge, D., and Barrett, V. J. (1987). Comparative analysis of two types of 5-hydroxytryptamine receptor mediating vasorelaxation: differential classification using tryptamines. Naunyn Schmiedebergs Arch. Pharmacol. 336: 365–373.

Saxena, P. R., and Lawang, A. (1985). A comparison of cardiovascular and smooth muscle effects of a 5-hydroxtryptamine and 5-carboxamidotryptamine, a selective agonist of 5-HT$_1$ receptors. Arch. Int. Pharmacodyn. 277: 235–252.

288

Saxena, P. R., and Verdouw, P. D. (1985). 5-Carboxamide tryptamine, a compound with high affinity for 5-HT$_1$ binding sites, dilates arterioles and constricts arteriovenous anastomoses. Br. J. Pharmacol. 84: 533–544.

Saxena, P. R., Duncker, D. J., Bom, A. H., Heiligers, J., and Verdouw, P. D. (1986). Effects of MDL 72222 and methiothepin on carotid vascular responses to 5-Hydroxytryptamine$_1$-like receptors. Naunyn Schmiedebergs Arch. Pharmacol. 333: 198–204.

Whiting, M. V., and Cambridge, D. (1990). Systemic and coronary vasodilator responses to 5-HT *in vivo* are not mediated by endothelial 5-HT$_1$-like receptors. Eur. J. Pharmacol. 183: 2117–2118.

Wright, C. E., and Angus, J. A. (1989). 5-Carboxamidotryptamine elicits 5-HT$_2$ and 5-HT$_3$-receptors mediated cardiovascular responses in the conscious rabbit: evidence for 5-HT release from platelets. J. Cardiovasc. Pharmacol. 13: 557–564.

Serotonin: Molecular Biology, Receptors and Functional Effects
ed. by J. R. Fozard/P. R. Saxena
© 1991 Birkhäuser Verlag Basel/Switzerland

The Subretrofacial Nucleus: A Major Site of Action for the Cardiovascular Effects of 5-HT$_{1A}$ and 5-HT$_2$ Agonist Drugs

A. K. Mandal, K. J. Kellar, and R. A. Gillis

Department of Pharmacology, Georgetown University School of Medicine, Washington, D.C. 20007, USA

Summary. The purpose of our study was to determine the CNS site(s) where activation of 5-HT$_{1A}$ and 5-HT$_2$ receptors influence arterial blood pressure. To achieve this purpose the prototypic 5-HT$_{1A}$ and 5-HT$_2$ receptor agonist drugs, 8-OH-DPAT and DOI, respectively, were microinjected into sites in the medulla oblongata known to control central sympathetic outflow in chloralose anesthetized cats while monitoring femoral arterial blood pressure and heart rate (ECG). Microinjection of 8-OH-DPAT (20 ng) into the subretrofacial nucleus in 6 animals produced large decreases in mean blood pressure. Heart rate was unaffected. Pretreatment with the 5-HT$_{1A}$ receptor antagonist, BMY-8227, prevented the response. Microinjection of DOI (100 and 300 ng, N = 6 animals) into the subretrofacial nucleus produced significant increases in mean arterial pressure, again, unaccompanied by any significant change in heart rate. These results indicate that the subretrofacial nucleus is a major site of CNS action for the blood pressure effects of 5-HT receptor agonist drugs.

Introduction

Evidence indicates that activation of 5-HT$_{1A}$ and 5-HT$_2$ receptors can produce pronounced changes in arterial blood pressure (Gradin and Lis 1985; Fozard et al. 1987; Ramage and Fozard 1987; McCall et al. 1987; Wouters et al. 1988; Doods et al. 1988; McCall and Harris 1988; Dabiré et al. 1989a; b; Alper et al. 1990; King and Holtman 1990; Clement and McCall 1990a). In the case of 5-HT$_{1A}$ receptors and arterial blood pressure effects, it is clear that these receptors exist in the CNS (McCall et al. 1987; Fozard et al. 1987; Laubie et al. 1989; Gillis et al. 1989). The exact location of these CNS 5-HT$_{1A}$ receptors is unclear. Data have been published indicating that the 5-HT$_{1A}$ receptors affecting arterial blood pressure are located on rostral ventrolateral sympathoexcitatory neurons (Laubie et al. 1989; Gillis et al. 1989; Clement and McCall 1990b), on central sympathetic neurons which lie antecedent to the ventrolateral sympathoexcitatory neurons (Clement and McCall 1990b), on the dorsal raphe nucleus (Connor and Higgins 1990), on the B1/B3 ventral medullary area (Valenta and Singer 1990; Helke and Phillips, 1990), and on the nucleus ambiguus (Izzo et al. 1988).

In the case of 5-HT$_2$ receptors and arterial blood pressure effects, it appears that these receptors are more widespread and occur in the periphery (Dabiré et al. 1989a; 1989b) as well as in the CNS (McCall et al. 1987; McCall and Harris 1988; King and Holtman 1990; Mandal et al. 1990a; Clement and McCall 1990a). Again, the location of CNS 5-HT$_2$ receptors affecting blood pressure is unclear. For example, Clement and McCall (1990a) recently reported that I.V. administration of the selective 5-HT$_2$ receptor agonist, 1-(2,5-dimethoxy-4-iodophenyl)-2-aminopropane (DOI), increased the firing rate of medullospinal sympathoexcitatory neurons located in the rostral ventrolateral medulla. However, microiontophoretic application of DOI failed to affect the firing of these sympathoexcitatory neurons.

The purpose of the present study was to determine the CNS site(s) where activation of 5-HT$_{1A}$ and 5-HT$_2$ receptors would influence arterial blood pressure. Since data of earlier studies indicate that both 5-HT$_{1A}$ and 5-HT$_2$ agonist drugs applied to the surface of the rostral ventrolateral medulla alters blood pressure (Gillis et al. 1989; King and Holtman 1990), we focussed on nuclei associated with this region of the hindbrain.

Materials and Methods
General

Experiments were performed on adult cats of either sex, weighing between 2.2 and 4.0 kg. Anesthesia was induced and maintained with a single intravenous dose of α-chloralose (75 mg/kg). Rectal temperature was monitored and maintained between 37.0° and 38.0°C with an infrared heating lamp connected to a thermistor. The femoral artery and vein were cannulated for the measurement of arterial blood pressure and for the systemic administration of drugs, respectively. Lead II of the electrocardiogram (ECG) was monitored, and heart rate was determined by measurement of the R-R interval. The trachea was cannulated for the purpose of instituting artificial respiration, if needed. Arterial blood pressure and the ECG were recorded simultaneously on a Hewlett-Packard eight-channel recorder (HP 7758B).

Control measurements were taken at approximately 5 minute intervals for at least 30 minutes before the administration of drugs in each experiment. Once a drug was administered, measurements were made at the time at which peak responses occurred and then at 5-minute intervals thereafter until cardiorespiratory activity had returned to the pre-drug level or had stabilized at a new level.

Microinjection Technique

To microinject drugs into the ventrolateral medulla, the following procedures were performed. After a longitudinal midline incision was made in the neck, the trachea and esophagus were transected and retracted cephalad. The prevertebral muscles were bluntly dissected away from the base of the skull, which then was removed. The dura and arachnoid membranes were then removed. Double-barrelled micro-pipettes (i.e., 0.3 mm; FHC, Inc., Brunswick, Me.) were pulled with a vertical pipette-puller (model 700C, David Kopf Instruments, Tujunga, Calif.), and the tips were cut to approximately 15 μm i.d. The pipette then was mounted on David Kopf electrode manipulator. Each barrel of the pipette was connected to polyethylene tubing and filled using negative pressure. One barrel was filled with either a 100 mM solution of L-glutamic acid or a solution of bicuculline methiodide (containing 10 ng) while the other was filled with either 8-OH-DPAT or DOI, which also contained a 1% solution of fast green (Sigma Chemical Co., St. Louis) for subsequent histological verification of the injection site.

Three microinjection sites were selected in evaluating the site of action of 8-OH-DPAT and DOI. One of these sites was the locus where microinjections of L-glutamic acid produce an increase in blood pressure, namely, the subretrofacial nucleus. The site has been described by McAllen (1986). The coordinates used for microinjection into the subretrofacial nucleus were 6.0–7.0 mm caudal to the foramen cecum, 4.0 mm lateral to the midline, and 1.0–1.5 mm below the ventral surface. With these coordinates, a micropipette was placed on each side. Then, using a series of pressure injections (30 psi, 400–1000 msec in duration) with a penumatic pump (model PV 800, World Precision Instruments, New Haven, Conn.), 100 mM L-glutamic acid in a volume of 20–60 nl was microinjected on both sides. If both microinjections produced a pressor response, then it was assumed that both micro-pipettes were in the subretrofacial nucleus. When the blood pressure returned to baseline value, either 8-OH-DPAT or DOI in a volume of 20–40 nl were microinjected bilaterally.

The second site is the caudal ventrolateral medulla, also known as the "bicuculline-sensitive hypotensive site" (Mandal et al. 1990b). The coordinates used for microinjection into the bicuculline-sensitive hypo-tensive site were 9.0 mm caudal to the foramen cecum, 4.0 mm lateral to the midline, and 1.5 mm below the ventral surface. Bilateral micro-injection of bicuculline (10 ng/side) was used first to locate the caudal ventrolateral site. If both microinjections produced a fall in blood pressure, then it was assumed that both micropipettes were in the "bicuculline-sensitive hypotensive site".

The third site selected within the ventrolateral medulla has been described as containing a high density of serotonin cell bodies and has

292

been termed the lateral wings of the B1/B3 cell groups or the parapyramidal region (Hökfelt et al., 1978; Helke et al., 1989). Many of the axons of these serotonergic cell bodies terminate in the intermediolateral cell column of the spinal cord, the major site of preganglionic sympathetic neurons. Pharmacological excitation of the parapyramidal cell bodies results in a pressor response (Howe et al. 1983; Pilowski et al. 1986; Minson et al. 1987). As defined by Marson and Loewy (1985) for cats, this cluster of serotonin neurons is just medial and superficial to the subretrofacial nucleus. The coordinates used for the parapyramidal region were 5.0–7.0 mm caudal to the foramen cecum, 3.5 mm lateral to the midline, and 0.3–0.8 mm below the ventral surface. Again, a volume of 20–60 nl of drug solution was microinjected using a series of pressure injections.

After completion of the experiment, the brain stem was removed and fixed in 10% formalin for at least 48 hours. The tissue was then blocked and transferred to a solution of 20% sucrose in phosphate buffered saline for at least 24 hours before sectioning. Fifty-micron sections were counterstained with neutral red to facilitate identification of nuclear groups, and the site of microinjection was easily identified by a bright green area; in some cases, tissue damage also marked the micropipette track.

Statistical Analysis

Data are presented as mean values along with the standard error of the mean. Data were analyzed using the two-tailed paired t-test. The criterion for significance was $p < 0.05$.

Results

Effect of Microinjection of 8-OH-DPAT into the Subretrofacial Nucleus on Mean Arterial Blood Pressure and Heart Rate

Six experiments were performed wherein 8-OH-DPAT was bilaterally microinjected into the subretrofacial nucleus. The dose of 8-OH-DPAT microinjected into each subrectrfacial nucleus was 20 ng (in 20 nl). Results obtained for effects on mean blood pressure and heart rate are presented in the histograms of Figure 1. As can be noted, 8-OH-DPAT microinjected bilaterally into the subretrofacial nucleus produced a significant decrease in mean arterial blood pressure. Microinjection of 8-OH-DPAT, however, had no significant effect on heart rate (Figure 1). The effects on blood pressure occurred immediately after microinjection of 8-OH-DPAT, reached a nadir at 3 to 5 minutes after injection, and lasted for 30 to 45 minutes. Two experiments were performed

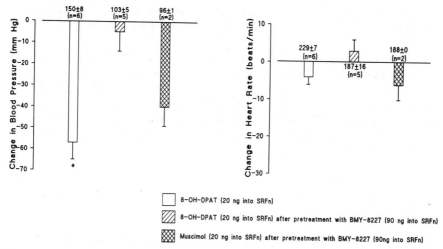

Figure 1. Effect of bilateral microinjection of 8-OH-DPAT into the subretrofacial nucleus (SRFn) on mean arterial blood pressure (left) and heart rate (right). Numbers at the base of each histogram indicate the base-line mean arterial pressure and heart rate prior to microinjection of drug. Numbers in parentheses indicate the number of animals studied. Vertical bars on the histograms indicate the S.E.M. The asterisk indicates statistical significance. Data are also presented indicating the effect of pretreatment with BMY-8227 on responses elicited by 8-OH-DPAT and responses elicited by muscimol.

wherein the vehicle for 8-OH-DPAT, namely, 0.9% saline, was bilaterally microinjected into the subretrofacial nucleus (20 nl per site). The results of saline vehicle microinjection on mean blood pressure and heart rate were no change and a decrease of 8 beats/min, respectively. In 5 experiments, 8-OH-DPAT was bilaterally microinjected into the subrectrofacial nucleus 5 mins after the 5-HT$_{1A}$ receptor antagonist, BMY-8227 (Yocca et al. 1986), was bilaterally microinjected into the same site (dose of 90 ng in a volume of 20 nl injected into each subretrofacial nucleus). As can be noted from the data presented in Figure 1, the expected decrease in mean arterial blood pressure with 8-OH-DPAT was absent in animals pre-treated with BMY-8227.

BMY-8227 itself caused a fall in mean arterial blood pressure. Thus, to assess whether BMY-8227 in the dose used to block the hypotensive effect of 8-OH-DPAT was acting non-specifically, we evaluated the ability of BMY-8227 to block an agent that produces hypotension after microinjection into the subretrofacial nucleus by affecting a non-serotonergic mechanism. This agent was muscimol (Gatti et al. 1987). In two experiments, muscimol (20 ng in 20 nl) was microinjected into each subretrofacial nucleus 5 minutes after pretreatment with BMY-8227, and the data appear in Figure 1. As can be noted, muscimol produced a large decrease in the mean blood pressure, similar in magnitude as that observed with 8-OH-DPAT microinjected into the subretrofacial nucleus of control animals.

294

Effect of Microinjection of 8-OH-DPAT into the Caudal Ventrolateral Medulla on Mean Arterial Blood Pressure and Heart Rate

Six experiments were performed wherein 8-OH-DPAT was bilaterally microinjected into the caudal ventrolateral medulla in a dose of 20 ng (in 20 nl) per site. The mean arterial blood pressure and heart rate of these animals averaged 118 ± 9 mm Hg and 179 ± 12 beats/min., respectively, prior to 8-OH-DPAT microinjection. Maximal changes in mean blood pressure and heart rate after 8-OH-DPAT microinjection were -17 ± 8 mm Hg ($p > 0.05$) and -3 ± 4 beats/min.

Effect of Microinjection of DOI into the Subretrofacial Nucleus on Mean Arterial Blood Pressure and Heart Rate

Six experiments were performed wherein DOI was bilaterally microinjected into the subretrofacial nucleus. The doses of DOI microinjected into each subretrofacial nucleus were 100 ng (in 20 nl) in 3 animals and 300 ng (in 20 nl) in 3 animals. Results obtained for effects on mean arterial blood pressure and heart rate are presented in the histograms of Figure 2. As can be noted, DOI microinjected bilaterally into the

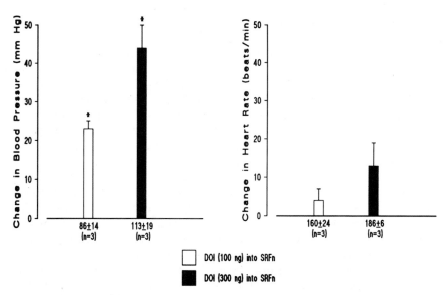

Figure 2. Effect of bilateral microinjection of DOI into the subretrofacial nucleus (SRFn) on mean arterial pressure (left) and heart rate (right). Numbers at the base of each histogram indicate the base-line mean arterial pressure and heart rate prior to microinjection of drug. Numbers in parentheses indicate the number of animals studied. Vertical lines on the histograms indicate the S.E.M. The asterisk indicates statistical significance.

subretrofacial nucleus produced a significant increase in mean arterial pressure. This increase occurred with both doses of DOI tested, and the pressor response appeared to be dose-related. Microinjection of DOI, however, had no significant effect on heart rate (Figure 2). The effect on blood pressure occurred immediately after microinjection of DOI, reached a peak at 1.5 to 3 min after injection, and lasted for 10 to 15 minutes.

Discussion

Two key sites in the hindbrain that influence arterial blood pressure through controlling central sympathetic outflow to the intermediolateral cell column are the subretrofacial nucleus and the caudal ventrolateral medulla (sometimes referred to as the bicuculline-sensitive hypotensive site – Mandal et al. 1990b). Neurons of the subretrofacial nucleus appear to comprise a major descending vasomotor pathway and receive afferent input from CNS neurons located in the baroreceptor pathway (McAllen 1986). Neurons of the caudal ventrolateral medulla appear to comprise a majority inhibitory input to subretrofacial nucleus (Blessing and Li 1989), and may also comprise part of the CNS portion of the baroreceptor pathway (Guyenet et al. 1987). 8-OH-DPAT microinjected into the subretrofacial nucleus in a dose as low as 20 ng produces a hypotensive effect that is in the same range as that observed after bilateral application of the drug to the ventral surface of the medulla and after I.V. administration in a dose of $10 \mu g/kg$ and $30 \mu g/kg$ (Gillis et al. 1989). On the other hand, 8-OH-DPAT in a dose of 20 ng microinjected bilaterally into the caudal ventrolateral medulla, exerted no significant hypotensive effect. Although there was a tendency for arterial pressure to decrease after microinjection of 8-OH-DPAT into the caudal ventrolateral medulla (-17 ± 8 mm Hg, N = 6 animals), the magnitude of the response was not statistically significant. In our earlier study of 3 animals (Mandel et al. 1990b), a response of similar magnitude (-21 ± 4 mm Hg) was observed after microinjection of 8-OH-DPAT into the caudal ventrolateral medulla.

Consistent with our data obtained with the 5-HT$_{1A}$ receptor agonist drug, 8-OH-DPAT, are our findings obtained with microinjection of another 5-HT$_{1A}$ receptor agonist drug, 5-methyl-urapidil, into the subretrofacial nucleus and the caudal ventrolateral medulla. Microinjection of 5-methyl-urapidil (25 ng bilaterally) into the subretrofacial nucleus decreased mean arterial pressure by 74 ± 14 mm Hg (Mandal et al. 1991). On the other hand, microinjection of the same dose of 5-methyl-urapidil into the caudal ventrolateral medulla decreased mean arterial pressure by only 21 ± 7 mm Hg (Mandal et al. 1991).

We feel confident in concluding that activation of 5-HT$_{1A}$ receptors in the subretrofacial nucleus results in hypotension. Our confidence is based on the pharmacological tools we have employed to test this hypothesis. 8-OH-DPAT is an agent with well-documented selectivity of action as an agonist for the 5-HT$_{1A}$ receptor. This is also true of 5-methyl-urapidil (Gross et al. 1987), another agent we have employed to test our hypothesis. Furthermore, BMY-8227, an agent that has been reported to block 5-HT$_{1A}$ receptors (Yocca et al. 1986), abolishes the hypotensive effect induced by microinjection of 8-OH-DPAT into the subretrofacial nucleus.

As in the case of 5-HT$_{1A}$ receptor agonists, we found that the 5-HT$_2$ receptor agonist drug, DOI, produced its effect on arterial blood pressure by acting in the subretrofacial nucleus. This was demonstrated by microinjection of DOI bilaterally into the subretrofacial nucleus in doses that were ineffective after topical application to the intermediate area (Mandal et al. 1990a), and observing significant increases in mean arterial blood pressure.

An area medial to the subretrofacial nucleus and reported to influence sympathetic outflow (Howe et al. 1983; Pilowsky et al. 1987) and contain 5-HT$_{1A}$ receptors (Thor and Helke 1990) is the parapyramidal region. This site could potentially be an important target for the actions of 5-HT$_{1A}$ receptor agonist drugs (8-OH-DPAT and 5-methyl-urapidil) as well as the 5-HT$_2$ receptor agonist drug, DOI. To test this hypothesis, we have microinjected 5-methyl-urapidil and DOI in the parapyramidial region in the same doses that were active in the subretrofacial nucleus. Results obtained indicated that neither 5-methyl-urapidil (Mandal et al. 1991) nor DOI (Mandal et al. 1990a) have any effect on mean arterial pressure after microinjection into this site. It should be noted that Helke and Phillips (1990) clearly observed a hypotensive effect after microinjection of 8-OH-DPAT into the parapyramidal area of the rat. It is therefore possible that a species difference exists for the site of action 8-OH-DPAT in the rostral ventrolateral medulla.

While these are the only data we are aware of wherein activation of 5-HT$_2$ receptors in the subretrofacial nucleus results in significant alterations in arterial blood pressure, Dabiré et al. 1989, were first to report that activation of 5-HT$_{1A}$ receptors in the subretrofacial nucleus influences arterial blood pressure and sympathetic nerve discharge. Their study was carried out in anesthetized dogs using 200 ng of 8-OH-DPAT in a volume of 200 nl and microinjected bilaterally. In Dabiré and colleagues' study, microinjection of 8-OH-DPAT into the nucleus tractus solitarius, raphe obscurus and raphe pallidus had no effect on blood pressure or sympathetic nerve activity.

Initially, we reported that serotonin applied to the ventrolateral medulla, specifically to the intermediate area where drug can readily diffuse to the subretrofacial nucleus, produces only a minor hypotensive effect (Gillis et al. 1989). However, after blockade of 5-HT$_2$ receptors at the intermediate area, 5-HT application produces a much greater hypotensive effect. We postulated that 5-HT application at the intermediate area would simultaneously activate both 5-HT$_{1A}$ and 5-HT$_2$ receptors at the intermediate area. Activation of both receptor types would result in little effect on arterial blood pressure because excitation of 5-HT$_{1A}$ receptors causes hypotension and excitation of 5-HT$_2$ receptors causes hypertension. Our new findings indicate that both types of receptors are probably located in the same nucleus, namely, the subretrofacial nucleus.

With the demonstration that activation of 5-HT$_{1A}$ receptors in the subretrofacial nucleus results in hypotension, we postulate that part of the hypotensive action of the new antihypertension agent, urapidil, is due to stimulation of 5-HT$_{1A}$ receptors in this nucleus. Evidence for this is as follows: (1) urapidil is known to bind to 5-HT$_{1A}$ recognition sites in brain tissue (Gross et al. 1987); (2) urapidil applied bilaterally to the intermediate area of the ventral surface of the medulla (a site where drug can readily diffuse to the subretrofacial nucleus) causes hypotension; this hypotensive effect is prevented by prior blockade of 5-HT$_{1A}$ receptors at the same site (Mandal et al. 1990b); (3) urapidil-induced hypotension after I.V. administration is counteracted by blockade of 5-HT$_{1A}$ receptors at the intermediate area of the ventrolateral medulla (Mandal et al. 1990b); and (4) 5-methyl-urapidil, a closely related analogue of urapidil, microinjected into the subretrofacial nucleus produces hypotension (Mandal et al. 1991).

Acknowledgements

Supported by a grant to Richard A. Gillis, from Byk-Gulden Pharmazeutica, Konstanz, F.R.G. Aloke K. Mandal is a recipient of a Medical Student Research Fellowship in Pharmacology-Clinical Pharmacology, Pharmaceutical Manufacturers Association Foundation, Inc., Washington, D.C. The authors express gratitude to Octavia Jones for her expert typing of the manuscript.

References

Alper, R. H. (1990). Hemodynamic and renin responses to ($\pm$)DOI, a selective 5-HT$_2$ receptor agonist, in conscious rats. Eur. J. Pharmacol. 1975: 323–332.

Blessing, W. W., and Li, Y. W. (1989). Inhibitory vasomotor neurons in the caudal ventrolateral region of the medulla oblongata. In Ciriello, J., Caverson, M. M. and Polossa, C. (Eds.), Progress in Brain Research (Vol. 81). New York: Elsevier, pp. 83–97.

Clement, M. E., and McCall, R. B. (1990a). Studies on the site of mechanism of the sympathoexcitatory action of 5-HT$_2$ agonists. Brain Res. 515: 299–302.

Clement, M. E., and McCall, R. B. (1990b). Studies on the site and mechanism of the sympatholytic action of 8-OH-DPAT. Brain Res. 521: 232–241.

Connor, H. E., and Higgins, G. A. (1990). Cardiovascular effects of 5-HT$_{1A}$ receptor agonists injected into the dorsal raphe nucleus of conscious rats. Eur. J. Pharmacol. 182: 63–72.

Dabiré, H., Chaouche-Teyara, K., Cherqui, C., Fournier, B., and Schmitt, H. (1989a). Characterization of DOI, a putative 5-HT$_2$ receptor agonist in the rat. Eur. J. Pharmacol. 168: 369–374.

Dabiré, H., Chaouche-Teyara, K., Cherqui, C., Fournier, B. and Schmitt, H (1989b). DOI is a mixed agonist-antagonist at post-junctional 5-HT$_2$ receptors in the pithed rat. Eur. J. Pharmacol. 170: 109–111.

Doods, H. N., Boddeke, H. W. G. M., Kalkman, H. O., Hoyer, D., Mathy, M. J., and Van Zwieten, P. A. (1988). Central 5-HT$_{1A}$ receptors and the mechanism of the central hypotensive effect of ($\pm$) 8-OH-DPAT, DP-5-CT, R28935 and urapidil. J. Cardiovasc. Pharmacol. 11: 432–441.

Fozard, J. R., Mir, A. K., and Middlemiss, D. N. (1987). Cardiovascular response to 8-hydroxy-2-(di-N-propylamino) tetralin (8-OH-DPAT) in the rat: site of action and pharmacological analysis. J. Cardiovasc. Pharmacol. 9: 328–347.

Gatti, P. J., Taveira Da Silva, A. M., and Gillis, R. A. (1987). Cardiorespiratory effects produced by microinjecting drugs that affect GABA receptors into nuclei associated with the ventral surface of the medulla. Neuropharmacol. 26: 423–431.

Gillis, R. A., Hill, K. J., Kirby, J., Quest, J. A., Hamosh, P., Norman, W. P., and Kellar, K. J. (1989). Effect of activation of CNS serotonin-1A receptors on cardiorespiratory function. J. Pharmacol. Exp. Ther. 248: 851–857.

Gradin, G. E., and Lis, E. V. (1985). Hypotensive action of 8-hydroxy-2-(di-N-propylamino) tetralin (8-OH-DPAT) in spontaneously hypertensive rats. Arch. Int. Pharmacodyn. 273: 251–259.

Gross, G., Hanft, G., and Kolassa, N. (1987). Urapidil and some analogues with hypotensive properties show high affinities for 5-hydroxytryptamine (5-HT) binding sites of the 5-HT$_{1A}$ subtype and for alpha-1-adrenoceptor binding sites. Naunyn-Schmideberg's Arch. Pharmacol. 336: 597–601.

Guyenet, P. G., Filtz, T. M., and Donaldson, S. R. (1987). Role of excitatory amino acids in rat vagal and sympathetic baroreflexes. Brain Res. 407: 272–284.

Helke, C. J., and Phillips, E. T. (1990). Ventral medullary hypotensive actions of a serotonin$_{1A}$ (5-HT$_{1A}$) agonist in the rat. FASEB J. 4: A1067.

Helke, C. J., Thor, K. B., and Sasek, C. A. (1989). Chemical neuroanatomy of the parapyramidal region of the ventral medulla in the rat. In Ciriello, J., Caverson, M. M., and Polosa, C. (eds), Progress in Brain Research (Vol. 81). New York: Elsevier, pp. 17–28.

Hökfelt, T., Ljungdahl, A., Steinbusch, H., Verhofstad, A., Nilsson, G., Brodin, E., Pernow, B., and Goldstein, M. (1978). Immunohistochemical evidence of substance P-like immunoreactivity in some 5-hydroxytryptamine-containing neurons in the rat central nervous system. Neuroscience 3: 517–538.

Howe, P. R. C., Kuhn, D. M., Minson, J. B., Stead, B. H., and Chalmers, J. P. (1983). Evidence for a bulbospinal pressor pathway in the rat brain. Brain Res. 270: 29–36.

Izzo, P. N., Jordan, D., and Ramage, A. G. (1988). Anatomical and pharmacological evidence supporting the involvement of serotonin in the central control of cardiac vagal motor neurones in the anaesthetized cat. J. Physiol (London) 406: 19P.

King, K. A., and Holtman, J. R. (1990). Characterizations of the effects of activation of ventral medullary serotonin receptor subtypes on cardiovascular activity and respiratory motor outflow to the diaphragm and larynx. J. Pharmacol. Exp. Ther. 252: 665–674.

Laubie, M., Drouillat, M., Dabiré, H., Cherqui, C., and Schmitt, H. (1989). Ventrolateral medullary pressor area: site of hypotensive and sympatho-inhibitory effects of ($\pm$) 8-OH-DPAT in anaesthetized dogs, Eur. J. Pharmacol. 160: 385–394.

Mandal, A. K., Kellar, K. J., and Gillis, R. A. (1991). The role of serotonin-1A receptor activation and alpha-adrenoceptor blockade in the hypotensive effect of 5-methyl-urapidil. J. Pharmacol. Exp. Ther. (Submitted).

Mandal, A. K., Keller, K. J., Norman, W. P., and Gillis, R. A. (1990a). Stimulation of serotonin$_2$ receptors in the ventrolateral medulla of the cat results in nonuniform increases in sympathetic outflow. Circ. Res. 67: 1267–1280.

Mandal, A. K., Zhong, P., Kellar, K. J., and Gillis, R. A. (1990b). Ventrolateral medulla: An important site of action for the hypotensive effect of drugs that activate serotonin-1A receptors. J. Cardiovasc. Pharmacol. 15: (suppl 7): S49–S60.

Marson, L., and Loewy, A. D. (1985). Topographic organization of substance P and monoamine cells in the ventral medulla of the cat. J. Auton. Nerv. Syst. 14: 271–285.

McAllen, R. M. (1986). Identification and properties of subretrofacial neurones: A decending cardiovascular pathway in the cat. J. Auton. Nerv. Syst. 17: 151–164.

McCall, R. B., and Harris, L. T. (1988). 5-HT$_2$ receptor agonists increase spontaneous sympathetic nerve discharge. Eur. J. Pharmacol. 151: 113–116.

McCall, R. B., Patel, B. N., and Harris, L. T. (1987). Effects of serotonin$_1$ and serotonin$_2$ receptor agonists and antagonists on blood pressure, heart rate, and sympathetic nerve activity. J. Pharmacol. Exp. Ther. 242: 1152–1157.

Minson, J. B., Chalmers, J. P., Caon, A. C., and Renaud, B. (1987). Separate areas of rat medulla oblongata with populations of serotonin and adrenaline-containing neurons alter blood pressure after L-glutamic acid stimulation. J. Auton. Nerv. Syst. 19: 39-50.

Pilowsky, P., Kapoor, V., Minson, J. B., West, M. J., and Chalmers, J. P. (1986). Spinal cord serotonin release and raised blood pressure after brainstem kainic acid injection. Brain Res. 366: 354–357.

Ramage, A. G., and Fozard, J. R. (1987). Evidence that the putative 5-HT$_{1A}$ receptor agonists, 8-OH-DPAT and ipsapirone, have a central hypotensive action that differs from that of clonidine in anaesthetized cats. Eur. J. Pharmacol. 138: 179–191.

Thor, K. B., Blitz-Siebert, A., and Helke, C. J. (1990). Discrete localization of high-density 5-HT$_{1A}$ binding sites in the midline raphe and parapyramidal regions of the ventral medulla oblongata of the rat. Neurosci. Letts. 108: 249–254.

Valenta, B., and Singer, E. A. (1990). Hypotensive effects of 8-hydroxy-2-(di-n-propylamino) tetralin and 5-mehtylurapidil following stereotaxic injection into the ventral medulla of the rat. Brit. J. Pharmacol. 99: 713–716.

Wouters, W., Tulp, M. Th. M., and Bevan, P. (1988). Flexinoxan lowers blood pressure and heart rate in cats via 5-HT$_{1A}$ receptors. Eur. J. Pharmacol. 149: 213–223.

Yocca, F. D., Smith, D. W., Hyalon, D. K., and Maayani, S. (1986). Dissociation of efficacy from affinity at the 5-HT$_{1A}$ receptor in rat hippocampal preparation. Neurosci. Abstr. 12: 422.

Serotonin: Molecular Biology, Receptors and Functional Effects
ed. by J. R. Fozard/P. R. Saxena
© 1991 Birkhäuser Verlag Basel/Switzerland

Cardiovascular Effects of Injection of 5-HT, 8-OH-DPAT and Flesinoxan into the Hypothalamus of the Rat

G. H. Dreteler[1], W. Wouters[1], and P. R. Saxena[2]

[1]*Departments of Pharmacology, Duphar B.V., P.O. Box 900, 1380 DA Weesp;*
[2]*Erasmus University Rotterdam, P.O. Box 1738, 3000 DR Rotterdam, The Netherlands*

Summary. The cardiovascular effects of injection of 5-HT and the 5-HT$_{1A}$ receptor agonists, 8-OH-DPAT and flesinoxan, into the hypothalamus of anesthetized spontaneously hypertensive rats (SHR) was investigated. Injection of 5-HT (10 μg) into the anterior hypothalamic (AH) region increased blood pressure by $16 \pm 2\%$, but did not change heart rate. Injection of 8-OH-DPAT (5 μg) into this area did not affect blood pressure or heart rate. Pressor responses ($16 \pm 3\%$) were also observed following 5-HT (10 μg) administration directly into the ventromedial hypothalamic (VMH) area. In contrast, however, injection of 8-OH-DPAT (1.25, 2.5 and 5 μg) into the VMH caused dose-dependent reductions in both blood pressure and heart rate (22 ± 4 and $15 \pm 3\%$, respectively at the highest dose). Surprisingly, injection of the 5-HT$_{1A}$ receptor agonist flesinoxan (10 μg) into the VMH did not affect blood pressure or heart rate. The results, therefore, suggest that 5-HT$_{1A}$ receptors are not involved in the cardiovascular effects elicited via the AH and VMH nuclei of the hypothalamus and that 8-OH-DPAT in the VMH seems to possess properties other than 5-HT$_{1A}$ receptor agonism.

Introduction

It has been shown that central 5-hydroxytryptamine (5-HT) neurones are important in cardiovascular regulation (Kuhn et al. 1980). However, despite extensive investigations, the exact role of central 5-HT in cardiovascular control is still not clear. The cardiovascular effects of central administration of 5-HT are variable and depend on the dose, animal species and site of drug administration. In rats, intracerebroventricular (i.c.v.) injections of 5-HT mainly produced pressor responses that were greatest when the drug was given into the third cerebral ventricle (Lambert et al., 1975; Lambert et al. 1978). Since, after i.c.v. administration, drugs are widely distributed throughout the brain, the exact central sites responsible for the cardiovascular effects of i.c.v. administered 5-HT are difficult to determine. However, Smits and Struyker-Boudier (1976) found that injection of 5-HT directly into the anterior hypothalamic/preoptic (AH/PO) nucleus, which is located in the diencephalon near the third ventricle, resulted in pressor responses that could be blocked by i.c.v. administration of methysergide. The importance of the hypothalamus in 5-HT mediated cardiovascular control was further stressed by the detection of rather high levels of 5-HT in the nucleus (Saavedra et al. 1974). Furthermore, the hypothalamus receives 5-HT neurones ascending from the midbrain raphe nuclei (Azmitia and Segal. 1978; Moore et al. 1978) and electrical stimulation of these raphe nuclei produced pressor responses as a result of 5-HT release in the AH/PO nucleus (Smits et al. 1978).

8-OH-DPAT and flesinoxan are potent, centrally acting agents, that lower blood pressure and heart rate in a variety of animal models (Gradin et al. 1985; Martin and Lis 1985; Fozard et al. 1987, Ramage and Fozard 1987; Wouters et al. 1988a; b; Laubie et al. 1988). The cardiovascular effects are mediated by stimulation of central 5-HT_{1A} receptors (Fozard et al. 1987; Wouters et al. 1988b, Saxena and Villalón 1990), since they can be blocked by the 5-HT receptor antagonist methiothepin (Fozard et al. 1987; Dreteler et al. 1990) and the putative 5-HT_{1A} and 5-HT_{1B} receptor antagonist pindolol (Wouters et al. 1988b; Connor and Higgins 1990). In rats, several brain nuclei have now been shown to be involved in 5-HT_{1A} receptor mediated cardiovascular control. Activation of 5-HT_{1A} receptors in the dorsal raphe by microinjection of 8-OH-DPAT or flesinoxan into this nucleus induced hypotension and bradycardia in conscious rats (Connor and Higgins 1990). The same blood pressure and heart rate effects were observed following injection of the 5-HT_{1A} receptor agonists 5-methylurapidil and 8-OH-DPAT into the raphe magnus and pallidus of the anesthetized rat (Valenta and Singer 1990). In contrast, activation of 5-HT_{1A} receptors in the raphe obscurus induced pressor responses (Dreteler et al. 1991).

Autoradiographic studies have revealed that 5-HT_{1A} binding sites are also present in the hypothalamus (Vergé et al. 1986). In the present study, a possible involvement of the hypothalamus in 5-HT_{1A} receptor mediated cardiovascular control has been investigated by using 8-OH-DPAT and flesinoxan, which were injected directly into the hypothalamus of anaesthetized rats.

Experimental Procedures

Male spontaneously hypertensive rats (275–325 g; Charles River, Sulzfeld, Germany) were anaesthetized with pentobarbital sodium (70 mg kg^{-1}; intraperitoneal). The left femoral artery was cannulated (PP10 tubing, outer diameter 0.61 mm and inner diameter 0.28 mm, connected to a length of medical PVC tubing) in order to allow blood pressure and heart rate measurements during the experiment. Subsequently, the animal was placed in a stereotaxic apparatus (David Kopf Instrument) and the skull of the animal was exposed. A stainless steel guide cannula (outer diameter: 0.8 mm; inner diameter 0.5 mm) with a length of 9.45 mm was inserted unilaterally into the brain at the stereotaxic co-ordinates according to Paxinos and Watson (1982): $A - 1.3$, $L + 0.6$, $V - 4.0$ for the anterior hypothalamic region (AH) and $A - 2.8$, $L + 1.0$, $V - 4.0$ for the ventromedial hypothalamus (VMH). Bregma was used as a point of reference for the determination of the anterior-posterior co-ordinate and the dorso-ventral co-ordinate was found by using the surface of the skull as zero. The cannula tip terminated 4.8 mm above the hypothalamus region to be injected. During the surgery and the experiment, the body temperature of the animal was kept at 37°C by means of a homeothermic blanket. Blood pressure was measured by connecting the femoral arterial cannula to a pressure transducer (Statham) and heart rate was derived from the pulse pressure.

After stabilisation of the blood pressure and heart rate, a needle (outer diameter: 0.47 mm; inner diameter: 0.15 mm) with a length of either 14.25 mm (AH) or 14.75 mm (VMH) attached to a 1 µl Hamilton syringe was inserted into the brain so that the tip of the needle was aimed at either the AH or the VMH region. The insertion of the needle consistently caused a drop in blood pressure and heart rate, which lasted approximately two minutes. When blood pressure and heart rate had returned to baseline values, 0.5 µl of a saline (0.9%) or drug solution was injected over a period of 1 minute. Injections of single doses of drug were performed in separate experiments. The effects on the cardiovascular variables were measured until 30 minutes after the hypothalamus injection. At the end of the experiment 0.5 µl of a Evans blue solution (5%) was injected into the hypothalamus region and the rat was perfused with a 10% formaldehyde solution. The brain was removed and transverse sections of a 100 µm were cut by means of a freeze microtome. Photographs were taken of the brain slices stained with Evans Blue and the injection sites were identified using the atlas of Paxinos and Watson (1982); see Figure 1.

Results

Effects of Injection of 5-HT and 8-OH-DPAT into the AH Nucleus

Baseline values for mean arterial blood pressure and heart rate for the various groups of animals are presented in Table 1. Microinjection of saline into the AH nucleus did not affect blood pressure and heart rate, but injection of 5-HT (10 µg) into the AH (Figure 1) caused a significant increase in blood pressure ($16 \pm 2\%$) but hardly affected heart rate (Figure 2). The effects were immediate in onset and maximum changes were reached about 1–5 minutes after drug injection. 8-OH-DPAT in a dose of 5 µg, injected directly into the AH, did not alter either blood pressure or heart rate (Figure 2).

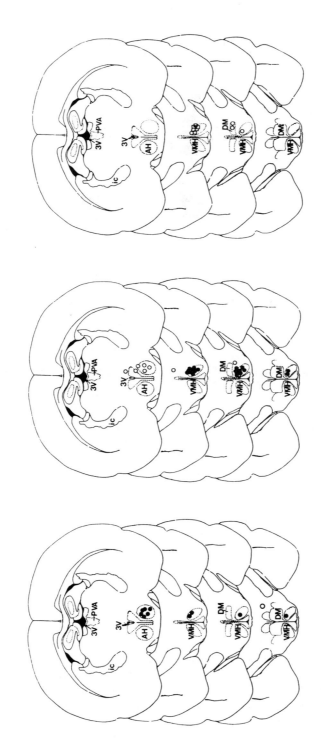

Figure 1. Cross-sections of rat brain at the level 1.8–3.3 mm caudal from bregma (adapted from Paxinos and Watson 1982) showing the histologically verified injection sites of 5-HT, 8-OH-DPAT and flesinoxan. ●: injection sites where positive effects were found; ○: injection sites where no effects were found; □: injection sites where flesinoxan did not induce blood pressure and heart rate effects but 8-OH-DPAT did. 3V, third ventricle; PVA, anterior paraventricular thalamic nucleus; ic, internal capsule; AH, anterior hypothalamic nucleus; DM, dorsomedial hypothalamic nucleus; VMH, ventromedial hypothalamic nucleus.

304

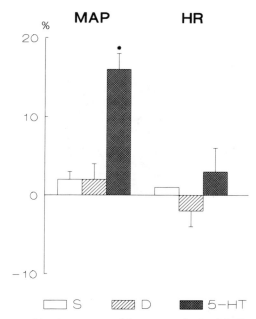

Figure 2. The effects on blood pressure and heart rate of microinjection of 0.5 µl saline (S; n = 6), 5-HT (10 µg; n = 6) or 8-OH-DPAT (D; 5 µg; n = 4) into the anterior hypothalamic nucleus (AH) of the anaesthetized SHR. *, p < 0.05 (student's t-test, unpaired) versus vehicle treated animals.

Table 1. Baseline values of mean arterial blood pressure (MAP) and heart rate (HR) in the various groups of rats that were microinjected with vehicle (0.5 µl) or drug into the AH nucleus. Values are mean ± s.e.m. from n observations

Treatment	Dose µg	Baseline value MAP	Baseline value HR	n
vehicle		156 ± 6	351 ± 12	6
5-HT	10	149 ± 4	317 ± 11	6
8-OH-DPAT	5	142 ± 2	321 ± 20	4

Effects of Microinjection of 5-HT, 8-OH-DPAT and Flesinoxan into the VMH Nucleus

Baseline values for the groups of rats microinjected with drug or vehicle directly into the VMH nucleus are presented in Table 2.

Injection of vehicle (0.5 µl) into the VMH did not cause any change in either blood pressure or heart rate (Figure 3). Injection of 5-HT (10 µg) into the VMH (Figure 1) caused an immediate significant increase in blood pressure reaching a maximum of 16 ± 3% (Figure 3)

Table 2. Baseline values of mean arterial blood pressure (MAP) and heart rate (HR) in the various groups of rats that were microinjected with vehicle (0.5 μl) or drug into the VMH nucleus. Values are mean ± s.e.m. from n observations

Treatment	Dose μg	Baseline value MAP	Baseline value HR	n
vehicle		144 ± 3	336 ± 7	4
5-HT	10	139 ± 3	330 ± 4	4
8-OH-DPAT	1.25	161 ± 5	377 ± 8	6
8-OH-DPAT	2.5	152 ± 5	369 ± 11	6
8-OH-DPAT	5	146 ± 5	347 ± 10	6
flesinoxan	10	170 ± 9	349 ± 9	8

after approximately 1–5 minutes. 5-HT also increased heart rate, although the sizes of the increments varied considerably (Figure 3).

Microinjection of 8-OH-DPAT (1.25, 2.5 and 5 μg) into the VMH (Figure 1) caused dose-dependent decreases in blood pressure and heart rate (Figure 4) with the largest reductions ($22 \pm 4\%$ for blood pressure and $15 \pm 3\%$ for heart rate) observed at the highest dose given. The effects were rapid in onset and peak changes were obtained 10–15 minutes after the injection.

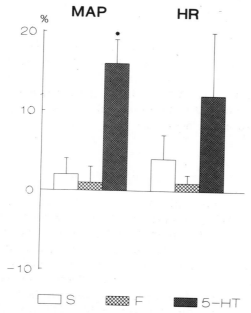

Figure 3. The effects on blood pressure and heart rate of microinjection of 0.5 μl saline (S; n = 4) flesinoxan (F; 10 μg; n = 8) or 5-HT (10 μg; n = 4) into the ventromedial hypothalamic (VMH) nucleus of the anaesthetized SHR. *, $p < 0.05$ (student's t-test, unpaired) versus vehicle treated animals.

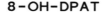

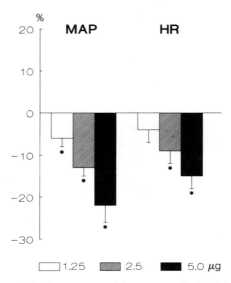

Figure 4. The effects on blood pressure and heart rate of microinjection of 8-OH-DPAT (1.25, 2.5 and 5 μg; n = 6 for each dose) into the ventromedial hypothalamic (VMH) nucleus of the anaesthetized SHR. *, p < 0.05 (student's t-test, unpaired) versus vehicle treated animals.

Microinjection of flesinoxan (10 μg) into the VMH (Figure 1) did not modify either blood pressure or heart rate (Figure 3). In five animals, 8-OH-DPAT was also microinjected into the same spot following the injection of flesinoxan (Figure 1). In all these five experiments, 8-OH-DPAT (5 μg), in contrast to flesinoxan, did cause a marked decrease in both blood pressure and heart rate.

Discussion

The aim of the present study was to investigate if the hypothalamus plays a role in 5-HT$_{1A}$ receptor mediated cardiovascular regulation. Thus, the 5-HT$_{1A}$ receptor agonists, 8-OH-DPAT and flesinoxan, were injected directly into two distinct areas within the hypothalamus i.e. the anterior hypothalamic (AH) and the ventromedial hypothalamic (VMH) region of anaesthetized rats. The effects on blood pressure and heart rate, following the drug injections, were studied.

In the present study, direct injection of 5-HT into the AH region caused increases in blood pressure and did not affect heart rate. These results are in agreement with the studies of Smits and Struyker-Boudier (1976), who first demonstrated that injection of 5-HT into the AH/PO region induced a pressor response, which was antagonized by i.c.v. injection of the unselective 5-HT receptor antagonist methysergide. In contrast to 5-HT, 8-OH-DPAT neither affected blood pressure nor heart rate when injected into the AH region. These results suggest that the AH nucleus, although integrated in central 5-HT-induced cardiovascular effects, does not play a role in $5\text{-}HT_{1A}$ receptor mediated cardiovascular control.

Electrical stimulation of the VMH nucleus of anaesthetized rats evokes either pressor or depressor responses, depending on the frequency of the stimulus (Faiers et al. 1976). Furthermore, $5\text{-}HT_{1A}$ receptors have been localized within the VMH (Vergé et al. 1986) and are suggested to be involved in the 5-HT induced hyperpolarization of VMH neurones (Newberry 1989). The data from the present study show that, as in the AH region, microinjection of 5-HT into the VMH region caused pressor effects. Direct administration of 8-OH-DPAT into the VMH induced dose-dependent decreases in both blood pressure and heart rate. Peak changes in these cardiovascular effects were not reached until 10–15 min following the injection, whereas peak changes in blood pressure and heart rate after intravenous administration of 8-OH-DPAT take about 5 min to occur (Dreteler et al. 1990). Furthermore, in order to obtain blood pressure and heart rate effects that are quantitatively comparable with the effects of intravenous administration (Dreteler et al. 1990), relatively high doses of 8-OH-DPAT were needed, when this drug was injected directly into the VMH. This contrasts with injections of $5\text{-}HT_{1A}$ receptor agonists into other brain nuclei like the dorsal raphe (Connor and Higgins 1990) and raphe pallidus and magnus (Valenta and Singer 1990), where very low doses of 8-OH-DPAT induced marked cardiovascular effects. Moreover, no changes in either blood pressure or heart rate were monitored following microinjection of another $5\text{-}HT_{1A}$ receptor agonist flesinoxan (10 μg) into the VMH region, whereas microinjection of 5 μg of flesinoxan into the dorsal raphe of conscious rats caused marked reductions in both blood pressure and heart rate (Connor and Higgins 1990). These results indicate that the $5\text{-}HT_{1A}$ receptor in the VMH is not directly involved in mediating the cardiovascular effects of 8-OH-DPAT or flesinoxan.

The discrepancy between the cardiovascular effects of 8-OH-DPAT and flesinoxan following injection into the VMH might be due to different pharmacokinetic properties of these drugs. However, recently, in several experimental models 8-OH-DPAT has been described to act on a 5-HT receptor subtype different from the $5\text{-}HT_{1A}$ receptor. 8-OH-DPAT contracts the dog isolated saphenous vein and increases the carotid arterial resistance in anaesthetized dogs by activation of a $5\text{-}HT_1$-like receptor that is not of the $5\text{-}HT_{1A}$ or $5\text{-}HT_{1D}$ receptor subtype (Perren et al. 1991). In rabbit isolated saphenous vein, 8-OH-DPAT mimicks the 5-HT induced contractions by acting on a $5\text{-}HT_1$-

like receptor, which does not seem to correlate with any of the known 5-HT$_1$ binding sites (van Heuven-Nolsen et al. 1990). Furthermore, Bom and colleagues (1989a) have shown that 8-OH-DPAT constricts arteriovenous anastomoses in anaesthetized pigs, an effect which can be antagonized by methiothepin. However, in the same model, the 5-HT$_{1A}$ receptor agonist ipsapirone appeared to be inactive with regard to the effect on arteriovenous anastomotic flow (Bom et al. 1988). In addition, the increase in arteriovenous anastomotic resistance by the 5-HT$_{1A}$ and 5-HT$_{1B}$ receptor agonist RU 24969, antagonizable with methiothepin, was resistant to blockade by pindolol, a putative antagonist of 5-HT$_{1A}$ and 5-HT$_{1B}$ receptors (Bom et al. 1989b). These results indicate that in anaesthetized pigs, 8-OH-DPAT and RU 24969 constrict arteriovenous anastomoses by acting at a 5-HT$_1$-like receptor, that does not seem to correspond to the 5-HT$_{1A}$ or 5-HT$_{1B}$ receptor subtype. Therefore, it is possible that the cardiovascular effects of 8-OH-DPAT administered into the VMH are, particularly in view of the lack of effectiveness of flesinoxan, mediated via a 5-HT receptor other than the 5-HT$_{1A}$ receptor.

In conclusion, it has been demonstrated that, although the AH and VMH nucleus do play a role in 5-HT-mediated cardiovascular control, the 5-HT$_{1A}$ receptor is not involved. Furthermore, it seems that 8-OH-DPAT may reduce blood pressure via a non-5-HT$_{1A}$ receptor in the VMH.

References

Azmitia, E. C., and Segal, M. (1978). An autoradiographic analysis of the differential ascending projections of the dorsal and median raphe nuclei in the rat. J. Comp. Neurol. 179: 641–668.

Bom, A. H., Saxena, P. R., and Verdouw, P. D. (1988). Further characterization of the 5-HT$_1$-like receptors in the carotid circulation of the pig. Br. J. Pharmacol. 94: 327P.

Bom, A. H., Verdouw, P. D., and Saxena, P. R. (1989a). Carotid haemodynamics in pigs during infusions of 8-OH-DPAT: reduction in arteriovenous shunting is mediated by 5-HT$_1$-like receptors. Br. J. Pharmacol. 96: 125–132.

Bom, A. H., Villalón, C. M., Verdouw, P. D., and Saxena, P. R. (1989b). The 5-HT$_1$-like receptor mediating reduction of porcine carotid arteriovenous shunting by RU 24969 is not related to either the 5-HT$_{1A}$ or the 5-HT$_{1B}$ subtype. Eur. J. Pharmacol. 171: 87–96.

Connor, H. E., and Higgins, G. A. (1990). Cardiovascular effects of 5-HT$_{1A}$ receptor agonists injected into the dorsal raphe nucleus of conscious rats. Eur. J. Pharmacol. 182: 63–72.

Dreteler, G. H., Wouters, W., and Saxena, P. R. (1990). Comparison of the cardiovascular effects of flesinoxan with that of 8-OH-DPAT in the rat. Eur. J. Pharmacol. 180: 339–349.

Dreteler, G. H., Wouters, W., Saxena, P. R., and Ramage, A. G. (1991). Pressor effects following microinjection of 5-HT$_{1A}$ receptor agonists into the raphe obscurus of the anaesthetized rat. Br. J. Pharmacol. 102: 317–322.

Faiers, A. A., Calaresu, F. R., and Mogenson, G. J. (1976). Factors affecting cardiovascular responses to stimulation of hypothalamus in the rat. Exp. Neurol. 51: 188–206.

Fozard, J. R., Mir, A. K., and Middlemiss, D. N. (1987). Cardiovascular response to 8-hydroxy-2-(di-n-propylamino)tetralin (8-OH-DPAT) in the rat: site of action and pharmacological analysis. J. Cardiovasc. Pharmacol. 9: 328–347.

Gradin, K., Pettersson, A., Hedner, T., and Persson, B. (1985). Acute administration of 8-hydroxy-2-(di-n-propylamino)tetralin (8-OH-DPAT) a selective 5-HT receptor agonist, causes a biphasic blood pressure response and a bradycardia in the normotensive Sprague-Dawley rat and in the spontaneously hypertensive rat. J. Neur. Transm. 62: 305–319.

Kuhn, D. M., Wolf, W. A., and Lovenberg, W. (1980). Review of the role of the central serotonergic neuronal system in blood pressure regulation. Hypertension 2: 243–255.

Lambert, G., Friedman, E., and Gershon, S. (1975). Centrally mediated cardiovascular responses to 5-HT. Life Sci. 17: 915–920.

Lambert, G. A., Friedman, E., Buchweitz, E., and Gershon, S. (1978). Involvement of 5-hydroxytryptamine in the central control of respiration, blood pressure and heart rate in the anesthetized rat. Neuropharmacol. 17: 807–813.

Laubie, M., Drouillat, M., Dabire, H., Cherqui, C., and Schmitt, H. (1988). Ventrolateral medullary pressor area: site of hypotensive and sympathoinhibitory effects of ($\pm$)-8-OH-DPAT in anaesthetised dogs. Eur. J. Pharmacol. 160: 385–394.

Martin, G. E., and Lis, E. V. (1985). Hypotensive action of 8-hydroxy-2-(di-n-propylamino)tetralin (8-OH-DPAT) in spontaneously hypertensive rats. Arch. Int. Pharmacodyn. 273: 251–261.

Moore, R. Y., Halaris, A. E., and Jones, B. E. (1978). Serotonin neurons of the midbrain raphe: ascending projections. J. Comp. Neurol. 180: 417–438.

Newberry, N. R. (1989). 5-HT$_{1A}$ receptors mediate a hyperpolarisation of rate ventromedial hypothalamic neurones in vitro. Br. J. Pharmacol. 98: 809P.

Paxinos, G., and Watson, C. (1982). The rat brain in stereotaxic co-ordinates. Academic Press, Sydney, Australia.

Perren, M. J., Feniuk, W., and Humphrey, P. P. A. (1991). Vascular 5-HT$_1$-like receptors which mediate contraction of the dog isolated vein and carotid arterial vasoconstriction in anaesthetised dogs are not of the 5-HT$_{1A}$ or 5-HT$_{1D}$ subtype. Br. J. Pharmacol. 102: 191–197.

Ramage, A. G., and Fozard, J. R. (1987). Evidence that the putative 5-HT$_{1A}$ receptor agonists, 8-OH-DPAT and ipsapirone, have a central hypotensive action that differs from that of clonidine in anaesthetised cats. Eur. J. Pharmacol. 138: 179–191.

Saavedra, J. M., Palkovits, M., Brownstein, M. J., and Axelrod, J. (1974). Serotonin distribution in the nuclei of the rat hypothalamus and preoptic region. Brain Res. 77: 157–165.

Saxena, P. R. and Villalón, C. M. (1990). Brain 5-HT$_{1A}$ receptor agonism: a novel mechanism for antihypertensive action. Trends Pharmacol. Sci. 11: 95–96.

Smits, J. F. M., Van Essen, H., and Struyker-Boudier, H. A. J. (1978). Serotonin mediated cardiovascular responses to electrical stimulation of the raphe nuclei in the rat. Life Sci. 23: 173–178.

Smits, J. F. and Struyker-Boudier, H. A. (1976). Intrahypothalamic serotonin and cardiovascular control in rats. Brain Res. 111: 422–427.

Valenta, B., and Singer, E. A. (1990). Hypotensive effects of 8-hydroxy-2-(di-n-propylamino)tetralin and 5-methylurapidil following stereotaxic microinjection into the ventral medulla of the rat. Br. J. Pharmacol. 99: 713–716.

Van Heuven-Nolsen, D., Tysse Klasen, T. H. M., Luo, Q., and Saxena, P. R. (1990). 5-HT$_1$-like receptors mediate contractions in the rabbit saphenous vein. Eur. J. Pharmacol. 191: 375–382.

Vergé, D., Daval, G., Marcinkiewicz, M., Patey, A., El Mestikawy, S., Gozlan, H., and Hamon, M. (1986). Quantitative autoradiography of multiple 5-HT$_1$ receptor subtypes in the brain of control or 5, 7-dihydroxytryptamine-treated rats. J. Neurosci. 6: 3474–3482.

Wouters, W., Hartog, J., and Bevan, P. (1988a). Flesinoxan. Cardiovasc. Drug Rev. 6: 71–83.

Wouters, W., Tulp, M. T. M., and Bevan, P. (1988b). Flesinoxan lowers blood pressure and heart rate in cats via 5-HT$_{1A}$ receptors. Eur. J. Pharmacol. 149: 213–223.

Serotonin: Molecular Biology, Receptors and Functional Effects
ed. by J. R. Fozard/P. R. Saxena
© 1991 Birkhäuser Verlag Basel/Switzerland

Pharmacological Characterisation of the Receptor Mediating 5-HT Evoked Motoneuronal Depolarization *In vitro*

P. M. Larkman and J. S. Kelly

Department of Pharmacology, University of Edinburgh, Edinburgh EH8 9JZ, UK

Summary. 5-hydroxytryptamine (5-HT) and noradrenaline (NA) evoked a slow depolarization of adult rat facial motoneurones (FM's) *in vitro*, associated with a reduction in a membrane potassium conductance. The 5-HT-evoked depolorization could not be mimicked by 8-OH-DPAT, dipropyl-5-CT or 2-methyl-5-HT: 5-CT however was able to evoke a depolarization of FM's previously shown to be depolarized by 5-HT. Methysergide selectively antagonised the 5-HT evoked depolarization but not the NA-evoked response. LY-53857 also antagonised the action of 5-HT while ketanserin depressed but could not fully abolish the depolarization even at high concentrations. Spiperone, methiothepin and ICS 205-930 were all ineffective. The results obtained exclude 5-HT_{1A} or 5-HT_3 receptor subtypes as mediators of the depolarization. Antagonism by methysergide, LY-53857 and to some extent ketanserin suggest but cannot differentiate between a $5\text{-HT}_{1C}/5\text{-HT}_2$ identity for this receptor.

Introduction

Central excitatory actions of 5-hydroxytryptamine (5-HT) have been demonstrated in a variety of *in vivo* and *in vitro* preparations using extracellular and intracellular recording methods (Roberts and Straughan 1967; McCall and Aghajanian 1979; Jahnsen 1980; Yoshimura and Nishi 1982; Davies et al. 1988). However, unlike the mechanism of central neuronal hyperpolarization involving an increase in membrane potassium ion conductance which is apparently common to all neurones inhibited by 5-HT, (Andrade and Nicoll 1987; Colino and Halliwell 1987; Joels and Gallagher 1988; Williams et al. 1988) multiple mechanisms of 5-HT-evoked depolarization have been demonstrated. Slow depolarization of hippocampal pyramidal neurones is mediated through reductions in a leak potassium conductance and a voltage sensitive potassium conductance, I_M (Andrade and Nicoll 1987; Colino and Halliwell 1987). In nucleus accumbens neurones an inwardly rectifying K^+ conductance is reduced (North and Uchimura 1989) while similar mechanisms are implicated in sympathetic preganglionic neurones (Ma and Dun 1986; Yoshimura and Nishi 1982), cortical neurones (Davies et al. 1987) and neonatal spinal motoneurones (Elliott and Wallis 1990). A second slow depolarization mediated by augmentation of an inward rectifier current, I_h, which is active at the resting potential has been shown to occur in neurones of the nucleus prepositus hypoglossi (PrH) (Bobker and Williams 1989) and dorsal lateral geniculate (LGNd) (Pape and McCormick 1989) as well as neonatal rat spinal motoneurones (Takashi and Berger 1990). A fast, rapidly desensitizing depolarization mediated through the integral ionophore/5-HT$_3$ receptor membrane protein has largely been studied in peripheral preparations (Derkach et al. 1989; Wallis and Dun 1988) but has been observed in cultured central neurones (Yakel et al. 1988).

In this study and elsewhere (Larkman et al. 1989; VanderMaelen and Aghajanian, 1980; Aghajanian and Rasmussen, 1989) adult rat facial motoneurones are shown to be depolarized by 5-HT and noradrenaline (NA) through a mechanism involving potassium channel closure. Using a range of putative 5-HT receptor ligands we present data aimed at characterizing the receptor mediating the 5-HT-evoked depolarization of facial motoneurones *in vitro* using intracellular electrophysiological techniques.

Materials and Methods

Adult Cob Wistar rats (160–250 g) were decapitated using a guillotine. The brainstem and attached cerebellum were quickly removed and placed in pre-oxygenated artificial cerebrospinal fluid (aCSF) at 4°C. A small block of the brainstem was prepared and attached to a plastic stage using cyanoacrylate glue with an agar block supporting one side. Slices were cut at approximately 350 μm using a Vibroslice (Cambden Inst.). The slices were quickly transferred to an interface-type chamber and perfused at a rate of 0.5 – 1 ml/min with aCSF at 37°C under a humidified atmosphere of 95%O$_2$/5%CO$_2$. Intracellular recordings were commenced only after at least 1h incubation. The composition of aCSF in mM was: NaCl 124, KCl 5, MgSO$_4$ 2, CaCl$_2$ 2, NaHCO$_3$ 26, HEPES 1.25, D-glucose 10, and the pH was 7.4 after equilibration with a 95%O$_2$/5%CO$_2$ gas mixture.

Intracellular recordings were made using microelectrodes pulled on a Flaming-Brown horizontal puller (Sutter Inst.) using 1.2mm outside diameter, thin walled, fibre, glass capillaries (Clark Electromedical). They were filled with 3M KCL and had DC resistances of 10–50 MΩ. A high input resistance bridge amplifier (Axoclamp IIA) in current clamp mode enabled simultaneous measurement of membrane potential and intracellular current injection through the recording electrode. Electrode resistance was monitored continuously and balanced when necessary. Output was observed on a digital storage oscilloscope (Gould 1425) and stored on video-tape using an analog-interfaced digital audio signal processor (Sony PCM 701ES) – video cassette recorder (Sony

SLF30) system (Lamb, 1985) to allow future computer analysis using a CED 502 interfaced to a PDP 11 (Crunelli et al. 1983).

All drugs were applied by addition to the superfusing aCSF after being diluted to the correct concentration from previously prepared stock solutions. 5-HT creatinine sulphate and noradrenaline HC1 were obtained from Sigma Chemical Co; 5-CT, methiothepin maleate and 2-methyl-5-HT were gifts from Glaxo; 8-OH-DPAT was supplied by Research Biochemicals Inc., ICS 205-930, methysergide and dipropyl-5-CT were gifts from Sandoz; spiperone and ketanserin were gifts from Janssen; LY-53857 was a gift from Eli Lilly.

Results

Intracellular current-clamp recordings obtained from adult rat facial motoneurones show superfusion of 5-HT (50 to 200 μM) and NA (10–100 μM) to evoke slow, monophasic, dose-dependent depolarizations of 2 to 14 mV (n = 70) and 2 to 10 mV (n = 20) respectively, which were rapidly reversed on removal of the neurotransmitter from the aCSF (Figure 1A, C). These depolarizations were associated with an increase in neuronal input resistance (Rm) and a lengthening of the time constant for charging of the neuronal membrane. The combination of these actions leads to increased excitability of the motoneurone which is reflected in the reduced amplitude of the injected current pulse required to reach spike threshold (Figure 1A, C).

Current-voltage plots obtained by plotting the peak voltage deflection against injected current pulse amplitude in both control and 5-HT exposed conditions are shown in Figure 1B. The increase in Rm is clearly indicated by the steeper slope of the current-voltage relationship in the presence of 5-HT. The point of intersection of current-voltage plots in the absence and presence of neurotransmitter gives a measure of the reversal potential (E_{rev}) for the action of the neurotransmitter which in this case is −90 mV for 5-HT (Figure 1B) and −84 mV for NA (not shown). When measured on the same population of motoneurones E_{5-HT} was −95 ± 11 mV and E_{NA} was −87 ± 5 mV (mean ± SD, n = 9). The use of 3M KCl filled microelectrodes which would be expected to reverse the chloride gradient across the membrane coupled with these values of E_{5-HT} and E_{NA} suggest that both depolarizations are mediated at least in part by a change in a conductance to K^+ ions.

The receptor mediating 5-HT-evoked depolarization of FM's was investigated in more detail using several ligands with putative 5-HT receptor affinity. The 5-HT receptor agonist 5-CT was tested for its ability to mimic the depolarization of FMs. 5-HT (150 μM) and 5-CT (100 μM) were both able to depolarize the same FM by 9mV (Figure 2). In both cases the depolarization was associated with an increase in Rm

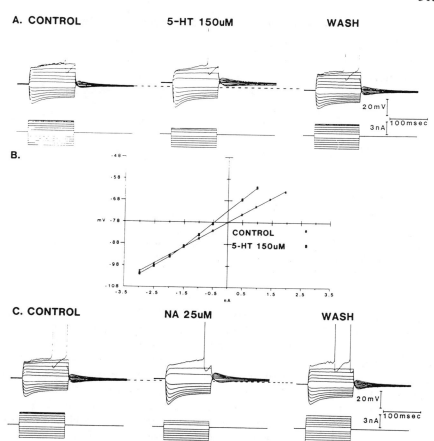

Figure 1. Electronic potentials and current-voltage plot from a facial motoneurone (FM) before, during and after 5-HT and NA superfusion. Voltage responses (upper traces) obtained by intracellular injection of current steps (lower traces) in the presence and absence of 5-HT (150 μM) (A) or NA (25 μM) (C). The FM had a resting potential of -78 mV. 5-HT evoked a depolarization of 5 mV associated with an increase in peak Rm from 7.5 to 11 MΩ while the NA-evoked response was a depolarization of 3 mV and a change in peak Rm from 7.4 to 12.6 MΩ. Spike amplitude is attenuated in these and subsequenct records by the sampling rates used in data analysis. In B) the control peak deflection current-voltage plot (asterisks) was linear throughout the tested range. The 5-HT (closed squares) relationship was linear up to the point of intersection with the control plot at -90 mV. (Lines drawn by at least squares linear regression analysis program.)

however the reversal potential for the action of 5-CT was more negative than that for 5-HT (not shown). In voltage clamp both ligands have been shown to evoke inward currents (n = 3). More detailed investigation will be required to determine whether these events are mediated through the same or different mechanisms.

The more selective 5-HT$_{1A}$ receptor agonists, 8-OH-DPAT (10 μM, n = 4) and dipropyl-5-CT (10 to 200 μM, n = 3), were both unable to

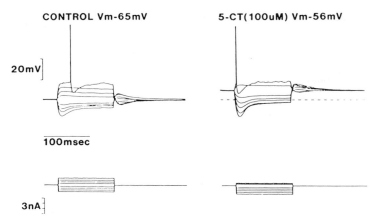

Figure 2. Electonic potentials showing superfusion of 5-CT (100 μM) to evoke a 9 mV depolarization of a FM resting at a potential of -65 mV. Peak Rm increased from 12.9 MΩ to 16.8 MΩ. The FM had previously been depolarized by 5-HT (150 μM) by 9 mV also associated with an increase in peak Rm of 11 to 17.5 MΩ (not shown).

mimic 5-HT evoked depolarization of FM's. In light of the partial agonist actions of 8-OH-DPAT on hippocampal neurones its ability to antagonise the 5-HT response of FM's was tested, however the 5-HT evoked depolarization in the presence of 8-OH-DPAT was indistinguishable from the control response. The 5-HT$_3$ receptor agonist, 2-methyl-5-HT (10–200 μM n = 3), was also unable to depolarize FM's.

That distinct receptors mediate the actions of 5-HT and NA was confirmed by experiments using the non-selective 5-HT$_1$/5-HT$_2$ receptor antagonist, methysergide, which selectively blocked 5-HT- but not NA-evoked depolarizations. Methysergide (100 μM) reduced the 5-HT evoked depolarization from 8.5 ± 5.5 mV to 1 ± 1 mV (n = 2) while at 10 μM the depolarization was reduced from 6.5 ± 2.5 mV to 3 ± 2 mV (n = 3) without altering the reversal potential of the 5-HT effect (Figure 3). Depolarization evoked by NA was unaffected by methysergide (50–100 μM) being 3.5 ± 0.7 mV before and 3.75 ± 0.4 mV after application (n = 2).

LY-53857 is an antagonist purported to show greater selectivity for 5-HT$_{1C}$/5-HT$_2$ receptors than its parent structure, methysergide. In this study when tested against the 5-HT-evoked depolarization it was found to effectively antagonize the response at concentrations of 50 to 100 μM. LY-53857 (50 μM) reduced a 5-HT evoked depolarization of 4.7 ± 0.6 mV to 2 ± 0 mV (n = 3) while increasing the concentration to 100 μM completely abolished the depolarization (n = 1).

Having previously ruled out 5-HT$_{1A}$ and 5-HT$_3$ receptors as mediating depolarization of FM's we attempted to differentiate between the 5-HT$_{1C}$ and 5-HT$_2$ receptors indicated by the actions of methysergide

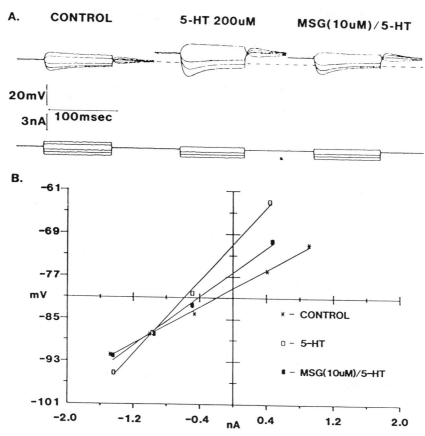

Figure 3. Antagonism of the 5-HT-evoked depolarization by methysergide (MSG) (10 μM).
A) 5-HT (200 μM) evoked an 8 mV depolarization from a resting potential of -79 mV
associated with an increase in peak Rm from 8.7 to 17.1 MΩ. Application of methysergide
(10 μM) did not alter the resting potential or peak Rm but reduced a subsequent 5-HT-
evoked depolarization to 3 mV with an increase in peak Rm to 11.5 MΩ.

B) Peak deflection current voltage plots from the records in A were linear and show the
reversal potential of the 5-HT response to be unaffected despite partial blockade by methysergide (-87.2 and -88.3 mV in the absence and presence of methysergide, respectively).

and LY-53857. The actions of ketanserin and spiperone two ligands
with selective affinity for 5-HT$_2$ receptors were tested. The effects of
ketanserin (1$-$100 μM) were not clear. Some suppression of the 5-HT-
evoked depolarization was obtained at all concentrations used though
full blockade of the kind observed with methysergide could not be
achieved even when the concentration was increased and superfusion
was extended for prolonged periods. Thus in Figure 4 prolonged application of ketanserin reduced a control 5-HT depolarization from 8 mV
to 5 mV after 20 minutes and to 3 mV after 60 minutes without altering

316

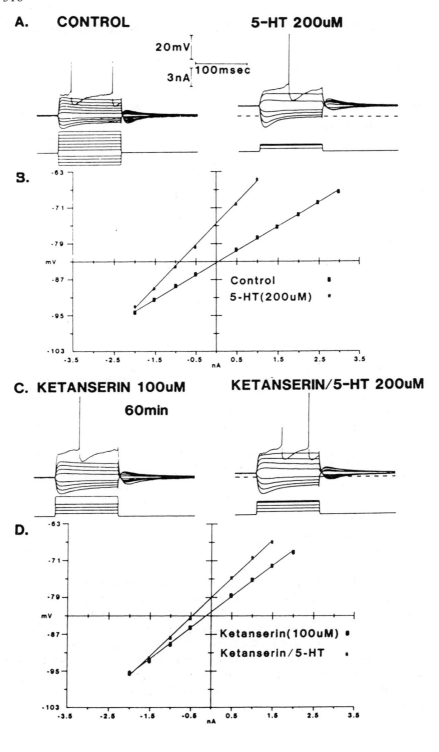

A. **CONTROL**   **5-HT 200uM**

20mV
3nA
100msec

B.

C. **KETANSERIN 100uM**   **KETANSERIN/5-HT 200uM**
60min

D.

the reversal potential seen in the corresponding current-voltage plots. A further reduction was not observed even with continued application. Somewhat surprisingly spiperone ($10-100 \mu M$, n = 3) failed to antagonise the 5-HT evoked depolarization, again after prolonged application. In addition to these results the $5-HT_1$ receptor antagonist, methiothepin ($10-100 \mu M$), and the $5-HT_3$ receptor antagonist, ICS 205-930 ($10 \mu M$), were both ineffective against the 5-HT-evoked depolarization of FM's.

Discussion

In common with other studies the data presented here show 5-HT and NA both evoke depolarization of FM's through a mechanism involving closure of potassium channels. The primary aim of this study was to attempt to define pharmacologically the receptor subtype mediating the 5-HT-evoked response. The ineffectiveness of the $5-HT_{1A}$ receptor ligands, 8-OH-DPAT, either as an agonist or antagonist, and dipropyl-5-CT, as an agonist, apparently rules out the involvement of this receptor subtype in 5-HT-evoked depolarization of FM's. The lack of agonism by 2-methyl-5-HT coupled with the absence of any antagonism by ICS 205-930 as well as a full blockade by methysergide suggest a $5-HT_3$ receptor is also not involved. Circumstantially the ionic mechanism of 5-HT-evoked depolarization is quite different from what would be expected through a $5-HT_3$ receptor mediated mechanism. Depolarization of FM's by the agonist 5-CT suggests a $5-HT_1$ type receptor may be involved. 5-CT also has agonist activity at $5-HT_4$ sites linked to cAMP production in colliculi neurones where it is significantly less active than 5-HT and 5-methoxytryptamine (5-MeOT) but of similar potency to zacopride. (Dumuis et al. 1988). It would appear however that this site can be ruled out for FM depolarization as ICS 205-930 is an antagonist at these $5-HT_4$ receptors. Further work is required to clarify the agonist action of 5-CT on FM's as current-voltage relationships suggest the action of this ligand and 5-HT may not be identical.

Suppression of the 5-HT evoked depolarization by methysergide, the $5-HT_1$ and $5-HT_2$ receptor antagonist, clearly distinguishes 5-HT and NA evoked responses while the effects of LY-53857 which has greater

---

Figure 4. The effects of ketanserin on 5-HT-evoked depolarization of a facial motoneurone.
A) A control 5-HT ($200 \mu M$) – evoked depolarization of 8 mV from a resting potential of -83 mV was associated with an increase in peak Rm form 5.4 to 9.4 MΩ (74%).
B) The peak deflection current voltage plots in the presence (asterisks) and absence (closed squares) of 5-HT intersect at -95 mV. C) Application of ketanserin ($100 \mu M$) for 60 minutes reduced the 5-HT evoked depolarization to 3 mV associated with a 25% increase in peak Rm to 8.5 MΩ. D) Control hyperpolarizing current steps apply to all voltage records. The point of intersection of current voltage plots in unchanged by the presence of ketanserin (-95 mV).

selectivity for 5-HT$_{1C}$/5-HT$_2$ sites (Cohen et al. 1983) suggests the identity of the mediating receptor falls into this category. Biochemical studies have shown ketanserin to possess high affinity for 5-HT$_2$ sites (Leysen et al. 1981); however, our results show that this ligand was only partially effective in preventing the 5-HT-evoked depolarization even at high perfusion concentrations. It could be suggested that if 5-HT depolarizes FM's through more than one mechanism and/or mediating receptor then ketanserin may block only a 5-HT$_2$ mediated component. Evidence from the current-voltage plots (Figure 4) showing a constant reversal potential at each stage of the ketanserin induced blockade suggests that more than one mechanism in this particular example is unlikely but cannot exclued the involvement of more than one receptor subtype. The effects of ketanserin on 5-CT evoked depolarizations have not been tested. Surprisingly given its high affinity for 5-HT$_2$ binding sites spiperone failed to antagonize 5-HT evoked depolarization of FM's.

Our data with both agonists and antagonists suggests a 5-HT$_{1C}$ receptor as being responsible for mediating the depolarization of FM's. Spiperone has poor affinity for this site, ketanserin has much higher affinity though not as high as for 5-HT$_2$ sites while 5-CT has an affinity for 5-HT$_{1C}$ sites which is much higher than for 5-HT$_2$ sites (Hoyer 1988). Clearly the possibility that 5-HT$_2$ or other receptor subtypes are also involved cannot be ruled out on the basis of electrophysiological evidence alone.

In vivo studies have shown that 5-HT mediated facilitation of glutamate-evoked firing of FM's can be antagonized by systemic and iontophoretic application of methysergide, cyproheptadine and cinanserin (McCall and Aghajanian 1979) while iontophoretic application of metergoline was also effective in the same study. Systemic application of 5-methoxy-N,N-dimethyltryptamine (5-MeODMT) evokes a depolarization of FM's *in vivo* which can be blocked with methysergide. (VanderMaelen and Aghajanian 1982). *In vivo* studies have also claimed iontophoretically applied ketanserin to be an antagonist of 5-HT-evoked facilitation of FM firing (Penington and Reiffenstein 1986). A more recent *in vitro* study has shown ritanserin applied in the superfusing aCSF to suppress excitatory actions of 5-HT while a depolarization can also be evoked by the 5-HT$_2$/5-HT$_{1C}$ agonist DOM (Rasmussen and Aghajanian 1990). These studies in combination with our data are still unable to clearly distinguish between a 5-HT$_{1C}$ and 5-HT$_2$ identity for this receptor.

Autoradiographic data show that while ketanserin binding sites in the FMN have been claimed to identify a 5-HT$_2$ site (Fischette et al. 1987), [^{3}H]-mesulergine binding to what is presumed to be the same site is only partially displaced by ketanserin and spiperone even though 5-HT can fully displace it (Pazos et al. 1985). Recent work using mRNA probes for 5-HT$_2$ 5-HT$_{1C}$ receptor messages has shown high densities of 5-HT$_2$ mRNA in the FMN largely located in cell bodies and proximal dendrites while 5-HT$_{1C}$ mRNA appears less abundant and more diffusely dispersed throughout the brainstem motor nuclei (Mengod et al. 1990a; b).

Our results show some similarities to the receptor type involved in depolarization of nucleus accumbens neurones (North and Uchimura 1989). Ketaserin, mianserin but also spiperone are all antagonists while 5-HT$_1$ and 5-HT$_3$ receptor ligands have no effect. A 5-HT$_2$ receptor is thus claimed to mediate this depolarization though surprisingly mCPP which would be expected to be an antagonist at these sites was ineffective. A similar conclusion was reached by a limited study on cortical neurones (Davies et al. 1987). Elliott and Wallis (1990) have shown using intracellular techniques that neonatal motoneurones are depolarized through what is likely to be a 5-HT$_2$ receptor-mediated closure of K$^+$ channels. However other work has shown that depolarization of presumably the same population of motoneurones can occur through augmentation of the inward rectifier through a receptor distinct from the 5-HT$_2$ receptor (Takahashi and Berger 1990). This indicates the involvement of a combination of mechanisms and receptors in 5-HT evoked depolarization of spinal motoneurones a situation which may also occur in FM's. A slow depolarization of enteric neurones involving a reduction in a potassium conductance has been claimed to be mediated through a novel receptor termed the 5-HT$_{1p}$ site. (Branchek et al. 1988). At this receptor the benzamide derivatives renzapride (BRL 24924) and the S isomer of zacopride are antagonists and agonists respectively but appear not to act through 5-HT$_4$ site as ICS 205-930, as in FM's, is not an antagonist. The testing of these compounds on FM's may be important in the further categorization of this 5-HT receptor. In the hippocampus a different receptor appears to be linked to potassium channel closure. The lack of effect of a wide range of ligands leaves classification open to debate though the results of some workers with ICS 205-930 as well as antagonist actions of renzapride has led to the suggestion that a 5-HT$_4$ may be responsible (Chaput et al. 1990).

Excitation mediated through augmentation of the inward rectifier I$_h$ in PrH and LGNd neurones appears to involve an as yet uncharacterized 5-HT$_1$-like receptor. As indicated previously 5-CT, but not 8-OH-DPAT and 2-methyl-5-HT, is an agonist while methysergide and spiperone but not ketanserin and mianserin are antagonists.

The conclusions of the present study suggest that, in broad agreement with similar actions of 5-HT on nucleus accumbens and cortical neurones as well as neonatal spinal motoneurones, the receptor involved in depolarization of FM's through K$^+$ channel closure belongs to the 5-HT$_2$ family, though a 5-HT$_{1C}$ or 5-HT$_2$ identity cannot be differentiated. Whether 5-CT acts through this or a distinct receptor remains to be determined. This serves to emphasize the heterogeneity of receptors involved in 5-HT evoked depolarization of central and peripheral neurones.

Acknowledgements

PML was in receipt of a SERC-CASE award in part funded by Glaxo, and is currently supported by a Merck Sharpe and Dohme fellowship. Support for the work was also provided by a Wellcome Trust award to J. S. Kelly.

References

Aghajanian, G. K., and Rasmussen, K. (1989). Intracellular studies in the facial nucleus illustrating a simple new method for obtaining viable motoneurones in adult rat brain slices. Synapse 3: 331–338.

Andrade, R., and Nicoll, R. A. (1987). Pharmacologically distinct actions of serotonin on single pyramidal neurones of the rat hippocampus recorded *in vitro*. J. Physiol. 394: 99–124.

Bobker, D. H., and Williams, J. T. (1989). Serotonin augments the cationic current Ih in central neurones. Neuron 2: 1535–1540.

Branchek, T. A., Mawe, G. M., and Gershon, M. D. (1988). Characterisation and localisation of a peripheral neural 5-hydroxytryptamine receptor subtype ($5-HT_{1P}$) with a selective agonist, 5-hydroxyindalpine. J. Neurosci. 8: 2582–2595.

Chaput, Y., Araneda, R. C., and Andrade, R. (1990). Pharmacological and functional analysis of a novel serotonin receptor in the rat hippocampus. Eur. J. Pharmacol. 182: 441–456.

Cohen, M. L., Fuller, R. W., and Kurz, K. D. (1983). LY-53857, a selective and potent serotonergic ($5-HT_2$) receptor antagonist, does not lower blood pressure in the spontaneously hypertensive rat. J. Pharmacol. Exp. Therap. 227: 327–332.

Colino, A., and Halliwell, J. V. (1987). Differential modulation of three separate K-conductances in hippocampal CA1 neurones by serotonin. Nature 328: 73–77.

Crunelli, V., Forda, S., Kelly, J. S., and Wise, J. C. M. (1983). A programme for the analysis of intracellular data recorded from *in vitro* preparations of central neurons. J. Physiol. 340: 13P.

Davies, M., Wilkinson, L. S., and Roberts, M. H. T. (1988). Evidence for excitatory 5-HT2-receptors on rat brainstem neurones. Br. J. Pharmacol. 94: 483–491.

Davies, M. F., Deisz, R. A., Prince, A., and Peroutka, S. J. (1987). Two distinct effects of 5-hydroxytryptamine on single cortical neurones. Brain Res. 423: 347–352.

Derkach, V., Surprenant, A., and North, R. A. (1989). $5-HT_3$ receptors are membrane ion channels. Nature 339: 706–709.

Dumius, A., Bouhelal, R., Sebben, M., Cory, R., and Bockaert, J. (1988). A non-classical 5-hydroxytryptamine receptor postively coupled with adenylate cyclase in the central nervous system. Mol. Pharmacol. 34: 880–887.

Elliott, P., and Wallis, D. I. (1990). The action of 5-hydroxytryptamine on lumbar motoneurones in neonatal rat spinal cord *in vitro*. J. Physiol. 426: 54P.

Fischette, C. T., Nock, B., and Renner, K. (1987). Effects of 5,7-dihydroxytryptamine on serotonin1 and serotonin2 receptors throughout the rat central nervous system using quantitative autoradiography. Brain Res. 421: 263–279.

Hoyer, D. (1988). Molecular pharmacology and biology of 5-HT1C receptors. TIPS 9: 89–94.

Jahsen, H. (1980). The action of 5-hydroxytryptamine on neuronal membranes and synaptic transmission in area CA1 of the hippocampus *in vitro*. Brain Res. 197: 83–94.

Joels, M., and Gallagher, J. P. (1988). Actions of serotonin recorded intracellularly in rat dorsal lateral septal neurons. Synapse 2: 45–53.

Lamb, T. D. (1985). An inexpensive digital tape recorder suitable for neurophysiological signals. J. Neurosci. Methods 15: 1–14.

Larkman, P. M., Pennington, N. J., and Kelly, J. S. (1989). Electrophysiology of adult rat facial motoneurones: the effects of serotonin (5-HT) in a novel *in vitro* brainstem slice. J. Neurosci. Methods 28: 133–146.

Leysen, J. E., Awouters, F., Kennis, L., Laduron, P. M., Vandenberk, J., and Jannsen, P. A. J. (1981). Receptor binding profile of R 41 468, a novel antagonist at 5-HT$_2$ receptors. Life Sci. 28: 1015–1022.

Ma, R. C., and Dun, N. J. (1986). Excitation of lateral horn neurons of the neonatal rat spinal cord by 5-Hydroxytryptamine. Dev. Brain Res. 24: 89–98.

McCall, R. B., and Aghajanian, G. K. (1979). Serotonergic facilitation of facial motoneurone excitation. Brain Res. 169: 11–27.

Mengod, G., Nguyen, H., Le, H., Waeber, C., Lubbert, H., and Palacios, J. M. (1990a). The distribution and cellular localization of the serotonin 1C receptor mRNA in the rodent brain examined by *in situ* hybridization histochemistry. Comparison with receptor binding distribution. Neuroscience 35: 577–591.

Mengod, G., Pompeiano, M., Martinez-Mir, M.I., and Palacios, J. M. (1990b). Localization of the mRNA for the 5-HT2 receptor by insitu hybridization histochemistry. Correlation with the distribution of receptor sites. Brain Res. 524: 139–143.

North, R. A., and Uchimura, N. (1989). 5-Hydroxytryptamine acts at 5-HT$_2$ receptors to decrease potassium conductance in rat nucleus accumbens neurones. J. Physiol. 417: 1–12.

Pape, H.-C., and McCormick, D. A. (1989). Noradrenaline and serotonin selectively modulate thalamic burst firing by enhancing a hyperpolarization-activated cation current. Nature 340: 715–718.

Pazos, A., Cortes, R., and Palacios, J. M. (1985). Quantitative autoradiographic mapping of serotonin receptors in the rat brain. II. Serotonin-2 receptors. Brain Res. 346: 231–249.

Penington, N. J., and Reiffenstein, R. J. (1986). Possible involvement of serotonin in the facilitatory effect of a hallucinogenci phenethylamine on single facial motoneurones. Can. J. Physiol. Pharmacol. 64: 1302–1309.

Rasmussen, K., and Aghajanian, G. K. (1990). Serotonin excitation of facial motoneurones: Receptor subtype classification. Synapse 5: 324–332.

Roberts, M. H. T., and Straughan, D. W. (1967). Excitation and depression of cortical neurones by 5-hydroxytryptamine. J. Physiol. 193: 269–294.

Takahashi, T., and Berger, A. J. (1990). Direct excitation of rat spinal motoneurones by serotonin. J. Physiol. 423: 63–76.

VanderMaelen, C. P., and Aghajanian, G. K. (1980). Intracellular studies showing modulation of facial motoneurone excitability by serotonin. Nature 287: 346–347.

VanderMaelen, C. P., and Aghajanian, G. K. (1982). Intracellular studies on the effects of systemic administration of serotonin agonists on rat facial motoneurons. Eur. J. Pharmacol. 78: 233–236.

Wallis, D. I., and Dun, N. J. (1988). A comparison of fast and slow depolarisations evoked by 5-HT in guinea-pig coeliac ganglion cells *in vitro*. Br. J. Pharmacol. 93: 110–120.

Williams, J. T., Colmers, W. F., and Pan, Z. Z. (1988). Voltage- and ligand- activated inwardly rectifying currents in dorsal raphe neurons *in vitro*. J. Neurosci. 8: 3499–3506.

Yakel, J. L., Trussell, L. O., and Jackson, M. B. (1988). Three serotonin responses in cultured mouse hippocampal and striatal neurones. J. Neurosci. 8: 1273–1285.

Yoshimura, M., and Nishi, S. (1982). Intracellular recordings from lateral horn cells of the spinal cord *in vitro*. J. Autonom. Nerv. Syst. 6: 5–11.

Serotonin: Molecular Biology, Receptors and Functional Effects
ed. by J. R. Fozard/P. R. Saxena
© 1991 Birkhäuser Verlag Basel/Switzerland

Stimulation of 5-HT₃ Receptors Inhibits Release of Endogenous Noradrenaline from Hypothalamus

P. Blandina[1,2], J. Goldfarb[1], and J. P. Green[1]

[1]*Department of Pharmacology, Mount Sinai Medical School of the City University of New York, New York, NY 10029, USA;* [2]*Dipartimento di Farmacologia Preclinica e Clinica, Università degli Studi di Firenze, 50134 Firenze, Italia*

Summary. Exposure to 20 mM KCl elicited a Ca^{++}-dependent release of endogenous NA (noradrenaline) from superfused rat hypothalamic slices. Two consecutive exposures, S_1 and S_2, respectively, produced NA release of similar magnitude ($S_2/S_1 = 1.01 \pm 0.07$, n = 5). 5-HT (5-hydroxytryptamine), 10 μM, inhibited KCl-evoked NA release by 50%, in the presence but not in the absence of the $5-HT_2/5-HT_{1C}$ receptor antagonist ritanserin. 2-ME-5-HT (2-methyl-5-hydroxytryptamine), a selective $5-HT_3$ receptor agonist, but not α-ME-5-HT (α-methyl-5-hydroxytryptamine), a $5-HT_1$-like and $5-HT_2$ receptor agonist, mimicked the 5-HT response in the presence and in the absence of ritanserin. ICS 205-930 ((3α-tropanyl)1H-indole-carboxylic acid ester), 1 nM, and s(−)zacopride, 3 nM, highly selective $5-HT_3$ receptor antagonists, inhibited the effects of both agonists. These observations provide direct evidence for a 5-HT-mediated modulation of endogeneous NA release from rat hypothalamus.

Introduction

$5-HT_3$ receptors, although first described over 30 years ago (the M receptor of Gaddum and Picarelli 1957), have become a focus of much attention, and understanding of the pharmacology and functions of these receptors has progressed rapidly since the introduction of selective antagonists, such as MDL 72222 (Fozard 1984), ICS 205-930 (Richardson et al. 1985), GR 38032 (Butler et al. 1988), BRL 24924 (Sanger 1987), zacopride (Smith et al. 1988), and LY277359 (Cohen et al. 1990).

$5-HT_3$ receptors were originally characterized in peripheral tissues such as the guinea pig ileum (Buchheit et al. 1985), rat vagus nerve (Ireland and Tyers 1987) and isolated rabbit heart (Fozard et al. 1979). The receptors, part of a cation-selective receptor-ion channel complex (Derkach et al. 1989), are located on neuronal elements. A prominent consequence of their activation is the release of other neurotransmitters, e.g. acetylcholine (ACh) and noradrenaline (NA). Until recently, little was known about the existence of these receptors in the central nervous system (CNS). However, the behavioral responses to $5-HT_3$ receptor antagonists, e.g., the prevention of cisplatin-induced emesis in the ferret

(Miner and Sanger 1986) and the modification of rodent and primate behavior in anxiolytic tests (Jones et al. 1988) suggested that 5-HT$_3$ receptors are also present in the CNS. The identification of 5-HT$_3$ binding sites in the brain of rat (Kilpatrick et al. 1987) and man (Barnes et al. 1988), supported this inference. In addition, 5-HT$_3$ receptor activation released preloaded [^{3}H]NA from hippocampal slices (Feuerstein and Hertting 1986) and endogeneous dopamine (DA) from striatal slices (Blandina et al. 1988, 1989b) and from the nucleus accumbens *in vivo* (Jiang et al. 1990). 5-HT$_3$ receptors have also been implicated in the modulation of [^{3}H]ACh release from rat cortical brain slices (Barnes et al. 1989).

The hypothalamus has binding sites and/or receptors of the 5-HT$_3$ (Kilpatrick et al. 1987), 5-HT$_2$ (Pazos et al. 1985) and 5-HT$_{1C}$ (Pazos and Palacios 1985; Molineaux et al. 1989) type. There is evidence of functional interactions between NA and 5-HT in the regulation of secretion of growth hormone (Collu et al. 1972; Conway et al. 1990). These reports, along with the prominent association of 5-HT$_3$ receptors with the modulation of transmitter release and the modulation of transmitter release by presynaptic 5-HT receptors (Göthert 1988), prompted us to investigate the effect of 5-HT on the release of endogenous NA from rat hypothalamic slices.

Materials and Methods

Hypothalamic slices (400 μ thick) from eight-week old, male Sprague-Dawley rats were superfused in a thermostatic chamber (37°C, 300 μl volume), at a rate of 0.5 ml/min with a medium the composition (mM) of which was: NaCl 113; NaHCO$_3$ 25; KCl 4.75; NaH$_2$PO$_4$ 1.18; CaCl$_2$ 2.52; MgSO$_4$ 1.19; glucose 10; nomifensine 0.01; tyrosine 0.05; pargyline 0.01. The medium was prewarmed and continuously gassed with 95% O$_2$ and 5% CO$_2$. In the medium containing 20 mM K$^+$, the increased KCl replaced an equimolar concentration of NaCl. 60 min after the start of superfusion, 1.5 ml fractions were collected into tubes containing 166 μl of a solution of 1N perchloric acid, 10 μM ascorbic acid, and 2.5 pg epinine (internal standard). Slices were twice stimulated by a 3 min exposure to a 20 mM K$^+$-medium at 12 (S$_1$) and 63 min (S$_2$) after the equilibrium period; the superfusate was collected for a total of 78 min. S$_1$ was used as control and the agonists were added to the medium 30 min before S$_2$. Antagonists were added to the medium at the onset of the superfusion. NA was determined by high-performance liquid chromatography-electrochemical detection (Waters, Milford, MA, U.S.A.) as described previously (Blandina et al. 1991).

The levels of NA in the extract were calculated by comparison of sample peak area with external standard peak area, both corrected by the internal standard peak area, and expressed as fmol/mg protein/3 min. Protein was measured by the method of Lowry et al. (1951). Evoked release at the S_1 and S_2 periods of stimulation was calculated as the total minus the spontaneous release. Spontaneous release was obtained by averaging the NA content in the four samples immediately before K^+ stimulation. Drug effects were evaluated by calculating the ratio of the evoked release (S_2/S_1) for the two stimulation periods. All values are expressed as means $\pm$ S.E., and the number of experiments (n) is also indicated. Comparisons between two means were carried out by Student's t test, paired or unpaired as appropriate. When experiments involved more than two treatment groups, the presence of significant treatment effects was first determined by a one-way analysis of variance (ANOVA). For all statistical tests, $p \leq 0.05$ was considered significant.

Results

The mean spontaneous NA release after 60 min perfusion was 171 ± 10 fmol/mg protein/3 min (n = 28). A first superfusion (S_1) for 3 min with medium containing 20 mM K^+ released 290 ± 18 fmol/mg protein/3 min of NA (n = 28). Previous studies have shown that the K^+-evoked release is Ca^{++}-dependent (Blandina et al. 1991). In 4 experiments, a second identical 20 mM K^+ superfusion (S_2), conducted 51 min after the first and with no drugs added to the superfusion medium, elicited a NA release of similar magnitude to S_1 and the mean S_2/S_1 ratio was 1.03 ± 0.08.

Ritanserin (1 μM), an antagonist at $5\text{-HT}_{1C}/5\text{-HT}_2$ receptors but not at 5-HT_3 receptors (Hoyer 1988), failed to modify either spontaneous or 20 mM K^+-evoked NA release as compared to controls. The spontaneous release in the presence of 1 μM ritanserin averaged 174 ± 12 fmol/ mg protein/3 min (n = 36), and the S_1 evoked release averaged 323 ± 24 fmol/mg protein/3 min (n = 36). The presence of ritanserin did not alter the ability to replicate the K^+-evoked release: the mean S_2/S_1 ratio was 0.98 ± 0.02 (n = 8) (Figure 1).

Neither 5-HT nor 2-ME-5-HT, alone or in combination with ritanserin (1 μM) altered spontaneous NA release. 5-HT, 10 μM, inhibited by about 50% the release of NA evoked by 20 mM K^+ in the presence (Figure 1), but not in the absence, of 1 μM ritanserin.

2-ME-5-HT inhibited NA release elicited by 20 mM K^+ in either the absence or the presence of 1 μM ritanserin, but 10 μM 2-ME-5-HT, in the presence of 1 μM ritanserin ($S_2/S_1 = 0.49 \pm 0.04$, n = 9) (Figure 1) was more effective than the same concentration in the absence of ritanserin ($S_2/S_1 = 0.66 \pm 0.04$, n = 6) (data not shown).

α-ME-5-HT (10 μM) did not alter spontaneous or 20 mM K^+-evoked NA release (data not shown). The 5-HT_3 receptor antagonists

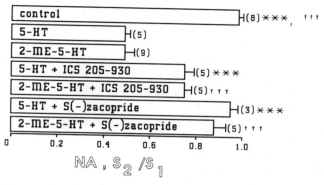

Figure 1. Influence of 5-HT ($10\,\mu M$) and 2-methyl-5-HT (2-ME-5-HT) ($10\,\mu M$) on 20 mM K^+-stimulated release of endogenous NA from hypothalamic slices. The slices were superfused with 20 mM K^+ medium for 3 min, at 12 (S_1) and 63 (S_2) min after the equilibrium period. $1\,\mu M$ ritanserin was present throughout all experiments. When used, ICS 205-930, 1 nM, or S($-$)zacopride, 3 nM, were present throughout the experiment. 5-HT or 2-ME-5-HT were added 30 min before the S_2 stimulation, and remained in the medium during and after the stimulation. Shown are means $\pm$ S.E. with (n). ***$p < 0.001$ vs. 5-HT, †††$p < 0.001$ vs 2-ME-5-HT (analysis of variance and unpaired Student's t test).

ICS 205-930 (1 nM) and S($-$)zacopride (3 nM) (Pinkus et al. 1990), alone or in combination with ritanserin ($1\,\mu M$), failed to alter spontaneous or 20 mM K^+-evoked release of NA. However, both ICS 205-930 (1 nM) and S($-$)zacopride (3 nM) antagonized the inhibition of the K^+-evoked NA release produced by 5-HT ($10\,\mu M$) and 2-ME-5-HT ($10\,\mu M$) (Figure 1).

Discussion

In the presence of ritanserin, which blocks 5-HT$_{1C}$ and 5-HT$_2$ receptors, 5-HT produced a decrease in the amount of endogenous NA released from hypothalamic slices exposed to 20 mM K^+. The 5-HT$_3$ receptor selective agonist, 2-ME-5-HT (Richardson et al. 1985), but not the 5-HT$_1$-like/5-HT$_2$ receptor agonist α-ME-5-HT (Ismaiel et al. 1990), mimicked the effect of 5-HT. The effects of 5-HT and 2-ME-5-HT were antagonized by 1 nM ICS 205-930, a concentration that blocks 5-HT$_3$ mediated responses (Donatsch et al. 1984; Butler et al. 1988; Round and Wallis 1986; Blandina et al. 1988, 1989b). S($-$)zacopride inhibited the effect of $10\,\mu M$ 5-HT and $10\,\mu M$ 2-ME-5-HT at a concentration, 3 nM, that corresponds to its affinity for 5-HT$_3$ receptors and binding sites (Pinkus et al. 1990). This response met the proposed criteria for a 5-HT$_3$ receptor.

Some, but not all, 5-HT$_3$ receptor responses show desensitization (Yakel and Jackson 1988; Todorovíc and Anderson 1990). If desensitization occurred at the beginning of superfusion with the agonists in our experiments, it would not be revealed under these conditions.

$5-HT_3$ receptor activation causes a rapid inward current, which depolarizes the cell membrane (Yakel and Jackson 1988; Todorovíc and Anderson 1990), an effect more likely to be associated with facilitation than inhibition of transmitter release. $5-HT_3$ receptor-mediated inhibition of release of [³H]ACh from cortex (Barnes et al. 1989) and of endogenous NA from hypothalamus may be indirect, mediated by the release of another substance(s) that, in turn, inhibits release of [³H]ACh and NA (see review by Chesselet 1984).

The fact that ritanserin must be present for the inhibitory effect of 5-HT to be manifest suggests that $5-HT_{1C}$ or $5-HT_2$ receptors (Hoyer 1988) also participate in the modulation of the NA release. The difference in the effectiveness of 10 μM 2-ME-5-HT in the presence and in the absence of ritanserin suggests the involvement of the $5-HT_{1C}$ receptor, since 2-ME-5-HT has very low affinity for the $5-HT_2$ receptor (Richardson et al. 1985; Ismaiel et al. 1990). The observation that α-ME-5-HT, a $5-HT_1$-like/$5-HT_2$ receptor agonist, did not significantly increase the spontaneous or the K^+-evoked release of NA suggests that $5-HT_{1C}$ receptor activation does not produce direct stimulatory effects. $5-HT_{1C}$ and $5-HT_3$ receptors might interact through their transducing pathways: phosphorylation of the $5-HT_3$ receptor could be caused by a protein kinase C secondary to $5-HT_{1C}$ receptor activation of phospholipase C (Hoyer et al. 1989). Alternatively, $5-HT_{1C}$ receptor activation could act independently, antagonizing other inhibitory mechanisms.

Because the amounts of endogenous transmitters released are very small, studies on release often involve measuring efflux of radioactive substances from tissues preloaded with exogenous labelled transmitters. It is assumed that release of exogenous transmitter represents or reflects release of endogenous transmitter. Indeed endogenous and preloaded radiolabelled transmitter release share some properties: both are released in a Ca^{++}-dependent manner by veratridine, $13-26$ mM K^+, and electrical stimulation (Chesselet 1984), and, at least for endogenous and exogenous DA release from rat striatal slices, the control by autoreceptors appears to be similar (Herdon and Nahorski 1987).

Despite these similarities, there is evidence that the release of exogenous transmitters does not always reflect release of endogenous transmitters. The K^+-evoked release of endogenous 5-HT from rat cortical slices was inhibited by concentrations of GABA 1000-times lower (Gray and Green 1987) than those reported to inhibit the K^+-evoked release of preloaded [³H]5-HT (Bowery et al. 1980; Schlicker et al. 1984). Histamine produced a greater increase in release of endogenous NA (Blandina et al. 1989a) than of preloaded [³H]NA (Subramanian and Mulder 1977) from hypothalamic slices. Two successive 50 mM K^+ stimuli (Herdon et al. 1985) or electrical field stimulations (Herdon and

Nahorski 1987) released similar amounts of endogenous DA from rat striatal slices, but the second stimulus released 50% less preloaded [^{3}H]DA than the first. In superfused rat brain slices the releasing effect of amphetamine was 3–4 times greater on endogenous DA than on preloaded [^{3}H]DA (Herdon et al. 1985). Amphetamine-induced release of endogenous NA and DA was compared to release of preloaded [^{3}H]NA and [^{3}H]DA in rat cerebral cortex: at concentrations up to 10 μM, amphetamine was slightly more effective in releasing the [^{3}H]catecholamines, but concentrations of amphetamine ranging from 100 to 1000 μM released greater amounts of endogenous than exogenous DA. Moreover, the basal release of the [^{3}H]catecholamines represented a percentage of the tissue levels 15–20 times higher than that of the endogenous amines (Arnold et al. 1977). 2-ME-5-HT produced a Ca^{++}-dependent release of endogenous DA from rat striatal slices (Blandina et al. 1989b) but failed to release preloaded [^{3}H]DA (Schmidt and Black 1989). EGTA, in a medium free of Ca^{++} and Mg^{++}, increased the release of endogenous DA (Okada et al., 1990) but not of preloaded [^{3}H]DA (Arias et al., 1984) from rat brain synaptosomes. In these experiments (and others not cited) the release of endogenous transmitters is clearly modulated differently from that of exogenous transmitters.

Acknowledgements

The work was supported by grants (DA01875 and DA04507) from the National Institute on Drug Abuse. The authors thank J. Walcott for technical assistance, Sandoz Ltd. for gifts of α-ME-5-HT, 2-ME-5-HT and ICS 205-930, Hoechst-Roussel for the gift of nomifensine, Janssen for the gift of ritanserin, and A. H. Robins for the gift of S(−)zacopride.

References

Arias, C., Sitges, M., and Tapia, R. (1984). Stimulation of [^{3}H]γ-aminobutyric acid release by calcium chelators in synaptosomes. J. Neurochem. 42: 507–514.

Arnold, E. A., Molinoff, P. B., and Rutledge, C. O. (1977). The release of endogenous norepinephrine and dopamine from cerebral cortex by amphetamine. J. Pharm. Exp. Ther. 202: 544–557.

Barnes, N. M., Costall, B., Ironside, J. W., and Naylor, R. J. (1988). Identification of 5-HT$_3$ recognition sites in human brain tissue using [^{3}H]zacopride. J. Pharm. Pharmac. 40: 668.

Barnes, J. M., Barnes, N. M., Costall, B., Naylor, R. J., and Tyers, M. B. (1989). 5-HT$_3$ receptors mediate inhibition of acetylcholine release in cortical tissue. Nature (London) 338: 762–763.

Blandina, P., Goldfarb, J., and Green J. P. (1988). Activation of a 5-HT$_3$ receptor releases dopamine from rat striatal slice. Eur. J. Pharm. 155: 349–350.

Blandina, P., Knott, P. J., Leung, L. K. H., and Green J. P. (1989a). Stimulation of the histamine H$_2$ receptor in the rat hypothalamus releases endogenous norepinephrine. J. Pharm. Exp. Ther., 249: 44–51.

Blandina, P., Goldfarb, J., Craddock-Royal, B., and Green, J. P. (1989b). Release of engodenous dopamine by stimulation of 5-hydroxytryptamine$_3$ receptors in rat striatum. J. Pharm. Exp. Ther. 251: 803–809.

Blandina, P., Goldfarb, J., Walcott, J., and Green, J. P. (1991). Serotonergic modulation of the release of endogenous norepinephrine from rat hypothalamic slices. J. Pharm. Exp. Ther. 256: 341–347.

Bowery, N. G., Hill, D. R., Hudson, A. L., Doble, A., Middlemiss, D. N., Shaw, J., and Turnbull, M. (1980). (−)Baclofen decreases neurotransmitter release in the mammalian CNS by an action at a novel GABA receptor. Nature (London) 283: 92–94.

Buchheit, K-H., Engel, G., Mutschler, E., and Richardson, B. P. (1985). Study of the contractile effect of 5-hydroxytryptamine (5-HT) in the isolated longitudinal muscle strip from guinea-pig ileum. Naunyn Schmiedeberg's Arch. Pharmac. 329: 36–41.

Butler, A., Hill, J. M., Ireland, S. J., Jordan, C. C., and Tyers, M. B. (1988). Pharmacological properties of GR38032F, a novel antagonist at 5-HT$_3$ receptors. Br. J. Pharm. 94: 397–412.

Chesselet, M. F. (1984). Presynaptic regulation of neurotransmitter release in the brain: facts and hypothesis. Neuroscience 12: 347–375.

Cohen, M. L., Bloomquist, W., Gidda, J. S., and Lacefield, W. (1990). LY277359 maleate: a potent and selective 5-HT$_3$ receptor antagonist without gastroprokinetic activity. J. Pharm. Exp. Ther. 254: 350–355.

Collu, R., Fraschini, F., Visconti, P., and Martini, L. (1972). Adrenergic and serotoninergic control of growth hormone secretion in adult male rats. Endocrinology 90: 1231.

Conway, S., Richardson, L., Speciale, S., Moherek, R., Mauceri, H., and Krulich, L. (1990). Interaction between norepinephrine and serotonin in the neuroendocrine control of growth hormone release in the rat. Endocrinology 126: 1022–1030.

Derkach, V., Surprenant, A., and North, R. A. (1989). 5-HT$_3$ receptors are membrane ion channels. Nature (London) 339: 706–709.

Donatsch, P., Engel, G., Richardson, B. P., and Stadler, P. A. (1984). ICS 205-930: A highly selective and potent antagonist at peripheral neuronal 5-hydroxytryptamine (5-HT) receptors. Br. J. Pharmac. 81: 34P.

Feuerstein, T. J., and Hertting, G. (1988). Serotonin (5-HT) enhances hippocampal noradrenaline (NA) release: Evidence for facilitatory 5-HT receptors within the CNS. Naunyn Schmiedeberg's Arch. Pharmac. 333: 191–197.

Fozard, J. R. (1984). MDL 72222: a potent and highly selective antagonist at neuronal 5-hydroxytryptamine receptors. Naunyn Schmiedeberg's Arch. Pharmac. 326: 36–44.

Fozard, J. R., Mobarok Ali, A. T. M., and Nevgrosh, G. (1979). Blockade of serotonin receptors on autonomic neurones by (−)cocaine and some related compounds. Eur. J. Pharmac. 59: 195–210.

Gaddum, J. H., and Picarelli, Z. P. (1957). Two kinds of tryptamine receptors. Br. J. Pharmac. 12: 323–328.

Göthert, M. (1988). Modulation of transmitter release by presynaptic serotonin receptors. In Herting, G., and Spatz, H.-C. (eds). Modulation of synpatic transmission and plasticity in nervous systems, NATO ASI Series, Vol. H19 Berlin-Heidelberg: Springer, pp. 55–68.

Gray, J., and Green, A. R. (1987). GABA$_B$-receptor mediated inhibition of potassium-evoked release of endogenous 5-hydroxytryptamine from mouse frontal cortex. Br. J. Pharm. 91: 517–522.

Herdon, H., Strupish, J., and Nahorski, S. (1985). Differences between the release of radiolabelled and endogenous dopamine from superfused rat brain slices: effects of depolarizing stimuli, amphetamine and synthesis inhibition. Brain Res. 348: 309–320.

Herdon, H., and Nahorski, S. R. (1987). Comparison between radiolabelled and endogeneous dopamine release from rat striatal slices: effects of electrical field stimulation and regulation by D$_2$-autoreceptors. Naunyn-Schmiedeberg's Arch. Pharm. 335: 238–242.

Hoyer, D. (1988). Molecular pharmacology and biology of 5-HT$_{1C}$ receptors. Trends Pharmacol. Sci. 9: 89–94.

Hoyer, D., Waeber, C., Schoeffter, P., Palacios, J. M., and Dravid A. (1989). 5-HT$_{1C}$ receptor mediated stimulation of inositol phosphate production in pig choroid plexus: A pharmacological characterization. Naunyn Schmiedeberg's Arch. Pharm. 339: 252–258.

Ismaiel, A. M., Titeler, M., Miller, K. J., Smith, T. S., and Glennon, R. A. (1990). 5-HT$_1$ and 5-HT$_2$ binding profiles of serotonergic agents α-methylserotonin and 2-methylserotonin. J. Med. Chem. 33: 755–758.

Ireland, S. J., and Tyers, M. B. (1987). Pharmacological characterization of 5-hydroxytryptamine-induced depolarization of the rat isolated vagus nerve. Br. J. Pharmacol. 90: 229–238.

Kilpatrick, G. J., Jones, B. J., and Tyers, M. B. (1987). Identification and distribution of 5-HT$_3$ receptors in rat brain using radioligand binding. Nature (London) 330: 746–748.

Jiang, I. H., Ashby Jr., C. R., Kasser, R. J., and Wang, R. Y. (1980). The effect of intraventricular administration of the 5-HT$_3$ receptor against 2-methylserotonin on the release of dopamine in the nucleus accumbens: an *in vivo* chronocoulometric study. Brain Res. 513: 156–160.

Jones, B. J., Costall, B., Domeney, A. M., Kelly, M. E., Naylor, R. J., Oakley, N. R., and Tyers, M. B. (1988). The potential anxiolytic activity of GR38032F, a 5-HT$_3$ receptor antagonist. Br. J. Pharmac. 93: 985–993.

Lowry, O. H., Rosebrough, N. J., Farr, A. L., and Randall, R. J. (1985). Protein measurement with the Folin phenol reagent. J. Biol. Chem. 193: 265–275.

Miner, W. D., and Sanger, G. J. (1986). Inhibition of cisplatin-induced vomiting by selective 5-hydroxytryptamine M-receptor antagonism. Br. J. Pharmac. 88: 479–499.

Molineaux, S. M., Jessel, T. M., Axel, R., and Julius, D. (1989). 5-HT$_{1C}$ receptor is a prominent serotonin receptor subtype in the central nervous system. Proc. Natl. Acad. Sci. USA 86: 6793–6797.

Okada, M., Mine, K., and Fujiwara, M. (1990). The Na$^+$-dependent release of endogenous dopamine and noradrenaline from rat brain synaptosomes. J. Pharm. Exp. Ther. 252: 1283–1288.

Pazos, A., and Palacios, J. M. (1985). Quantitative autoradiographic mapping of serotonin receptors in the rat brain. I. Serotonin-1 receptors. Brain Res. 346: 205–230.

Pazos, A., Cortés, R., and Palacios, J. M. (1985). Quantitative autoradiographic mapping of serotonin receptors in the rat brain. II. Serotonin-2 receptors. Brain Res. 346: 231–249.

Pinkus, L. M., Sarbin, N. S., and Gordon, J. C., and Munson Jr., H. R. (1990). Antagonism of [^{3}H]zacopride binding to 5-HT$_3$ recognition sites by its (R) and (S) enantiomers. Eur. J. Pharm. 179: 231–235.

Richardson, B. P., Engel, G., Donatsch, P., and Stadler, P. A. (1985). Identification of 5-hydroxytryptamine M-receptor subtypes and their specific blockade by a new class of drugs. Nature (London) 316: 126–131.

Round, A., and Wallis, D. I. (1986). The depolarizing action of 5-hydroxytryptamine on rabbit vagal afferent and sympathetic neurones in vitro and its selective blockade by ICS 205-930. Br. J. Pharmac. 88: 485–494.

Sanger, G. J. (1987). Increased gut cholinergic activity and antagonism of 5-hydroxytryptamine M-receptor subtypes by BRL 24924: Potential clinical importance of BRL 24924. Br. J. Pharm. 91: 77–87.

Schlicker, E., Classen, K., and Göthert, M. (1984). GABA receptor-mediated inhibition of serotonin release in the rat brain. Naunyn Schmiedeberg's Arch. Pharm. 326: 99–105.

Schmidt, C. J., and Black, C. K. (1989). The putative 5-HT$_3$ agonist phenylbiguanide induces carrier-mediated release of [^{3}H]dopamine. Eur. J. Pharm. 167: 309–310.

Smith, W. L., Sancilio, L. F., Owera-Atepo, J. B., Naylor, R. J., and Lambert, L. (1988). Zacopride: A potent 5-HT$_3$ antagonist. J. Pharm. Pharmacol. 40: 301–302.

Subramanian, N., and Mulder, A. H. (1967). Modulation by histamine of the efflux of radiolabeled catecholamines from rat brain slices. Eur. J. Pharm. 43: 143–152.

Todorovíc, S., and Anderson, E. G. (1990). 5-HT$_2$ and 5-HT$_3$ receptors mediate two distinct depolarizing responses in rat dorsol root ganglion neurons. Brain Res. 511: 71–79.

Yakel, J. L., and Jackson, M. B. (1988). 5-HT$_3$ receptors mediate rapid responses in cultured hippocampus and a clonal cell line. Neuron 1: 615–621.

Serotonin: Molecular Biology, Receptors and Functional Effects
ed. by J. R. Fozard/P. R. Saxena
© 1991 Birkhäuser Verlag Basel/Switzerland

Antagonism of Serotonin Agonist-Elicited Increases in Serum Corticosterone Concentration in Rats

R. W. Fuller

Lilly Research Laboratories, Eli Lilly and Company, Lilly Corporate Center, Indianapolis, IN 46285, USA

Summary. Serum corticosterone and adrenocorticotropin (ACTH) concentrations are increased in rats by numerous direct-acting and indirect-acting serotonin receptor agonists. Antagonism of these increases by serotonin receptor antagonists has been useful in verifying their serotonergic nature and in defining central vs. peripheral sites of action. In addition, the use of agonists and antagonists with selectivity toward particular subtypes of serotonin receptors helps elucidate the identity of receptors that mediate agonist-elicited increases in serum corticosterone concentration. At least two serotonin receptor subtypes, 5-HT_{1A} and $5\text{-HT}_{2/1C}$ receptors, appear capable of mediating increases in serum corticosterone concentrations. The use of 8-hydroxy-2-(di-*n*-propylamino)tetralin (8-OH-DPAT) and quipazine as agonists at these receptors, respectively, provides a convenient means of assessing central serotonin receptor-blocking efficacy of newly developed serotonin receptor antagonists in rats. Data from studies in humans are beginning to appear, suggesting that neuroendocrine markers in humans will also be useful in characterizing serotonin receptor antagonists clinically.

Introduction

The pituitary-adrenocortical axis in rats is activated by various direct-acting and indirect-acting serotonin receptor agonists, the latter group of drugs including the serotonin precursor L-5-hydroxytryptophan (Popova et al. 1972; Fuller et al. 1976), serotonin uptake inhibitors such as fluoxetine (Fuller et al. 1976) and serotonin-releasing drugs such as p-chloroamphetamine (Fuller and Snoddy 1980). These effects are probably initiated by activation of central serotonin receptors leading to CRF (corticotropin-releasing factor) release, inasmuch as CRF is released from isolated rat hypothalamus *in vitro* by these serotonergic agents (Calogero et al. 1989), and synaptic contacts between serotonergic terminals and CRF-containing neurons in the paraventricular nucleus of rat hypothalamus have been demonstrated (Liposits et al. 1987). Direct and indirect serotonin receptor agonists have been shown to increase not only serum corticosterone concentration (Fuller and Snoddy 1990) but also serum ACTH concentration (Gartside and Cowen 1990; Gilbert et al. 1988) and CRF concentration in portal blood to the pituitary gland (Gibbs and Vale 1983) *in vivo* in rats.

Which of the multiple serotonin receptor subtypes that have been identified in brain (Peroutka, 1988; Frazer et al. 1990) mediate(s) pituitary-adrenocortical activation by serotonin agonists is a topic of current investigation. One means of verifying the involvement of serotonin receptors in the activation of pituitary-adrenocortical function by serotonin receptor agonists in rats is through the use of serotonin receptor antagonists, and if the antagonists chosen discriminate among serotonin receptor subtypes they provide a useful way in investigating which receptors mediate the effects of each agonist.

Enhanced Pituitary-Adrenocortical Function in Rats

Many direct- and indirect-acting serotonin receptor agonists increase serum corticosterone concentration in rats (e.g., Fuller 1981; Koenig et al. 1988; Nash et al. 1988a; Bagdy et al. 1989; Calogero et al. 1990; King et al. 1989; Fuller and Snoddy 1990). Figure 1 shows the increase in serum corticosterone in rats by two direct-acting serotonin agonists; the increases elicited by 8-hydroxy-DPAT and quipazine were dose-dependent (Figure 1A). Although antagonist studies to be described indicate that separate receptor subtypes mediate the corticosterone increase by these two agonists, their effects were not additive when maximally effective doses of the two agonists were combined (Figure 1B).

Antagonism of Agonist Effects in Rats

Figure 2 shows the dose-dependent antagonism by spiperone of the quipazine-elicited increase in serum corticosterone concentration in rats. The ED_{50} value (dose that would antagonize the increase by 50%) was calculated from these data to be 1.5 mg/kg. The ED_{50} values for numerous antagonists have been calculated in this way, and the order of potency among selected antagonists was metergoline > pirenperone > setoperone > methiothepin > mianserin > ketanserin > ritanserin > spiperone > clozapine > trazodone. Compounds that were not effective in blocking the quipazine-induced increase in serum corticosterone concentration at the doses tested included fluphenazine, perphenazine, haloperidol, ICS 205-930 and MDL 72222. Some of the antagonists that were effective have selective affinity for $5\text{-}HT_2$ receptors, implicating the $5\text{-}HT_2$ receptor as a mediator of this effect of quipazine. Although distinguishing between $5\text{-}HT_2$ receptors and $5\text{-}HT_{1C}$ receptors is difficult, the efficacy of spiperone, which has low affinity for $5\text{-}HT_{1C}$ receptors (Hoyer 1988; Peroutka 1988), is one finding that favors the involvement of $5\text{-}HT_2$ receptors instead of $5\text{-}HT_{1C}$ receptors in the serum corticosterone-increasing effect of quipazine.

332

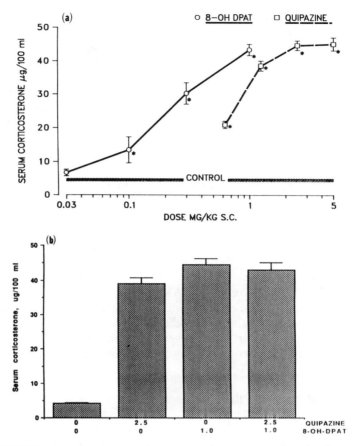

Figure 1.(a) Dose-dependent increases in serum corticosterone concentration in rats elicited by 8-OH-DPAT (○) or by quipazine (□). (b) Lack of additivity of the corticosterone increases elicited by maximally effective doses of 8-OH-DPAT and quipazine. The mg/kg s.c. doses of quipazine and 8-OH-DPAT are shown below each column. In both graphs, mean values ± standard errors for 5 rats per group are shown, measurements being made 1 hr after the s.c. injection of 8-OH-DPAT or quipazine. In Figure 1(b), the doses were 1.0 mg/kg s.c. for 8-OH-DPAT and 2.5 mg/kg s.c. for quipazine. Serum corticosterone concentration was measured spectrofluorometrically (Solem and Brinck-Johnsen, 1965). The data in Figure 1(a) are from Fuller and Snoddy (1990).

In contrast to quipazine, 8-OH-DPAT increases serum corticosterone concentration in a manner that is not sensitive to antagonism by metergoline (Figure 3). The ED_{50} value for metergoline in antagonizing the quipazine-induced increase in serum corticosterone concentration was 0.03 mg/kg i.p. (Fuller and Snoddy 1979), and even at a dose 100 times higher, metergoline did not antagonize the effect of 8-OH-DPAT (Figure 3). The effect of 8-OH-DPAT, a serotonin receptor agonist with high and selective affinity for the $5\text{-}HT_{1A}$ receptor subtype (Hoyer 1988;

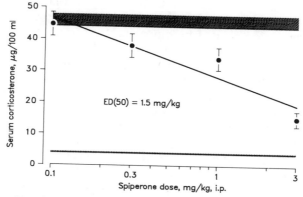

Figure 2. Dose-dependent antagonism by spiperone of the quipazine-induced increase in serum corticosterone concentration in rats. Quipazine maleate (10 mg/kg i.p.) was injected 1 hr before rats were killed and 1 hr after spiperone, which was injected i.p. at the doses indicated. The horizontal shaded areas indicate mean ± standard error ranges for vehicle-treated rats (bottom) and rats receiving quipazine alone (top). Mean values ± standard errors for 5 rats per group are shown in all cases.

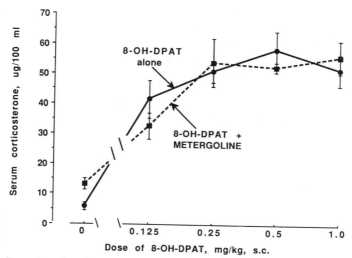

Figure 3. Dose-dependent increase in serum corticosterone concentration in rats elicited by 8-OH-DPAT in metergoline-pretreated and in non-pretreated rats. 8-Hydroxy-DPAT was injected s.c. 1 hr before rats were killed and 1 hr after an injection of metergoline (3 mg/kg i.p.) or water. Mean values ± standard errors for 5 rats per group are shown.

Peroutka 1988), is blocked by drugs such as pindolol and penbutolol, which have highest affinity for the 5-HT$_{1A}$ receptor among serotonin receptor subtypes (Koenig et al. 1987; Gilbert et al. 1988; Przegalinski et al. 1989; Fuller and Snoddy 1990). Thus strong evidence exists that 5-HT$_{1A}$ receptor activation mediates the increase in serum corticosterone concentration elicited by 8-OH-DPAT and that 5-HT$_2$ (or possibly 5-HT$_{1C}$) receptor activation mediates the increase in serum corticosterone concentration elicited by quipazine.

The involvement of specific receptor subtypes in corticosterone-increasing effects of other serotonin agonists has also been investigated. Buspirone, gepirone and ipsapirone are 5-HT$_{1A}$ receptor agonists that increase plasma ACTH (Gilbert et al. 1988) and corticosterone (Koenig et al. 1988) concentrations in rats as 8-OH-DPAT does. The effects of gepirone and ipsapirone are counteracted by 5-HT$_{1A}$ receptor antagonists (Gilbert et al. 1988; Koenig et al. 1988). Some investigators have reported antagonism of buspirone effects by 5-HT$_{1A}$ receptor antagonists (Gilbert et al. 1988), but others have found pindolol not to antagonize corticosterone elevation by buspirone (Koenig et al. 1988; R. W. Fuller and H. D. Snoddy, unpublished data). Buspirone affects dopaminergic neurons and is metabolized to 1-(2-pyrimidinyl)piperazine, an α_2 receptor antagonist (Fuller and Perry 1989) which can increase plasma corticosterone in rats (Matheson et al. 1989); additional mechanisms besides 5-HT$_{1A}$ receptor activation may be involved in enhancement of pituitary-adrenocortical function by buspirone.

Increases in serum corticosterone concentration elicited by MK-212 in rats are blocked by clozapine, melperone, setoperone, and LY-53857, suggesting an involvement of 5-HT$_2$ receptors (Lorens and Van de Kar 1987; Nash et al. 1988b). On the other hand, King et al. (1989) have argued that 5-HT$_{1C}$ receptors mediate the MK-212-elicited increases in rat plasma ACTH concentration based on blockade of the effects by mesulergine and metergoline but not spiperone or ketanserin (at doses up to 1.25 mg/kg).

Use in Characterizing Serotonin Receptor Antagonists

The antagonism of agonist-elicited increases in serum ACTH or corticosterone concentration is a useful assay in the development of serotonin receptor antagonists. Such antagonism appears to measure the blockade of serotonin receptors within the central nervous system (Fuller et al. 1986; Haleem et al. 1989). The robustness of the agonist-elicited increases facilitates determination of ED$_{50}$ values for antagonists so that quantitative comparison among compounds is possible.

Drugs used in the treatment of schizophrenia block dopamine D$_2$ receptors. There is interest currently in drugs that block not only dopamine D$_2$ receptors but also serotonin 5-HT$_2$ receptors. Clozapine has that characteristic, and some researchers believe this dual action may be related to the relative lack of acute and long-term extrapyramidal side effects associated with clozapine and that blockade of 5-HT$_2$ receptors

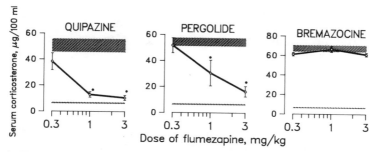

Figure 4. Flumezapine antagonizes the effects of quipazine and pergolide but not bremazocine on serum corticosterone concentration in rats. Quipazine maleate (10 mg/kg i.p.), pergolide mesylate (0.3 mg/kg i.p.) or bremazocine (0.1 mg/kg s.c.) was injected 1 hr before rats were killed and 1 hr after treatment with flumezapine. Mean values ± standard errors for 5 rats per group are shown. The data are from Fuller and Mason (1986).

helps in the treatment of negative symptoms of schizophrenia (Meltzer 1989). In rats, dopamine D_2 receptor agonists increase serum ACTH and corticosterone concentrations (Fuller and Snoddy 1984). Antagonism of the same endocrine response to different agonists, namely the quipazine-induced and the pergolide-induced increases in serum ACTH or corticosterone in rats, can be used in characterizing compounds that block 5-HT_2 and D_2 receptors. Figure 4 shows that flumezapine antagonizes both the quipazine-induced and pergolide-induced increases in serum corticosterone concentrations with ED_{50} values of 0.48 and 0.93 mg/kg i.p., respectively (Fuller and Mason 1986). Flumezapine does not impair pituitary-adrenocortical responsivity, as shown by its lack of effect on serum corticosterone increases elicited by bremazocine, a *kappa* opioid agonist (Fuller and Leander 1984). The antagonism of quipazine and pergolide effects by flumezapine can therefore be attributed to block of 5-HT_2 and D_2 receptors, respectively.

Antagonism of Agonist Effects in Humans

m-Chlorophenylpiperazine, a direct-acting serotonin agonist that is a metabolite of the antidepressant drug trazodone, is being used as a pharmacologic probe in humans to evaluate functionality of central serotonin receptors. It elicits increases in plasma ACTH and cortisol, among other changes, thought to be mediated by activation of serotonin receptors in brain. Mueller et al. (1986) reported that metergoline blocked the m-chlorophenylpiperazine-induced increases in plasma hormones in humans, strengthening the notion that serotonin receptors mediate the effect. Neither m-chlorophenylpiperazine nor metergoline has high selectivity among serotonin receptor subtypes (Hoyer 1988), so the identity of the specific receptors that mediate these effects of m-chlorophenylpiperazine in humans remains unclear.

MK 212, a centrally active serotonin agonist, increases serum cortisol in humans (Lowy and Meltzer 1988), and its effects are blocked by clozapine (Meltzer et al. 1987), supporting the idea that clozapine can block 5-HT$_2$ receptors at clinically used doses. Lesch et al. (1990) showed that the ipsapirone-induced increase in plasma ACTH and cortisol concentrations in humans was blocked by ($\pm$)pindolol. Because the 5-HT$_{1A}$ receptor is the only known serotonin receptor for which ipsapirone and pindolol share a common high affinity, these results implicate the activation of 5-HT$_{1A}$ receptors in ACTH and cortisol responses to ipsapirone in humans. As more specific agonists and antagonists become available for study in humans, the identification of serotonin receptor subtypes that can affect pituitary-adrenocortical function in this species can be better understood.

Acknowledgement

I am grateful to Harold D. Snoddy for producing all the experimental results shown in this paper.

References

Bagdy, G., Calogero, A. E., Murphy, D. L., and Szemeredi, K. (1989). Serotonin agonists cause parallel activation of the sympathoadreno-medullary system and the hypothalamo-pituitary-adrenocortical axis in conscious rats. Endocrinology 125: 2664–2669.

Calogero, A. E., Bagdy, G., Szemeredi, K., Tartaglia, M. E., Gold, P. W., and Chrousos, G. P. (1990). Mechanisms of serotonin receptor agonist-induced activation of the hypothalamic-pituitary-adrenal axis in the rat. Endocrinology 126: 1888–1894.

Calogero, A. E., Bernardini, R., Margioris, A. N., Bagdy, G., Gallucci, W. T., Munson, P. J., Tamarkin, L., Tomai, T. P., Brady, L., Gold, P. W., and Chrousos, G. P. (1989). Effects of serotonergic agonists and antagonists on corticotropin-releasing hormone secretion by explanted rat hypothalamus. Peptides 10: 189–200.

Frazer, A., Maayani, S., and Wolfe, B. B. (1990). Subtypes of receptors for serotonin. Ann. Rev. Pharmacol. Toxicol. 30: 307–348.

Fuller, R. W. (1981). Serotonergic stimulation of pituitary-adrenocortical function in rats. Neuroendocrinology 32: 118–127.

Fuller, R. W., Kurz, K. D., Mason, N. R., and Cohen, M. L. (1986). Antagonism of a peripheral vascular but not an apparently central serotonergic response by xylamidine and BW 501C67. Eur. J. Pharmacol. 125: 71–77.

Fuller, R. W., and Leander, J. D. (1984). Elevation of serum corticosterone in rats by bremazocine, a K-opioid agonist. J. Pharm. Pharmacol. 36: 345–346.

Fuller, R. W., and Mason, N. R. (1986). Flumezapine, an antagonist of central dopamine and serotonin receptors. Res. Commun. Chem. Pathol. Pharmacol. 54: 23–34.

Fuller, R. W., and Perry, K. W. (1989). Effects of buspirone and its metabolite, 1-(2-pyrimidinyl)piperazine, on brain monoamines and their metabolites in rats. J. Pharmacol. Exp. Ther. 248: 50–56.

Fuller, R. W., and Snoddy, H. D. (1984). Central dopamine receptors mediating pergolide-induced elevation of serum corticosterone in rats. Characterization by the use of antagonists. Neuropharmacology 23: 1389–1394.

Fuller, R. W., and Snoddy, H. D. (1980). Effect of serotonin-releasing drugs on serum corticosterone concentration in rats. Neuroendocrinology 31: 96–100.

Fuller, R. W., and Snoddy, H. D. (1979). The effects of metergoline and other serotonin receptor antagonists on serum corticosterone in rats. Endocrinology 105: 923–928.

Fuller, R. W., and Snoddy, H. D. (1990). Serotonin receptor subtypes involved in the elevation of serum corticosterone concentration in rats by direct- and indirect-acting serotonin agonists. Neuroendocrinology 52: 206–211.

Fuller, R. W., Snoddy, H. D., and Molloy, B. B. (1976). Pharmacologic evidence for a serotonin neural pathway involved in hypothalamus-pituitary-adrenal function in rats. Life Sci. 19: 337–346.

Gartside, S. E., and Cowen, P. J. (1990). Mediation of ACTH and prolactin responses to 5-HT by 5-HT$_2$ receptors. Eur. J. Pharmacol. 179: 103–109.

Gibbs, D. M., and Vale, W. (1983). Effect of the serotonin reuptake inhibitor fluoxetine on corticotropin-releasing factor and vasopressin secretion into hypophysial portal blood. Brain Res. 280: 176–179.

Gilbert, F., Brazell, C., Tricklebank, M. D., and Stahl, S. M. (1988). Activation of the 5-HT$_{1A}$ receptor subtype increases rat plasma ACTH concentration. Eur. J. Pharmacol. 147: 431–439.

Haleem, D. J., Kennett, G. A., Whitton, P. S., and Curzon, G. (1989). 8-OH-DPAT increases corticosterone but no other 5-HT$_{1A}$ receptor-dependent responses more in females. Eur. J. Pharmacol. 164: 435–443.

Hoyer, D. (1988). Functional correlates of serotonin 5-HT$_1$ recognition sites. J. Receptor Res. 8: 59–81.

King, B. H., Brazell, C., Dourish, C. T., and Middlemiss, D. N. (1989). MK-212 increases rat plasma ACTH concentration by activation of the 5-HT$_{1C}$ receptor subtype. Neurosci. Lett. 105: 174–176.

Koenig, J. I., Gudelsky, G. A., and Meltzer, H. Y. (1987). Stimulation of corticosterone and β-endorphin secretion in the rat by selective 5-HT receptor subtype activation. Eur. J. Pharmacol. 137: 1–8.

Koenig, J. I., Meltzer, H. Y., and Gudelsky, G. A. (1988). 5-Hydroxytryptamine$_{1A}$ receptor-mediated effects of buspirone, gepirone and ipsapirone. Pharmacol. Biochem. Behav. 29: 711–715.

Lesch, K.-P., Sohnle, K., Poten, B., Schoellnhammer, G., Rupprecht, R., and Schulte, H. M. (1990). Corticotropin and cortisol secretion after central 5-hydroxytryptamine-1A (5-HT$_{1A}$) receptor activation: effects of 5-HT receptor and β-adrenoceptor antagonists. J. Clin. Endocrinol. Metab. 70: 670–674.

Liposits, Zs., Phelix, C., and Paull, W. K. (1987). Synaptic interaction of serotonergic axons and corticotropin releasing factor (CRF) synthesizing neurons in the hypothalamic paraventricular nucleus of the rat. A light and electron microscopic immunocytochemical study. Histochemistry 86: 541–549.

Lorens, S. A., and Van de Kar, L. D. (1987). Differential effects of serotonin (5-HT$_{1A}$ and 5-HT$_2$) agonists and antagonists on renin and corticosterone secretion. Neuroendocrinology 45: 305–310.

Lowy, M. T., and Meltzer, H. Y. (1988). Stimulation of serum cortisol and prolactin secretion in humans by MK-212, a centrally active serotonin agonist. Biol. Psychiat. 23: 818–828.

Matheson, G. K., Gage-White, D., White, G., Guthrie, D., Rhoades, J., and Dixon, V. (1989). The effects of gepirone and 1-(2-pyrimidinyl)-piperazine on levels of corticosterone in rat plasma. Neuropharmacology 28: 329–334.

Meltzer, H. Y. (1989). Clinical studies on the mechanism of action of clozapine: the dopamine-serotonin hypothesis of schizophrenia. Psychopharmacology 99: S18–S27.

Meltzer, H. Y., Bastani, B., Kwon, K., Ramirez, L., and Nash, J. F. (1987). Clozapine as a 5-HT$_2$ antagonist in man. Soc. Neurosci. Abstr. 13: 800.

Mueller, E. A., Murphy, D. L., and Sunderland, T. (1986). Further studies of the putative serotonin agonist, m-chlorophenylpiperazine: Evidence for a serotonin receptor mediated mechanism of action in humans. Psychopharmacology 89: 388–391.

Nash, J. F., Meltzer, H. Y., and Gudelsky, G. A. (1988a). Antagonism of serotonin receptor mediated neuroendocrine and temperature responses by atypical neuroleptics in the rat. Eur. J. Pharmacol. 151: 463–469.

338

Nash, J. F., Meltzer, H. Y., and Gudelsky, G. A. (1988b). Elevation of serum prolactin and corticosterone concentrations in the rat after the administration of 3,4-methylenedioxymethamphetamine. J. Pharmacol. Exp. Ther. 245: 873–879.

Peroutka, S. J. (1988). Serotonin receptor subtypes. ISI Atlas of Science, pp. 1–4.

Popova, N. K., Maslova, L. N., and Naumenko, E. V. (1972). Serotonin and the regulation of the pituitary-adrenal system after deafferentation of the hypothalamus. Brain Res. 47: 61–67.

Przegalinski, E., Budziszewska, B., Warchol-Kania, A., and Blaszczynska, E. (1989). Stimulation of corticosterone secretion by the selective 5-HT$_{1A}$ receptor agonist 8-hydroxy-2-(di-n-propylamino)tetralin (8-OH-DPAT) in the rat. Pharmacol. Biochem. Behav. 33: 329–334.

Solem, J. H., and Brinck-Johnsen, T. (1965). An evaluation of a method for determination of free corticosteroids in minute quantities of mouse plasma. Scand. J. Clin. Invest. (Suppl. 80) 17: 1–14.

Serotonin: Molecular Biology, Receptors and Functional Effects
ed. by J. R. Fozard/P. R. Saxena
© 1991 Birkhäuser Verlag Basel/Switzerland

Influence of 5-HT$_{1A}$ and 5-HT$_2$ Receptor Agonists on Blood Glucose and Insulin Levels

F. Chaouloff, V. Baudrie, and D. Laude

Laboratoire de Pharmacologie, Groupe de Neuropharmacologie, CNRS, CHU Necker-E.M., 156 rue de Vaugirard, 75015 Paris, France

Summary. It is now 20 years that the role of central 5-HT in the regulation of circulating glucose and insulin levels has been a matter of interest. Nowadays, the possibility to investigate the metabolic influence of ligands thought to be selective for 5-HT receptor subtypes provides new insights in the relationships between 5-HT and the energetic metabolism. Then, the purpose of this paper is to review the respective consequences of 5-HT$_{1A}$ or 5-HT$_2$ receptor activation (by compounds such as 8-OH-DPAT, buspirone, ipsapirone, diisopropyl-5-carboxytryptamine, RU 24969, DOI or alpha-methyl-5-HT) on blood glucose and insulin levels, with special reference to the tight relationships between the serotonergic system and the sympathoadrenal system. Lastly, the putative relevance of these findings to physiology will be discussed.

Introduction

Since the early seventies, the hypothesis for a control of circulating glucose and insulin levels by serotonin (5-HT) systems has been the matter of numerous works. However, twenty years ago, tryptophan and 5-hydroxytryptopan (i.e. 5-HT precursors), monoamine oxidase inhibitors, 5-HT uptake inhibitors, unspecific 5-HT receptor antagonists, and 5-HT itself were the sole appropriate tools to investigate the above hypothesis. Logically, the use of these unspecific compounds (and different animal species) has led to conflicting conclusions (Gagliardino et al. 1971; Furman 1974; Wong and Tyce 1978). Nowadays, the characterization of the various 5-HT receptor types and subtypes, and the development of increasingly selective 5-HT receptor agonists and antagonists, permit to further define the relationships between the serotonergic system and metabolism. Thus, we have already shown that the i.v. administration of the selective 5-HT$_{1A}$ receptor agonist 8-hydroxy-2-(di-n-propylamino)tetralin (8-OH-DPAT) triggers hyperglycemia and inhibition of insulin release (Chaouloff and Jeanrenaud 1987). Pharmacological analyses allowed us to suggest the involvement of both 5-HT$_{1A}$ and alpha$_2$-adrenoceptors in the metabolic effects of 8-OH-DPAT (Chaouloff and Jeanrenaud 1987). Other experiments with adrenalectomised rats and with rats which were pretreated either with a catecholamine synthesis inhibitor or a ganglionic blocker suggested that adrenomedullary catecholamine release mediated the 8-OH-DPAT-induced hyperglycemia (Chaouloff et al. 1990a). Accordingly, we and others reported that 8-OH-DPAT dose-dependently increases plasma adrenaline levels (Bagdy et al. 1989; Chaouloff et al. 1990a). In view of the marked sympathoinhibitory (and vagoexcitatory) effects of 5-HT$_{1A}$ receptor agonists (Ramage and Fozard 1987), the adrenaline-releasing effects of 8-OH-DPAT appeared to be somewhat surprising. On that basis, we have investigated whether the adrenaline-releasing effect of 8-OH-DPAT was due to specific 5-HT$_{1A}$ receptor stimulation. Indeed, appropriate pharmacological tools led us to conclude that 5-HT$_{1A}$ receptors mediate the adrenaline-releasing and hyperglycemic effects of 8-OH-DPAT (Chaouloff et al. 1990b). Furthermore, the 5-HT$_{1A}$ receptor agonists buspirone and ipsapirone were found to increase in a dose-dependent manner plasma adrenaline and glucose levels (Chaouloff et al. 1990c). The above elements indicate that 5-HT$_{1A}$ receptor agonists have differential effects on the sympathetic nervous system. Thus, these compounds inhibit sympathetic activity, while they activate adrenomedullary catecholamine release. Of interest is the finding that few physiological situations have already been reported to have such differential consequences upon sympathoadrenal activity (for a review, Landsberg and Krieger 1989).

Then, the purpose of the following studies was to further define the relationships between the serotonergic and the sympathetic systems, and to evaluate their consequences on plasma glucose and insulin levels.

Materials and Methods

Male rats (IFFA CREDO, Les Oncins, France), weighing 250–300 g, were used in the following studies. Rats were anaesthetised for 30 min by means of methohexital (40 mg/kg i.p.; Brietal, Lilly, France), and implanted with an intracardiac catheter through the right jugular vein (Chaouloff et al. 1990b). The animals were used 5–6 h after awakening. All the drugs were given i.v., the volumes of injection being 1 or 2 ml/kg. Plasma adrenaline levels were assessed by HPLC coupled with an electrochemical detection, following adsorption onto aluminium oxide. Plasma glucose and corticosterone levels were respectively measured by means of an automated glucose analyser and radioimmunoassays.

Results and Discussion

5-HT$_{1A}$ Receptors, Adrenaline Release, and Blood Glucose Levels

The first series of experiments were devoted to the influence of ganglionic transmission upon the adrenaline-releasing and hyperglycemic effects of 8-OH-DPAT (0.25 mg/kg; BioBlock, Illkirch, France). For that purpose, adrenal acetylcholine release was either impaired by pentobarbital anaesthesia, or the adrenomedullary transmission blocked by the nicotinic receptor antagonist hexamethonium.

Pentobarbital anaesthesia was found to blunt both the adrenaline-releasing and hyperglycemic effects of 8-OH-DPAT (Chaouloff et al. 1990d). This result is interesting for many reasons; firstly, it indicates that a procedure that is commonly used in pharmacological studies may directly affect the neuroendocrinological consequences of 5-HT_{1A} receptor activation. Secondly, it suggests that ganglionic transmission is a prerequisite for 8-OH-DPAT-induced adrenaline release (as confirmed later in hexamethonium-pretreated rats; see below). Thirdly, it brings a new insight in the relationships between the serotonergic system and blood pressure. Thus, the administration of 8-OH-DPAT in conscious rats triggers a rapid and transient increase in blood pressure, followed by a long-lasting hypotension (Fozard et al. 1987). Whereas the latter is due to the sympathoinhibitory effect of 8-OH-DPAT, the mechanisms responsible for the initial increase in blood pressure are still not fully clarified. Interestingly, the latter increase in blood pressure is abolished in pentobarbital anaesthetised animals (Fozard et al. 1987), strongly suggesting that the adrenaline-releasing effect of 8-OH-DPAT is responsible for the initial increase in blood pressure. Our suggestion has been recently confirmed in adrenalectomised rats (Bouhelal and Mir 1990).

A more direct evidence for adrenal preganglionic acetylcholine release being a key mechanism in 8-OH-DPAT-induced adrenaline release (and hyperglycemia) was provided by the use of nicotinic blockade. Thus, in animals pretreated with hexamethonium (5 or 10 mg/kg; Sigma, Paris, France), both the adrenaline-releasing and hyperglycemic effects of 8-OH-DPAT were dose-dependently diminished (Chaouloff et al. 1990e). Moreover, the corticosterone-releasing effects of 8-OH-DPAT (Koenig et al. 1987) were also diminished by prior hexamethonium treatment (Chaouloff et al. 1990e). This could indicate – as reported in the ether stress model (Rivier and Vale 1983) – that adrenaline release is responsible for the high corticosterone levels measured in 8-OH-DPAT-treated rats. Nonetheless, the significant increase in basal corticosterone levels in hexamethonium-pretreated rats (Chaouloff et al. 1990e) impedes any definitive conclusion.

The results reported above clearly indicated that a 5-HT_{1A} receptor activated, either directly or indirectly, acetylcholine release from preganglionic nerves. The question which remained was that concerning the location of this receptor. Is this receptor located in the adrenal gland, or in the spinal cord (i.e. are we dealing with a peripheral or a central 5-HT_{1A} receptor)? In a study led in collaboration with Dr G. R. Martin (Wellcome Research Labs, Beckenham, UK), we have measured plasma adrenaline and glucose levels in rats which were injected with diisopropyl-5-carboxamidotryptamine (DP-5-CT; Wellcome Research Labs, Beckenham, UK). This 5-HT$_{1A}$ receptor agonist is believed to be unable to cross the blood-brain barrier (Mir et al. 1987; Doods et al. 1988). However, due to the high affinity of DP-5-CT for another 5-HT receptor, namely the 5-HT$_{1D}$ receptor (Hoyer 1988), we have also analysed the respective influences of the 5-HT$_{1B}$/5-HT$_{1D}$ receptor agonist CGS 12066B (Hoyer 1988), and of the 5-HT$_{1A}$/5-HT$_{1B}$/5-HT$_{1D}$ receptor agonist RU 24969 (Hoyer 1988). Whereas RU 24969 (Roussel-Uclaf, Romainville, France) increased dose-dependently (0.5–4.5 mg/kg) plasma adrenaline and glucose levels, the latter remained unaffected by CGS 12066B (1.5 and 4.5 mg/kg; BioBlock, Illkirch, France) (Laude et al. 1990). At doses already shown to be effective in peripheral models (0.1 and 0.3 mg/kg), DP-5-CT was devoid of any adrenaline-releasing property, thus indicating that peripheral 5-HT$_{1A}$ receptors, if any, do not control adrenaline release (Laude et al. 1990). Moreover, that an extremely high dose of DP-5-CT (1 mg/kg) triggers adrenaline release (Laude et al. 1990), probably indicates that at this dose, a significant number of molecules enters the central nervous system.

5-HT$_2$ Receptors, Sympathetic Activity, and Blood Glucose and Insulin Levels

The activation of 5-HT$_2$ receptors elicits sympathoexcitation (McCall and Harris 1988) and adrenal catecholamine release (Bagdy et al. 1989). Thus, we have analysed the consequences of such relationships between 5-HT$_2$ receptors and the sympathetic nervous system upon blood glucose and insulin levels. For that purpose, we have used the 5-HT$_{1C}$/5-HT$_2$ receptor agonist DOI (Glennon 1987), and we have added a pharmacological analysis to assess the identity of the receptors involved in DOI metabolic effects. Our results indicate that DOI (BioBlock, Illkirch, France) dose-dependently (0.125-2 mg/kg) increases plasma glucose levels without affecting those of insulin (thus suggesting inhibition of insulin release) (Chaouloff et al. 1990f). DOI-induced hyperglycemia was prevented or diminished by 5-HT$_{1C}$/5-HT$_2$ receptor blockers (Hoyer 1988), e.g. LY 53857 (0.2 mg/kg; Eli Lilly, Indianapolis, IN, U.S.A.), ketanserin (0.25 mg/kg; Janssen, Beerse, Belgium), and ritanserin (0.2 mg/kg; Janssen). Interestingly, the 5-HT$_{1C}$ receptor agonists/5-HT$_2$ receptor antagonists mCPP and TFMPP (Conn and Sanders-Bush 1987) (both used at a 2 mg/kg dose; BioBlock, Illkirch, France) did not affect plasma glucose levels, but both dose-dependently diminished DOI-induced hyperglycemia, indicating the involvement of 5-HT$_2$ receptors (Chaouloff et al. 1990f). It is likely that a significant part of these metabolic effects of DOI are mediated by the sympathetic nervous system (as revealed by the metabolic consequences of either prior ganglionic blockade, or pretreatment with an alpha 2-adrenoceptor blocker; Chaouloff et al. 1990f). Lastly, the use of the peripherally acting $5\text{-}HT_{1C}/5\text{-}HT_2$ receptor agonist alpha-methyl-5-HT (Hoyer 1988), which triggers a LY 53857-sensitive hyperglycemia and insulin release, has allowed us to suggest that (i) both peripheral and central $5\text{-}HT_2$ receptors are involved in DOI-induced hyperglycemia, (ii) central $5\text{-}HT_2$ receptors are responsible for DOI-induced inhibition of insulin release (Chaouloff et al. 1990f).

This study shows that the stimulation of central $5\text{-}HT_{1A}$ and $5\text{-}HT_2$ receptors causes hyperglycemia and inhibition of insulin release. It is likely that these metabolic events are consequences of the tight relationships that link the serotonergic system and the sympathetic nervous system.

Are the Findings Relevant to Physiology?

The above results lead to the fundamental question of their relevance to physiological processes. Indeed, the above results are too recent to allow any clearcut answer to that question. However, of interest are the following preliminary findings. Thus, we recently observed that neither $5\text{-}HT_{1A}$ nor $5\text{-}HT_2$ receptor antagonists were able to affect 2-deoxyglucose (2-DG)-induced hyperglycemia (submitted for publication). Neuroglucopenia induced by 2-DG elicits a set of reflexes which results in high blood glucose levels, and inhibition of insulin release. It is noteworthy that 2-DG triggers sympathoinhibition and adrenal catecholamine release (Landsberg and Krieger 1989), thus suggesting the mediation by central $5\text{-}HT_{1A}$ receptors. As reported above, our results clearly indicate that the latter hypothesis is unlikely. Another negative finding is that which concerns the insulin stress model. Insulin administration, which is known since 1972 to increase central 5-HT synthesis, elicits adrenal catecholamine release (Kalil et al. 1986). Thus, we have investigated whether $5\text{-}HT_{1A}$ receptor blockade by ($-$)-propranolol, or $5\text{-}HT_{1C}/5\text{-}HT_2$ receptor blockade by LY 53857, may prevent insulin-induced adrenaline release. Indeed, the latter was found to remain unaltered by these pretreatments (in preparation). These negative findings give support to the idea that the above pharmacological findings may not have any physiological implication, although it is certainly too early to draw any firm conclusions. At the present time, we are investigating other models of adrenaline release which could involve the stimulation of 5-HT receptors. The results from these works will allow us to clearly appreciate the extent to which the above pharmacological data are related to physiological events.

344

References

Bagdy, G., Calogero, A. E., Murphy, D. L., and Szemeredi, K. (1989). Serotonin agonists cause parallel activation of the sympatho-adrenomedullary system and the hypothalamo-pituitary-adrenocortical axis in conscious rats. Endocrinology 125: 2664–2669.

Bouhelal, R., and Mir, A. K. (1990). Role of the adrenal gland in the metabolic and cardiovascular effects of 8-OH-DPAT in the rat. European J. Pharmacol. 181: 89–95.

Chaouloff, F., Baudrie, V., and Laude, D. (1990a). Adrenaline-releasing effects of the 5-HT$_{1A}$ receptor agonists 8-OH-DPAT, buspirone, and ipsapirone in the conscious rat. Br. J. Pharmacol. 99: 39P.

Chaouloff, F., Baudrie, V., and Laude, D. (1990b). Evidence that 5-HT$_{1A}$ receptors are involved in the adrenaline-releasing effects of 8-OH-DPAT in the conscious rat. Naunyn-Schmiedeb. Arch. Pharmacol. 341: 381–384.

Chaouloff, F., Baudrie, V., and Laude, D. (1990c). Evidence that the 5-HT$_{1A}$ receptor agonists buspirone and ipsapirone activate adrenaline release in the conscious rat. European J. Pharmacol. 177: 107–110.

Chaouloff, F., Baudrie, V., and Laude, D. (1990d). Pentobarbital anaesthesia prevents the adrenaline-releasing effect of the 5-HT$_{1A}$ receptor agonist 8-OH-DPAT. European J. Pharmacol. 180: 175–178.

Chaouloff, F., and Jeanrenaud, B. (1987). 5-HT$_{1A}$ and alpha 2-adrenergic receptors mediate the hyperglycemic and hypoinsulinemic effects of 8-hydroxy-2-(di-n-propylamino)tetralin in the conscious rat. J. Pharmacol. Exp. Ther. 243: 1159–1166.

Chaouloff, F., Laude, D., and Baudrie, V. (1990e). Ganglionic transmission is a prerequisite for the adrenaline-releasing and hyperglycemic effect of 8-OH-DPAT. European J. Pharmacol. 185: 11–18.

Chaouloff, F., Laude, D., and Baudrie, V. (1990f). Effects of the 5-HT$_{1C}$/5-HT$_2$ receptor agonists DOI and alpha-methyl-5-HT on plasma glucose and insulin levels in the rat. European J. Pharmacol. 187: 435–443.

Conn, P. J., and Sanders-Bush, E. (1987). Relative efficacies of piperazines at the phosphoinositide hydrolysis-linked serotonergic (5-HT$_2$ and 5-HT$_{1C}$) receptors. J. Pharmacol. Exp. Ther. 242: 552–557.

Doods, H. N., Bodeke, H. W. G. M., Kalkman, H. O., Hoyer, D., Mathy, M. J., and Van Zwieten, P. A. (1988). Central 5-HT$_{1A}$ receptors and the mechanism of the central hypotensive effect of (+)-8-OH-DPAT, DP-5-CT, R 28935, and urapidil. J. Cardiovasc. Pharmacol. 11: 432–437.

Fozard, J. R., Mir, A. K., and Middlemiss, D. N. (1987). Cardiovascular response to 8-hydroxy-2-(di-n-propylamino)tetralin (8-OH-DPAT) in the rat: site of action and pharmacological analysis. J. Cardiovasc. Pharmacol. 9: 328–347.

Furman, B. L. (1974). The hypoglycaemic effect of 5-hydroxytryptophan. Br. J. Pharmacol. 50: 575–580.

Gagliardino, J. J., Zieher, L. M., Iturriza, F. C., Hernandez, R. E., and Rodriguez, R. R. (1971). Insulin release and glucose changes induced by serotonin. Horm. Metab. Res. 3: 145–150.

Glennon, R. A. (1987). Central serotonin receptors as targets for drug research. J. Med. Chem. 30: 1–12.

Hoyer, D. (1988). Functional correlates of 5-HT$_1$ recognition sites. J. Receptor Res. 8: 59–81.

Khalil, Z., Marley, P. D., and Livett, B. G. (1986). Elevation in plasma catecholamines in response to insulin stress is under both neuronal and nonneuronal control. Endocrinology 119: 159–167.

Koenig, J. I., Gudelsky, G. A., and Meltzer, H. Y. (1987). Stimulation of corticosterone and B-endorphin secretion in the rat by selective 5-HT receptor subtype activation. European J. Pharmacol. 137: 1–8.

Landsberg, L., and Krieger, D. R. (1989). Obesity, metabolism, and the sympathetic nervous system. Am. J. Hypertension 2: 125S–132S.

Laude, D., Baudrie, V., Martin, G. R., and Chaouloff, F. (1990). Effects of the 5-HT$_1$ receptor agonists DP-5-CT, CGS 12066B, and RU 24969 on plasma adrenaline and glucose levels in the rat. Naunyn-Schmiedeb. Arch. Pharmacol. 342: 378–381.

McCall, R. B., and Harris, L. T. (1988). 5-HT$_2$ receptor agonists increase spontaneous sympathetic nerve discharge. European J. Pharmacol. 151: 113–116.

Mir, A. K., Hibert, M., and Fozard, J. R. (1987). Cardiovascular effects of N,N-dipropyl-5-carboxamidotryptamine, a potent and selective 5-HT$_{1A}$ receptor ligand. In Nobin, A., Owman, C., and Arneklo-Nobin, B. (eds.), Neuronal Messengers on Vascular Function. New York: Elsevier, pp. 21–29.

Ramage, A. G., and Fozard, J. R. (1987). Evidence that the putative 5-HT$_{1A}$ agonists, 8-OH-DPAT and ipsapirone, have a central hypotensive action that differs from that of clonidine in anaesthetized cats. European J. Pharmacol. 138: 179–191.

Rivier, C., and Vale, W. (1983). Modulation of stress-induced ACTH release by corticotropin-releasing factor, catecholamines, and vasopressin. Nature 305: 325–326.

Wong, K. L., and Tyce, G. M. (1978). Effect of the administration of L-5-hydroxytryptophan and a monoamine oxidase inhibitor on glucose metabolism in rat brain. J. Neurochem. 31: 613–620.

Serotonin: Molecular Biology, Receptors and Functional Effects
ed. by J. R. Fozard/P. R. Saxena

Novel *in vivo* Models of 5-HT$_{1A}$ Receptor-Mediated Activity: 8-OH-DPAT-induced Spontaneous Tail-Flicks and Inhibition of Morphine-Evoked-Antinociception

M. J. Millan, K. Bervoets, S. Le Marouille-Girardon, C. Grévoz, and F. C. Colpaert

Neurobiology Division, FONDAX, Groupe de recherche servier, 7 rue Ampere, 92800 Puteaux, France

Summary. Subcutaneous administration of the high efficacy 5-HT$_{1A}$ receptor agonists, 8-OH-DPAT, RU 24969, 5-MeODMT and lisuride induced spontaneous tail-flicks (STFs) in lightly-restrained rats; that is, tail-flicks in the absence of extraneous stimulation. This response was *not* induced in mice. 5-HT$_{1A}$ receptor partial agonists, such as buspirone, gepirone and flesinoxan and 5-HT$_{1A}$ receptor antagonists, such as BMY 7378, NAN-190, (−)-alprenolol and methiothepin failed to induce STFs in rats and blocked the action of 8-OH-DPAT. Agonists and antagonists at other 5-HT receptor types neither induced nor antagonised STFs. In mice, 5-HT$_{1A}$ agonists and partial agonists, but not antagonists, blocked morphine-induced antinociception (MIA) without affecting nociceptive thresholds when administered alone. Similarly, in rats, 5-HT$_{1A}$ receptor partial agonists, but not antagonists, inhibited MIA. Agonists at other 5-HT receptor types did not influence MIA.

It is concluded that a high efficacy agonist action at 5-HT$_{1A}$ receptors, but not other 5-HT receptor types, elicits STFs in rats but not in mice. Further, 5-HT$_{1A}$ receptor agonists and partial agonists but not antagonists inhibit MIA both in mice and rats. These tests may prove useful in the detection of novel ligands at 5-HT$_{1A}$ receptors.

Introduction

A multiplicity of binding sites for serotonin (5-HT) exists in the CNS (Glennon 1990; Hoyer 1988). The current classification of 5-HT$_{1A}$, 5-HT$_{1B}$, 5-HT$_{1C}$, 5-HT$_{1D}$, 5-HT$_2$, 5-HT$_3$ and 5-HT$_4$ receptors is based largely on binding and *in vitro* studies and an important challenge is both to determine the functional correlates and pathophysiological significance of this multiplicity and to develop novel ligands interacting selectively with particular receptor subtypes. Correspondingly, there is a need for *in vivo* models permitting the detection of activity at particular 5-HT receptor types. 5-HT$_{1A}$ receptors are involved in the control of, for example, appetite, mood and cardiovascular function (Dourish et al. 1987). They have also been implicated in the modulation of nociception (Berge et al. 1989; Besson 1990) and are localized in tissues, such as the dorsal horn of the spinal cord (Daval et al. 1987), which play a key role in nociceptive processes. However, available data are conflicting with 5-HT$_{1A}$ agonists documented to elicit either analgesia or hyperalgesia, or to be inactive (Berge et al. 1989; Besson 1990; Franklin et al. 1989; Millan et al. 1989; Solomon and Gebhart 1988). Further, there is an intriguing report that the prototypical 5-HT$_{1A}$ receptor agonist, 8-OH-DPAT, can attenuate morphine-induced antinociception (MIA) in mice (Berge et al. 1985).

Recently, we discovered that high efficacy 5-HT$_{1A}$ receptor agonists such as 8-OH-DPAT induce spontaneous tail flicks (STFs) in rats: this response may account for their apparent hyperalgesic properties. The present paper summarizes our findings (Millan et al. 1989, in press; Millan and Colpaert 1990a; b) concerning both STFs and MIA in rodents. Several interrelated models are characterized, the employment of which may facilitate the discovery and classification of 5-HT$_{1A}$ receptor agonists, partial agonists and antagonists.

Material and Methods

Male Wistar rats of 200–220 g and male NMRI mice of 20–22 g with free access to chow and water were employed.

For evaluation of STFs, rats were introduced into horizontal plastic cylinders with the tail hanging freely over the lab bench. After 5 min adaptation, STFs elicited in 5 min were counted. A STF was defined as the raising of the tail to a level higher than that of the body axis. The action of 8-OH-DPAT alone was evaluated 10 min after its administration and that of all other drugs alone at 30 min. For antagonist studies with 8-OH-DPAT, drugs were given 30 min before evaluation of STFs. For antagonist studies with RU 24969, lisuride and 5-MeODMT, these were given 30 min, and antagonists, 40 min, before testing.

For algesiometric tests, mice and rats were gently restrained under paper wadding. For heat, a beam of light was focussed on the tip and the latency withdrawal recorded. For pressure, an incremental weight was applied to a point 3 cm from the tip and the latency to withdrawal recorded. Basal latencies to respond to heat were ca. 3.0 sec and there was a cut-off of 8 sec. For pressure, the corresponding values were 80 g and 250 g for mice, and 150 g and 500 g for rats. The action of 5-HT$_{1A}$ ligands alone and of morphine was evaluated 30 min following treatment. For antagonist studies, morphine was administered 30 min, and 5-HT$_{1A}$ ligands, 40 min prior to testing. Dose-response curves to morphine were performed in the presence of an invariant dose of 5-HT$_{1A}$ ligands. Maximal possible antinociception (% MPA) was calculated according to the following formula: (Drug or Vehicle + Morphine) − Drug or Vehicle Alone ÷ (Cutoff-Drug or Vehicle Alone) × 100.100% and 0% correspond to cut-off and basal latencies, respectively.

All drugs were administered subcutaneously. At least 4 animals were used for each data point and at least 3 doses evaluated per dose-response curve. Data were analysed by 1-way ANOVA followed by Newman-Keuls test with the limit of significance set at 0.05. Dose-response curves were analysed for ID_{50}s (dose reducing agonist action by 50%) plus 95% confidence limits.

Results

Figure 1 shows that the high efficacy $5-HT_{1A}$ receptor agonists, 8-OH-DPAT, RU 24969, 5-MeODMT and lisuride, but not the $5-HT_{1B/1C}$ agonist, TFMPP nor the $5-HT_{1C/2}$ agonist, DOI dose-dependently induced STFs in rats. The $5-HT_3$ receptor agonist, 2-methyl-5-HT was inactive at doses of up to 40.0 mg/kg (not shown). $5-HT_{1A}$ receptor partial agonists and antagonists failed to elicit STFs (Table 1). Methiothepin and BMY 7378 completely blocked STFs elicited by these high efficacy $5-HT_{1A}$ receptor agonists (Figure 1). STFs elicited by 8-OH-DPAT were not only blocked by methiothepin ($ID_{50} = 0.01$ mg/kg) and BMY 7378 (0.4), but also by the $5-HT_{1A}$ partial agonists, buspirone (0.4), gepirone (2.0), ipsapirone (0.3), LY 165,163 (1.1), MDL 72832 (0.03), flesinoxan (5.6), and the $5-HT_{1A}$ receptor antagonists, (−)-pindolol (0.9), (−)-alprenolol (0.5), (±)-isamoltane (4.2), spiperone (0.06), spiroxatrine (0.2) and NAN-190 (0.03). Antagonists at $5-HT_{1C/2}$ receptors, ritanserin and ICI 169,369 and at $5-HT_3$ receptors, GR 38032F and MDL 72222, in each case at doses of up to 10.0 mg/kg, were ineffective (not shown). Morphine and the opioid antagonist, naloxone, at doses of up to 10.0 mg/kg, failed either to induce STFs or to modify the action of 8 OH-DPAT (not shown).

In view of the induction of STFs, high efficacy $5-HT_{1A}$ receptor agonists could *not* be examined in rats for their influence upon nociception and MIA. However, in mice, STFs were *not* observed and, in this species, agonists could be evaluated. The $5-HT_{1A}$ agonists, partial agonists and antagonists mentioned above failed to significantly modify basal latencies to respond to heat or pressure in mice or rats across a range of doses encompassing those employed in the above-described studies (not shown). However, in the presence of the $5-HT_{1A}$ receptor agonists and partial agonists, the dose-response curve for MIA was clearly shifted in parallel to the right without any loss of maximal effect (Figure 2 and Table 2). In distinction, in the presence of the $5-HT_{1A}$ receptor antagonists, no influence upon the dose-response curve for MIA was observed (Figure 2 and Table 2). Further, TFMPP, DOI and 2-methyl-5-HT did not modify MIA (not shown). In addition, in mice, employing an invariant dose of morphine (5.0 mg/kg) which induced

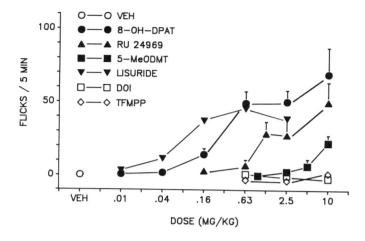

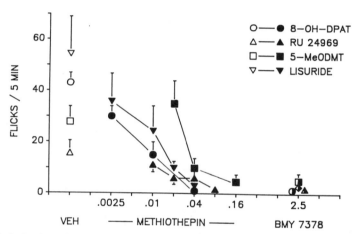

Figure 1. Induction of spontaneous tail-flicks (STFs) by 5-HT$_{1A}$ receptor agonists and their inhibition by the antagonists, methiothepin and BMY 7378.

Upper panel: Dose-dependent induction of STFs by the 5-HT$_{1A}$ agonists, 8-OH-DPAT, RU 24969, 5-MeODMT and lisuride but not by the 5-HT$_{1B/1C}$ agonist, TFMPP nor the 5-HT$_{1C/2}$ agonist, DOI. Mean ± SEM shown. The minimum effective does (significantly different from vehicle was 0.04, 0.16, 1.25 and 5.0 for lisuride, 8-OH-DPAT, RU 24969 and 5-MeODMT, respectively. ANOVA as follows. 8-OH-DPAT, F $(6,117) = 25.5$, $p < 0.001$; RU 24969, F $(5,26) = 7.7$, $p < 0.01$; 5MeODMT, F $(4,41) = 12.3$, $p < 0.001$; TFMPP, F $(4,31) = 1.8$, $p > 0.05$ and DOI, F $(3,18) = 0.9$, $p > 0.05$.

Lower panel: Inhibition of STFs by methiothepin and BMY 7378. Mean ± SEM shown. The ID$_{50}$ (95% confidence limits) for methiothepin was as follows; 8-OH-DPAT, 0.010 $(0.003 - 0.161)$, lisuride, 0.006 $(0.002 - 0.012)$, RU 24969, 0.014 $(0.006 - 0.042)$ and 5-MeODMT, 0.033 $(0.011 - 0.100)$.

Table 1. Inability of 5-HT$_{1A}$ receptor partial agonists (PAG) and antagonists (ANT) to induce spontaneous tail-flicks in rats.

Activity	Drug	Dose range (mg/kg, s.c.)	Flicks/5 min (at max. effect dose)
PAG	buspirone	0.08–20.0	1.7 ± 0.9 (0.31)
PAG	gepirone	0.08–40.0	1.2 ± 1.2 (5.0)
PAG	ipsapirone	0.04–10.0	1.4 ± 1.4 (10.0)
PAG	LY 165,163	0.04–10.0	0.9 ± 0.7 (0.16)
PAG	flesinoxan	0.04–10.0	5.4 ± 4.0 (10.0)
PAG	MDL 72832	0.16–10.0	2.0 ± 0.7 (0.16)
ANT	BMY 7378	0.04–10.0	2.2 ± 1.1 (0.63)
ANT	NAN-190	0.01–0.16	0 ± 0 (10.0)
ANT	methiothepin	0.0025–0.16	1.4 ± 0.6 (0.04)
ANT	(−)-alprenolol	2.5–40	0.5 ± 0.5 (2.5)
ANT	(−)-pindolol	0.31–5.0	0.8 ± 0.6 (1.25)
ANT	(±)-isamoltane	0.16–40	1.4 ± 0.7 (40.0)
ANT	spiperone	0.01–0.16	1.2 ± 0.8 (0.04)
ANT	spiroxatrine	0.01–2.5	0.3 ± 0.3 (0.63)

For each drug, data were analysed by 1-way ANOVA: F values ranged from 0.16–1.81 and, in no case, were significant (> 0.05). N was 4–8 per close.

Table 2. Influence of 5-HT$_{1A}$ receptor agonists, partial agonists and antagonists, as well as the buspirone metabolite, 1-PP, upon the dose-response relationship for induction of antinociception by morphine.

Drug	AD$_{50}$ (95% confidence limits)			
	Mice		Rats	
	Heat	Pressure	Heat	Pressure
vehicle	1.8 (1.2–2.5)	1.4 (1.1–1.8)	0.8 (0.6–1.1)	1.2 (0.9–1.6)
8-OH-DPAT	4.9 (3.2–7.6)	5.3 (4.3–6.4)	NT	NT
lisuride	5.2 (4.0–6.8)	4.7 (3.8–5.0)	NT	NT
busipirone	4.4 (3.1–6.3)	2.8 (1.9–4.1)	6.3 (6.3–10.4)	6.7 (3.7–12.2)
LY 165,163	7.2 (5.0–10.2)	6.1 (3.7–10.1)	4.3 (2.3–8.4)	3.6 (2.0–6.6)
flesinoxan	8.1 (5.3–12.1)	5.2 (3.4–7.9)	5.0 (2.1–11.7)	8.2 (5.4–12.6)
BMY 7378	1.2 (0.7–1.9)	1.3 (0.8–2.3)	0.6 (0.3–1.4)	0.9 (0.5–1.9)
methiothepin	0.9 (0.5–1.8)	1.4 (0.8–2.1)	1.2 (0.7–2.2)	1.4 (0.9–2.4)
1-PP	1.3 (0.9–2.0)	1.5 (1.0–2.3)	1.1 (0.7–2.8)	1.8 (1.1–3.2)

NT signifies not tested, owing to induction of STFs. The values are AD$_{50}$ s (dose of morphine yielding half-maximal antinociception). Note, there is no overlap of confidence limits as compared to vehicle for agonists. Doses as follows: 8-OH-DPAT (0.63), lisuride (0.63), buspirone (10.0), LY 165,163 (10.0), flesinoxan (10.0), BMY 7378 (2.5), methiothepin (0.16) and 1-PP (10.0).

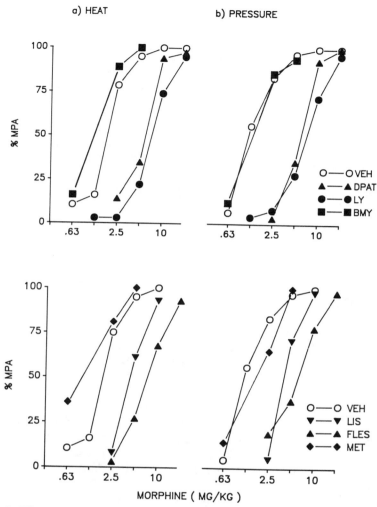

a) HEAT   b) PRESSURE

Figure 2. Effect of 5-HT$_{1A}$ receptor agonists and partial agonists upon the dose-response relationship for the induction of antinociception by morphine in mice. The dose-response curve to morphine was performed in the presence of vehicle (VEH), 0.63 mg/kg 8-OH-DPAT (DPAT), 10.0 mg/kg LY 165,163 (LY), 2.5 mg/kg BMY 7378 (BMY), 0.63 mg/kg lisuride (LIS), 10.0 mg/kg flesionoxan (FLES) and 0.16 mg/kg methiothepin (MET). The mean is shown (SEM omitted for clarity).

maximal antinociception, dose-response curves for the inhibition of MIA by 5-HT$_{1A}$ agonists and partial agonists could be established. ID 50s for reduction of MIA against heat/pressure were as follows: 8-OH-DPAT (0.5/0.6), lisuride (0.4/0.8), RU 24969 (0.7/8.4), buspirone (7.2/6.6), LY 165,163 (4.2/4.4) and flesinoxan (3.2/5.2).

352

Discussion

8-OH-DPAT, lisuride, 5-MeODMT and RU 24969, each of which induced STFs in rats, are classified as high efficacy 5-HT$_{1A}$ receptor agonists based on the *in vitro* inhibition of forskolin-stimulated adenyl cyclase in hippocampal membranes and, *in vivo*, the induction of the 5-HT behavioural syndrome (Dourish et al. 1987; Glennon, 1990; Glennon and Lucki 1990; Hoyer 1988; Tricklebank 1985). In contrast, several 5-HT$_{1A}$ ligands classified as partial agonists (see Table 1), based on their behaviour in these and other paradigms (see above refs and Hartog and Wouters 1988; Mir et al. 1988; Smith and Peroutka 1986), failed to elicit STFs. These findings imply that, for induction of STFs, a *high* efficacy agonist action at 5-HT$_{1A}$ receptors is necessary. The argument that 5-HT$_{1A}$ receptors are critical for induction of STFs is supported by the observation that agonists at 5-HT$_{1B/1C}$ (TFMPP), 5-HT$_{1C/2}$ (DOI) and 5-HT$_3$ receptors (2-methyl-5-HT) (Glennon 1990; Hoyer 1988) were not able to induce STFs. Further, STFs elicited by 8-OH-DPAT could not be modified by selective antagonists (Glennon 1990; Hoyer 1988) at 5-HT$_{1C/2}$ (ritanserin and ICI 169,369) or 5-HT$_3$ (GR 38032 F and MDL 72222) receptors (not shown). In distinction, as detailed in Results and shown in Figure 1, STFs were powerfully inhibited by the mixed 5-HT$_{1A}$ receptor antagonist, methiothepin, and several types of drugs possessing antagonist activity at 5-HT$_{1A}$ receptors (Hartog and Wouters 1988; Glennon 1990; Hoyer 1988; Mir et al. 1988; Smith and Peroutka 1986; Tricklebank 1985). First, the β-blockers, ($-$)-alprenolol, ($-$)-pindolol and ($\pm$)-isomoltane, which are mixed 5-HT$_{1A/1B}$ receptor antagonists; in contrast, pure β-blockers (ICI 118,551 and betaxolol) did not affect the action of 8-OH-DPAT (not shown). Second, the D$_2$ receptor antagonists, spiperone and spiroxatrine, which act as mixed 5-HT$_{1A/2}$ antagonists; in contrast, selective D$_2$ antagonists (pimozide and raclopride) did not modify the action of 8-OH-DPAT (not shown). Third, the selective 5-HT$_{1A}$ partial agonists, buspirone, gepirone, ipsapirone, LY 165,163, flesinoxan and MDL 72832. Fourth, the putative, selective 5-HT$_{1A}$ antagonists, BMY 7378 (Yocca et al. 1987) and NAN-190 (Glennon et al. 1988). Collectively, these findings allow for the attribution of STFs to an activation of 5-HT$_{1A}$ receptors. Their location remains unclear but the ability of 8-OH-DPAT to elicit STFs in rats deprived of CNS 5-HT neurones by treatment with 5,7-DHT indicates that they are post-synaptic to 5-HT pathways (Millan et al. unpublished).

The discovery of STFs suggests that reports of hyperalgesia with 5-HT$_{1A}$ agonists in rats may be artifactual (Berge et al. 1989; Millan et al. 1989; Solomon and Gebhart 1988). Indeed, 5-HT$_{1A}$ receptor agonists (in mice), 5-HT$_{1A}$ partial agonists and 5-HT$_{1A}$ antagonists all failed to modify nociceptive thresholds to heat or pressure in either mice or rats (see Results). These data suggest that 5-HT$_{1A}$ receptors do not play a major role in control of nociception *per se*. Further, although it has been claimed that 8-OH-DPAT can induce antinociception in the hot-plate (Berge et al. 1989), this action is mediated by α_2 rather than 5-HT$_{1A}$ receptors (Millan, unpublished). Nevertheless, the present data *do* indicate an important and novel role of 5-HT$_{1A}$ receptors, that is, in the inhibition of MIA. This action was very robust in that all 5-HT$_{1A}$ agonists tested inhibited MIA dose-dependently and resulted in a shift in the dose-response curve for MIA to the right. Further, the antago-nism of MIA can be prevented by BMY 7378 (Millan and Colpaert 1990b) and is not reproduced by agonists at other 5-HT receptor types. Importantly, 5-HT$_{1A}$ receptor agonists neither evoke withdrawal in morphine-dependent mice nor antagonise the induction of Straub tail by morphine: this indicates that antagonism of MIA reflects neither pharmacokinetic factors nor intrinsic opioid antagonist properties (Mil-lan and Colpaert 1990b). The location of the 5-HT$_{1A}$ receptors responsi-ble for inhibition of MIA will be of importance to establish. It is conceivable that inhibition of MIA reflects activation of raphe-localized 5-HT$_{1A}$ autoreceptors resulting in a diminution of serotoninergic trans-mission (Sprouse and Aghajanian, 1987). However, there is increasing evidence that 5-HT does not invariably mediate antinociception (Berge et al. 1985; Besson 1990; Franklin et al. 1989) and the present data suggest that the role of 5-HT in the modulation of nociception may require reappraisal, at least as regards the possibility of contrasting roles of individual receptor types. Indeed, the present finding that activation of 5-HT$_{1A}$ receptors inhibits MIA provides an interesting distinction to 5-HT$_2$ and 5-HT$_3$ receptors, each of which can mediate antinociception (Glaum et al. 1988; Solomon and Gebhart 1988). It is possible that stimulus quality and intensity, as is the case with, for example, K-opioid agonists (Millan 1990) may emerge to be important variables determin-ing the efficacy of 5-HT receptor ligands.

In summary, induction of STFs by 5-HT$_{1A}$ receptor agonists, and their antagonism by 5-HT$_{1A}$ partial agonists and antagonists, may represent novel tests for the detection of drugs interacting with 5-HT$_{1A}$ receptors. Although the paradigm of STFs cannot differentiate 5-HT partial agonists from antagonists, this can be acheived by an evaluation of the influence upon MIA. In contrast to STFs, 5-HT$_{1A}$ partial agonists act like agonists in inhibiting MIA, whereas 5-HT$_{1A}$ receptor antago-nists are ineffective. To recapitulate, 5-HT$_{1A}$ receptor agonists induce STFs in rats and inhibit MIA in mice; 5-HT$_{1A}$ partial agonists inhibit STFs in rats and inhibit MIA in mice and rats; 5-HT$_{1A}$ receptor antagonists inhibit STFs in rats and fail to influence MIA in mice or rats. Opioid agonists and antagonists can also be distinguished in that they neither induce nor inhibit STFs. These simple, robust models may prove of wide utility in the characterization of novel 5-HT$_{1A}$ ligands.

354

Acknowledgement

V. Green is thanked for secretarial assistance.

References

Berge, O.-G., Fasmer, O. B., Ogren, S. O., and Hole, K. (1985). The putative serotonin receptor agonist, 8-OH-DPAT, antagonises the antinociceptive effect of morphine. Neurosci. Lett. 54: 71–75.

Berge, O. -G., Post, C., and Archer, T. (1989). The behavioural pharmacology of serotonin in pain processes, In Bevan, P., Cools, A. R., and Archer, T. (Eds) The Behavioural Pharmacology of 5-HT. New York, Lawrence Earlbaum Association Inc, pp. 301–320.

Besson, J.-M. (1990). Serotonin and Pain (Ed.). Amsterdam, Excerpta Medica, pp. 339.

Daval, G., Verge, D., Basbaum, A., Bourgoin, S., and Hamon, M. (1987). Autoradiographic evidence of serotonin binding sites on primary afferent fibres in the dorsal horn of the rat spinal cord. Neurosci. Lett. 83: 71–76.

Dourish, T. P., Hutson, P. H., and Ahlenius, S. (Eds.), (1987). Brain $5\text{-}HT_{1A}$ receptors. Chichester, Ellis Horwood.

Franklin, K. B. J. (1989). Analgesia and the neural substrate of reward. Neurosci. Biobehav. Rev. 133: 149–154.

Glaum, S. R., Proudfit, H. K., and Anderson, E. G. (1988). Reversal of the antinociceptive effects of intrathecally administered serotonin in the rat by a selective $5\text{-}HT_3$ antagonist. Neurosci. Lett. 95: 313–317.

Glennon, R. A. (1990). Serotonin receptors: Clinical implications. Neurosci. Biobehav. Rev. 14: 35–47.

Glennon, R. A., and Lucki, I. (1989). Behavioural models of 5-HT receptor activation. In: Sanders-Bush, E. (Ed.), The Serotonin Receptors. New York, Humana Press, pp. 253–292.

Glennon, R. A., Naiman, N. A., Peirson, M. E., Titeler, M., Lyon, R. A., and Weisberg, E. (1988). NAN-190: an arylpiperazine analog that antagonises the stimulus effects of the $5\text{-}HT_{1A}$ agonist 8-hydroxy-2-(di-n-propylamino)tetralin (8-OH-DPAT). Eur. J. Pharmacol. 154: 339–341.

Hartog, J., and Wouters, W. (1988). Flesinoxan hydrochloride. Drugs Future 13: 31–33.

Hoyer, D. (1988). Functional correlates of 5-HT 1 recoginition sites. J. Rec. Res. 8: 59–81.

Millan, M. J. (1990). K-opioid receptors and analgesia. Trends Pharmacol. 11: 71–76.

Millan, M. J., Bervoets, K., and Colpaert, F. C. (1989). Apparent hyperalgesic action of the $5\text{-}HT_{1A}$ agonist, 8-OH-DPAT in the rat reflects induction of spontaneous tail-flicks. Neurosci. Lett. 107: 227–232.

Millan, M. J., and Colpaert, F. C. (1990a). Attenuation of opioid-induced antinociception by $5\text{-}HT_{1A}$ partial agonists in the rat. Neuropharmacology 29: 315–318.

Millan, M. J., and Colpaert, F. C. (1990b). Characterization of the blockade of morphine-induced antinociception by an agonist action at $5\text{-}HT_{1A}$ receptors. In: Besson, J. -M., (Ed.), Serotonin and Pain. Amsterdam, Excerpta Medica. pp. 263–284.

Mir, A. K., Hibert, M., Tricklebank, M. D., Middlemiss, D. N., Kidd, E. J., and Fozard, J. R. (1988). MDL 72832: a potent and stereoselective ligand of central and peripheral $5\text{-}HT_{1A}$ receptors. Eur. J. Pharmacol. 149: 107–120.

Smith, L. M., and Peroutka, S. J. (1986). Differential effects of 5-hydroxytryptamine 1A selective drugs on the 5-HT behavioural syndrome, Pharmacol. Biochem. Behav. 24: 1513–1519.

Solomon, R. E., and Gebhart, G. F. (1988). Mechanism of the effects of intrathecal serotonin on nociception and blood pressure in the rat. J. Pharmacol. Exp. Ther. 245: 905–912.

Sprouse, J. A., and Aghajanian, G. K. (1987). Electrophysiological responses of serotoninergic dorsal raphe neurons to $5\text{-}HT_{1A}$ and $5\text{-}HT_{1B}$ agonists. Synapse 1: 3–9.

Tricklebank, M. D. (1985). The behavioural response to 5-HT receptor agonists and subtypes of the central 5-HT receptor. Trends Pharmacol. 7: 403–407.

Yocca, F. M., Hyslop, D. K., Smith, D. W., and Maayani, S. (1987). BMY 7378, a buspirone analog with high affinity and low intrinsic activity at the $5\text{-}HT_{1A}$ receptor in rat and guinea-pig hippocampal membranes. Eur. J. Pharmacol. 137: 293–294.

Serotonin: Molecular Biology, Receptors and Functional Effects
ed. by J. R. Fozard/P. R. Saxena
© 1991 Birkhäuser Verlag Basel/Switzerland

Evidence that the Unilateral Activation of 5-HT$_{1D}$ Receptors in the Substantia Nigra of the Guinea-Pig Elicits Contralateral Rotation

G. A. Higgins[2], C. C. Jordan[1] and M. Skingle[1]

[1]*Department of Neuropharmacology, Glaxo Group Research Ltd., Ware, Herts SG12 0DP, UK;* [2]*Addiction Research Foundation, 33 Russell Street, Toronto, Ontario, Canada M5S 2S1*

Summary. The 5-HT$_1$ receptor agonists, 5-carboxamidotryptamine (5-CT) and sumatriptan, induced a marked contralateral rotation following unilateral infusion into the substantia nigra of the guinea-pig. In contrast the 5-HT$_{1A}$ receptor agonist, 8-OH-DPAT, produced only a very small response, whilst DOI and 2-methyl-5-HT were without effect. At doses lower than those required to produce contralateral rotation, 5-CT and sumatriptan induced a small and consistent ipsilateral rotation.

The contralateral rotation induced by 5-CT was attenuated following pretreatment with the poorly selective 5-HT receptor antagonists, methiothepin and metergoline, but not by ritanserin or ondansetron. An involvement of dopaminergic systems in the rotational response to 5-CT was implied by the antagonism of 5-CT-induced rotation by haloperidol. Taken together these data suggest that 5-CT-induced contralateral rotation may be mediated by 5-HT$_{1D}$ receptor activation and may therefore represent the first model to study this receptor subtype *in vivo.*

Introduction

Central 5-HT$_{1D}$ receptors have been identified in several species including guinea-pig and man (for review see Waeber et al. 1990a). In each species studied, the distribution of this receptor subtype is similar with the highest density in regions associated with the control of movement such as the dopamine nigro-striatal system. Furthermore, a possible involvement of 5-HT$_{1D}$ receptors in movement disorders such as Huntington's chorea have been implicated by the autoradiography studies of Waeber and Palacios (1989). In the rat, a species known to have a distribution of 5-HT$_{1B}$ recognition sites similar to the distribution of 5-HT$_{1D}$ sites in guinea-pig and man (Pazos et al. 1985; Waeber et al. 1989a; b), it has been shown that unilateral injection of 5-HT into the substantia nigra (SN) causes contralateral rotation (Oberlander et al. 1981). The 5-HT$_{1B}$ and 5-HT$_{1D}$ receptors in the SN of the rat and calf respectively are negatively coupled to adenylate cyclase and it has been suggested that these two sites subserve the same function in different species (Hoyer and Middlemiss 1989). Therefore it was of potential interest for us to investigate the effects of 5-HT receptor agonists at 5-HT$_{1D}$ sites in the SN of the guinea-pig.

Materials and Methods

Male, Dunkin-Hartley guinea-pigs (300–500 g, Interfauna) were anaesthetised with chloral hydrate (400 mg/kg i.p.) and ketamine (40 mg/kg i.p.) and placed in a Kopf stereotaxic frame. Guide cannulae consisting of a stainless steel tube (23 gauge) in a perspex holder, were implanted unilaterally, 8 mm above the left substantia nigra (SN) (Anterior +6.0 mm (interaural line), Lateral +2.2 mm (midline), Ventral −4.0 mm (skull surface)) with the mouthpiece set 3.2 mm below zero; coordinates according to Luparello (1967). Stainless steel stylets (30 gauge) were used to keep the guide cannulae patent during the 7 day recovery period. These extended 1 mm beyond the cannula tip.

Seven days after surgery, the guinea-pigs were lightly restrained, the stylets removed and a microinjection needle (30 gauge) was inserted into the guide cannula such that it terminated 8 mm below the cannula tip. Drug solutions or saline vehicle were slowly infused over a 120 s period in a dose volume of 2 μl. A further 30 s was allowed for diffusion away from the cannula tip. The injection units were then removed and the stylets replaced. Unless otherwise stated, the guinea-pigs were used in one experiment only.

Drugs for subcutaneous or intraperitoneal injection were administered in a dose volume of 2 ml/kg, except chloral hydrate which was injected in a volume of 10 ml/kg. The pH of all solutions was in the range 4–7. For intra-nigral injections, a dose volume of 2 μl was used. Immediately after discrete injection the guinea-pigs were placed individually in hemispherical bowls (33 cm diameter; 30 cm high) typically for a 3 h period unless otherwise stated. Rotational behaviour was automatically measured by connecting the animals via a harness to a transducer which, in turn, was linked to an on-line computer (Olivetti M28) for the automatic recording of both ipsilateral-(I) and contralateral (C) rotations.

For antagonist studies, the test compound or vehicle was administered subcutaneously in the nape of the neck 30 min before the nigral injection of 5-CT (10 μg). All studies were carried out in naive animals.

Upon completion of testing, histological verification of each injection site was made following injection with 2 μl of India ink. Placements outside the SN were excluded from the study.

To determine whether antagonism of 5-CT-induced rotation was due to sedation rather than pharmacological antagonism, each effective compound was assessed for overt signs of behavioural depression and locomotor activity (using infra-red beam photocells) at doses which antagonised 5-CT.

Results

Microinfusion of 5-CT ($2-25\,\mu g$) into the left SN of the guinea-pig induced a strong contralateral (C) rotation (Figure 1). Although this response was almost immediate in onset, at higher doses ($10-25\,\mu g$) the most intense period of activity was 1–4 h post injection (Figure 2). This apparent delay may be due to the induction of ataxia which impaired locomotor performance. During this time period, the animals showed a clear asymmetrical posture with the head directed away from the injected side.

At lower doses of 5-CT ($0.08-2\,\mu g$), a small but nonetheless re-producible, ipsilateral (I) rotation was observed (Figure 1). The magnitude of this effect was much smaller than the contralateral response (e.g., 5-CT $0.4\,\mu g$ I $= 39 \pm 16$ rotations 3 h^{-1}; 5-CT 25 μg C $= 436 \pm 109$ rotations 3 h^{-1}).

Significant contralateral rotation was induced by the nigral infusion of sumatriptan ($10-25\,\mu g$) or 8-OH DPAT ($10-25\,\mu g$) into the SN, although the response to 8-OH-DPAT was much smaller than that elicited by 5-CT and sumatriptan (Figure 1). In common with 5-CT, sumatriptan at a dose ($2\,\mu g$) subthreshold for inducing contralateral rotation, produced a small ipsilateral response as did 8-OH-DPAT ($25\,\mu g$).

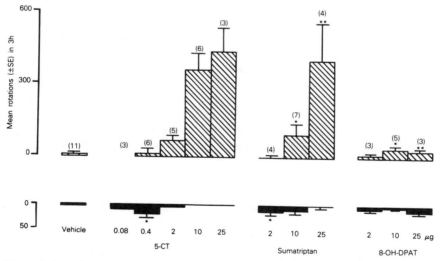

Figure 1. Rotational behaviour elicited by 5-HT$_1$ receptor agonists following unilateral intra-nigral injection in the guinea-pig. Contralateral ($\boxed{}$) and ipsilateral ($\blacksquare$) rotations were recorded for 3 h. Significant differences (*$p < 0.05$; **$p < 0.01$) compared to vehicle were calculated using the unpaired "t"-test. Bracketed figures above each histobar refer to the number of animals used in each treatment group.

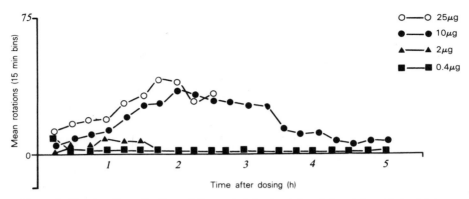

Figure 2. The duration of action of the contralateral rotation induced by unilateral intra-nigral injection of 5-CT. Contralateral rotations were recorded every 15 min for up to 5 h following unilateral injection of 5-CT. The following doses refer to the total dose given to each guinea-pig: (■) 0.4 μg, (▲) 2 μg, (●) 10 μg and (○) 25 μg.

In contrast to the effects of the 5-HT_1 receptor agonists, the 5-HT_3 receptor agonist, 2-Me-5-HT, failed to induce any significant rotational behaviour following intra-nigral infusion (25 μg = I = 7 ± 3 rotations 3 h^{-1}; C = 11 ± 1 rotations 3 h^{-1}, n = 4). The selective 5-HT_{1C}/5-HT_2 receptor agonist (±) DOI (25 μg) was relatively ineffective although some rotational behaviour was observed (I = 58 ± 28 rotations 3 h^{-1}; C = 32 ± 30 rotations 3 h^{-1}, n = 4). However there was no obvious preference for either direction and these scores may therefore reflect the behavioural hyperactivity which became apparent approximately 60 min after DOI infusion.

The contralateral rotation induced by intra-nigral infusion of 5-CT (10 μg) was dose-dependently blocked after pretreatment with methiothepin (1 mgkg^{-1} s.c.) and metergoline (5–10 mgkg^{-1} s.c.) (Figure 3). Following saline pretreatment, the rotational response to 5-CT was exclusively contralateral. However at doses of methiothepin and metergoline which significantly blocked the contralateral response, some ipsilateral rotation was observed (Figure 3).

Both ritanserin (1 mgkg^{-1} s.c.) and ondansetron (0.5 mgkg^{-1} s.c.) failed to antagonise 5-CT-induced contralateral rotation; however haloperidol (0.3 mgkg^{-1} s.c.) produced a significant antagonism of this response (Figure 4).

Peak rotational responses to intra-nigral 5-CT occurred at 90–120 min post administration. The effect of the antagonists, given alone, on locomotor activity was monitored for 120 min. At doses which significantly antagonised 5-CT-induced contralateral rotation, methiothepin (1 mgkg^{-1} s.c.), metergoline (10 mgkg^{-1} s.c.) and haloperiodol (0.3 mgkg^{-1} s.c.) did not significantly reduce spontaneous activity in the

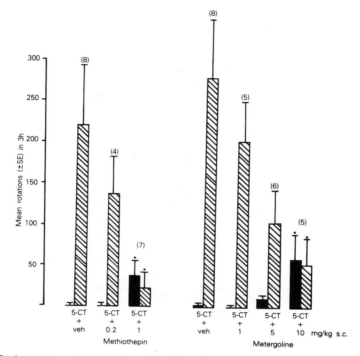

Figure 3. Antagonism of 5-CT-induced rotation by methiothepin and metergoline. Antagonists were given s.c. 30 min before unilateral intra-nigral injection of 5-CT (10 μg). Contralateral (▨) and ipsilateral (■) rotation were recorded for 3 h post 5-CT injection. Significant differences (*$p < 0.05$) compared to 5-CT/vehicle were calculated using the unpaired "t"-test. Bracketed figures above each histobar refer to the number of animals used in each treatment group.

guinea-pig (data not shown). Furthermore, no overt signs of sedation were seen in any of these groups except in the metergoline- and methiothepin-pretreated groups where 2/7 of the animals displayed hind limb abduction with a lowered body position.

Discussion

The present study demonstrates that unilateral microinfusion of the non-selective 5-HT$_1$ receptor agonist 5-CT into the SN of the guinea-pig elicits a marked contralateral rotation. Various lines of evidence support an involvement of 5-HT$_{1D}$ receptors in this response. First, the rank order of potency for agonists to induce this response i.e., 5-CT > sumatriptan > 8-OH-DPAT is in good agreement with their relative potency at inhibiting adenylate cyclase activity (Schoeffter et al. 1988; Schoeffter and Hoyer 1989) and relative affinity for 5-HT$_{1D}$

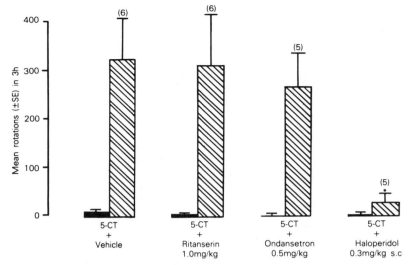

Figure 4. Effect of ritanserin, ondansetron and haloperiodol on 5-CT-induced rotation following intra-nigral injection. Ritanserin (1 mg/kg), ondansetron (0.5 mg/kg), haloperidol (0.3 mg/kg) or vehicle were given s.c. 30 min before unilateral intra-nigral injection of 5-CT (10 μg). Contralateral ($\boxtimes$) and ipsilateral ($\blacksquare$) rotations were recorded for 3 h. Significant differences (*$p < 0.05$) compared to 5-CT/vehicle were calculated using unpaired "t"-test. Bracketed figures above each histobar refer to the number of animals used in each treatment group.

binding sites (Waeber et al. 1990a). Failure of the 5-HT$_{1C}$/5-HT$_2$ receptor agonist, ($\pm$) DOI (Glennon et al. 1984), and 5-HT$_3$ receptor agonist, 2-methyl 5-HT, to elicit rotation when given intranigrally, suggests that these receptors are not involved.

Second, the non-selective 5-HT$_1$/5-HT$_2$ receptor antagonists, methiothepin and metergoline, at doses which failed to produce any marked behavioural effects, attenuated the contralateral rotations induced by 5-CT. In contrast, the 5-HT$_{1C}$/5-HT$_2$ receptor antagonist ritanserin, was inactive at a dose (1 mg/kg s.c.) which has previously been used to antagonise DOI-induced "wet-dog" shakes in the guinea-pig, an effect mediated by central 5-HT$_2$ receptors (Cole et al. 1990). Similarly activation of the 5-HT$_3$ receptors can be excluded since ondansetron, at a high dose (0.5 mg/kg s.c.), was ineffective.

Finally, only discrete injections made directly into the SN elicited rotations, microinfusions of 5-CT just outside this structure failed to induce significant rotation. This strongly suggests an involvement of the SN in the rotation behaviour. As binding studies have shown that 5-HT receptors in the SN are primarily the 5-HT$_{1D}$ subtype, it is likely that this receptor mediates the response.

The present study does not give any insight into the precise location of the 5-HT$_{1D}$ sites or neural circuitry involved in the rotational behaviour. However the elegant lesioning studies of Waeber et al. (1990b) suggest that 5-HT$_{1D}$ receptors are located on the terminals of the striatal neurones projecting to the pars reticulata of the SN. Unilateral 6-hydroxydopamine lesions of the nigral dopaminergic cells failed to decrease nigral 5-HT$_{1D}$ binding sites. In contrast when the striatonigral pathway was lesioned by injecting quinolinic acid into the caudate-putamen, marked reductions in binding in SN were noted on the side ipsilateral to the lesion. These authors tentatively suggest that 5-HT$_{1D}$ receptors may control the release of neurotransmitters other than 5-HT, via a presynaptic action. In our study the involvement of dopamine in the 5-CT-induced rotational response is implicated by the antagonism of 5-CT by the dopamine receptor antagonist, haloperidol. It would be of interest to determine whether a selective dopamine uptake blocker augmented the rotational response to intranigral 5-CT. Lesion studies in the rat suggest that the response to intranigral 5-HT does not require the participation of nigrostriatal neurons (Oberlander et al. 1981). Therefore the locus of interaction between the dopamine and 5-HT system is unclear although it is evident that nigral 5-HT$_{1D}$ receptors are not localised on dopamine cell bodies.

An interesting finding from these studies was the ipsilateral rotation observed at doses of 5-CT and sumatriptan lower than those required to induce contralateral rotation. The small magnitude of this response made pharmacological characterisation using antagonists virtually impossible. However, following pretreatment with metergoline and methiothepin, a significant ipsilateral rotation was unmasked in animals treated with a dose of 5-CT that in control animals produced exclusively contralateral rotation. This may indicate that the ipsilateral rotation produced by 5-CT is not mediated via 5-HT$_1$/5-HT$_2$ receptors. Since both methiothepin and metergoline have affinity for the dopamine D$_2$-binding site (Leysen et al. 1981) it could be argued that they prevent 5-CT-induced rotation indirectly by blocking D$_2$-binding sites (cf. haloperidol). However, unlike methiothepin and metergoline, haloperidol attenuates 5-CT-induced contralateral rotation but does not reveal the ipsilateral rotation responses. This would suggest that haloperidol is acting at a different site(s) to metergoline and methiothepin. It would be interesting to examine the effects of central 5,7-dihydroxytryptamine (5,7-DHT) lesions on both the ipsi- and contralateral rotation response to 5-CT. Such studies may provide further insight into the mechanisms underlying these rotational responses. Following intra-nigral infusion in the rat, 5-HT-induced contralateral rotation is potentiated by 5,7-DHT lesions (Oberlander et al. 1981) suggesting the involvement of postsynaptic 5-HT receptors within the substantia nigra. Based on both the functional and anatomical similarity between 5-HT$_{1D}$ and 5-HT$_{1B}$ receptors (see Introduction) this equivalent response in the rat may represent an *in vivo* 5-HT$_{1B}$ receptor model and also imply the involvement of postsynaptic 5-HT$_{1D}$ receptors in the guinea-pig model described here.

In conclusion, 5-CT-induced contralateral rotation in the guinea-pig may represent the first animal model for the study of 5-HT$_{1D}$ receptors *in vivo*. Although the results for the agonists strongly implicate an action at the 5-HT$_{1D}$ site, definitive classification is not possible without selective 5-HT$_{1D}$ antagonists. A functional involvement of 5-HT$_{1D}$ receptors with dopaminergic systems seems evident, a finding supported by the anatomical distribution of this receptor subtype (Waeber et al. 1989a, 1989b). This may suggest a role for selective 5-HT$_{1D}$ receptor compounds in the treatment of movement disorders.

References

Cole, N., Higgins, G. A., and Skingle, M. (1990). Inhibition of DOI-induced wet dog shakes in the guinea-pig by 5-HT$_2$ receptor antagonists. J. Psychopharmacol. 4: 307P.

Glennon, R. A., Titeler, M., and McKenny, J. D. (1984). Evidence for 5-HT$_2$ involvement in the mechanism of action of hallucinogenic agents. Life Sci. 35: 2505–2511.

Hoyer, D., and Middlemiss, D. N. (1989). Species differences in the pharmacology of terminal 5-HT autoreceptors in mammalian brain. Trends Pharmacol. Sci. 10: 130–132.

Leysen, J. E., Awouters, F., Kennis, L., Laduron, P. M., Vandenberk, J., and Janssen, P. A. J. (1981). Receptor binding profile of R41 468; a novel antagonist at 5-HT$_2$ receptors. Life Sci. 28: 1015–1022.

Luparello, T. J. (1967). Stereotaxic atlas of the forebrain of the guinea-pig. Basel (Switzerland) New York: S. Karger.

Oberlander, C., Hunt, P. F., Dumont, C., and Boissier, J. R. (1981). Dopamine independent rotational response to unilateral intra-nigral injection of serotonin. Life Sci. 28: 2595–2601.

Pazos, A., Cortes, R., and Palacios, J. M. (1985). Quantitative autoradiographic mapping of serotonin receptors in the rat brain. II. Serotonin$_2$ receptors. Brain Res. 346: 231–249.

Schoeffter, P., and Hoyer, D. (1989). How selective is GR43175? interactions with functional 5-HT$_{1A}$, 5-HT$_{1B}$, 5-HT$_{1C}$ and 5-HT$_{1D}$ receptors. Naunyn-Schmiedeberg's Arch Pharmacol. 340: 135–138.

Schoeffter, P., Waeber, C., Palacios, J. M., and Hoyer, D. (1988). The 5-hydroxytryptamine 5-HT$_{1D}$ receptor subtype is negatively coupled to adenylate cyclase in calf substantia nigra. Naunyn-Schmiedeberg's Arch Pharmacol. 337: 602–608.

Waeber, C., Schoeffter, P., Palacios, J. M., and Hoyer, D. (1989a). 5-HT$_{1D}$ receptors in guinea-pig and pigeon brain. Radioligand binding and biochemical studies. Naunyn-Schmiedeberg's Arch Pharmacol. 340: 479–485.

Waeber, C., Dietl, M. M., Hoyer, D., and Palacios, J. M. (1989b). 5-HT$_1$ receptors in the vertebrate brain. Regional distribution examined by autoradiography. Naunyn-Schmiedeberg's Arch Pharmacol. 340: 486–496.

Waeber, C., and Palacios, J. M. (1989). Serotonin-1 receptor binding sites in the human basal ganglia are decreased in Huntington's chorea but not in Parkinson's disease: a quantitative *in vitro* autoradiography study. Neurosci. 32: 337–347.

Waeber, C., Schoeffter, P., Hoyer, D., and Palacios, J. M. (1990a). The serotonin 5-HT$_{1D}$ receptor: A progress review. Neurochemical Res. 15: 567–582.

Waeber, C., Zhang, L., and Palacios, J. M. (1990b). 5-HT$_{1D}$ receptors in the guinea-pig brain: pre- and postsynaptic localizations in the striatonigral pathway. Brain Res. 528: 197–206.

Part III

Pathophysiological Roles and Prospects for New Therapies

Serotonin: Molecular Biology, Receptors and Functional Effects
ed. by J. R. Fozard/P. R. Saxena

5-Hydroxytryptamine and the Human Heart

A. J. Kaumann

*SmithKline Beecham Pharmaceutical, The Frythe, Welwyn, Hertfordshire AL6 9AR, UK;
Clinical Pharmacology Unit, Department of Medicine, University of Cambridge,
Addenbrooke's Hospital, Cambridge CXB2 2QQ, UK*

Summary. 5-Hydroxytryptamine (5-Ht) causes a variety of effects on human heart through
several 5-HT receptor subtypes. Both $5\text{-}HT_1$-like receptors and $5\text{-}HT_2$ receptors in large
coronary arteries mediate spasm. It is likely that vagal $5\text{-}HT_3$ receptors mediate reflex
bradycardia while sinoatrial $5\text{-}HT_4$-like receptors mediate tachycardia. Right atrial $5\text{-}HT_4$-like
receptors mediate increases in contractile force and relaxation through a cyclic AMP-dependent pathway.

Introduction

Evidence obtained over 30 years ago has established that 5-hydrox-
ytryptamine (5-HT) can both increase and decrease heart rate in man.
The mechanisms of these heart rate changes are still unknown but new
evidence allows plausible explanations to be forwarded. Recent experiments have shown that 5-HT is a potent positive inotropic agonist on
isolated preparations of human right atrium. 5-HT also contracts large
coronary arteries isolated from human heart. These findings will now be
discussed in detail, in particular, with regard to the 5-HT receptor
subtypes involved (Figure 1).

Tachycardia

Hollander et al. (1957) were the first to report that the intravenous
injection of 5-HT increases pulse rate in man. Tachycardia was subsequently observed by Le Messurier et al. (1959), Harris et al. (1960) and
Parks et al. (1960). Harris et al. (1960) also saw that bradycardia
preceded the tachycardia in some subjects but the most consistent effect
of 5-HT was to raise heart rate. Intravenously infused 5-HT also
increases respiration and forearm blood flow but these effects are
preceded by the initiation of tachycardia (Le Messurier et al. 1959;
Parks et al. 1960), suggesting a direct interaction of 5-HT with the
sinoatrial node.

What is the nature of human sinoatrial 5-HT receptors? Direct
evidence is lacking but a receptor that resembles $5\text{-}HT_4$ receptors in

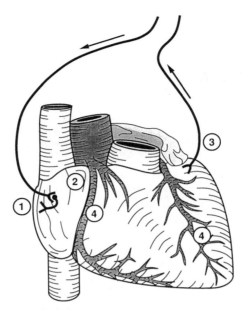

Figure 1. Putative 5-HT receptors in the human heart.
1 Sinoatrial node: 5-HT_4-like receptors?
2 Right atrium: 5-HT_4-like receptors.
3 Afferent vagus: 5-HT_3 receptors?
4 Large coronary artery smooth muscle: 5-HT_1-like receptors and 5-HT_2 receptors.

rodent brain (Dumuis et al. 1989; Bockaert et al. 1990) has been discovered in human right atrium (Kaumann et al. 1989a and 1990a) and it is likely that human sinoatrial 5-HT receptors are also 5-HT_4-like. 5-HT and certain benzamides, including renzapride, cause tachycardia in the pig which is blocked by high concentrations of 3α-tropanyl-1H-indole-3-carboxylic acid (ICS 205-930) (Villalón et al. 1990). Renzapride is a potent agonist in stimulating adenylyl cyclase in rodent brain and the effect is blocked by ICS 205-930 but not by other 5-HT receptor antagonists selective for 5-HT_{1A}, 5-HT_{1C}, 5-HT_{1D}, 5-HT_2 or 5-HT_3 receptors (Dumuis et al. 1988a and 1989; Bockaert et al. 1990). In view of the properties of renzapride and ICS 205-930, Villalón et al. (1990) suggested that 5-HT-induced porcine tachycardia is mediated through 5-HT_4 receptors.

Kaumann (1990) found that 5-HT, renzapride, cisapride and 5-carboxamido-tryptamine (5-CT) caused tachycardia in piglet right atria that was blocked by ICS 205-930. The rank order of potency was 5-HT > renzapride > cisapride > 5-CT and both renzapride and cisapride were partial agonists. Although this pattern resembles that expected of 5-HT_4 receptors, it is not identical. On embryonic mouse colliculi neurons both cisapride and renzapride are as potent as and more efficacious that 5-HT in stimulating adenylyl cyclase (Dumuis et al. 1989). In addition, the affinity of ICS 205-930 appears to be around 5 times higher for piglet sinoatrial 5-HT receptors than for rodent cerebral 5-HT_4 receptors (Dumuis et al. 1988b). In view of these differences piglet sinoatrial 5-HT receptors are for the time being designated 5-HT_4-like receptors (Kaumann 1990). Piglet sinoatrial 5-HT_4-like receptors greatly resemble human right atrial 5-HT receptors (Kaumann 1990). ICS 205-930 has similar affinity for human right atrial 5-HT receptors of both species (human, $-\log K_B = 6.7$, Kaumann et al. 1990a; porcine, $-\log K_B = 6.9$, Kaumann 1990) and renzapride is also a partial agonist in human (Kaumann et al. 1990b). Thus, human right atrial 5-HT receptors appear to resemble more piglet sinoatrial 5-HT_4-like receptors than rodent cerebral 5-HT_4 receptors and are, therefore, called 5-HT_4-like. The benzamide cisapride, the potent and most efficacious agonist on cerebral 5-HT_4 receptors, causes only moderate tachycardia in man (Bateman 1986), consistent with the existence of human sinoatrial 5-HT_4-like receptors. The future advent of potent and selective 5-HT_4 receptor antagonists will be useful in uncovering further differences or similarities between cardiac 5-HT_4-like receptors and cerebral 5-HT_4 receptors.

Bradycardia

The occasional slowing of heart rate caused by intravenously adminis-tered 5-HT in man (Harris et al. 1960) could conceivably be due to activation of 5-HT_3 receptors located on afferent vagal fibres thereby initiating a Bezold-Jarisch reflex. There is evidence for the location of such vagal 5-HT_3 receptors in the cat heart (Saxena et al. 1985; Mohr et al. 1986). The bradycardia in experimental animals observed with 5-HT is abolished by vagotomy, atropine and the 5-HT_3 receptor antagonist MDL 72222 (Fozard 1984; Mohr et al. 1986), consistent with the involvement of 5-HT_3 receptors on the afferent vagus and an activation of sinoatrial muscarinic receptors via the efferent vagus (Figure 1). The 5-HT-induced cardiac slowing can be accompanied by hypotension in man (Harris et al. 1960), rat (Fozard 1984) and cat (Saxena et al. 1985; Mohr et al. 1986). The hypotension is, at least in part, due to a decreased efferent sympathetic nerve acitvity, observed during the Bezold-Jarisch reflex in cat (Mohr et al. 1986) and a similar component may play a role in man.

Positive Inotropism

5-HT increases both contractile force and cyclin AMP through un-known 5-HT receptors in molusc hearts (Sawada et al. 1984), where

5-HT is an excitatory neurotransmitter. Similarly, it is well known that the mammalian excitatory neurotransmitter (−)-noradrenaline increases cardiac force and cyclin AMP through β-adrenoceptors, and (−)-adrenaline also causes these effects. For example in human right atrium (−)-noradrenaline can increase contractile force and cyclic AMP through β_1-adrenoceptors while (−)-adrenaline can do so through β_2-adrenoceptors (Kaumann et al. 1989b). Although 5-HT does not appear to be a neurotransmitter in human heart, a vestige of ancestral 5-HT receptors coupled to adenylyl cyclase could exist. Experiments did indeed reveal that 5-HT caused a positive inotropic effect and increased cyclic AMP in isolated human atrium (Kaumann et al. 1989a, 1990a; b). However, it quickly became apparent that cardiac 5-HT receptors of moluscs and man differ (Kaumann et al. 1990c). For example, LSD is a powerful inotropic agonist in molusc heart (effective at 10 fmol.l^{-1}; Greenberg 1960) but has no stimulant or blocking effects in human atrium (400 nmol.l^{-1}; Kaumann et al. 1990c).

5-HT is 5 times more potent than (−)-noradrenaline as an inotropic and lusitropic stimulant in human atrium and the inotropic efficacy of 5-HT is around 0.6 that of (−)-noradrenaline under conditions of blockade of neuronal uptake of the amines (Kaumann et al. 1990a). Like (−)-noradrenaline and (−)-adrenaline (Kaumann et al. 1989b), 5-HT also stimulates cyclic AMP-dependent protein kinase in human atrium (Kaumann et al. 1989a, 1990a; b; c), consistent with a subsequenct phosphorylation of phospholamban and troponin leading to accelerated cardiac relaxation (England 1983). Thus, the accelerated human atrial relaxation observed with (−)-noradrenaline through β_1-adrenoceptors, (−)-adrenaline through β_2-adrenoceptors (Hall et al. 1990) and 5-HT through 5-HT receptors occurs via a common final biochemical pathway.

What is the nature of human atrial 5-HT receptors? The positive inotropic effects of 5-HT are resistant to blockade by β-adrenoceptor and α-adrenoceptor antagonists, ruling out an indirect effect of 5-HT via release of (−)-noradrenaline. The effects are also resistant to the blockade of 5-HT$_{1A}$, 5-HT$_{1C}$, 5-HT$_{1D}$, 5-HT$_1$-like, 5-HT$_2$ and 5-HT$_3$ receptors (Kaumann et al. 1989a, 1990a; c). The effects are antagonised competitively with moderate potency by ICS 205-930 and renzapride (Kaumann et al. 1990a; b). Renzapride has also positive inotropic effects and is thus a partial agonist (Kaumann et al. 1990b). These properties and the increase in atrial cyclic AMP caused by 5-HT resemble those of 5-HT$_4$ receptors of rodent brain (Dumuis et al. 1989; Bockaert et al. 1990). However, as in piglet sinoatrial node, renzapride is less potent and less efficacious that 5-HT on human atrium. On the other hand, on embryonic mouse colliculi neurones renzapride is as potent as 5-HT and more efficacious than 5-HT in stimulating adenylyl cyclase (Dumuis et al. 1989). Thus, human right atrial 5-HT$_4$-like receptors are not necessarily identical to rodent cerebral 5-HT$_4$ receptors. Human right atrial 5-HT$_4$-like receptors appear similar to 5-HT receptors located in cholinergic neurones of guinea-pig ileum, on which renzapride is also a less potent agonist than 5-HT (Craig and Clarke 1990).

Human right atrial 5-HT-like receptors appear to be in concert with other receptors positively coupled to adenylyl cyclase (Kaumann et al. 1989c and 1990d). It has recently been discovered that chronic β_1-adrenoceptor blockade in patients causes inotropic hyperresponsiveness to ($-$)-adrenaline mediated through β_2-adrenoceptors without modifying β_1-adrenoceptor-mediated responses in human atrium (Hall et al. 1988 and 1990). Interestingly, chronic β_1-adrenoceptor blockade also enhances 5-HT-induced postitive inotropic responses in human right atrium (Kaumann et al. 1990d). Because inotropic responses to cyclic AMP are not modified and the apparent affinity of β_2-adrenoceptor agonists appears unchanged it has been proposed that chronic blockade of β_1-adrenoceptors enhances coupling of β_2-adrenoceptors to the adenylyl cyclase (Hall et al. 1990) and this may also happen with 5-HT$_4$-like receptors. The tonic effects of ($-$)-noradrenaline, mediated through β_1-adrenoceptors, appear to suppress partially the coupling of other receptors (including β_2 and 5-HT$_4$-like) to the adenylyl cyclase (Kaumann et al. 1989c). The mechanism of this crosstalk amongst receptors must involve modification of the function (and perhaps expression) of G$_s$ protein isoforms that couple receptors to the adenylyl cyclase and/or ion channels (Birnbaumer et al. 1990). Any model that accounts for changes in the function of G$_s$-coupled receptors due to chronic β_1-adrenoceptor blockade must assume that the receptors coexist in the same atrial cell. Such evidence has recently been produced by Sian Harding and Kaumann (unpublished experiments), showing that β_1-adrenoceptors, β_2-adrenoceptors and 5-HT$_4$-like receptors coexist in a single disaggregated human atrial cell amd mediate shortening of cell contraction with the corresponding agonists.

Coronary Spasm

5-HT is an important vasoactive compound released from aggregating platelets that can alter coronary artery function. Depending on the integrity of vascular endothelium, 5-HT can either relax or constrict coronary arteries. Relaxation has been observed with both 5-HT (Cocks and Angus 1983) and aggregating platelets (Cohen et al. 1983) in canine coronary arteries with intact endothelium. However, it is not yet known whether platelet-derived 5-HT can cause endothelium-dependent relaxation in human coronary artery (Vanhoutte and Shimokawa 1989). Toda and Okamura (1990) have only found evidence for 5-HT receptors mediating relaxation in coronary arteries of dog but not in coronary arteries of man and Japanese monkey.

Human large coronary arteries are contracted by 5-HT in a complex fashion. In arteries obtained from hearts of young persons (9–15 yr) 5-HT causes tonic contractions, while in arteries of older individuals 5-HT can elicit both rhythmic (phasic) and tonic contractions (Godfraind and Miller 1983; Godfraind et al. 1984). In human coronary arteries 5-HT is a more potent spasmogen than acetylcholine, noradrenaline or PGF_{2alpha} (Godfraind and Miller 1983; Godfraind et al. 1984). The high potency of 5-HT is further enhanced in the presence of low, barely contracting, concentrations of the thromboxane A mimetic U44069 in canine coronary arteries (Mullane et al. 1982). This finding needs still to be verified in human coronary arteries because it may be relevant to spasm elicited by 5-HT and related agonists.

Which receptors mediate the 5-HT-induced contractions of human coronary arteries? Frenken and Kaumann (1985) found evidence for two 5-HT receptors mediating 5-HT-induced contractions in endothelium-denuded canine coronary arteries, one blockable by ketanserin (i.e. $5-HT_2$) the other resistant to ketanserin. This finding was confirmed for human coronary arteries and presented to the Amsterdam 5-HT meeting by the author in October 1988. Since then, evidence has been accumulating that confirms the involvement of both $5-HT_2$ receptors and $5-HT_1$-like receptors (Connor et al. 1989; Kaumann and Brown 1989 and 1990; Toda and Okamura 1990). $5-HT_1$-like receptor mediated contractions in human coronary arteries have also been reported with sumatriptan (Connor et al. 1989), 5-carboxyamidotryptamine and methysergide (Toda and Okamura 1990). The 5-HT-induced contractions, resistant to ketanserin, are antagonised by methiothepin (a compound that blocks both $5-HT_2$ and $5-HT_1$-like receptors) (Kaumann and Brown 1990b). In a study by Kaumann and Brown (1990b) with coronary arteries obtained from 10 patients receiving cardiac transplants, 5-HT caused rhythmic and/or tonic contractions. The rhythmical contractions caused by 5-HT were usually resistant to blockade by ketanserin but blocked by methiothepin, consistent with mediation through $5-HT_1$-like receptors. In 3 patients 5-HT caused only ketanserin-resistant contractions but blocked by methiothepin, again consistent with a mediation via $5-HT_1$-like receptors. In the other 7 patients both $5-HT_1$-like receptors and $5-HT_2$ receptors mediated 5-HT-induced spasm. In one coronary artery from an ischaemic heart patient 5-HT-induced contractions of a prestenotic segment which were predominantly of $5-HT_1$-like nature while contractions of the poststenotic segment which were predominantly of $5-HT_2$ nature (see also Kaumann and Brown 1990a). With regard to the participation of $5-HT_1$-like and $5-HT_2$ receptors no systematic difference was detected between coronary arteries with or without atheromatous plaque. The experiments in these ten patients suggest that on average 5-HT$_1$-like receptors contribute more than 5-HT$_2$ receptors to 5-HT-induced spasm in human coronary arteries. However, a high number of patients needs to be studied because other authors have suggested that the participation of 5-HT$_2$ receptors could be more important than that of 5-HT$_1$-like receptors in 5-HT-induced spasm (Connor et al. 1989; Toda and Okamura 1990).

The discovery of 5-HT$_1$-like receptors may have some bearing in patients with coronary spasm and Prinzmetal angina. Freedman et al. (1984) showed that ketanserin was ineffective in patients with Prinzmetal angina, ruling out a role of 5-HT. However, 5-HT could actually cause coronary spasm through 5-HT$_1$-like receptors because these are not blocked by ketanserin.

References

Bateman, D. N. (1986). The action of cisapride on gastric emptying and the pharmacodynamics and pharmocokinetics of oral diazepam. Eur. J. Clin. Pharmacol. 30: 205–208.

Birnbaumer, L., Abramowitz, J., and Brown, A. M. (1990). Receptor-effector coupling by G proteins. Biochem. Biophys. Acta. 1031: 103–224.

Bockaert, J., Sebben, M., and Dumuis, A. (1990). Pharmacological characterization of 5-hydroxytryptamine$_4$ (5-HT$_4$) receptors positively coupled to adenylate cyclase in adult guinea pig hippocampal membranes: Effect of substituted benzamide derivatives. Mol. Pharmacol. 37: 408–411.

Cocks, T. M., and Angus, J. A. (1983). Endothelium-dependent relaxation of coronary arteries by noradrenaline and serotonin. Nature 305: 627–630.

Cohen, R. A., Shepherd, J. T., and Vanhoutte, P. M. (1983). Inhibitory role of the endothelium in the responses of isolated coronary arteries to platelets. Science 221: 173–174.

Connor, H. E., Feniuk, W., and Humphrey, P. P. A. (1989). 5-Hydroxytryptamine contracts human coronary arteries predominantly via 5-HT$_2$ receptor activation. Eur. J. Pharmacol. 161: 91–94.

Craig, D. A., and Clarke, D. (1990). Pharmacological characterisation of a neuronal receptor for 5-hydroxytryptamine in guinea pig ileum with properties similar to the 5-hydroxytryptamine$_4$ receptor. J. Pharmacol. Exp. Ther. 252: 1378–1386.

Dumuis, A., Bouhelal, R., Sebben, M., Cory, R., and Bockaert, J. (1988a). A nonclassical 5-hydroxytryptamine receptor positively coupled with adenylate cyclase in the central nervous system. Mol. Pharmacol. 34: 880–887.

Dumuis, A., Bouhelal, R., Sebben, M., and Bockaert, J. (1988b). A 5-HT receptor in central nervous system, positively coupled with adenylate cyclase, is antagonised by ICS 205-930. Eur. J. Pharmacol. 146: 187–188.

Dumuis, A., Sebben, M., and Bockaert, J. (1989). The gastrointestinal prokinetic benzamide derivatives are agonists at the non-classical 5-HT receptor (5-HT$_4$) positively coupled to adenylate cyclase in neurones. Naunyn-Schmiedeberg's Arch. Pharmacol. 340: 403–410.

England, P. (1983). Phosphorylation of cardiac muscle contractile proteins. In Cardiac Metabolism ed. Drake-Holland, A. J., and Noble, MIM pp. 365–389 John Wiley & Sons Limited, New York.

Fozard, J. (1984). MDL 72222, a potent and highly selective antagonist at neuronal 5-hydroxytryptamine receptors. Naunyn-Schmiedeberg's Arch. Pharmacol. 326: 36–44.

Freedman, S. B., Chierchia, S., Rodriguez, P., Bugiardim, R., Smith, B., and Maseri, A. (1984). Ergonovine-induced myocardial ischemia: no role for serotonergic receptors? Circulation 70: 178–183.

Frenken, M., and Kaumann, A. J. (1985). Ketanserin causes surmountable antagonism of 5-hydroxytryptamine-induced contractions in large coronary arteries of dog. Naunyn-Schmiedberg's Arch. Pharmacol. 328: 301–303.

Goodfraind, T., and Miller, R. C. (1983). Specificity of action of Ca^{++} entry blockers. A comparison of their actions in rat arteries and in human coronary arteries. Circ. Res. Suppl. 52: 81–93.

Goodfraind, T., Finet, M., Socrates Lima, J., and Miller, R. C. (1984). Contractile activity of human coronary arteries and human myocardium *in vitro* and their sensitivity to calcium entry blockade by nifedipine. J. Pharmacol. Exp. Ther. 230: 514–518.

Greenberg, M. J. (1960). Structure-activity relationship of tryptamine analogues on the heart of venus mercenaria. Br. J. Pharmacol. 15: 375–388.

Hall, J. A., Kaumann, A. J., and Brown, M. J. (1990). Selective β_1-Adrenoceptor Blockade Enhances Positive Inotropic Responses to Endogenous Catecholamines Mediated Through β_2-Adrenoceptors in Human Atrial Myocardium. Circ. Res. 66: 1610–1623.

Hall, J. A., Kaumann, A. J., Wells, F. C., Brown, J. J. (1988). β_2-Adrenoceptor mediated inotropic responses of human atria: receptor subtype regulation by atenolol. Br. J. Pharmacol. 193: 116P.

Harris, P., Fritts, H. W., and Cournand, A. (1960). Some circulatory effects of 5-hydroxytryptamine in man. Circulation 21: 1134–1139.

Hollander, W., Michelson, A. L., and Wilkins, R. W. (1957). Serotonin and antiserotonins. I. Their circulatory, respiratory and renal effects in man. Circulation 16: 246–255.

Kaumann, A. J. (1990). Piglet sinoatrial 5-HT receptors resemble human atrial $5-HT_4$-like receptors. Naunyn-Schmiedeberg's Arch. Pharmacol. 342: 619–622.

Kaumann, A. J., and Brown, A. M. (1990). Allosteric modulation of arterial $5-HT_2$ receptors. P. R. Saxena, D. I. Wallis, W. Wouters and P. Bevan (eds), Cardiovascular Pharmacology of 5-Hydroxytryptamine, pp. 127–142. Kluwer Academic Publishers, Dordrecht (Netherlands).

Kaumann, A. J., and Brown, A. M. (1990b). Human coronary artery spasm induced by 5-hydroxytryptamine: role of receptor subtypes, plaque and stenosis. Nauny-Schmiedeberg's Arch. Pharmacol. 1341: R90.

Kaumann, A. J., Sanders, L., Brown, A. M., Murray, K. J., and Brown, M. J. (1989a). A receptor for 5-hydroxytryptamine in human atrium. Br. J. Pharmacol. 98: 664P.

Kaumann, A. J., Hall, J. A., Murray, K. J., Wells, F. C., and Brown, M. J. (1989b). A comparison of the effects of adrenaline and noradrenaline on human heart: the role of β_1- and β_2-adrenoceptors in the stimulation of adenylate cyclase and contractile force. Eur. Heart J. 10 (Suppl. B): 29–37.

Kaumann, A. J., Murray, K. J., Brown, A. M., Hall, J. A., Sanders, L., and Brown, M. J. (1989c). Transregulation of G_S protein-coupled receptors by chronic β_1-adrenoceptor blockade in human atrium. J. Mol. Cell. Cardiol. 21 (Suppl. III) S14.

Kaumann, A. J., Sanders, L., Brown, A. M., Murray, K. J., and Brown, M. J. (1990a). A 5-hydroxytryptamine receptor in human atrium. Br. J. Pharmacol. 100: 879–885.

Kaumann, A. J., Sanders, L., Brown, A. M., Murray, K. J., and Brown, M. J. (1990b). Human atrial 5-HT receptors: similarity to rodent neuronal $5-HT_4$ receptors. Br. J. Pharmacol. 100: 319P.

Kaumann, A. J., Murray, K. J., Brown, A. M., Frampton, J. E., Sanders, L., and Brown, M. J. (1990c). Heart 5-HT receptors. A novel 5-HT receptor in human atrium. Serotonin: From Cell Biology to Pharmacology & Therapeutics. ed. Paoletti, R., Vanhoutte, P., Brunello, N., and Maggi, F. M. pp. 347–354. Dordrecht, Boston, London: Kluwers.

Kaumann, A. J., Sanders, L., and Brown, M. J. (1990d). Chronic β_1-adrenoceptor blockade enhances positive inotropic responses to 5-hydroxytryptamine in human atrium. J. Mol. Cell. Cardiol. Vol. 22 (Suppl. III) S2.

Le Messurier, D. H., Schwartz, C. J., and Whelan, R. F. (1959). Cardiovascular effects of intravenous infusions of 5-hydroxytryptamine in man. Br. J. Pharmacol. 14: 246–250.

Mohr, B., Bom, A. H., Kaumann, A. J., Thamer, V. (1987). Reflex inhibition of efferent renal sympathetic nerve activity by 5-hydroxytryptamine and nicotine is elicited by different epicardial receptors. Pflugers Arch 409: 145–151.

Mullane, K. M., Bradley, G., and Moncada, S. (1982). The interactions of platelet-derived mediators on isolated canine coronary arteries. Eur. J. Pharmacol. 84: 115–118.

Parks, V. J., Sandison, A. G., Skinner, S. L., and Whelan, R. F. (1960). The stimulation of respiration by 5-hydroxytryptamine in man. J. Physiol. 151: 342–251.

Saxena, P. R., Mylecharane, E. J., and Heiligers, J. (1985). Analysis of the heart rate effects of 5-hydroxytryptamine in the cat; Mediation of tachycardia by 5-HT$_1$-like receptors. Naunyn Schmiedebergs Arch. Pharmacol. 330: 121–129.

Sawada, M., Ichinose, M., Ito, I., Maeno, T., and Mcadoo, D. J. (1984). Effects of 5-hydroxytryptamine on membrane potential, contractility, accumulation of cyclic AMP, and Ca^{2+} movements in anterior aorta and ventricle of aplysia. J. Neurophysiol. 51: 361–374.

Toda, N., and Okamura, T. (1990). Comparison of the responses to carboxamidotryptamine and serotonin in isolated human, monkey and dog coronary arteries. J. Pharmacol. Exp. Ther. 253: 676–682.

Vanhoutte, P. M., and Shimokawa, H. (1989). Endothelium-derived relaxing factor and coronary vasospam. Circulation 80: 1–9.

Villalón, C. M., den Boer, M. O., Heiligers, J. P. C., and Saxena, P. R. (1990). Mediation of 5-hydroxytryptamine-induced tachycardia in the pig by the putative 5-HT$_4$ receptor. Br. J. Pharmacol. 100: 665–667.

Serotonin: Molecular Biology, Receptors and Functional Effects
ed. by J. R. Fozard/P. R. Saxena
© 1991 Birkhäuser Verlag Basel/Switzerland

The Effects of BRL 24924 (Renzapride) on Secretion of Gastric Acid and Pepsin in Dogs

B. Johansen, J. M. Lyngsø, and K. Bech

Department of Surgical Gastroenterology, Odense University Hospital, 5000 Odense C, Denmark

Summary. The effects of serotonin (5-HT) and the 5-HT$_3$ receptor antagonist, BRL 24924, on gastric acid and pepsin secretion were evaluated in conscious dogs with a gastric fistula. In accordance with previous studies serotonin inhibited gastric secretion. BRL 24924 possessed both inhibitory and stimulatory effects depending on the background stimulation. The inhibitory effects of serotonin on gastric secretion were counteracted by BRL 24924 only during infusion of pentagastrin, which by itself releases serotonin. 5-HT$_3$ receptors are of minor importance in the control of gastric secretion.

Introduction

Serotonin (5-Hydroxytryptamine, 5-HT) inhibits gastric secretion of acid and pepsin *in vivo* in animals and man (Thompson 1971, Bech 1988), which can be blocked by β-adrenoceptor antagonists and methysergide (Bech 1986a, 1988, Bech and Johansen 1989). The role of 5-HT$_3$ receptors in the regulation of gastric secretion has not been evaluated previously. BRL 24924 is a 5-HT$_3$ receptor antagonist with significant actions on gastric emptying and emesis evoked by radiation and cytotoxic drugs (Andrews and Hawthorn 1987, Hawthorn et al. 1988, Staniforth and Corbett 1987).

The purpose of the present study was to analyze the effects of serotonin and BRL 24924 on gastric secretion of acid and pepsin in non-anaesthetized dogs.

Materials and Methods

Conscious dogs (Beagles, n = 6) with a gastric fistula were used. The experiments were carried out during a fasted state which was established by registration of motility by intra-luminal pressure transducers showing patterns of phase I or II in the interdigestive motility just before commencement of experiments.

Experiments Performed

Single dose experiments were performed with intravenous infusion of the drugs together with isotonic NaCl with KCl (10 mmol/l). An initial series of control studies was performed using stimulation (2 hours) with pentagastrin (0.25 and 1.0 µg/kg/h, Peptavlon®, ICI), bethanechol (80 µg/kg/h, Urecholine®, Merck, Sharpe and Dohme) or histamine (15 µg/kg/h, Histamin™, SAD). During this continuous stimulation, experiments with 5-HT (10 µg/kg/min, 5-Hydroxytryptamine creatinine sulphate, Sigma) with or without BRL 24924 (0.5 and 1.0 mg/kg, (±)-(endo)-4-amino-5-chloro-2-methoxy-N-(1-azabicyclo[3.3.1]non-4-yl, Beecham) were carried out. BRL 24924, by itself, was analyzed for effects on gastric secretion. Atropine (Atropin™, DAK) was used in the dose of 0.1 mg/kg.

Gastric Acid Secretion

Gastric secretion was collected every 15 min and the volume measured to the nearest 0.5 ml. The acid content per ml was measured by *in vitro* titration to pH 7.0 (Radiometer, TTA 81 Autopipetting Titration Station). Acid secretion was expressed as mmol H^+/15 min.

Gastric Pepsin Secretion

The pepsin content was determined immediately after the experiments by a haemoglobin substrate method (enzymatic degradation technique) (Bech and Andersen 1985, Bech 1986a). An aliquot (0.1 ml) of the gastric juice was diluted in 4.9 ml 0.01 N HCl, and 0.5 ml of this solution was added to a haemoglobin solution (2.0 ml of 1.55 mmol/l haemoglobin with 0.5 ml 0.3 N HCl). The reaction was performed in a water bath (25°C) and stopped after exactly 10 min with 5 ml 0.3 N trichloroacetic acid. The content was filtered (Schleicher and Schül® 589 filters) and measured at 280 nm on a spectrophotometer (Beckman 25). Crystalline pepsin from pig stomach (Koch-Light Ltd) with an activity of 2200 units/mg was used for standard solutions and curves. Pepsin secretion was expressed as 10^3 Units/15 min.

Statistical Evaluations

Results were obtained every 15 min; the data are presented as median values (range) each for six dogs. The Wilcoxon test for paired differences was used, considering $P < 0.05$ significant.

Results

Effects of Stimulants

Pentagastrin, bethanechol and histamine (Figures 1–3) increased gastric secretion by volume and acid output inducing a steady state level in the second hour of infusion. For pepsin secretion, a stimulatory effect was seen during infusion of pentagastrin (32.5×10^3 Units/15 min) and bethanechol (19.9×10^3 Units/15 min) whereas an inhibitory effect was found for histamine inducing a level of 1.7×10^3 Units/15 min in comparison to 9.5×10^3 Units/15 min obtained during fasting.

Effects of 5-HT

5-HT inhibited pentagastrin and bethanechol stimulated gastric acid secretion by decreasing volume and titratable acidity (Figures 1–2). For pepsin secretion, similar inhibitory effects were obtained for 5-HT

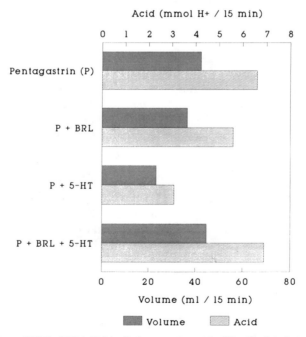

Figure 1. Effects of BRL 24924 (0.5 mg/kg) on pentagastrin (P) stimulated gastric secretion (volume and acid) with and without 5-HT (10 μg/kg/min).

Bethanechol

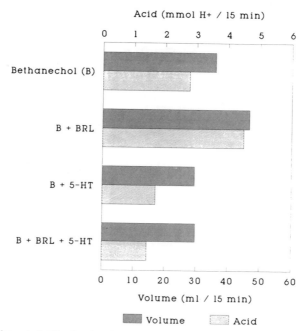

Acid (mmol H+ / 15 min)

Volume (ml / 15 min)

Volume   Acid

Figure 2. Bethanechol (B) stimulated secretion (volume and acid) during infusion of BRL 24924 (0.5 mg/kg) with and without 5-HT (10 μg/kg/min).

during stimulation by pentagastrin and bethanechol inducing 10.6 and 7.2×10^3 Units/15 min, respectively. During histamine stimulation (Figure 3), 5-HT increased non-significantly the volume of gastric secretion and the pepsin secretion $(8.9 \times 10^3$ Units/15 min) but was without effects on acid secretion.

Effects of BRL 24924

During pentagastrin stimulation, BRL 24924 (0.5 mg/kg) decreased non-significantly acid secretion (volume and acid) (Figure 1) and significantly pepsin secretion $(5.4 \times 10^3$ Units/15 min). Similar results were obtained for the dose of 1 mg/kg (not shown). When the dose of pentagastrin was reduced to 0.25 μg/kg/h, BRL 24924 showed significant *stimulatory* effects on volume and acid output but was without effects on pepsin secretion (not shown).

During infusion of bethanechol, BRL 24924 (0.5 mg/kg) stimulated significantly gastric acid secretion (volume and acid, Figure 2) and

Histamine

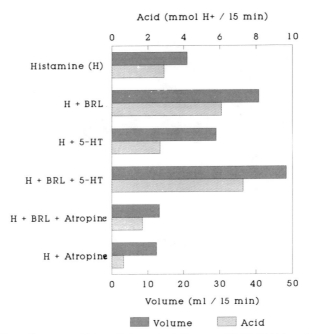

Figure 3. Effects of atropine (0.1 mg/kg) on the influence of BRL 24924 on histamine (H) stimulated gastric secretion (volume and acid) with and without 5-HT (10 μg/kg/min).

pepsin secretion (31.1×10^3 Units/15 min). Similar results were obtained for BRL 24924 at a dose of 1 mg/kg (not shown).

For histamine stimulation (Figure 3), BRL 24924 (0.5 and 1 mg/kg) increased significantly secretory volume, acid and pepsin (11.8×10^3 Units/15 min) secretion, an effect which was counteracted by atropine.

BRL 24924 counteracted the acid inhibitory effect of 5-HT only during pentagastrin stimulation (Figure 1).

Discussion

We found BRL 24924 to stimulate gastric secretion as measured by volume of output, acid and pepsin secretion. Atropine blocked the stimulatory action of BRL 24924 obtained during a background stimulation by histamine. 5-HT inhibited gastric secretion and BRL 24924 counteracted this inhibitory effect only during infusion of pentagastrin.

In accordance with previous studies in dogs (Bech 1986a and 1988, Bech and Johansen 1989) we found stimulatory effects (volume, acid and pepsin) by infusion of pentagastrin, bethanechol and histamine with the exception of an inhibitory effect of histamine on pepsin secretion. Furthermore, serotonin inhibited gastric secretion, which is in agreement with data obtained earlier (Bech and Andersen 1985, Bech 1986b, Bech and Johansen 1989). Contrary to results obtained earlier, pepsin secretion and volume output were increased by infusion of serotonin, 10 μg/kg/min, during a background stimulation by histamine. This difference could be due to individual responses of the dogs to histamine (15 μg/kg/h) which has been noted before (Bech 1986a). In this study the median acid output was 2.9 mmol H^+/15 min whereas it was 4.6 mmol H^+/15 min in the previous study.

During pentagastrin stimulation we found dual actions of BRL 24924 showing a stimulatory effect on acid secretion by a low dose of pentagastrin whereas the opposite effect was found for a high dose of pentagastrin. This divergence in effect or biphasic response could be due to a blockade of serotonin effects as pentagastrin by itself stimulates the output of 5-HT (Izzat and Waton 1987). In agreement with the effects of BRL 24924, methysergide, a non-selective 5-HT receptor antagonist, induced a potentiation of pentagastrin stimulated gastric secretion (Bech and Andersen 1985, Bech 1986b).

BRL 24924 has been suggested to possess two possible modes of action. 1. to act as a 5-HT$_3$ receptor antagonist without cholinergic activity. 2. to act by releasing acetylcholine via 5-HT$_4$ receptors.

The possible mechanism of action for the 5-HT$_3$ receptors on gastric secretion is not known. The serotonin-counteracting effect of BRL 24924, as found during pentagastrin stimulation, could be due to an inhibition of the serotonin effect by an action on the 5-HT$_3$ receptors. If so, the serotonin response of gastric secretion should be blocked by other 5-HT$_3$ antagonists and should be mimicked by 5-HT$_3$ agonists which we have shown not to be the case.

5-HT$_2$ receptors have been shown not to play any role in the regulation of gastric secretion (Bech 1988; Bech and Johansen 1989). Selective 5-HT$_1$ receptor antagonists have not been examined. It is notable that β-adrenoceptor antagonists were very potent blockers of serotonin effects on gastric secretion and as β-adrenoceptor antagonists may act as 5-HT$_1$ receptor antagonists, this matter needs further analysis.

With respect to a possible action via acetylcholine a prejunctionally mediated release of acetylcholine has been shown to mediate the contractile actions of BRL 24924 on the guinea-pig ileum (Sanger 1987). Furthermore, BRL 24924 had no effects on contractions evoked by exogenous acetylcholine (Sanger 1987). The secretory stimulation obtained during infusion of histamine could be explained by the ability of

BRL 24924 to increase cholinergic activity as atropine blocked this stimulation.

A 5-HT$_4$ receptor has been proposed in brain and gut (Clarke et al. 1989; Dumis et al. 1989). This receptor site for 5-HT triggers cholinergically mediated contractions in guinea-pig ileum and is resistant to conventional 5-HT$_1$ and 5-HT$_2$ receptor antagonists. The site has also been clearly distinguished from a 5-HT$_3$ receptor (Dumuis et al. 1989).

In conclusion, in the present study, BRL 24924 acts as a 5-HT$_3$ receptor antagonist but it may also stimulate gut cholinergic activity via 5-HT$_4$ receptors.

Acknowledgements

This study was supported by grants from the Danish Medical Research Foundation (No. 12-8472), Nordic Insulin Foundation Committee and the foundations of P. Carl Petersen, Direktør Ib Henriksen, Novo, Carlsberg, I. O. Buck, Lægevidenskabens Fremme and Fyns Amt. The authors thank Karen Bentsen for technical assistance, Beecham Pharmaceuticals (BRL 24924) and Merck, Sharpe and Dohme (bethanechol) for supplying drugs.

References

Andrews, P. L. R., and Hawthorn, J. (1987) Evidence for an extra-abdominal site of action for the 5-HT$_3$-receptor antagonist BRL 24924 in the inhibition of radiation evoked emesis in the ferret. Neuropharmacology 26(9): 1367–1370.

Bech, K., and Andersen, D. (1985) Effect of serotonin on pentagastrin stimulated gastric acid secretion and gastric antral motility in dogs with gastric fistula. Scand. J. Gastroenterol. 20: 1115–1123.

Bech, K. (1986a) Exogenous serotonin and histamine-stimulated gastric acid and pepsin secretion in dogs. Scand. J. Gastroenterol. 21: 1205–1210.

Bech, K. (1986b) Effect of serotonin on bethanechol stimulated gastric acid secretion and gastric antral motility in dogs. Scand. J. Gastroenterol. 21: 655–661.

Bech, K. (1988) The role of somatostatin and serotonin in the β-adrenoceptor regulation of gastric function. Dan. Med. Bull. 35(2): 122–140.

Bech, K., and Johansen, B. (1989) Effects of serotonin on gastric secretion in vivo. In: Vanhoutte, P. M., Paoletti R. (ed.), Serotonin: from cell biology to pharmacology and therapeutics. Amsterdam, Kluwer, pp. 229–233.

Clarke, D. E., Craig, D. A., and Fozard, J. R. (1989) The 5-HT$_4$ receptor: naughty, but nice. Trends Pharmacol. Sci. 10: 385–386.

Dumuis, A., Sebben, M., and Bockaert, J. (1989) The gastrointestinal prokinetic derivatives are agonists at the non-classical 5-HT receptor (5-HT$_4$) positively coupled to adenylate cyclase in neurons. Naunyn-Schmiedeberg's Arch. Pharmacol. 340: 403–410.

Hawthorn, J., Ostler, K. J., and Andrews, P. L. R. (1988) The role of abdominal visceral innervation and 5-hydroxy-tryptamine M-receptors in vomiting induced by the cytotoxic drugs cyclophosphamide and cis-platin in the ferret. Quart. J. Exp. Physiol. 73: 7–21.

Izzat, A., and Waton, N. G. (1987) Release of 5-hydroxy-tryptamine by pentagastrin and its role in the "fade" of stimulated gastric secretion in cats. J. Physiol. 383: 499–507.

Sanger, G. J. (1987) Increased gut cholinergic activity and antagonism of 5-hydroxytryptamine M-receptors by BRL 24924: potential clinical impotance of BRL 24924. Br. J. Pharmac. 91: 77–87.

Staniforth, D. H., and Corbett, R. (1987) The effect of BRL 24924 on upper gastrointestinal tract activity. Br. J. Clin. Pharmac. 24: 263–264.

Thompson, J. M. (1971) Serotonin and the alimentary tract. Res. Commun. Chem. Pathol. Pharmacol. 2: 687–781.

Serotonin: Molecular Biology, Receptors and Functional Effects
ed. by J. R. Fozard/P. R. Saxena
© 1991 Birkhäuser Verlag Basel/Switzerland

Constipation Evoked by 5-HT$_3$ Receptor Antagonists

G. J. Sanger, K. A. Wardle, S. Shapcott, and K. F. Yee

SmithKline Beecham Pharmaceuticals, Medicinal Research Centre, Coldharbour Road, The Pinnacles, Harlow, Essex CM19 5AD, UK

Summary. In human volunteers, mild constipation has been reported as a side-effect of high doses of granisetron, ondansetron and ICS 205-930. This could be mimicked using conscious guinea-pigs, in which the output of faecal pellets were counted over a 12 h period. In guinea-pig isolated colon, faecal pellets are expelled spontaneously. This expulsion was prevented by morphine or clonidine; granisetron or ICS 205-930 (10^{-7} or 10^{-6} M) reduced the rate of pellet expulsion. In contrast to the conscious animal experiments, ondansetron (10^{-6} M) was without activity *in vitro*. This study suggests that the mechanisms by which 5-HT$_3$ receptor antagonists cause constipation are complex, and that not all antagonists behave in the same manner.

Introduction

Several 5-HT$_3$ receptor antagonists exist (see King and Sanger 1989), but to investigate the involvement of 5-HT$_3$ receptors in normal gut function it is important to study only those antagonists which do not also activate 5-HT$_4$ receptors. Examples of the latter include zacopride and BRL 24924, both of which are effective gut motility stimulants because of 5-HT$_4$ receptor activation (see Sanger and King 1988).

In normal human volunteers, granisetron and ondansetron have little or no effect on the motility of the stomach and small intestine (Staniforth 1989; Talley et al. 1989; Gore et al. 1990), although there have been some reports linking ICS 205-930 with an increased upper gut motility (Meleagros et al. 1987; Akkermans et al. 1988). By contrast, there have been consistent reports of a change in human volunteer lower bowel function after administration of selective 5-HT$_3$ receptor antagonists. These include mild constipation (Gore et al. 1990; Stacher et al. 1989; Upward et al. 1990), colonic stasis (Gore et al. 1990; Talley et al. 1989) and colonic hypermotility (Stacher et al. 1989). The latter, if non-propulsive, would lead to constipation (Connel 1962). These reports suggest that 5-HT$_3$ receptors may play a role in the normal physiology of lower bowel function. In view of the significance of such a suggestion, we have, therefore, studied the actions of granisetron, ondansetron and ICS 205-930 on guinea-pig lower bowel function.

Materials and Methods

Faecal Pellet Output from Conscious Guinea-pigs

Male albino guinea-pigs (Duncan Hartley, 350–550 g) were transferred to a constant low-level light room, 7 days before experimentation. They were allowed free access to food and water. 24 hours before each experiment, the guinea-pigs were weighed and split randomly into 4 groups of 6 animals. These were transferred, in pairs, to wire-bottom metabolic cages, again with free access to food and water. On the day of the experiment each group of guinea-pigs were injected intraperitoneally (i.p.) with volumes of room temperature saline (0.35–0.55 ml) and with 0.1, 1.0 or 10 mg kg^{-1} of either granisetron, ICS 205-930 or ondansetron. The number of faecal pellets produced by each pair of guinea-pigs at 1, 3, 5, 8, 10 and 12 h post dosing were then counted.

Faecal Pellet Output from Guinea-pig Isolated Colon

Adult male albino guniea-pigs (250–450 g weight) were killed by cervical dislocation and mid to distal colon (about 25 cm) was removed. These were immediately placed in a horizontal organ bath containing Krebs solution (mM: NaCl 121.5, $CaCl_2$ 2.5, KH_2PO_4 1.2, KCl 4.7, $MgSO_4$ 1.2, $NaHCO_3$ 25.0, glucose 5.6) bubbled with 5% CO_2 and O_2 and maintained at 37°C; the Krebs solution was then perfused through the bath at a constant rate of 50 ml min^{-1}, throughout the experiment.

The colon was secured within the bath by 4 frog heart clips attached at regular intervals along a stainless steel bar fitted to the base of the organ bath; care was taken to ensure that faecal pellet movement was not obstructed by the clips. Movement of the faecal pellets always occurred in an anal direction and this was detected using isotonic force transducers attached to the colon via frog heart clips; for these experiments, the times taken for each successive pellet to arrive at a transducer approximately half way along the colon, were recorded. After approximately consistent control readings were obtained (30–40% of pellets; n = 3 minimum), the bath was drained and re-filled with Krebs solution containing the compound under investigation. If the compound failed to prevent or reduce pellet movement, the colon was allowed to empty. If the compound reduced or prevented pellet movement, it was left in contact with the colon for 60 min. If the colon had not discharged its contents within this period, the bath was drained, re-filled and then perfused for 60 min with Krebs solution containing the same compound plus naloxone (10^{-7} M). Pellets in transit under the transducer at the time of compound administration were disregarded.

Statistical Analysis

For the experiments *in vivo*, the results are expressed as means $\pm$ standard error of the mean (sem) and the effects of the 5-HT$_3$ receptor antagonists were compared with the effects of saline (dosed on the same day) using the Student's unpaired t-test. For the *in vitro* experiments, the effects of the various treatments were compared with the faecal pellet transit times measured prior to any drug addition to the bathing solution. Analysis was carried out using a weighted analysis of variance method on log transformed data. The mean $\pm$ sem were calculated and re-expressed in the original scale (i.e. antilog of mean = geometric mean).

Results

Faecal Pellet Output from Conscious Guinea-pigs

Granisetron, ondansetron and ICS 205-930 0.1–10 mg kg^{-1} i.p. reduced the number of faecal pellets excreted by guinea-pigs over a 12 h period. For granisetron and ondansetron, there was evidence of dose-dependence in their action, whereas for ICS 205-930, dose-dependence was not clear (Figure 1).

Faecal Pellet Output from Guinea-pig Isolated Colon

The addition of solvent (Krebs solution) to the bathing solution did not affect the faecal pellet transit time. Similarly, the addition of naloxone 10^{-7} M was without effect (Figure 2). This consistency in the time of pellet movement was probably due to the shortening of the colon as each pellet was expelled. In this way, successive pellets did not travel increasingly long distances in order to reach the transducer.

Granisetron 10^{-7} or 10^{-6} M and ICS 205-930 10^{-6} M reduced the faecal pellet transit time; for granisetron this action appeared to be concentration-dependent. By contrast, ondansetron 10^{-6} M was without significant effect (Figure 2).

The ability of granisetron 10^{-6} M to slow faecal pellet transit was reversed by the concomitant addition of naloxone 10^{-7} M to the bathing solution (Figure 3). Similarly, this concentration of naloxone reversed the abilities of morphine 10^{-7} M and clonidine 10^{-8} M to delay faecal pellet transit (Figure 3).

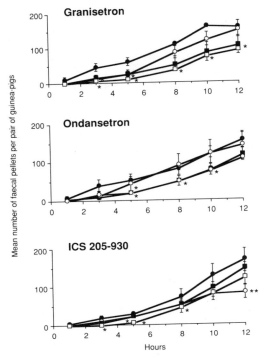

Figure 1. Inhibition of faecal pellet defecation by granisetron, ondansetron and ICS 205-930. In each experiment, the 5-HT$_3$ receptor antagonists were given by the i.p. route (● saline; ○ 0.1 mg kg^{-1}; ■ 1 mg kg^{-1}; □ 10 mg kg^{-1}) and the number of faecal pellets excreted were counted at intervals over a 12 h observation period. The results are given as means ± s.e.m. and were analysed statistically using a Student's unpaired t-test; *$p < 0.05$ compared with saline, n = three pairs of guinea-pigs per group.

Discussion

Granisetron, ICS 205-930 and ondansetron at doses of 0.1 and/or 1 and 10 mg kg^{-1} i.p. reduced the output of faecal pellets from conscious guinea-pigs, measured over a 12 h period. These doses are higher than those which maximally antagonise the Bezold-Jarisch reflex in anaesthetised rats (Richardson et al. 1985; Butler et al. 1988; Sanger and Nelson 1989). For granisetron, the doses are also higher than those which maximally prevented cytotoxic drug-induced vomiting in ferrets (Bermudez et al. 1988), whereas for ondansetron and ICS 205-930, the reported anti-emetic doses in ferrets (i.v. route) are similar to those which caused constipation in guinea-pigs (i.p. route) (Costall et al. 1986; 1987). Whilst it may not be strictly correct to compare the efficacy of 5-HT$_3$ receptor antagonists between different species (see Lattimer et al. 1989), a similar difference in relative doses is also seen in human

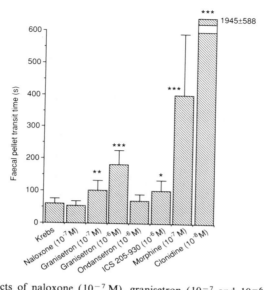

Figure 2. The effects of naloxone (10^{-7} M), granisetron (10^{-7} and 10^{-6} M), ondansetron (10^{-6} M), ICS 205-930 (10^{-6} M), morphine (10^{-7} M) and clonidine (10^{-8} M) on endogenous faecal pellet transit time in guinea-pig isolated mid-to-distal colon. Results were analysed using a weighted analysis of variance method in the log transformed data and are expressed as antilog (means ± sem); $*0.1 > p > 0.05$, $**p < 0.05$, $***p < 0.001$, compared with pre-treatment control values.

volunteers. Thus, the dose of granisetron which causes constipation ($\geqslant 80\ \mu g\ kg^{-1}$ i.v.) is greater than the maximally-effective anti-emetic dose ($20-40\ \mu g\ kg^{-1}$ i.v.; Upward et al. 1990). The ability of high doses of 5-HT$_3$ receptor antagonist to cause constipation in guinea-pig may, therefore, be analogous to the constipation in human volunteers (see Introduction), justifying further work on the mechanisms by which constipation is evoked in the guinea-pig.

In guinea-pig isolated mid-to-distal colon, faecal pellet transit was delayed by granisetron or ICS 205-930 and delayed or prevented by morphine or clonidine. For granisetron, morphine and clonidine, this action was reversed by a concentration of naloxone (10^{-7} M) which by itself, had no effect on faecal pellet transit. Opioids may, therefore, be fundamentally involved in the regulation of abnormal colonic motility and this is supported by reports of naloxone reversing constipation in 2 patients with idiopathic chronic constipation (Kreek et al. 1983) and increasing faecal wet weights in geriatric patients (Kreek et al. 1984).

In marked contrast with the actions of granisetron and ICS 205-930 *in vitro* and with the constipating action of all three antagonists *in vivo*, ondansetron 1.0 μM did not delay faecal pellet transit in the isolated colon preparation. Although it is not possible to draw firm conclusions,

386

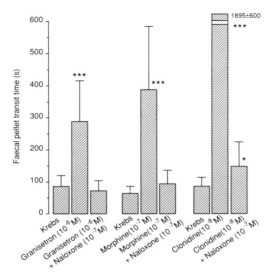

Figure 3. The effects of granisetron (10^{-6} M), morphine (10^{-7} M) and clonidine (10^{-8} M) on endogenous faecal pellet transit time in guinea-pig isolated mid-to-distal colon. Each of these compounds significantly ($*0.1 > p > 0.05$, $**p < 0.05$, $***p < 0.001$) increased transit time, an effect reversed by the concomitant addition of naloxone (10^{-7} M). Results were analysed using a weighted analysis of variance method in the log transformed data and are expressed as antilog (mean ± sem).

it is suggested that this lack of activity of ondansetron is not due to a failure to penetrate into the colonic nerve plexuses. Thus, in segments of intact guinea-pig colon, ondansetron antagonised 5-HT-evoked contractions (pA_2 8.7, unpublished) in a manner similar to that found in the longitudinal muscle-myenteric plexus preparation of guinea-pig colon (pA_2 7.1; Grossman et al. 1989). For ondansetron to cause constipation *in vivo*, the compound must, therefore, act either at another region of the gut (e.g. caecum or rectum, but implying a regional variance in intestinal "recognition" of different 5-HT$_3$ receptor antagonists), or antagonise a 5-HT$_3$ receptor outside the gut but at a site which can control lower bowel function (e.g. spinal ganglia or spinal cord, where 5-HT$_3$ receptors have been detected; see Hamon et al. 1989). Although not proven, it is possible to speculate that granisetron and ICS 205-930 could also antagonise the 5-HT$_3$ receptor at this second site of action, implying that their actions on the isolated colon represent an additional action of secondary importance.

From the data presented, it cannot be concluded that granisetron and ICS 205-930 slowed faecal pellet transit *in vitro* by antagonising a 5-HT receptor. However, in view of the selectivity of (for example) granisetron (Sanger and Nelson 1989) it is hard to find an alternative action which is also shared by ICS 205-930. If these 2 compounds do act by antagonising a 5-HT receptor, then a simple adherence to the Bradley et al. (1986) 5-HT receptor classification would suggest that the receptor cannot be 5-HT$_3$. Thus, ondansetron was inactive. However, it should be recognised that precedents for exceptions already exist in the original classification. MDL 72222 fails to antagonise at the 5-HT$_3$ receptor in the guinea-pig isolated ileum and yet both the compound and the model are accepted tools in 5-HT$_3$ receptor research. Similarly, in an anaesthetised rat model of the visceral pain reflex, granisetron and ICS 205-930 both behaved as effective analgesics, whereas ondansetron was inactive (Moss and Sanger 1990). Differences in gastrointestinal activity do, therefore, exist between the selective 5-HT$_3$ receptor antagonists, although at present, it is not possible to conclusively attribute these differences to a non-5-HT$_3$ receptor, to a subtype of the 5-HT$_3$ receptor or to peculiarities of the models used.

References

Akkermans, L. M. A., Vos, A., Hoekstra, A., Roelofs, J. M. M., and Horowitz, M. (1988). Effect of ICS 205-930 (a specific 5-HT$_3$ receptor antagonist) on gastric emptying of a solid meal. Gastroenterol. 94: A4.

Bermudez, J., Boyle, E. A., Miner, W. D., and Sanger, G. J. (1988). The anti-emetic and anti-nauseant potential of the 5-hydroxytryptamine$_3$ receptor antagonist BRL 43694. Br. J. Cancer. 58: 644–650.

Bradley, P. B., Engel, G., Fenuik, W., Fozard, J. R., Humphrey, P. P. A., Middlemiss, D. N., Mylecharane, E. J., Richardson, B. A., and Saxena, P. R. (1986). Proposals for the classification and nomenclature of functional receptors for 5-hydroxytryptamine. Neuropharmacol. 25: 563–576.

Butler, A., Hill, J. M., Ireland, S. J., Jordon, C. C., and Tyers, M. B. (1988). Pharmacological properties of GR 38032F, a novel antagonist of 5-HT$_3$ receptors. Br. J. Pharmacol. 94: 397–412.

Connell, A. M. (1962). The motility of the pelvic colon. Part 2. Paradoxical motility in diarrhoea and constipation. Gut. 3: 342–348.

Costall, B., Domeney, A. M., Naylor, R. J., and Tattersall, F. D. (1986). 5-Hydroxytryptamine M-receptor antagonism to prevent cisplatin-induced emesis. Neuropharmacol. 25: 959–961.

Costall, B., Domeney, A. M., Gunning, S. J., Naylor, R. J., Tattersall, F. D., and Tyers, M. B. (1987). GR 38032F: A potent and novel inhibitor of cisplatin-induced emesis in the ferret. Br. J. Pharmacol. 91: 90P.

Gore, S., Gilmore, I. T., Haigh, C. G., Brownless, S. M., Stockdale, H., and Morris, A. I. (1990). Colonic transit in man is slowed by ondansetron (GR 38032F), a selective 5-hydroxytryptamine receptor (type 3) antagonist. Aliment. Pharmacol. Therap. 4: 139–144.

Grossman, C. J., Bunce, K. T., and Humphrey, P. P. A. (1989). Investigation of the 5-HT receptors in guinea-pig descending colon. Br. J. Pharmacol. 97: 451P.

Hamon, M., Gallissot, M. C., Menard, F., Gozlan, H., Bourgoin, S., and Vergé, D. (1989). 5-HT$_3$ receptor binding sites are on capsaicin-sensitive fibres in the rat spinal cord. Eur. J. Pharmacol. 164: 315–322.

King, F. D., and Sanger, G. J. (1989). 5-HT$_3$ receptor antagonists. Drugs of the Future. 14: 875–889.

Kreek, M. J., Schaffer, R. A., Hahn, E. F., and Fishman, I. (1983). Naloxone, a specific opioid antagonist, reverses chronic idiopathic constipation. Lancet. 1: 261–262.

Kreek, M. J., Paris, P., Bartol, M. A., and Muller, D. (1984). Effects of short term oral administration of the specific opioid antagonist naloxone on faecal evacuation in geriatric patients. Gastroenterol. 86: 114.

388

Lattimer, N., Rhodes, K. F., and Saville, V. L. (1989). Possible differences in 5-HT$_3$-like receptors in the rat and the guinea-pig. Br. J. Pharmacol. 96: 270P.

Meleagros, L., Kreymann, B., Ghatei, M. A., and Bloom, S. R. (1987). Mouth-to-caecum transit time in man is reduced by a novel serotonin M-receptor antagonist. Gut. 28: A1373.

Moss, H. E., and Sanger, G. J. (1990). The effects of granisetron, ICS 205-930 and ondansetron on the visceral pain reflex induced by duodenal distension. Br. J. Pharmacol. 100: 497–501.

Richardson, B. P., Engel, G., Donatsch, P., and Stadler, P. A. (1985). Identification of serotonin M-receptor subtypes and their blockade by a new class of drugs. Nature. 316: 126–131.

Sanger, G. J., and King, F. D. (1988). From metoclopramide to selective gut motility stimulants and 5-HT$_3$ receptor antagonist. Drug Design and Delivery 3: 273–295.

Sanger, G. J., and Nelson, D. R. (1989). Selective and functional 5-hydroxtryptamine$_3$ receptor antagonism by BRL 43694 (granisetron). Eur. J. Pharmacol. 159: 113–124.

Stacher, G., Gaupmann, G., Schneider, C., Stacher-Janotta, G., Steiner-Mittelbach, G., Abatzi, Th-A., and Steinringer, H. (1989). Effects of a 5-hydroxytryptamine$_3$ receptor antagonist (ICS 205-930) on colonic motor activity in healthy men. Br. J. Clin. Pharmacol. 28: 315–322.

Staniforth, D. H. (1989). Oro-caecal transit time in man unaffected by 5-HT$_3$ antagonism: a comparison of BRL 24924 and BRL 43694. Br. J. Clin. Pharmacol. 27: 701P–702P.

Talley, N. J., Philips, S. F., Miller, L. J., Haddad, A., Miller, L. J., Twomey, C., Zinsmeister, A. R., and Ciociola, A. (1989). Effect of selective 5-HT$_3$ antagonist (GR 38032F) on small intestinal transit time and release of gastrointestinal peptides. Dig. Dis. Sci. 34: 1511–1515.

Upward, J. W., Arnold, B. D. C., Link, C., Pierce, D. M., Allen, A., and Tasker, T. C. G. (1990). The clinical pharmacology of granisetron (BRL 43694), a novel specific 5-HT$_3$ antagonist. Eur. J. Cancer. 26 (Suppl. 1): S12–S15.

Serotonin: Molecular Biology, Receptors and Functional Effects
ed. by J. R. Fozard/P. R. Saxena
© 1991 Birkhäuser Verlag Basel/Switzerland

Subchronic D-Fenfluramine Treatment Enhances the Immunological Competance of Old Female Fischer 344 Rats

L. M. Petrovic[1], S. A. Lorens[2], M. George[2], T. Cabrera[2],
B. H. Gordon[3], R. J. Handa[1], D. B. Campbell[3], and J. Clancy, Jr.[1]

[1]*Department of Cell Biology, Neurobiology and Anatomy, and* [2]*Pharmacology,
Loyola University Chicago, Stritch School of Medicine, Maywood, IL 60153, USA;* [3]*Servier
Research and Development, Fulmer Hall, Windmill Road, Fulmer, Slough SL3 6HH, UK*

Summary. Young (5 mo) and old (21 mo) female Fischer 344 (F344) rats received the serotonin (5-HT) releaser and reuptake inhibitor, d-fenfluramine (d-Fen; 0.6 mg/kg/day, p.o.), in their deionized drinking water for 30–38 days. The control animals had access *ad libitum* to deionized water. In comparison to the young animals, the old rats showed a significantly higher percentage of splenic large granular lymphocytes (%LGL) and greater concanavalin A (Con-A) stimulated T cell proliferation. The old d-Fen treated rats, furthermore, showed a significant: 1) further increase in %LGL; 2) enhancement of both basal and recombinant interleukin-2 (rIL-2) stimulated natural killer cytotoxicity; and, 3) augmentation of lipo-polysaccharide (LPS) induced B cell mitogenesis. d-Fenfluramine did not affect any of the immunological parameters studied in the yound rats, and did not affect the animals' body weights, 24 h fluid intakes, or the levels of medial frontal cortex monoamines and their metabolites. These data suggest that prolonged 5-HT release and/or reuptake inhibition enhances immune competence in old but not in young female F344 rats.

Introduction

It is well established that male rats exhibit age-related immunological dysfunctions (see Lorens et al. 1990). Evidence also has been advanced that 5-HT is involved in the regulation of the immune system (Slauson et al. 1984; Jackson et al. 1985; Khan et al. 1986; Hellstrand and Hermodsson 1987; Fillion et al. 1989). Although age-related changes in CNS 5-HT and 5-hydroxyindoleacetic acid (5-HIAA) levels are not observed in the rat until senescence (28 mo), alterations in the 5-HT response to stress and in 5-HT receptor densities have been observed much earlier (see references cited in Lorens et al. 1990).

The objective of the present experiment was to determine whether sustained increased 5-HT release and/or reuptake inhibition, produced by daily administration of d-fenfluramine (d-Fen) (Garattini et al. 1988), would affect splenic immune functions in young and/or old female F344 rats.

Materials and Methods

Animals

Barrier reared young (5 mo) and old (21 mo) female F344 rats were obtained from the NIA colony at Harlan Sprague-Dawley Inc. (Indianapolis, IN). The animals were housed individually and maintained on a 12 h light/dark cycle (lights on at 07:00 h) in a temperature (20–22°C) and humidity (52–55%) controlled AAALAC approved facility.

Experimental Protocol

The animals received d-fenfluramine hydrochloride (d-Fen; Servier, France; 0.6 mg/kg/day, p.o.) in their deionized drinking water for 30–38 days. The control animals had access *ad libitum* to unadulterated deionized water. Body weights were measured every three days and 24 h fluid intakes were determined daily. The concentrations of d-Fen were adjusted according to each rat's body weight and 24 h fluid consumption. Vaginal smears were performed daily for two weeks prior to sacrifice in order to ascertain that the old rats were anestrus, and to enable the sacrificing of the young rats during either diestrus or estrus.

The rats were euthanized instantaneously using a guillotine, and thoroughly examined for gross pathological conditions. The brain was removed and dissected over ice as detailed previously (Lorens et al. 1990). The rostral 2.0 mm of the medial frontal cortex (MFC) and the rostral 4.0 mm of the suprarhinal dorsolateral cortex (DLC) were obtained and stored at −70°C. The concentrations of the monoamines and their metabolites in the MFC were analyzed by HPLC-EC using a modification of the procedure detailed in Lorens et al. (1990). The concentrations of d-Fen and its metabolite, d-norfenfluramine (d-nor-Fen) were quantified in the DLC according to Richards et al. (1989). The spleen was excised using a sterile field, and spleen cell suspensions, prepared immediately as previously described (Clancy et al. 1983; Lorens et al. 1990), were used to measure: 1) the percentage of splenic large granular lymphocytes (%LGL); 2) natural killer cell (NK) cytotoxicity, before and after rIL-2 stimulation; 3) Con-A T lymphocyte activation; and, 4) LPS stimulation of B lymphocyte proliferation.

Cytocentrifuged Giemsa stained spleen cell preparations were analyzed microscopically, and the %LGL was evaluated by counting 200–400 mononuclear cells.

The NK mediated ^{51}Cr release assay was performed in triplicate using YAC-1 lymphoma targets as detailed previously (Lorens et al. 1990). Interleukin-2 stimulated YAC-1 killing also was performed as described previously (Vujanovic et al. 1988; Lorens et al. 1990), using human rIL-2 (Cetus, Emeryville, CA), containing 1.25×10^6 units of rIL-2/mg protein. Lytic units were calculated using a computer program (Clinical Immunological Services Program Resources Inc., Frederick, MD). One lytic unit (LU) was defined as the number of leukocytes in 10^6 effectors to lyse 30% of 5×10^3 targets [LU (30%)/10^6].

Spleen cell suspensions were cultured in microtiter plates (Corning Glass Works, Corning, NY) in the presence of 3.0 ug/ml Con-A (Pharmacia, Uppsala, Sweden) and 10.0 mg/ml LPS (Sigma, St. Louis, MO) for 48 h at 37°C. The cells then were pulsed for 6 h with 1.0 uCi of ^{3}H-thymidine (DuPont-New England Nuclear Research, Boston, MA), harvested, and thymidine incorporation determined with a liquid scintillation counter. The data are reported as the mean of triplicate wells.

Data Analysis

Data reduction and analyses were performed by an IBM PC-XT utilizing PC ANOVA (version 1.0) software (Human Systems Dynamics, Northridge, CA). The immunological and neurochemical data first were examined by an analysis of variance (ANOVA) using a two factor (age × drug) design followed, when appropriate, by the Newman-Keuls' multiple range test for post hoc comparisons (Winer 1971). The body weight and fluid consumption data were analyzed using a three-factor (age × drug × time) repeated measure (time) ANOVA. Brain drug and metabolite levels were analyzed using a t-test for independent means (two-tail).

Results

Examination of the brain, cranial and oral cavities, thorax, abdomen and musculocutaneous tissue did not reveal any gross pathological conditions, except in two old vehicle treated rats. These animals had pituitary adenomas and were eliminated from the study prior to the analysis of the data.

Analysis of the body weight data showed significant age [$F(1, 28) = 196.2, p < 0.00001$], time [$F(9, 252) = 3.1, p < 0.002$], age × time [$F(9,252) = 33.8, p < 0.00001$], and age × drug × time [$F(9, 252) = 3.5, p < 0.006$] effects, but no significant drug, drug × age, or drug × time effects. At the beginning of the experiment the young rats weighed (180–228 g) significantly ($p < 0.01$) less than the old rats (263–318 g). During the course of this experiment, the young rats gained an average of 11 g, while the old rats lost an average of 16 g. Importantly, no differences were observed between the d-Fen and vehicle treated young or old groups. At the time of sacrifice, the body weights (mean ± S.E.M.) were: young vehicle group, 208 ± 4 g; young d-Fen group, 210 ± 4 g; old vehicle group, 272 ± 10 g; old d-Fen group, 266 ± 9 g.

The daily fluid consumption of the four groups remained constant throughout the experiment and averaged 21 ml/day. The ANOVA of the fluid intake data did not show any significant age, drug, time or interaction effects.

The DLC concentrations (mean ± S.E.M. ng/g) of both d-Fen (young, 56 ± 19; old, 226 ± 79) and d-norFen (young, 516 ± 63; old, 831 ± 11) were significantly higher in the old rats ($t = 2.9$ and 2.6, respectively, df = 14, $p < 0.02$). Importantly, no significant group differences were observed in the MFC concentrations (Table 1) of dopamine (DA) and its metabolites 3,4-dihydroxyphenylacetic acid (DOPAC) and homovanillic acid (HVA); norepinephrine (NE) and its metabolite 3-methoxy-4-hydroxyphenylglycol (MHPG); or, 5-HT and its metabolite 5-HIAA.

Significant age [$F(1, 28) = 151.5, p < 0.001$], drug [$F(1, 28) = 15.3, p < 0.0001$], and age × drug interaction [$F(1, 28) = 10.2, p < 0.004$] effects on %LGL were obtained. The post hoc analysis showed that not only was the %LGL significantly ($p < 0.01$) greater in the old compared to the young rats, but it was higher ($p < 0.01$) in the old rats treated with d-Fen (Table 2).

The ANOVA of the baseline (non-stimulated) NK cytotoxicity (LU) data showed a significant age effect [$F(1, 27) = 11.8, p < 0.002$]. The Newman-Keuls' test indicated that the d-Fen old group differed ($p < 0.05$) from both of the young groups (Table 2).

Significant age [$F(1, 27) = 26.4, p < 0.0001$], drug [$F(1, 27) = 18.6, p < 0.0004$], and age × drug [$F(1, 27) = 23.1, p < 0.0002$] effects on rIL-2/LU were obtained. The post hoc analysis showed that rIL-2/LU in the

Table 1. Effect of subchronic d-fenfluramine (d-Fen) treatment on the concentrations of monoamines and their metabolites in the medial frontal cortex

GROUP	N	DA	DOPAC	HVA	NE	MHPG	5-HT	5-HIAA
YOUNG								
Vehicle	11	161	74	44	446	324	1070	824
		±3	±4	±5	±14	±33	±55	±34
d-Fen	10	149	75	44	442	288	987	849
		±7	±5	±4	±11	±22	±47	±50
OLD								
Vehicle	5	154	65	35	455	309	983	889
		±7	±7	±7	±22	±28	±50	±25
d-Fen	5	137	68	30	423	320	1032	891
		±10	±6	±5	±19	±29	±85	±31

Data presented as group mean ± S.E.M. N = number of rats/group.
One young d-Fen sample was lost due to technical error.

Table 2. Effects of subchronic d-fenfluramine (d-Fen) treatment on splenic percent large granular lymphocytes (%LGL) and basal (LU) and rIL-2 stimulated (rIL-2/LU) natural killer cytotoxicity in young (5 mo) and old (21 mo) F344 female rats

GROUP	N	%LGL	LU	rIL-2/LU
YOUNG				
Vehicle	11	3.9	3.0	23.4
		±0.4	±0.5	±2.7
d-Fen	11	5.4	2.5	21.9
		±1.0	±0.3	±2.5
OLD				
Vehicle	5	22.4*	4.8	24.4
		±4.0	±2.5	±2.5
d-Fen	5	36.8**	6.6*	51.8**
		±2.2	±1.9	±2.7

Data presented as group mean ± S.E.M. N = number of rats/group.
Superscripts refer to significant individual group differences determined by an ANOVA followed by a Newman-Keuls' test:
*Significantly (%LGL, $p < 0.01$; LU, $p < 0.05$) greater than both of the young groups.
**Significantly ($p < 0.01$) greater than the other three groups.

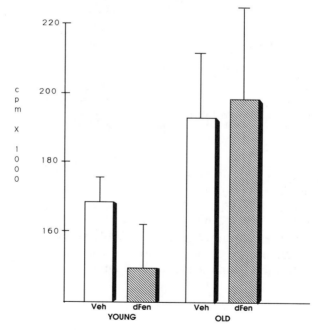

Figure 1. Concanavalin A (Con-A) induced splenic T cell mitogenesis in young (5 mo) and old (21 mo) female F344 rats treated for 30–38 days with vehicle (open bars) or d-Fen (shaded bars). Data are presented as group mean ± S.E.M. cpm × 10³. CPMs in the absence of Con-A ranged between 1–3 × 10³. The ANOVA showed that Con-A stimulated T lymphocyte proliferation was significantly greater in the old than in the young animals. Although the marginal mean (195 cpm × 10³) for the old rats was 24% higher than in the young rats, the post hoc analysis did not reveal any significant individual group differences.

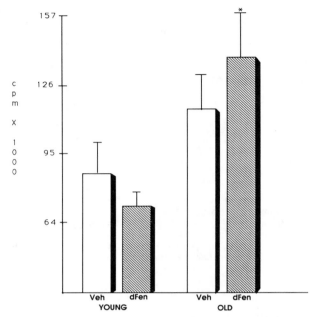

Figure 2. Lipopolysaccharide (LPS) induced splenic B cell mitogenesis in young (5 mo) and old (21 mo) female F344 rats treated for 30–38 days with vehicle (open bars) or d-Fen (shaded bars). Data are presented as group mean ± S.E.M. cpm × 10³. CPMs in the absence of LPS ranged between 1–3 × 10³. The ANOVA revealed that the LPS stimulated B lymphocyte proliferation was significantly greater in the old than in the young animals. The marginal mean (126 cpm × 10³) for the old rats was 58% higher than in the young rats. The post hoc analysis showed that the LPS stimulated B cell mitogenesis in the old rats treated with d-Fen was greater (*$p < 0.05$) than that in the young d-Fen treated rats.

d-Fen treated old group was significantly ($p < 0.01$) greater than in the other three groups (Table 2).

T (Figure 1) and B (Figure 2) lymphocyte proliferation induced by Con-A [$F(1, 28) = 4.8, p < 0.04$] and LPS [$F(1, 28) = 10.7, p < 0.003$], respectively, were significantly greater in the old animals. The old d-Fen treated rats, furthermore, showed greater ($p < 0.05$) LPS induced B cell mitogenesis than the young d-Fen treated animals (Figure 2).

Discussion

The old female F344 rats, like old male F344 rats (Lorens et al. 1990), showed a higher relative level of splenic LGLs than young rats. The present study demonstrated, moreover, that splenic LGL levels were further elevated in the old but not in the young female rats treated subchronically with d-Fen. Subchronic d-Fen also significantly increased basal and rIL-2 stimulated NK activity, as well as LPS induced B cell proliferation, in the old but not in the young female rats.

Brain levels of d-Fen and d-norFen were significantly higher in the old than in the young animals, indicating that d-Fen metabolism and clearance is slower in aged female rats. The absence of d-Fen effects on immune functions in young animals thus may be due to inadequate drug and metabolite levels being obtained. However, our recent studies (unpublished) employing a higher dose (1.8 mg/kg, p.o.) of d-Fen in young female rats likewise failed to alter immune function although brain d-Fen and metabolite levels were significantly higher in the young than in the old rats. These results suggest that the immunological consequences of subchronic d-Fen treatment are age-dependent.

The d-Fen dose and route of administration employed did not alter the levels of MFC monoamines and their metabolites. Thus, the immunological effects observed in the present study are not associated with substantial changes in CNS monoamine turnover. It remains to be determined, moreover, whether the effects of d-Fen on splenic immune functions are centrally and/or peripherally mediated.

It is now well documented that the immune system plays a key role in normal physiological functions as well as in disease and stressful conditions (Cunnick et al. 1988; Rabin et al. 1988; Bellinger et al. 1989). In rats, the aging process is accompanied by a decline in some immune functions (Bellinger et al. 1989; Lorens et al. 1990). However, whereas murine (Muzzioli et al. 1986) studies have documented an age related decline in NK activity, there may be no effect (Facchini et al. 1986) or an actual increase in NK activity in aging humans (Krishnaraj and Blandford 1986).

The results from several studies are controversial regarding the nature of 5-HT effects on various compartments of the immune system. The additional variables of species, sex and age differences have been largely ignored in investigations of the role of 5-HT. A study done with spleen cells from 2–3 month old CBA mice (Jackson et al. 1985) suggested that 5-HT and its precursor, 5-hydroxytryptophan (5-HT), when given *in vivo* subcutaneously 30–60 min before immunization, significantly impaired the *in vitro* T cell dependent antibody response to sheep red blood cells (SRBC). Conversely, inhibition of 5-HT synthesis by parachlorophenylalanine two days before antigen administration markedly enhanced the SRBC response. Because selective depletion of central 5-HT following treatment with 5,7-dihydroxytryptamine (100 ug intracisternally) had no effect on the murine splenic SRBC response, the authors suggested that the immunomodulatory effect of 5-HT was mediated by peripheral rather than by central neural pathways. It also has been reported (Khan et al. 1986) that phytohemagglutinin (PHA) induced human T lymphocyte proliferation *in vitro* was inhibited by 20 ug/ml of 5-HT. This confirmed a previous report that 10 mM 5-HT significantly inhibited PHA induced proliferation of T lymphocytes obtained from 22–39 year old humans (Slauson et al. 1984). None of the *in vivo* murine or *in vitro* human studies appear to have evaluated the effect of age or sex, nor the duration of the inhibitory effects.

In contradistinction to these suppressive effects, the addition of 1–100 uM 5-HT significantly enhanced *in vitro* NK cytotoxicity of percoll-fractionated human peripheral blood lymphocytes (PBL) (Hellstrand and Hermodsson 1987). Such enhancement appeared to be 5-HT$_1$ receptor specific and dependent on a non-interferon or IL-2 like factor released from plastic adherent cells isolated along with the PBL. The adherent cell derived enhancing factor(s) required at least 1.0 h or 10 uM 5-HT for their induction and subsequent activation of CD16+/CD3-NK effector cells (Hellstrand and Hermodsson 1990a). Also, enhanced NK effectors persisted in an activated state for at least 16 h. Finally, recent studies indicate that the 5-HT$_{1A}$ receptor on monocytes can be stimulated by 5-HT to release a prostaglandin independent factor(s) which stimulates baseline NK activity (Hellstrand and Hermodsson 1990b). This same receptor on monocytes also can be stimulated to release factor(s) which synergize with rIL-2 to activate NK cells. No attempt was made to discern any age or sex differences between the test subjects.

The stimulatory effect of d-Fen on baseline and rIL-2 stimulated NK activity in old (21 mo) female F344 rats confirm and extend the *in vitro* human studies with *in vivo* data. While adherent cell dependency of the presently reported enhanced rat effector cells was not addressed, discrete ages and one gender group were employed. Future studies will be necessary in order to determine whether the 5-HT$_{1A}$ receptor on rat spleen macrophages is affected by d-Fen as well as whether there is a direct or indirect effect of d-Fen on rat NK cells or the alpha (p55) and/or beta (p70/75) components of their IL-2 receptor (Fung and Greene 1990).

There is a growing body of evidence that both humoral and cell-mediated immune responses are more effective in females than in males (Grossman 1990). Furthermore, a study (Yonezawa et al. 1989) which used young (6–7 mo) and old (25–26 mo) Wistar rats of both sexes showed that the 5-HT content of platelets seems to be sex and age-related, since age-related decreases were observed in males while no age-related difference in platelet 5-HT content was seen in females. In light of the above, we are in the process of determining whether subchronic d-Fen treatment will produce effects in male F344 rats similar to those obtained in the present study using female F344 rats.

References

Bellinger, D. L., Ackerman, K. D., Felten, S. Y., Lorton, D., and Felten, D. L. (1989). Noradrenergic sympathetic innervation of thymus, spleen, and lymph nodes: aspects of development, aging and plasticity in neural immune interaction. In: Hadden, J. W., Masek, K., and Nistico, G., (eds.), Interactions among CNS, Neuroendocrine and Immune Systems. Rome-Milan: Pythagora, pp. 35–66.

Clancy, J. Jr., Mauser, L., and Chapman, A. (1983). Level and temporal pattern of naturally cytolytic cells during acute graft-vs-host disease in the rat. Cell. Immunol. 79: 1–10.

Cunnick, J. E., Lysle, D. T., Armfield, A., and Rabin, B. S. (1988) Shock-induced modulation of lymphocyte responsiveness and natural killer cell activity: differential mechanisms of induction. Brain Behav. Immunity. 2: 102–113.

Facchini, A., Mariani, E., Mariani, A. R., Papa, S., Vitale, M., and Manzoli, F. A. (1986). NK cells during human aging. In: Facchini, A., Haaijman, J. J., and Labo, G. (eds.), Immunoregulation in aging. Rijswijk, The Netherlands: EURAGE, pp. 229–237.

Fillion, M. P., Prudhomme, N., Haour, F., Fillion, G., Bonnet, M., Lespinats, G., Masek, K., Flegel, M., Corvaia, N., and Launay, J. M. (1989). Hypothetical role of the serotonergic system in neuroimmunomodulation: preliminary molecular studies. In: Hadden, J. W., Masek, K., and Nistico, G. (eds.), Interactions Among CNS, Neuroendocrine and Immune Systems. Rome-Milan: Pythagora, pp. 235–250.

Fung, M. R., and Greene, W. C. (1990). The human interleukin-2 receptor: insights into subunit structure and growth signal transduction. Sem. Immunology 2: 119–128.

Garattini, S., Bizzi, A., Caccia, S., Mennini, T., and Samanin, R. (1988). Progress in assessing the role of serotonin in the control of food intake. Clin. Neuropharmacol. 11: S8–S32.

Grossman, C. J. (1990). Are there underlying immune-neuroendocrine interactions responsible for immunological sexual dimorphism? Prog. Neuroendocrinimmunol. 3: 75–82.

Hellstrand, K., and Hermodsson, S. (1987). Role of serotonin in the regulation of human natural killer cell cytotoxicity. J. Immunol. 139: 869–875.

Hellstrand, K., and Hermodsson S. (1990a). Enhancement of human natural killer cell cytotoxicity by serotonin: role of non-T/CD16$^+$ NK cells, accessory monocytes, and 5-HT$_{1A}$ receptors. Cell. Immunol. 127: 199–214.

Hellstrand, K., and Hermodsson, S. (1990b). Monocyte-mediated suppression of IL-2-induced NK-cell activation. Scand. J. Immunol. 32: 183–192.

Jackson, J. C., Cross, R. J., Walker, R. F., Markesbery, W. R., Brooks, W. R., and Roszman, T. L. (1985). Influence of serotonin on the immune response. Immunology 54: 505–512.

Khan, I. A., Bhardwaj, G., Malla, N., Wattal, C., and Agrawal, S. C. (1986). Effect of serotonin on T lymphocyte proliferation *in vitro* in healthy individuals. Int. Archs Allergy appl. Immun. 81: 378–380.

Krishnaraj, R., and Blandford G. (1986). An analysis of human NK cell features that change with age. In: Facchini, A., Haaijman, J. J., and Labo, G. (eds.), Immunoregulation in aging. Rijswijk, The Netherlands: EURAGE, pp. 239–245.

Lorens, S. A., Hata, N., Handa, R. J., Van de Kar, L. D., Guschwan, M., Goral, J., Lee, J. M., Hamilton, M. E., Bethea, C. L., and Clancy, Jr., J. (1990). Neurochemical, endocrine and immunological responses to stress in young and old Fischer 344 male rats. Neurobiol. Aging 11: 139–150.

Muzzioli, M., Provinciali, M., and Fabris, N. (1986). Age-dependent modification of natural killer activity in mice. In: Facchini, A., Haaijman, J. J., and Labo, G. (eds.), Immunoregulation in aging. Rijswijk, The Netherlands: EURAGE, pp. 221–227.

Rabin, B. S., Ganguli, R., Cunnick, J. E., and Lysle, D. T. (1988). The central nervous system-immune system relationship. Clinics Lab. Med. 8: 253–268.

Richards, R. P., Gordon, B. H., Ings, R. M. J., Campbell, D. B., and King, L. J. (1989). The measurement of d-fenfluramine and its metabolite, d-norfenfluramine in plasma and urine with an application of the method to pharmacokinetic studies. Xenobiotica 19: 547–553.

Slauson, D. O., Walker, C., Kristensen, F., Wang, Y., and de Weck, A. L. (1984). Mechanisms of serotonin-induced lymphocyte proliferation inhibition. Cell. Immunol. 84: 240–252.

Vujanovic, N. L., Herberman, R. B., and Hiserodt, J. L. (1988). Lymphokine-activated killer cells in rats: analysis of tissue and strain distribution, ontogeny, and target specificity. Cancer Res. 48: 878–883.

Winer, B. J. (1971). Statistical principles in experimental design. New York: McGraw-Hill.

Yonezawa, Y., Kondo, H., and Nomaguchi, T. A. (1989). Age-related changes in serotonin content and its release reaction of rat platelets. Mech. Ageing Dev. 47: 65–75.

Serotonin: Molecular Biology, Receptors and Functional Effects
ed. by J. R. Fozard/P. R. Saxena

Serotonin as a Vascular Smooth Muscle Cell Mitogen

T. A. Kent[1,2,3], A. Jazayeri[1], and J. M. Simard[4,5]

Departments of Neurology[1], Psychiatry[2], Pharmacology[3], Surgery/Division of Neurosurgery[4], and Physiology[5], University of Texas Medical Branch, Galveston, TX 77550, U.S.A.

Summary. Serotonin (5-HT) stimulated the uptake of ^{3}H-thymidine into cultured vascular smooth cells from the rat aorta. Thrombin also stimulated mitogenesis, and potentiated the effect of 5-HT. However, the percent increase in 5-HT-induced mitogenesis was unchanged after the addition of thrombin. The mitogenic effect of 5-HT was not inhibited by the dihydropyridine calcium channel blocker, nifedipine (1 μM), but the action of thrombin was. These results suggest that 5-HT and thrombin act through different, but interacting, signal transduction mechanisms to produce mitogenesis.

Introduction

It is becoming apparent that a number of substances that are vasoactive also stimulate mitogenesis in vascular smooth muscle cells (SMC) (Owens 1989). Conversely, a number of substances originally identified as growth factors also have vasoactive properties (Berk and Alexander 1989). The mechanisms of action of these various factors most likely differ, and their effects on *in vivo* smooth muscle cell function is not yet known. It is hypothesized that the development of atheromatous lesions is linked to the action of these serum or endothelial derived growth factors (Ross 1986). Serotonin(5-HT) is a known vasoconstrictor in peripheral and cerebral blood vessels, and is locally released at sites of thrombosis. It has also been shown to increase the incorporation of ^{3}H-thymidine into cultured vascular SMC from the aorta of various species (Nemeck et al. 1986).

We have recently reported that there are at least two 5-HT receptor subtypes on cultured vascular SMC from the rat aorta, with pharmacological profiles suggestive of their identity as 5-HT$_{1B}$ and 5-HT$_2$ binding sites (Jazayeri et al. 1989). Presence of the 5-HT$_2$ site was consistent with the known pharmacology in the rat aorta, in which the 5-HT acts as a low potency vasoconstrictor with effects thought to be mediated via action at this receptor subtype (Peroutka 1984).

The presence of the high affinity 5-HT$_{1B}$ site was somewhat surprising, as an *in vivo* function for this subtype in these cells has not been described. Although in many whole vessel studies more than one receptor subtype may be vasoactive (Hamel et al. 1989), such a function has not been shown for the 5-HT$_{1B}$ site in rat aorta. Conversely, it has been previously shown that the 5-HT$_{1B}$ receptor, when transfected into fibroblasts, mediates mitogenesis induced by 5-HT (Seuwen et al. 1988). Because of the association of this receptor subtype with growth, we have begun a series of experiments with cultured vascular SMC to better understand the mechanisms underlying 5-HT induced mitogenesis, with the goal of clarifying the potential roles of multiple receptor subtypes and the interplay between vasoconstriction and mitogenesis. This study was intended to define potential roles of low and high affinity 5-HT receptors in mitogenesis. In addition, we explored the mechanisms of mitogenesis by studying the role of calcium channels and the interaction between 5-HT and another SMC mitogen, thrombin.

Materials and Methods

Cultured vascular SMC from WKY rat aorta were used in these studies (Jazayeria and Meyer 1988). For comparison, cultured SMC from the guinea pig basilar artery were also used in some studies. These cells were identified by their characteristic growth pattern (hillock and valley), electron microscopic appearance (presence of thick filaments), and positive staining to monoclonal alpha smooth muscle actin (Sigma).

Passaged cells (passage 10–30) were plated either in 96 or 24 well plates and seeded to approximately 75% confluence. Cells were maintained in RPMI + 10% FCS for 24 hours. The media was then replaced with a serum-free quiescing media (Q media) (Kavanaugh et al. 1988) containing transferrin (5 μg/ml), ascorbic acid (110 μg/ml) and insulin (1 μg/ml) in DMEM and Ham's F12 media (1:1, v:v). This media had been previously tested for effects on cell viability using trypan blue exclusion, and yielded >90% viable cells after 48 hours.

After 24 hours, the quiescing media was removed and growth factors (5-HT, 10^{-12} to 10^{-5} M and/or thrombin 0.0001 Units/ml to 10 U/ml) in the quiescing media or fresh quiescing media alone was added to all wells. After 20 hours of incubation, ^{3}H-thymidine (2 μCi/ml; Dupont NEN; 81 Ci/mmol) was added for four additional hours. The cells were then washed with Hank's buffered saline solution. ^{3}H-thymidine incorporated into acid insoluble protein was extracted by first washing with 10% tetraethylammonium Cl$^-$, then ethanol/ether (2:1), and the cell debris was extracted into 0.5 N NaOH. The solution was placed into scintillation vials, and counted (Beckman 3801) at approximately 50% efficiency.

We determined the effect of mitogens on cell proliferation. 500,000 cells were seeded into 35 mm tissue culture dishes and treated in an identical fashion as above, except ^{3}H-thymidine was not added. 24

hours after addition of growth factors, cells were trypsinized and gently triturated to reduce clumping. Trypan blue was added, and the number of viable cells (those which excluded the dye) was counted in triplicate in a hemocytometer.

The effect of co-incubation with nifedipine was studied. All experiments were performed in the dark. The nifedipine (1 μM final concentration) was dissolved in polyethylene glycol and polyethylene glycol was included in the Q media for all experimental conditions.

Results

5-HT stimulated ^{3}H-thymidine incorporation by 5 fold with an EC_{50} of 3×10^{-7} (Figure 1). The effect of 5-HT was curvilinear. The decrease that occurred at high concentrations may represent a true decrease in mitogenesis, or may be due to an acceleration of mitogenesis at high 5-HT concentrations that is missed by addition of ^{3}H-thymidine for the last four hours. High concentrations of 5-HT might also be toxic. By comparison, 5-HT was more potent but less effective as a mitogen in

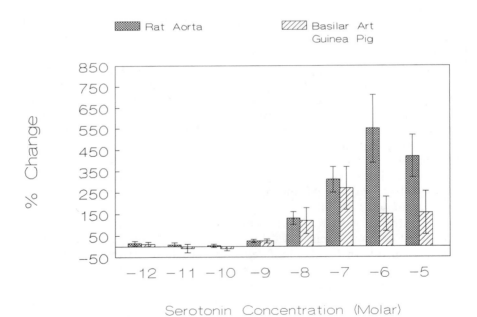

Figure 1. Percent change (mean ± S.D.) in ^{3}H-thymidine incorporation (y axis) in cultured Wistar-Kyoto rat aorta and guinea pig basilar artery smooth muscle cells as a function of increasing serotonin concentration (x-axis). Serotonin was more effective but less potent in rat aorta vs guinea pig basilar artery. The mechanism of decrease in ^{3}H-thymidine incorporation seen at high serotonin concentration is not yet known.

guinea pig basilar artery ($EC_{50} = 2 \times 10^{-8}$ M; % increase = 250) (Figure 1). These differences in potency parallel differences in potency of the vasoconstrictive effect of 5-HT in whole vessel studies (Peroutka 1984; Chang et al. 1988), although the differences do not necessarily reflect action at the same receptors which mediate contraction.

In WKY aortic cells, thrombin stimulated mitogenesis by about the same magnitude as 5-HT, reaching a maximum at 10 U/ml (Figure 2). When combined with 5-HT, 1 U/ml of thrombin not only increased mitogenesis at baseline (no 5-HT), but overall increased ^{3}H-thymidine incorporation (Figure 3). However, the percent change from baseline (no 5-HT) was the same with or without thrombin (Figure 3, insert). Our data indicate that low concentrations of 5-HT (10^{-12}–10^{-10} M) may reduce the thrombin induced increase in ^{3}H-thymidine incorporation ($p < 0.05$; ANOVA and post-hoc Newman-Keuls), but the magnitude of this effect was small.

In preliminary studies we have studied 5-HT$_1$ and 5-HT$_2$ antagonists but found that the viability of the cells in Q media was affected, rendering results suspect. Some modification in the Q media may be necessary for these pharmacological studies.

Neither 5-HT (10^{-7} M) nor 5-HT + thrombin (1 U/ml) increased cell number. There were $570,000 \pm 38,000$ cells (mean $\pm$ S.D.) counted in

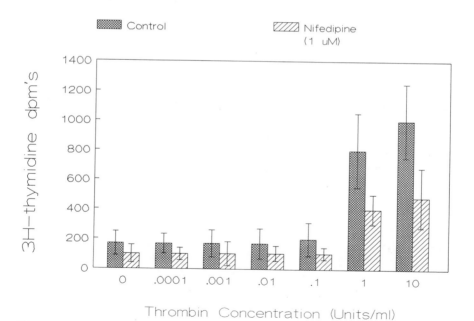

Figure 2. Percent change in ^{3}H-thymidine incorporation (y axis) in rat aorta smooth muscle cells as a function of increasing concentrations of thrombin. Nifedipine (single hatch bars) reduced ^{3}H-thymidine incorporation ($p < 0.5$; t-test performed at 1 and 10 Units/ml).

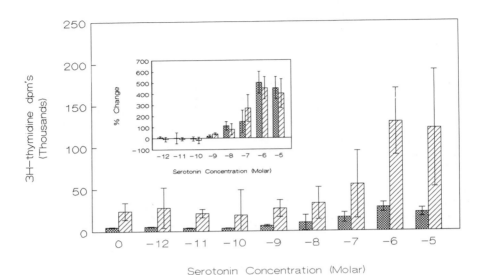

Figure 3. Potentiating effect of thrombin (1 Unit/ml) on the increase in ³H-thymidine incorporation induced by various serotonin concentrations in rat aorta smooth muscle cells. ³H-thymidine disintegrations per minute (dpm) was increased throughout. The percentage increase from baseline (no added serotonin) was identical with or without thrombin (insert). There was a slight, but statistically significant decrease in ³H-thymidine incorporation at low serotonin concentrations ($p = 0.05$; ANOVA and post-hoc Newman-Keuls).

the wells with Q media, $510,000 \pm 63,000$ after incubation with 5-HT, and $590,000 \pm 35,000$ after incubation with thrombin and 5-HT ($p > 0.5$; t-test).

Nifedipine did not inhibit the mitogenic effect of 5-HT (Figure 4). However, it reduced the maximal thrombin mitogenic effect by approximately 60% (Figure 2). When the combination of 5-HT and thrombin was studied, ³H-thymidine incorporation was reduced by 40% (data not shown).

Discussion

Our results confirm the action of 5-HT as a SMC mitogen in aorta, and extend this finding to a cerebral vessel, the basilar artery of the guinea pig. Differences in potency were noted, which parallel *in vivo* differences in vasoconstrictive potency of 5-HT and provide circumstantial evidence for the involvement of different 5-HT receptor subtypes in mitogenesis in the different vessels.

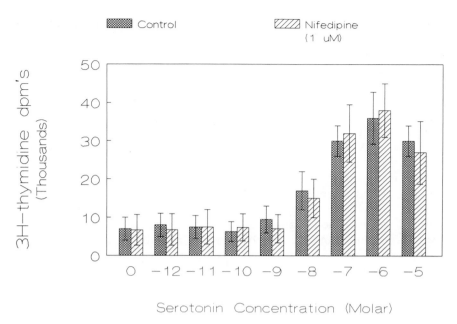

Figure 4. Effect of nifedipine on the serotonin-induced increase in ^{3}H-thymidine incorporation in rat aorta smooth muscle cells. No effect was evident.

Radioligand receptor binding studies and *in situ* hybridization experiments provide further evidence for the presence of different receptor subtypes on these cells. We have demonstrated the presence of a high affinity 5-HT$_{1B}$ site and a 5-HT$_2$ site on the aortic SMC (Jazayeri et al. 1989). We have also found a curvilinear Scatchard plot of ^{3}H-5-HT binding on the guinea pig basilar artery SMC (Simard et al. 1990) with a high affinity (< 1 nM Kd) and lower affinity (Kd ca. 20 nM) site, whose pharmacological profile is suggestive of a 5-HT$_1$-like subtype (e.g. sensitive to methiothepin, 5-carboxyamidotryptamine, yohimbine, and insenstive to propranolol). Moreover, our recent *in situ* hybridization studies with ^{32}P-labelled oligonucleotide probe complementary to rat 5-HT$_2$ receptor mRNA (Dupont NEN) demonstrated abundant hybridization to the rat aortic SMC, but little to the guinea pig basilar artery (unpublished data). Potential species differences in mRNA sequence are probably not responsible for this difference, since cultured SMC from the guinea pig aorta also showed abundant hybridization.

The relatively low potency of the 5-HT effect in cultured rat aortic SMC suggests that the mitogenic effect is not mediated by the 5-HT$_{1B}$ subtype. The exact concentration of 5-HT in solution is not known, however, and could be lower than the starting concentration. Definitive identification of the subtypes responsible for the mitogenic effect of 5-HT will require the use of other agonists and antagonists.

Dihydropyridines have been shown to inhibit ^{3}H-thymidine incorporation in cultured vascular SMC stimulated by other growth factors, notably platelet derived growth factor (Nilsson et al. 1985), suggesting that calcium is involved in mitogenesis, although other non-specific effects of these agents may be important. Their beneficial effect on retarding the development of atheromas (Lichtlen et al. 1990) has been attributed at least partially to these inhibitory effects on mitogenesis. Our results suggest that the inhibition is not universal, since the stimulation by 5-HT in WKY SMC was not affected, while the effect of thrombin was. *In vivo*, it is likely that the actions of multiple growth factors, in conjunction or succession, may be important in mediating the hyperplastic growth and migration of SMC seen in atheromatous lesions.

Should results comparable to those reported here be replicated in human vessels, the small but statistically significant inhibition of thrombin-induced mitogenesis by low concentrations of 5-HT may be worth exploiting via an agonist at this receptor subtype. In addition, antagonists at the 5-HT receptor subtype mediating the increase in mitogenesis may act to inhibit potential mitogenic effects of locally high concentrations of 5-HT seen after thrombosis.

The mechanism of mitogenic effect of thrombin has not been clarified, although it has been suggested that multiple signal transduction pathways become activated (Seuwen et al. 1989; Seuwen and Pouyssegur 1990). Interactions with 5-HT could occur at a receptor level or through intracellular messengers. Our data indicate that the signal pathways acitvated by thrombin are different from those utilized by 5-HT, because of differences observed in sensitivity to a dihydropyridine calcium channel antagonist and because maximal stimulation by 5-HT could be further enhanced by thrombin.

Conversely, our data indicate that thrombin and 5-HT probably interact, as suggested by the finding that after stimulation with thrombin, the same percent increase in ^{3}H-thymidine incorporation follows the addition of 5-HT. We envision two possible explanations for this type of interaction: 1) there is an increase in the number of cells able to react to 5-HT, e.g. more cells are "primed" by thrombin to react to 5-HT, which then respond in the usual manner as before, or 2) thrombin increased the number of chromosomes per cell (polyploidy), and then each chromosome pair responds to 5-HT in the same manner as those not simulated by thrombin. Polyploidy is a known property of SMC in culture (Owens 1989), and it has been suggested that an increase in chromosome number may occur as a result of stimulation with certain mitogens. Our data showing that the number of cells was not increased by these growth factors, while incorporation of ^{3}H-thymidine was increased, supports this hypothesis. It is possible that an increase in cell number may be evident if studied at longer time intervals.

Acknowledgements

This work was supported by grants from The Moody Foundation (89-61), The National Headache Foundation, The American Heart Association/Texas Affiliate, an American Heart Association Medical Student Fellowship, the Research Foundation of the American Association of Neurological Surgeons, and the National Heart, Lung and Blood Institute (HL42646).

References

Berk, B. C., and Alexander, R. W. (1989). Vasoactive effects of growth factors. Biochem. Pharmacol. 38: 219–225.

Chang, J. Y., Hardebo, J. E., and Owman, C. H. (1988). Differential vasomotor action of noradrenaline, serotonin, and histamine in isolated basilar artery from rat and guinea-pig. Acta Physiol. Scand. 132: 91–102.

Hamel, E., Robert, J. P., Young, A. R., and MacKenzie, E. T. (1989). Pharmacological properties of the receptor(s) involved in the 5-hydroxytryptamine-induced contraction of the feline middle cerebral artery. J. Pharmacol. Exp. Ther. 249: 879–889.

Jazayeri, A., and Meyer, W. J. (1988). Beta-adrenergic receptor binding characteristics and responsiveness in cultured Wistar-Kyoto rat arterial smooth muscle cells. Life Sci. 43; 721–725.

Jazayeri, A., Meyer, W. J., and Kent, T. A. (1989). 5-HT$_{1B}$ and 5-HT$_2$ serotonin binding sites in cultured Wistar-Kyoto rat aortic smooth muscle cells. Eur. J. Pharmacol. 169: 183–187.

Kavanaugh, W. M., Williams, L. T., Ives, H. E., and Coughlin, S. R. (1988). Serotonin-induced deoxyribonucleic acid synthesis in vascular smooth muscle cells involves a novel pertussis toxin-sensitive pathway. Mol. Endocrinol. 2: 599–605.

Lichtlen, P. R., Hugenholtz, P. G., Rafflenbeul, W., Hecker, H., Jost, S., and Deckers, J. W. (1990). Retardation of angiographic progression of coronary artery disease by nifedipine. Lancet 335: 1109–1113.

Nemecek, G. M., Coughlin, S. R., Handley, D. A., and Moskowitz, M. A. (1986). Stimulation of aortic smooth muscle cell mitogenesis by serotonin. Proc. Natl. Acad. Sci. USA. 83: 674–678.

Nilsson, J., Sjolund, M., Palmberg, L, Von Euler, A. M., Jonzon, B., and Thyberg, J. (1985). The calcium antagonist nifedipine inhibits arterial smooth muscle cell proliferation. Atherosclerosis 58: 109–122.

Owens, G. K. (1989). Control of hypertrophic versus hyperplastic growth of vascular smooth muscle cells. Am. J. Physiology 257: H1755–1764.

Peroutka, S. J. (1984). Vascular serotonin receptors. Biochem. Pharmacol. 33: 2349–51.

Ross, R., and Glomset, J. A. (1976). The pathogenesis of atherosclerosis. New Engl. J. Med. 295: 420–425.

Seuwen, K., Magnaldo, I., and Pouyssegur, J. (1988). Serotonin stimulates DNA synthesis in fibroblasts acting through 5-HT$_{1B}$ receptors coupled to a G$_i$ protein. Nature 335: 254–256.

Seuwen, K., Chambard, J. C., L'Allemain, G., Magnaldo, I., Paris, S., and Pouyssegur, J. (1989). Thrombin as a growth factor: Mechanisms of signal transduction. In: Meyer, P., Marche, P. (eds.), Blood Cells and Arteries in Hypertension and Atherosclerosis. New York, Raven Press.

Seuwen, K., and Pouyssegur, J. (1990). Serotonin as a growth factor. Biochem. Pharmacol. 39: 985–990.

Simard, J. M., Kent, T. A., and Jazayeri, A. (1990). Serotonin, protein kinase C and Ca^{2+} channels in basilar artery cells. The Second IUPHAR Satellite Meeting on Serotonin (abstract P21).

Serotonin: Molecular Biology, Receptors and Functional Effects
ed. by J. R. Fozard/P. R. Saxena
© 1991 Birkhäuser Verlag Basel/Switzerland

Serotonin, the Endothelium and the Coronary Circulation

F. C. Tanner[1], V. Richard[1], M. Tschudi[1], Z. Yang[1], and T. F. Lüscher[1,2]

Department of Research, Laboratory of Vascular Research[1], and Department of Medicine, Division of Cardiology[2], University Hospital, CH-4031 Basel, Switzerland

Summary. Serotonin releases endothelium-derived relaxing factors and also causes contraction of vascular smooth muscle cells. The latter may be important at sites of damaged endothelium and promote thrombus formation and vasospasm. In epicardial porcine coronary arteries, endothelium-dependent relaxations to serotonin are impaired by the inhibitor of nitric oxide formation L-N^G-monomethyl arginine (L-NMMA). Since pertussis toxin reduces endothelium-dependent relaxations to serotonin, a pertussis toxin sensitive G protein is involved. In intramyocardial porcine coronary arteries, L-NMMA is a weak inhibitor of the endothelium-dependent relaxation to serotonin and pertussis toxin has no effect. Thus, the importance of nitric oxide as well as that of pertussis toxin sensitive G protein decreases from large to small porcine coronary arteries.

In epicardial porcine coronary arteries, oxidized, but not native LDL inhibit endothelium-dependent relaxations to serotonin in a dose-dependent manner. As the relaxations to the nitric oxide donor SIN-1 are completely maintained, a reduced vascular responsiveness to endothelium-derived nitric oxide can be excluded. In contrast to serotonin, the endothelium-dependent relaxations to bradykinin are unaffected by oxidized LDL. As the same inhibitory pattern is obtained with inhibitors of endothelium-derived nitric oxide, the effect of oxidized LDL appears to be related to a reduced production and/or release of the endogenous nitrate. This may promote platelet aggregation and vasospasm, both known clinical events in patients with coronary artery disease.

In human left anterior descending coronary arteries, threshold and low concentrations of endothelin-1 potentiate contractions to serotonin. As the production of endothelin-1 by endothelial cells is stimulated by thrombin, this effect of the peptide may contribute to the vasospastic events associated with platelet aggregation.

Introduction

Upon release into the circulation from the enterochromaffine cells of the gut serotonin is taken up and stored by platelets or is inactivated by the liver. Only small concentrations of the monoamine are therefore present in plasma (Tyce 1985; Gillis 1985). At sites of functionally or morphologically damaged endothelium, however, platelet aggregation is favored, which leads to the local release of high amounts of serotonin. The effects of serotonin in the circulation are complex (Vanhoutte 1985). It evokes vasoconstriction in large and vasodilation in small arteries (Vanhoutte 1985). Thus, it causes contraction of epicardial coronary arteries accompanied with increased total blood flow in the coronary circulation. When it is infused *in vivo*, it causes changes in blood pressure, which are variable and depend on the species and the experimental conditions (Page 1954).

In the porcine coronary artery, contractions to serotonin are mediated by 5-HT_2 receptors on vascular smooth muscle cells (Van Nueten et al. 1981). Relaxations to the monoamine are caused indirectly by 5-HT_1 receptors on endothelial cells with subsequent release of endothelium-derived relaxing factor and activation of soluble guanylate cyclase in smooth muscle cells (Houston and Vanhoutte 1988). At sites where the endothelium is damaged, contraction of blood vessels and aggregation of platelets are favored by a reduced formation of endothelium-derived relaxing factor and eventually by a facilitated access of serotonin to the smooth muscle cells. Thus, serotonin may play an important role in vasospasm and thrombus formation, both events occurring in atherosclerotic coronary arteries (Vincent et al. 1983; Shimokawa et al. 1983).

Here we summarize studies which have been designed to characterize the endothelium-derived relaxing factor released by serotonin in large and in small porcine coronary arteries and to analyse its role in pathophysiological conditions such as atherosclerosis and coronary vasospasm.

Serotonin in Coronary Arteries

In epicardial porcine coronary arteries suspended in organ chambers for isometric tension recording, endothelium-dependent relaxations to serotonin (performed in the presence of ketanserin) are completely prevented by the scavenger of nitric oxide hemoglobin and markedly reduced by the inhibitor of guanylate cyclase methylene blue (Richard et al. 1991). They are also impaired by the inhibitor of nitric oxide formation L-N^G-monomethyl arginine (L-NMMA) indicating that the monoamine stimulates the release of nitric oxide from L-arginine in endothelial cells. Since the effect of L-NMMA can be reversed by L-arginine, but not D-arginine, the nitric oxide formation is specifically inhibited under these conditions (Richard et al. 1991; Palmer et al. 1989). Pertussis toxin, which irreversibly inactivates certain G proteins, can interfere with the release of endothelium-derived relaxing factors upon stimulation with some, but not all agonists mediating endothelium-dependent responses (Richard et al. 1991; Flavahan and Vanhoutte 1989). In the porcine coronary artery, it inhibits endothelium-dependent relaxations to serotonin. These relaxations, however, are not completely prevented by the toxin, since the monoamine still relaxes the blood vessels to a moderate extent under these conditions. In addition, L-NMMA causes a further inhibition of the response to

408

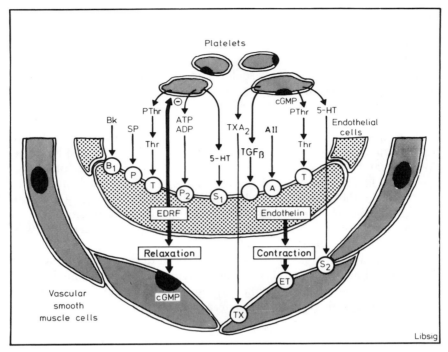

Figure 1. Role of the endothelium in the vascular effects of serotonin. The intact endothelium releases endothelium-derived relaxing factors upon stimulation with serotonin and other substances set free by activated platelets. The resulting vasodilation and impaired platelet aggregation prevents the progression of the initiated thrombus to vascular occlusion. At sites of damaged endothelium these beneficial effects are reduced and contractions to serotonin and thromboxane A_2 predominate. Vasospasm and thrombus formation are the consequences.

serotonin in the presence of pertussis toxin. This indicates that only a part of the formation of nitric oxide involves a pertussis toxin sensitive G protein (Richard et al. 1991).

In intramyocardial porcine coronary arteries studied in a myograph system, endothelium-dependent relaxations to serotonin are of similar potency as those in epicardial coronary arteries (Tschudi et al. 1991). In contrast, L-NMMA is much less effective in intramyocardial as compared to epicardial coronary arteries suggesting that the relative importance of endothelium-derived nitric oxide decreases from large to small vessels in this vascular bed (Tschudi et al. 1991). As pertussis toxin does not affect the endothelium-dependent relaxations to serotonin in intramyocardial coronary arteries, the signal transduction pathway after stimulation of serotonergic receptors on endothelial cells differs in large and small arteries of the coronary circulation (Tschudi et al. 1991).

These results indicate that serotonin causes the release of endothelium-derived nitric oxide in both epicardial and intramyocardial porcine

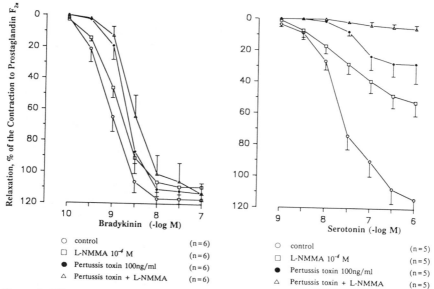

Figure 2. Effect of L-NMMA (10^{-4} M) and pertussis toxin (100 ng/ml) on endothelium-dependent relaxations to serotonin and bradykinin in epicardial porcine coronary arteries. The relaxation to serotonin is inhibited by L-NMMA and by pertussis toxin. L-NMMA has an additional inhibitory effect in the presence of pertussis toxin. In contrast, neither L-NMMA nor pertussis toxin affect the endothelium-dependent relaxation to bradykinin.

coronary arteries and that the importance of endothelium-derived nitric oxide as well as the involvement of a pertussis toxin sensitive G protein in the signal transduction pathway decreases from large to small coronary arteries.

Lipoproteins and Atherosclerosis

In contrast to previous concepts of atherosclerosis, the endothelium remains morphologically intact in the course of the early stages of atherogenesis (Faggiotto et al. 1984). Functionally, however, pronounced alterations of the endothelium occur (Vincent et al. 1983; Shimokawa et al. 1983). Low density lipoproteins (LDL) are a major risk factor for coronary artery disease, and oxidized LDL have been shown to be present in atherosclerotic plaques of rabbits and humans (Ylä-Herttuala et al. 1989). We therefore investigated whether native or oxidized LDL interfere with functional properties of the endothelium of isolated epicardial porcine coronary arteries.

Oxidized LDL inhibit endothelium-dependent relaxations to serotonin in a dose-dependent manner while native LDL do not have any effect even at a comparable concentration (Tanner, Boulanger and

410

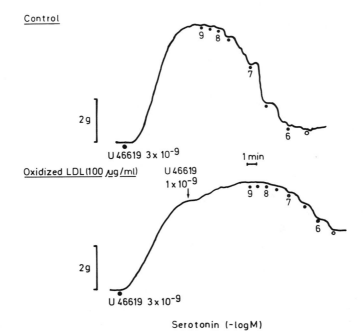

Figure 3. Effect of oxidized LDL (100 μg/ml) on endothelium-dependent relaxations to serotonin in epicardial porcine coronary arteries. The oxidized LDL were added to the organ chambers for two hours and then removed by extensive washing. After ten minutes the vessels were contracted and then exposed to cumulative concentrations of serotonin. Note the impaired relaxation to serotonin in the vessel exposed to oxidized LDL as compared to control.

Lüscher, unpublished). Since the relaxations of porcine coronary arteries without endothelium to the nitric oxide donor SIN-1, the active metabolite of molsidomine, are completely maintained, a reduced vascular responsiveness to endothelium-derived relaxing factor can be excluded. The potency of endothelium-dependent relaxations of most blood vessels varies depending on the level of precontraction of the

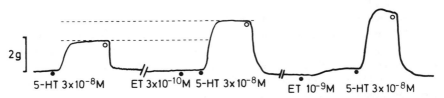

Figure 4. Contractile responses to serotonin (3×10^{-8} M) in a human left anterior descending coronary artery under control conditions (left) and after incubation with a threshold (3×10^{-10} M) or low (10^{-9} M) concentration of endothelin-1. Note the augmented response to serotonin in the presence of endothelin-1 as compared to control conditions.

arteries and oxidized LDL has been shown to potentiate contractions to several agonists including potassium chloride (Galle et al. 1990). In our experiments, however, the contractions to the thromboxane analoge U 46619 and those to potassium chloride are not different between arteries having been exposed or not exposed to oxidized LDL and thus cannot explain the effect of the lipoprotein. Binding of serotonin to the LDL particle also cannot be considered as an explanation since the LDL was removed from the organ chambers by extensive washing ten minutes before the arteries were precontracted. Thus, oxidized LDL specifically interferes with endothelial function of epicardial coronary arteries of the pig. This effect must be related to a reduced production of endothelium-derived nitric oxide, since the endothelium-dependent relaxations to bradykinin are not inhibited by oxidized LDL and are also unaffected by L-NMMA and pertussis toxin (Richard et al. 1990; Tanner, Boulanger and Lüscher, unpublished).

In hypercholesterolemic porcine coronary arteries, the endothelium-dependent relaxations to serotonin, but not those to bradykinin are impaired (Shimokawa and Vanhoutte, 1988). As the same pattern of action has been obtained in arteries exposed to oxidized LDL, the reduced endothelium-dependent relaxations of hypercholesterolemic arteries may be explained by an accumulation of oxidized LDL in the blood vessel wall. This interpretation is in line with the fact that a decreased release of endothelium-derived relaxing factor accounts for the impaired endothelium-dependent relaxations in hypercholesterolemia (Shimokawa and Vanhoutte 1989), and it is also consistant with the hypothesis that oxidized LDL plays an important role particularly in early stages of atherogenesis (Quinn et al. 1987). When compared to hypercholesterolemia, established atherosclerosis further impairs the endothelium-dependent relaxations to serotonin and also reduces those to bradykinin (Shimokawa and Vanhoutte 1988). At this stage of the disease, the concomitant release of vasoactive prostaglandins from the endothelium and/or intimal thickening as a functional barrier may also contribute.

Vasospasm

Endothelin-1 is a 21-amino acid peptide that is produced and released by endothelial cells upon stimulation with thrombin and platelet-derived substances such as TGFβ (Yanagisawa et al. 1988; Boulanger et al. 1990). It has potent vasoconstrictor properties that exceed those of the other cardiovascular hormones. Thus, it has been implicated in coronary vasospasm and unstable angina (Kurihara et al. 1989). Measurements of the circulating levels of endothelin-1, however, revealed concentrations that are below those inducing contractions of isolated blood vessels (Ando et al. 1989). Threshold levels of the peptide could still play an important role sensitizing the blood vessel wall to the contractile effect of serotonin. Indeed, threshold and low concentrations of endothelin-1 have been shown to augment contractions to serotonin in human left anterior descending coronary arteries (Yang et al. 1990). This effect of endothelin-1 may have important clinical implications. In unstable angina, local platelet activation occurs, predominantly at sites where oxidized LDL has accumulated in the vessel wall. Thrombin, which is formed after activation of the coagulation cascade, is present at high concentrations and stimulates the production of endothelin-1. Serotonin is released from platelets, and its vasoconstrictor action is amplified by the endothelium-derived peptide. Thus, endothelin-1 may importantly contribute to the effects of serotonin in acute ischemic syndromes associated with platelet activation.

Conclusions

Serotonin plays an important role in the local regulation of coronary vascular tone by the release of endothelium-derived relaxing factor and by direct contractile effects on vascular smooth muscle cells. In epicardial porcine coronary arteries, the endothelium-derived relaxing factor stimulated by the monoamine is nitric oxide, whereas in intramyocardial coronary arteries another factor must also be released. Oxidized, but not native LDL reduce endothelium-dependent relaxations to serotonin by an impaired release of nitric oxide and thus favour platelet aggregation and vasospasm. Threshold concentrations of endothelin-1 released upon platelet aggregation potentiate contractions to serotonin and with that further contribute to acute ischemic syndromes. Thus, serotonin is involved in the pathogenesis of potentially fatal events associated with coronary artery disease.

References

Ando, K., Hirata, Y., Shichiri, M., Emori, T., and Marumo, F. (1989). Presence of immunoreactive endothelin-1 in human plasma FEBS Lett. 245: 164–166.
Boulanger, C., and Lüscher, T. F. (1990). Release of endothelin-1 from the porcine aorta: Inhibition by endothelium-derived nitric oxide. J. Clin. Invest. 85: 587–590.
Faggiotto, A., Ross, R., and Harker, L. (1984). Studies of hypercholesterolemia in the nonhuman primate. I. Changes that lead to fatty streak formation. Arteriosclerosis 4: 323–340.
Flavahan, N. A., and Vanhoutte, P. M. (1989). Pertussis toxin inhibits endothelium-dependent relaxations to certain agonists in porcine coronary arteries. J. Physiol. 408: 549–560.
Galle, J., Bassenge, E., and Busse, R. (1990). Oxidized low density lipoproteins potentiate vasoconstrictions to various agonists by direct interaction with vascular smooth muscle. Circ. Res. 66: 1287–1293.
Gillis, C. N. (1985). Peripheral metabolism of Serotonin. In Vanhoutte, P. M. (ed), Serotonin and the Cardiovascular System. New York: Raven Press, p. 27.

Houston, D. S., and Vanhoutte, P. M. (1988). Comparison of serotonergic receptor subtypes on the smooth muscle and endothelium of the canine coronary artery. J. Pharmacol. Exp. Ther. 244: 1–10.

Kurihara, H., Yoshizumi, M., Sugiyama, T., Yamaoki, K., Nagai, R., Takaku, F., Satoh, H., Inui, J., Yanagisawa, M., Masaki, T., and Yasaki, Y. (1989). The possible role of endothelin-1 in the pathogenesis of vasospasm. J. Cardiovasc. Pharmacol. 13 (suppl. 5): 129–132.

Page, I. H. (1954). Serotonin (5-hydroxytryptamine). Physiol. Rev. 34: 563.

Palmer, R. M. J., Rees, D. D., Ashton, D. S., and Moncada, S. (1989). L-arginine is the physiological precursor for the formation of nitric oxide in endothelium-dependent relaxation. Biochem. Biophys. Res. Comm. 153: 1251–1256.

Quinn, M. T., Parthasarathy, S., Fong, L. G., and Steinberg, D. (1987). Oxidatively modified low density lipoproteins: A potential role in recruitment and retention of monocyte/ macrophages during atherogenesis. Proc. Natl. Acad. Sci. USA. 84: 2995–2998.

Richard, V., Tanner, F. C., Tschudi, M., and Lüscher, T. F. (1990). Differential activation of the endothelial L-arginine pathway by bradykinin, serotonin and clonidine in porcine coronary arteries. Am. J. Physiol. 259: H1433–H1439.

Shimokawa, H., Tomoike, H., Nabeyama, S., Yamamoto, H., Araki, H., Nakamura, M., Ishii, Y., and Tanaka, K. (1983). Coronary artery spasm induced in atherosclerotic miniature swine. Science 221: 560–562.

Shimokawa, H., and Vanhoutte, P. M. (1988). Dietary cod-liver oil improves endothelium-dependent responses in hypercholesterolemic and atherosclerotic porcine coronary arteries. Circulation 78: 1421–1430.

Shimokawa, H., and Vanhoutte, P. M. (1989). Impaired endothelium-dependent relaxation to aggregating platelets and related vasoactive substances in porcine coronary arteries in hypercholesterolemia and in atherosclerosis. Circ. Res. 64: 900–914.

Tschudi, M., Richard, V., Bühler, F. R., and Lüscher, T. F. (1991). The endothelial L-arginine pathway in intramyocardial porcine coronary resistance arteries. Am. J. Physiol. In press.

Tyce, G. M. (1985). Biochemistry of Serotonin. In Vanhoutte, P. M. (ed.), Serotonin and the Cardiovascular System. New York: Raven Press, p. 1.

Vanhoutte, P. M. (1985). Peripheral serotonergic receptors and hypertension. In Vanhoutte, P. M. (ed.), Serotonin and the Cardiovascular System. New York: Raven Press, p. 123.

Van Nueten, J. M., Janssen, P. A. J., van Beek, J., Xhonneux, R., Verbeuren, T. J., Vanhoutte, P. M. (1981). Vascular effects of ketanserin (R 41 468), a novel antagonist of 5-HT2 serotonergic receptors. J. Pharmacol. Exp. Ther. 218: 217.

Vincent, G. M., Anderson, J. L., and Marshall, H. W. (1983). Coronary vasospasm producing coronary thrombosis and myocardial infarction. N. Engl. J. Med. 309: 220–223.

Yanagisawa, M., Kurihara, H., Kimura, S., Tomobe, Y., Kobayashi, M., Mitsui, Y., Yazaki, Y., Goto, K., and Masaki, T. (1988). A novel potent vasoconstrictor peptide produced by vascular endothelial cells. Nature 332: 411–415.

Yang, Z., Richard, V., von Segesser, L., Bauer, E., Stulz, P., Turina, M., and Lüscher, T. F. (1990). Threshold concentrations of endothelin-1 potentiate contractions to norepinephrine and serotonin in human arteries. A new mechanism of vasospasm? Circulation 82: 188–195.

Ylä-Herttuala, S., Palinski, W., Rosenfeld, M. E., Parthasarathy, S., Carew, T. E., Butler, S., Witztum, J. L., and Steinberg, D. (1989). Evidence for the presence of oxidatively modified low density lipoprotein in atherosclerotic lesions of rabbit and man. J. Clin. Invest. 84: 1086–1095.

Serotonin: Molecular Biology, Receptors and Functional Effects
ed. by J. R. Fozard/P. R. Saxena
© 1991 Birkhäuser Verlag Basel/Switzerland

Effects of Selective 5-HT$_2$ Receptor Antagonists on some Haemodynamic Changes Produced by Experimental Pulmonary Embolism in Rabbits

J. L. Amezcua[1], E. Hong[2], and R. A. Bobadilla[3]

[1]Escuela Militar de Graduados de Sanidad, Universidad del Ejército y Fuerza Aérea, Mexico City; [2]Sección de Terapéutica Experimental, CINVESTAV, I.P.N., Mexico City; [3]Escuela Superior de Medicina del Instituto Politécnico Nacional, Mexico City (partially supported by a COFAA grant)

Summary. Heparin, aspirin and the selective 5-HT$_2$ antagonists ketanserin and pelanserin partially prevented the platelet-fall, rise en right ventricular pressure and systemic hypotension induced by the intravenous administration of an autologous clot or a suspension of collagen fibrils in rabbits. Whereas platelet-depletion nearly fully prevented the haemodynamic effects of the experimental thromboembolism. Intravenous 5-HT resembled the effects of pulmonary embolism and was only partially antagonized by the 5-HT$_2$ receptor antagonists. These results suggest an additive role for platelet-derived vasoactive factors and thrombin in these experimental models and point out a role for 5-HT receptors eliciting reflex responses in the lung during the embolization.

Introduction

The haemodynamic consequences of vascular thromboembolism are not the mere result of the mechanical presence of the clot but involve the direct and reflex responses evoked by several vasoactive mediators (Malik 1983). Among these substances serotonin (5-HT) is an outstanding candidate to play an important pathophysiological role as it is accumulated in the circulating platelets and released upon stimulation which invariably occurs as the thrombotic cascade is triggered (Huval 1983). Thromboxane A2 (TXA$_2$) and thrombin are also vasoactive substances present in the thrombotic environment and their effects are likely to combine with those of 5-HT in the modulation of the overall vascular response to thrombus formation (De Clerck 1982). The complexity of the 5-HT actions in the vasculature is still a matter of vast interest since they vary according to animal species, the vascular bed, the presence or absence of endothelium and the contribution of reflex responses elicited by 5-HT acting on peripheral nerve endings (Cohen 1990, McQueen 1990). The systematization of these complex array of responses seems now possible by the recognition of different subtypes of 5-HT receptors involved in different vascular responses and the availability of selective agonists and antagonists for some of these receptors. We have studied the relative role of thrombin and of platelet-derived vasoactive factors such as TXA_2 and 5-HT on two different models of pulmonary thromboembolism in rabbits.

Materials and Methods

General Procedure

Male, white, New Zealand rabbits weighing 2.5 to 3.5 kg were anaesthetised with pentobarbitone (30–35 mg/kg) injected through the ear vein. The chest was opened and the respiration was mechanically assisted via a tracheal cannula and indwelling cannulae were placed in the right ventricle via the right jugular vein, the carotid artery and the left jugular vein. The right ventricular pressure (RVP) and the systemic blood pressure (SBP) were continously recorded by means of Statham transducers attached to a Grass polygraph. A basal recording of 20 min was obtained before any drug was given.

Pretreatments

The animals were divided into several groups according to the following pretreatments:

1. A platelet antibody (20 mg of ram gamma globulin, i.v.) prepared as described below, 24 h before the experiment.
2. Aspirin (10 mg/kg, i.v.)
3. Heparin (200 U/kg, i.v.)
4. Ketanserin (300 μg/kg, i.v.)
5. Pelanserin (450 μg/kg, i.v.)

The pretreatments 2 to 5 were given 20 min before the challenges.

Challenges

The challenges consisted of either an autologous clot (60 mg/kg i.v.) or a suspension of collagen fibrils (90 μg/kg i.v., infused over a period of 2 min). At least six animals of each pretreatment group were given each challenge. The observation of the haemodynamic changes was continued over at least 30 min, but the results are expressed as the maximal change observed in each indicator.

Platelet Counts

An arterial blood sample was taken immediately and 5 min after the challenge and platelets were counted in a Coulter Counter (Coulter Electronics Inc. model ZBI).

Preparation of Autologous Clot

1 ml of arterial blood was taken in a 1 ml glass syringe immediately after anaesthetising the rabbits in those cases which were going to be challenged with an autologous clot. After keeping the blood 10 min in the syringe it was pulled out the syringe and cut into slices of about 1 mm width to make up the calculated dose for each animal.

Collagen Fibril Preparation

The Achilles tendons of some rabbits were excised, cleaned of muscle and finely chopped with scissors. One gram of tendon was placed in 75 ml of a solution of acetic acid 0.25 M in distilled water and stirred in a magnetic plate at 4°C during 24 hours. The resulting suspension was centrifuged at 1400X g for 20 min and the supernatant was used. The strength of the solution thus obtained was titrated against a commercial preparation of collagen (Horm) in washed platelet bioassay in a Payton aggregometer. The protein concentration was determined by the method of Lowry (Lowry, 1951) and in pilot experiments a standard dose of 60 mg/kg was established to be a sublethal dose capable of producing consistent haemodynamic changes and a significant fall in platelet count. The suspension was kept at 4° and freshly prepared every week.

Preparation of the Platelet-Antibody

Washed rabbit platelets were prepared according to the method of Van Loghem (Van Loghem 1959). The concentration of washed platelets was adjusted to 500,000 μl. The priming dose of platelet antigen was given subcutaneously to a 25 kg ram and consisted of 3 ml of washed platelets thoroughly mixed with 3 ml of Freund adjuvant. Boosting doses were similarly prepared but injected intraperitoneally once a week up to 5 weeks, and then every month until the end of the experiments. Blood samples were monthly taken after the fifth dose and the gamma globulin fraction was precipitated from the serum with a saturated solution of ammonium sulphate and thrice dialysed against a borate buffer pH 7.6. The gamma globulin fraction was freeze-dried. Pilot experiments in rabbits showed that 20 mg of protein thus obtained was capable of reducing the platelet count of rabbits by more than 90% without significantly changing the leucocyte or the erythrocyte counts.

Supplementary Experiments

Additional experiments were done in rabbits pretreated with ketanserin ($300 \mu g/kg$ i.v., $n = 6$) and pelanserin ($450 \mu g/kg$ i.v., $n = 6$) that were challenged with 5-HT ($250 \mu g/kg/min$, i.v., 2 min).

Statistics

The differences among the means were assessed with a *t*-test for multiple constrasts as described by Brown and Hollander (Brown and Hollander, 1977) and a difference was considered significant at a level of $p < 0.05$. The results are expressed as the mean $\pm$ s.e.m.

Results

Pulmonary embolization produced by either an autologous clot or a collagen suspension caused a reversible but sharp rise in RVP followed a few seconds later by a fall in SBP in control animals. It is remarkable that the only pretreatment capable of inhibiting nearly completely the haemodynamic changes produced by the experimental thromboembolism in both models was the depletion of platelets (see Figures 1 to 4). Heparin was more effective in preventing these haemodynamic changes in the autologous clot model, specially the fall in systemic blood pressure, whereas aspirin was more effective in the collagen model than in the autologous clot model (see Figures 1 to 4). However, the selective $5\text{-}HT_2$ antagonists ketanserin and pelanserin produced protection in

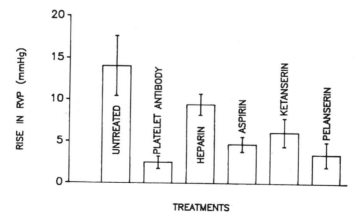

Figure 1. Rise in the right ventricular pressure (RVP) produced by an intravenous injection of a collagen suspension in rabbits in the presence of different pretreatments. The results are the mean $\pm$ s.e.m. of six experiments in each group.

418

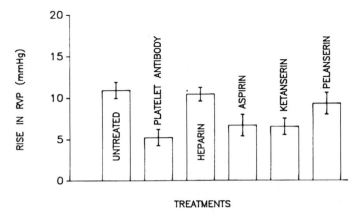

Figure 2. Rise in the right ventricular pressure (RVP) produced by an intravenous injection of an autologous clot in rabbits in the presence of different pretreatments. The results are the mean ± s.e.m. of six experiments in each group.

both models. The effects of these antagonists were not significantly different from those of aspirin. The platelet count falled $71.3 \pm 8.1\%$ with the collagen challenge in the animals without pretreatment and $37 \pm 4.5\%$ with the autologous clot. This drop in platelet count was prevented more effectively by heparin, aspirin, ketanserin and pelanserin in the collagen model, whereas this indicator behaved in a rather bizarre manner in the clot model i.e. neither heparin nor aspirin had any effect, but ketanserin and pelanserin significantly prevented the fall in the platelet count.

The infusion of 5-HT closely resembled the effects of pulmonary embolization in both the fall in platelet count and the changes in RVP

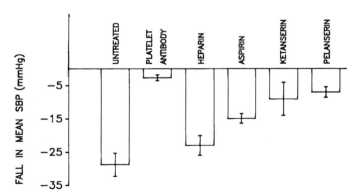

Figure 3. Fall in the mean systemic blood pressure (SBP) produced by an intravenous injection of a suspension of collagen fibrils in the presence of different pretreatments. The results are the mean ± s.e.m. of six experiments in each group.

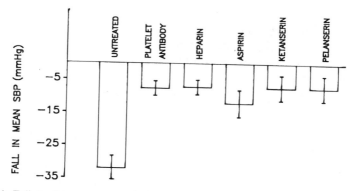

Figure 4. Fall in the mean systemic blood pressure (SBP) produced by an intravenous injection of an autologous clot in the presence of different pretreatments. The results are the mean ± s.e.m. of six experiments in each group.

and SBP, these effects of 5-HT were inhibited about 50% by the pretreatment with either ketanserin and pelanserin.

Discussion

This study illustrates the crucial role of platelet-derived vasoactive mediators in the production of the haemodynamic changes produced by a thrombotic challenge in the lung. One of the models resembles the situation of the most common type of pulmonary embolism i.e., that produced by a migrating blood clot which predominately activates the clotting factors and secondarily promotes platelet recruitment. By contrast, the collagen model resembles the starting events involved in arterial thrombosis which is initiated by the adhesion of platelets to an area of intimal damage followed by their aggregation and release reaction. The fact that neither aspirin nor the $5\text{-}HT_2$ antagonists could fully prevent the effects of the thomboembolism suggests that TXA_2 and 5-HT may add their effects to produce the amplification of the vascular responses following thrombus formation and/or impactation. In addition, the protection observed with heparin suggests that the vascular effects of the clotting factors, particularly those of thrombin are likely to participate in these events, as thrombin has been shown to have vasoactive properties which vary according to the vascular bed (White 1980) and to the presence or absence of endothelium (De Mey 1981).

Moreover, the intravenous infusion of 5-HT simulates the effects of thromboembolism, and its partial inhibition by $5\text{-}HT_2$ receptor antagonist indicates a role for these receptors, most likely at the vascular level. However, if we consider the fall of the systemic blood pressure that occurs in our models in spite of the virtually complete clearance of

5-HT in the lung (Vane 1969), we need to suggest that triggering of a hypotensive reflex response by 5-HT. Indeed, the presence of 5-HT receptors in the so called juxta-pulmonary capillary nerve terminals has been reported (McQueen 1990) and its stimulation is followed by brady-cardia and systemic hypotension in rabbits (Armstrong 1986). These actions of 5-HT are likely to be mediated by 5-HT$_3$ receptors rather than the 5-HT$_2$ since ketanserin and palenserin the two selective antag-onists of this type used in this study only caused a partial inhibition of the haemodynamic responses to exogenous or platelet-derived 5-HT, whereas MDL 72222 has been shown to be a selective inhibitor of the reflex responses to intravenous 5-HT (McQueen 1990). This study accentuates the need for further investigation into the role of these receptors in the pathophysiology of pulmonary thromboembolism.

References

Armstrong, D. U., Kay, I. S., and Russel, N. J. W. (1986). MDL 72222 antagonizes the reflex tachypneoic response to miliary pulmonary embolism anaesthetized rabbits. J. Physiol. 381: 13p.

Brown, B., and Hollander M. (1977). Statistics. A biomedical introduction. John Wiley and Sons, Sydney.

Cohen, M. L. (1990). Receptors for 5-HT in the cardiovascular system. In: Saxena, P. R., Wallis, D. I., Wouters, W., and Bevan, P. (eds.), Cardiovascular Pharmacology of 5-hydroxytryptamine. Dordrecht: Kluwer, pp. 295–302.

DeClerck, F., and Van Neuten, J. (1982). Platelet-mediated vascular contraction: inhibition of the serotonergic component by ketanserin. Thromb. Res. 27: 713–727.

De Mey, J. G., and Vanhoutte, P. M. (1981). Role of the intima in the relaxation of the canine femoral artery caused by thrombin. Arch. Int. Pharmacodyn. Ther. 250: 314–320.

Huval, W., Mathieson, M., Stemp, L., Dunham, B., Jones, A., and Shepro, D. (1983). Therapeutic benefits of 5-HT inhibition following pulmonary embolism. Ann. Surg. 197: 220–225.

Lowry, O. R., Rosenborough, N., Farr, L., and Randall, R. (1951). Protein measurement with the Folin-phenol reagent. J. Biol. Chem. 193: 265–275.

Malik, A. B. (1983). Pulmonary microembolism. Physiol. Rev. 63: 1115–1207.

McQueen, D. S. (1990). Cardiovascular reflexes and 5-hydroxytryptamine. In: Saxena, P. R., Wallis, D. I., Wouters, W., and Bevan, P. (eds.), Cardiovascular Pharmacology of 5-hydroxytryptamine. Dordrecht: Kluwer, pp. 233–246.

Vane, J. R. (1969). The release and fate of vasoactive hormones in the circulation. Br. J. Pharmacol. 53: 209–242.

Van Loghem, J. (1959). Serological and genetical studies on platelet antigens. Vox. Sang. 4: 161–169.

White, R. P., Chapbrau, E. Dugdale, M., and Robertson, J. T. (1980). Stroke. 11: 363–365.

Serotonin: Molecular Biology, Receptors and Functional Effects
ed. by J. R. Fozard/P. R. Saxena
© 1991 Birkhäuser Verlag Basel/Switzerland

The Vasoconstrictor Action of Sumatriptan on Human Isolated Dura Mater

P. P. A. Humphrey[1], W. Feniuk[1], M. Motevalian[2], A. A. Parsons[2], and E. T. Whalley[2]

[1]Pharmacology Division, Glaxo Group Research Limited, Park Road, Ware, Hertfordshire SG12 0DP; [2]Department of Physiological Sciences, University of Manchester, Oxford Road, Manchester, M13 9PT, UK

Summary. Sumatriptan is a selective agonist for a 5-HT$_1$-like receptor sub-type which mediates contraction of some cranial blood vessels *in vitro*. It has been shown to be a very effective novel agent for the acute treatment of migraine and this has stimulated much interest in its clinical mode of action. It has been suggested that the pain of migraine results from a sterile neurogenic inflammatory response of intracranial blood vessels innervated by the trigeminal nerve. Furthermore, sumatriptan has been shown to inhibit plasma protein extravasation from blood vessels of the dura mater, induced by antidromic stimulation of the trigeminal nerve in the anaesthetised rat and guinea-pig. This effect could be explained by a localised vasoconstriction of the meningeal vessels. We now provide evidence that sumatriptan will potently constrict the blood vessels within the human isolated perfused dura mater. Preliminary work on characterising the 5-HT receptor involved indicates that it is a 5-HT$_1$-like receptor. These studies add weight to the view that the anti-migraine action of sumatriptan results from selective cranial vasoconstriction.

Introduction

Sumatriptan is under development as a novel drug treatment for acute migraine (Humphrey et al. 1989a). Chemically, it is structurally similar to the neurotransmitter amine, 5-hydroxytryptamine (5-HT; serotonin) but its pharmacological profile is much more restricted (Humphrey et al. 1988; 1989b). Thus sumatriptan has a very specific and selective agonist action on some 5-HT$_1$ sub-types, notable the 5-HT$_1$-like receptor which mediates vasoconstriction of cranial blood vessels but is devoid of activity at 5-HT$_2$, 5-HT$_3$ and 5-HT$_4$ receptors (Feniuk et al. 1990).

Sumatriptan was developed on the basis that 5-hydroxytryptamine itself will alleviate the symptoms of migraine (Humphrey et al. 1990a). However, 5-HT produces many effects which are very undesirable including bronchoconstriction, platelet aggregation and activation of respiratory and cardiovascular reflexes. Since such effects are mediated via 5-HT$_2$ and 5-HT$_3$ receptor activation, these actions are not shared by sumatriptan. Although there is circumstantial evidence to implicate a depletion of 5-HT brain levels during a migraine attack, it seems unlikely that sumatriptan is acting as a "replacement" therapy (Humphrey 1990). Thus sumatriptan does not readily cross the blood brain barrier (Humphrey et al. 1990b; Sleight et al. 1990). It, therefore, seems reasonable at this stage to accept the more parsimonious explanation that sumatriptan is simply acting as a vasoconstrictor agent. Thus other vasoconstrictors too are known to be effective in treating acute migraine attacks although their unwanted side effects limit their use. Since $5\text{-}HT_1$-like receptors mediating vasoconstriction are located predominantly on cranial blood vessels, sumatriptan has a more localised vasoconstrictor action (Feniuk et al. 1989; Perren et al. 1989; Saxena and Ferrari 1989). The question remains as to which particular cranial vessels are dilated during a migraine attack?

Neuroanatomical and other considerations have led to the view that the meningeal circulation may be particularly important (Blau 1978; Moskowitz 1990). Thus the vasculature of the dura mater, together with the large conduit vessels of the carotid and vertebral circulations are particularly pain-sensitive. These vessels are densely innervated by "C" fibre sensory afferents which run in the trigeminal nerve, the Vth cranial nerve known to be implicated in head pain of vascular origin. Moskowitz (1984) has suggested that vascular headaches such as migraine are initiated by a sterile neurogenic inflammatory condition associated with vasodilatation, extravasation and activation of pain pathways. It has been argued that vasoconstriction of the affected vessels may be sufficient to alleviate the pain (Humphrey 1990). It is already known that the human basilar artery, which contains $5\text{-}HT_1$-like receptors, is contracted by sumatriptan (Parsons et al. 1989). We have now investigated the effect of sumatriptan on the perfused, isolated vasculature of the human dura mater.

Methods

Human dura mater was obtained from cadavers, up to 18 hours post-mortem, and placed in Krebs-Henseleit solution at $37°C$ and gassed with 95% O_2 and CO_2. The major meningeal artery was cannulated and the preparation transferred to a large jacketed organ bath (50 ml). The dura was then perfused with Krebs-Henseleit solution ($37°C$) at a constant flow rate using a peristaltic pump. The perfusion rate was maintained at 2 ml/min and changes in perfusion pressure were recorded from a side arm on the perfusion circuit, connected to a pressure transducer. The immersion solution (Krebs-Henseleit) was also continuously changed at a flow rate of 10 ml/min. Perfusion pressure was recorded on a Grass polygraph chart recorder. Initial mean resting perfusion pressure was 59 ± 5 mmHg (values are mean $\pm$ s.e. mean, n = 105). Agonists were injected into the perfusion stream as a bolus, using a standard 50 μl volume.

Following an initial equilibration period of approximately 30 minutes, the reactivity of the vasculature to an initial challenge dose of potassium chloride (0.5 mmole) was studied. The initial challenge with potassium chloride increased perfusion pressure in 57% of the tissues. Those not responding were discarded. The composition of the Krebs-Henseleit solution was (mM) NaCl (118), KCl (4.75) $CaCl_2.6H_2O$ (2.55), $MgSO_4.7H_2O$ (1.2), KH_2PO_4 (1.19), $NaHCO_3$ (25) and glucose (11).

Studies with Agonists

Dose-effect curves to bolus injections of 5-HT were first obtained in all preparations. Increases in perfusion pressure (reflecting vasoconstriction) were measured as a percentage of the mean resting perfusion pressure. Once perfusion pressure had returned to normal, a second dose-effect curve to either sumatriptan or 5-carboxamidotryptamine (5-CT) was obtained, 60–90 minutes later. Dose-effect curves for each of the agonists was plotted and pED_{50} values calculated for each preparation ($-\log_{10}$ dose in mole producing 50% of the maximal effect). Since full dose-effect curves were not always obtained, effects produced by 10^{-6} mole 5-HT and 10^{-7} mole sumatriptan or 5-CT were considered to be maximal. Equipotent dose ratios (5-HT = 1) were derived by dividing pED_{50} values for the test drug by the pED_{50} values for the 5-HT in a given experiment.

Studies with Antagonists

Preliminary studies showed that dose-effect curves to 5-HT, sumatriptan and 5-CT were reproducible. Control dose-effect curves to 5-HT, sumatriptan or 5-CT were determined and then repeated in the presence of known concentrations of antagonist in the Krebs-Henseleit perfusing (but not bathing) the dural preparation. The antagonists examined were either ketanserin (100 nM), ondansetron (100 nM) or methiothepin (100 nM); each antagonist was in contact for 30 minutes prior to the start of and during an agonist dose-effect curve. Full dose-effect curves to the agonists could not always be obtained in the presence of an antagonist and, therefore, dose-ratios were calculated at the level of 30% of the maximum effect in the absence of antagonist. Specificity studies were carried out using prostaglandin $F_{2\alpha}$ ($PGF_{2\alpha}$)-induced increases in perfusion pressure.

Drugs Used

5-Hydroxytryptamine creatinine sulphate (Sigma); 5-carboxamidotryptamine (Glaxo); ketanserin (Janssen); methiothepin maleate

424

(Hoffman La Roche); ondansetron (Glaxo); prostaglandin $F_{2\alpha}$ (Sigma) and sumatriptan (Glaxo).

Results

Studies with Agonists

5-Hydroxytryptamine, sumatriptan and 5-CT all produced dose-dependent increases in perfusion pressure in human isolated perfused dura mater (see Figure 1). Comparison of pED_{50} values (Table 1) showed that 5-CT was approximately two times more potent than 5-HT whilst sumatriptan was approximately three times weaker than 5-HT. The maximum effect produced by sumatriptan appeared to be less than that produced by 5-HT or 5-CT, though it was not statistically significant (Table 1).

Studies with Antagonists

The effect of several 5-HT receptor blocking drugs was assessed against 5-HT-induced increases in perfusion pressure in human isolated perfused dura mater. The increases in perfusion pressure produced by 5-HT

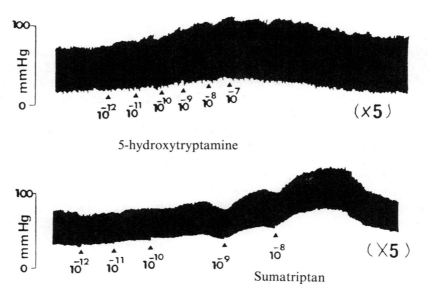

Figure 1. Original recordings showing change in perfusion pressure of perfused human dura on dosing with of 5-hydroxytryptamine or sumatriptan. Agonists were added, as bolus doses in a standard 50 μl volume, to the perfusion fluid.

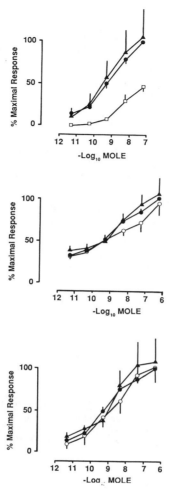

Figure 2. The effect of various 5-HT receptor blocking drugs on vasoconstrictor responses to 5-hydroxytryptamine in human isolated dural preparations. Concentration-effect (c-e) curves to 5-HT were repeated to provide two control curves (filled symbols). The third c-e curve in the presence of an antagonist shows that vasoconstrictor responses to 5-HT were antagonised by the 5-HT$_1$-like receptor blocking drug, methiothepin (top panel; 100 nM, open symbols) but not by the 5-HT$_2$ receptor blocking drug, ketanserin (middle panel; 100 nM, open symbol) or the 5-HT$_3$ receptor blocking drug, ondansetron (lower panel; 100 nM, open symbols). Each point represents the mean of between 3–13 values and the vertical bars represent the s.e. mean.

were not affected by the presence of either 5-HT$_2$ receptor blocking drug ketanserin (100 nM) or the 5-HT$_3$ receptor blocking drug, ondansetron (100 nM) (Figure 2). Neither ketanserin nor ondansetron alone had any significant effect on basal perfusion pressure.

In contrast methiothepin (100 nM) caused a marked attenuation of the increases in perfusion pressure produced by 5-HT (Figure 2).

Table 1. Agonist potencies of 5-HT, 5-CT and sumatriptan in isolated perfused human dura mater

Agonist	pED_{50}	Maximum % change in perfusion pressure	Equipotent dose ratio
5-Hydroxytryptamine	8.9 ± 0.3	103 ± 10	1
5-Carboxamidotryptamine	9.2 ± 0.3	114 ± 14	0.5
			$(0.05 - 5.0)$
Sumatriptan	8.4 ± 0.6	$62 \pm 12*$	2.5
			$(0.3 - 25)$

pED_{50} is the $-\log$ dose (mole) of agonist producing 50% of its maximum vasoconstrictor action.
Values are mean $\pm$ s.e.m. or geometric mean (95% confidence limits) from at least five experiments.
An equipotent dose ratio of <1 indicates that the compound is more active than 5-HT or of >1 indicates that the compound is weaker than 5-HT.
*Not statistically significant from maximum response to 5-HT at the $p = 0.05$ level by Students "t" test.

Table 2. Antagonist effect of methiothepin (100 nM) on 5-hydroxytryptamine-, 5-carboxamidotryptamine-, sumatriptan- and prostaglandin $F_{2\alpha}$-induced increases in perfusion pressure of human isolated dura mater

Agonist	Dose-Ratio	n
5-Hydroxytryptamine	30 $(10-79)$	7
5-Carboxamidotryptamine	25 $(7.4 - 85)$	5
Sumatriptan	93 $(37 - 234)$	3
Prostaglandin $F_{2\alpha}$	1.6 $(0.3 - 9.8)$	3

Dose ratio represents the mean x-fold rightward shift of the agonist dose-effect curve produced by the antagonist. Values are geometric means (95% confidence limits) from n preparations. Note that vasoconstrictor responses to 5-hydroxytryptamine, 5-carboxamidotryptamine and sumatriptan were antagonised to a similar degree by methiothepin, while vasoconstrictor responses to prostaglandin $F_{2\alpha}$ were barely affected.

Methiothepin also antagonised the effect of sumatriptan and 5-CT to a similar extent as for 5-HT (Table 2). Methiothepin itself (100 nM) had little or no effect on basal perfusion pressure and it did not modify increases in perfusion pressure produced by $PGF_{2\alpha}$ (Table 2).

Discussion

The vasculature of the dura mater, one of the meninges surrounding the brain, has been strongly implicated as the source of head pain of the migraine type (see Introduction). It has a dense perivascular nerve plexus innervating it, which transmits information into the brain at the level of the pons via the trigeminal nerve. It has been argued that vasodilation of dural blood vessels, in a pathological scenario involving a sterile neurogenic inflammatory response (see Buzzi and Moskowitz 1990), could lead to a substantial reduction in the threshold for activation of sensory pain fibres which would be reversed by a vasoconstrictor agent (see Humphrey et al. 1990b). As the anti-migraine drug, sumatriptan, has a selective vasoconstrictor profile, (Humphrey et al. 1989a; b, 1990a; den Boer et al. 1991), it was important to determine whether it was capable of constricting the vasculature of human dura mater. For obvious practical purposes, these studies had to be carried out *in vitro*. It also presented us with the opportunity to investigate the effects of 5-HT itself and to determine the receptor type(s) involved.

Sumatriptan caused a dose-dependent vasoconstriction in isolated preparations of the human dura mater. This finding extends the number of isolated cranial blood vessels known to be constricted by sumatriptan (Connor et al. 1989; Parsons et al. 1989). Importantly too, in the light of interest in the dura mater as a key locus in the pathophysiology of migraine, it raises the possibility that the dural vasculature is an important site of the anti-migraine action of sumatriptan.

5-Hydroxytryptamine and 5-carboxamidotryptamine also produced vasoconstriction of the dural vasculature. The relative potencies of these agonists are very similar to that observed in other cranial vessels already mentioned, with 5-carboxamidotryptamine being about twice as potent as 5-HT itself and sumatriptan some three-fold weaker than 5-HT. This rank order of agonist potency is indicative of a $5-HT_1$-like receptor involvement (see Bradley et al. 1986; Feniuk et al. 1985; Humphrey et al. 1988). More convincing evidence for this comes from studies with antagonists. Thus the vasoconstrictor action of 5-HT (and sumatriptan; data not shown) was not antagonised by either the $5-HT_2$ receptor blocking drug, ketanserin, or the $5-HT_3$ receptor blocking drug, ondansetron. However, the vasoconstrictor action of all three agonists was antagonised by the classical $5-HT_1$-like receptor antagonist, methiothepin. Approximate estimates of its affinity in the form of estimated pA_2 values for the single concentration of antagonist studied are in the range 8.5–9.0. Although with the experimental protocol employed it could be argued that full equilibrium conditions were not achieved, these values are close to pA_2 values obtained in other $5-HT_1$-like containing vessels (Connor et al. 1989; Parsons et al. 1989). The antagonism produced by methiothepin was judged to be specific for the 5-HT receptor agonists since vasoconstrictor responses to prostaglandin $F_{2\alpha}$ were barely affected. These findings are entirely consistent with the view that the vasculature of the human dura mater contains predominantly $5-HT_1$-like receptors which explains the vasoconstrictor activity of sumatriptan in these vessels.

The question remains as to whether vasoconstriction per se can explain the anti-migraine action of sumatriptan. Certainly it has long been known that intra-cranial vessels are pain-sensitive and that their distension will initiate pain (Graham and Wolff 1938; Nichols et al. 1990). Vasodilators are well known to initiate migraine attacks in migraineurs and vasoconstrictors will relieve the pain (see Lance 1973). It seems reasonable, therefore, to believe the distension and inflammation of vessels inside a fixed cranium could lead to a marked reduction in the threshold for activation of pain fibre endings in the vessel wall. Logically a vasoconstrictor might be expected to reverse this process reducing the transmural pressure thereby raising the threshold for neuronal firing. A localised vasoconstrictor action would also be expected to reduce or inhibit any concomitant extravasation from the dilated vascular site (see Williams and Peck 1977).

Indeed sumatriptan is active in an animal model of dural extravasation, though it has been postulated that a neuronal mechanism is involved (Buzzi and Moskowitz 1990). Thus Moskowitz has proposed that sumatriptan acts by a direct inhibitory action on trigeminal nerve endings as it does on some sympathetic nerves (Humphrey et al. 1988; Buzzi and Moskowitz 1990; Moskowitz 1990). Further experiments will be needed to resolve this controversy but nevertheless, all the evidence seems to point to the site of the anti-migraine action of sumatriptan being at the level of the vessel wall of the affected cephalic blood vessels, be it on the smooth muscle or the innervating nerve endings. It seems likely that the meningeal vasculature will be shown to be an important site. Unfortunately, we were unable to examine the effect of sumatriptan on neuronal function in this study because no neuronal vasomotor responses could be demonstrated in response to electrical stimulation, probably because of neuronal degeneration in the dural vessels after death.

References

Blau, J. N. (1978). Migraine: A vasomotor instability of the meningeal circulation. Lancet 1136–1139.

Bradley, P. B., Engel, G., Feniuk, W., Fozard, J. R., Humphrey, P. P. A., Middlemiss, D. N., Mylecharane, E. J., Richardson, B. P., and Saxena, P. R. (1986). Proposals for the classification and nomenclature of functional receptors for 5-hydroxytryptamine. Neuropharmacology 25: 563–576.

Buzzi, M. G., and Moskowitz, M. A. (1990). The anti-migraine drug sumatriptan (GR43175) selectively blocks neurogenic plasma extravasation from blood vessels in dura mater. Br. J. Pharmacol. 99: 202–206.

Connor, H. E., Feniuk, W., and Humphrey, P. P. A. (1989). Characterization of 5-HT receptors mediating contraction of canine and primate basilar artery by use of GR43175, a selective 5-HT$_1$-like receptor agonist. Br. J. Pharmacol. 96: 379–387.

Den Boer, M. O., Villaón, C. M., Heiligers, J. P. C., Humphrey, P. P. A., and Saxena, P. R. (1991). The role of 5-HT$_1$-like receptors in the reduction of porcine arteriovenous shunting by sumatriptan. Br. J. Pharmacol. 102: 323–330.

Feniuk, W., Humphrey, P. P. A., Perren, M. J., and Watts, A. D. (1985). A comparison of 5-hydroxytryptamine receptors mediating contraction in rabbit aorta and dog saphenous vein: evidence for different receptor types obtained by use of selective agonists and antagonists. Br. J. Pharmacol. 86: 697–704.

Feniuk, W., Humphrey, P. P. A., and Perren, M. J. (1989). The selective carotid arterial vasoconstrictor action of GR43175 in anaesthetized dogs. Br. J. Pharmacol. 96: 83–90.

Feniuk, W., Humphrey, P. P. A., Perren, M. J., Connor, H. E., and Whalley, E. T. (1990). The rationale for the use of sumatriptan in the treatment of migraine. J. Neurology. 238: S57–S61.

Graham, J. R., and Wolff, H. G. (1938). Mechanism of migraine headache and action of ergotamine tartrate. Arch. Neurol. Psychiat. Chicago 39: 737–763.

Humphrey, P. P. A. (1990). 5-Hydroxytryptamine and the pathophysiology of migraine. J. Neurology. 238: S38–S44.

Humphrey, P. P. A., Feniuk, W., Perren, M. J., Connor, H. E., Oxford, A. W., Coates, I. H., and Butina, D. (1988). GR43175, a selective agonist for the 5-HT$_1$-like receptor in dog isolated saphenous vein. Br. J. Pharmacol. 94: 1123–1132.

Humphrey, P. P. A., Feniuk, W., Perren, M. J., Oxford, A. W., and Brittain, R. T. (1989a). Sumatriptan succinate. Drugs of the Future 14(1): 35–39.

Humphrey, P. P. A., Feniuk, W., Perren, M. J., Connor, H. E., and Oxford, A. W. (1989b). The pharmacology of the novel 5-HT$_1$-like receptor agonist, GR43175. Cephalalgia 9(Suppl. 9): 23–33.

Humphrey, P. P. A., Apperley, E., Feniuk, W., and Perren, M. J. (1990a). A rational approach to identifying a fundamentally new drug for the treatment of migraine. In Saxena, P. R., Wallis, D. I., Wouters, W., and Bevan, P. (eds.), Cardiovascular pharmacology of 5-hydroxytryptamine. Dordrecht: Kluwer, pp. 417–431.

Humphrey, P. P. A., Feniuk, W., Perren, M. J., Beresford, I. J. M., Skingle, M., and Whalley, E. T. (1990b). Serotonin and migraine. Annals of New York Academy of Sciences. 600: 587–598.

Lance, J. W. (1973). The mechanism and management of headache. Second edition, London: Butterworths.

Moskowitz, M. A. (1984). Neurobiology of vascular head pain. Annals of Neurology 16: 157–168.

Moskowitz, M. A., and Buzzi, M. G. (1991). Neuroeffector functions of sensory fibres: implications for headache mechanisms and drug actions. J. Neurology. 238: S18–S22.

Nichols, F. T., Mawad, M., Mohr, J. P., Stein, B., Hilal, S., and Michelsen, W. J. (1990). Focal headache during balloon inflation in the internal carotid and middle cerebral arteries. Stroke 21: 555–559.

Parsons, A. A., Whalley, E. T., Feniuk, W., Connor, H. E., and Humphrey, P. P. A. (1989). 5-HT$_1$-like receptors mediate 5-hydroxytryptamine-induced contraction of human isolated basilar artery. Br. J. Pharmacol. 96: 434–449.

Perren, M. J., Feniuk, W., and Humphrey, P. P. A. (1989). The selective closure of feline carotid arteriovenous anastomoses (AVAs) by GR43175. Cephalalgia 9(Suppl.9): 41–46.

Saxena, P. R., and Ferrari, M. D. (1989). 5-HT$_1$-like receptor agonists and the pathophysiology of migraine. Trends Pharmacol. Sci. 10: 200–204.

Sleight, A. J., Cervenka, A., and Peroutka, S. J. (1990). In vivo effects of sumatriptan (GR43175) on extracellular levels of 5-HT in the guinea pig. Neuropharmacology 29: 511–513.

Williams, T. J., and Peck, M. J. (1977). Role of prostaglandin-mediated vasodilatation in inflammation. Nature 270: 530–532.

Serotonin: Molecular Biology, Receptors and Functional Effects
ed. by J. R. Fozard/P. R. Saxena

Influence of 5-HT₃ Receptor Antagonists on Limbic-Cortical Circuitry

B. Costall and R. J. Naylor

Postgraduate Studies in Pharmacology, The School of Pharmacy, University of Bradford, Bradford, West Yorkshire BD7 1DP, UK

Summary. Evidence is presented that 5-HT₃ receptor antagonists such as ondansetron are highly effective in influencing behavioural disturbance in the absence of effect on normal behaviour. The 5-HT₃ receptor antagonists demonstrate a remarkable potency and breadth of action over a wide dose range in rodent and primates to inhibit aversive behaviour, antagonise the hyperactivity caused by mesolimbic dopamine excess and facilitate performance in cognitive tests. These effects are achieved in the absence of withdrawal phenomana following cessation of chronic treatment. The studies indicate a widespread role for 5-HT₃ receptors in the regulation of mesolimbic-cortical functioning, and the importance of the 5-HT₃ receptor antagonists as investigative tools to establish the role of 5-HT₃ receptors in anxiety, schizophrenia and cognitive dysfunction, and their treatment.

Introduction

Forebrain regions of the limbic and cortical circuitry concerned with functions as diverse as motor activity, anxiety and cognition receive an extensive 5-HT innervation from the raphe nuclei. There is considerable evidence that the 5-HT input may contribute to the control of these behaviours, although the precise nature of the contribution is frequently not clear (see review by Costall et al. 1990). This may partly have been caused by the limitations of the investigative techniques used, e.g. brain lesions causing non-selective damage, and the use of 5-HT agonists or antagonists which invariably lacked specificity or selectivity of action. In particular, in the early studies the extent to which drug action depended on 5-HT receptor subtypes was not appreciated; an up to data assessment indicates the existance of at least seven 5-HT receptor subtypes (see reviews by Bradley et al. 1986; Costall and Naylor 1990).

The aim of the present chapter is to review evidence indicating that antagonist action at the 5-HT₃ receptor subtype may influence limbic driven activity, aversive (anxiety) responding and cognition. The major findings of the studies are that 5-HT₃ receptor antagonists are potent and effective in moderating a disturbed behaviour in the limbic/cortical systems in the absence of effect on normal behaviour.

The Identification of 5-HT₃ Receptors in the Brain

5-HT$_3$ receptor agonists and antagonists were first identified by their actions in peripheral preparations such as the rabbit heart and vagus nerve (Fozard 1984; Richardson et al. 1985; Ireland and Tyers 1987) and the initial behavioural studies using 5-HT$_3$ receptor antagonists such as ondansetron, ICS 205-930 and zacopride were undertaken in the absence of any evidence that 5-HT$_3$ receptor antagonists were actually present within the brain. However, Kilpatrick et al. (1987), using the highly potent and selective 5-HT$_3$ receptor antagonist, GR 65630, first demonstrated the presence of 5-HT$_3$ receptor/recognition sites in the cortex and limbic areas of rat brain. This was confirmed by Barnes et al. (1988a) and other workers using other ligands (see Costall et al. 1990b), and subsequently in man (Barnes et al. 1988b). The demonstration of a relatively high density of 5-HT$_3$ receptors in the cortical and limbic regions was important and provided the necessary sites of action for 5-HT$_3$ receptor antagonists to modify behaviour.

The Effect of 5-HT₃ Receptor Antagonists on 'normal' Behaviour

The possibility that 5-HT$_3$ receptor antagonists may modify behaviour was first investigated using ondansetron in 'normal' mice, rats and primates (Costall et al. 1987; Jones et al. 1988). The results of the studies were conclusive, revealing that ondansetron, administered over an extensive dose range, did not overtly modify behaviour in the three species. This has been a consistent finding with other 5-HT$_3$ receptor antagonists (see review by Costall et al. 1990b; Carboni et al. 1989a; b) and many indicate in normal animals the absence of a 5-HT$_3$ tone. It clearly distinguishes the 5-HT$_3$ receptor antagonists from other psychopharmacological agents which have the potential to cause sedation, motor or cognitive impairment (see below).

The Anxiolytic Profile of 5-HT₃ Receptor Antagonists in Animal Models of Anxiety

Animal tests for anxiety are based on the assumptions that there are drugs which act therapeutically to reduce anxiety, that the clinical effect of certain drugs is fairly specific to anxiety, that anxiety is present in animals, and is modified in a similar way by anti-anxiety drugs. Early experiments using the conflict test indicated that behaviours suppressed by punishment are released after 5-HT synthesis inhibition or lesioning of the 5-HT pathway (see review by Costall et al. 1990b). This gave credence to the hypothesis that a reduction in 5-HT function can lead to an anxiolytic effect, although 5-HT$_1$ and 5-HT$_2$ receptor antagonists have inconsistent effects in animal models of anxiety (see review by Costall et al. 1990b). The possibility that 5-HT$_3$ receptors may contribute to the genesis of anxiety became amenable to investigation with the introduction of ondansetron and related agents.

The water-lick conflict has been used extensively as a screening procedure for potential anxiolytic activity (Vogel et al. 1971) and was the first test procedure used to assess the action of ondansetron. However, ondansetron failed to display an anxiolytic profile in this test (Jones et al. 1988). It might have been concluded that ondansetron and perhaps other 5-HT$_3$ receptor antagonists had no anxiolytic potential. However, it was reasoned that the water lick conflict test had been validated by known anxiolytic agents from the benzodiazepine series; anxioltyic agents with a different mode of action may not necessarily be active in this test. Therefore other tests for anxiolytic potential were investigated and in the mouse light/dark discrimination test, ondansetron and subsequently other 5-HT$_3$ receptor antagonists were found to have an anxiety profile of action (Jones et al. 1988; Costall et al. 1990b).

The test was initially described by Crawley and Goodwin (1980), and its basis is that mice placed into a two compartment black and white test box demonstrate an aversion to the brightly illuminated area and show a preference for exploration in the darkly illuminated black section. The aversion to the white area is inhibited by benzodiazepines, mice spending more time in the white section exhibiting increased exploratory rearings and line crossings, with corresponding decreases in these behaviours in the black. Mice also showed an increased latency of first movement from the white to the black section (Costall et al. 1990b). The important finding was that a similar behavioural profile was induced by the 5-HT$_3$ receptor antagonists; the 5-HT$_3$ receptor antagonists were distinguished from the benzodiazepines by a much greater potency and absence of sedation or muscle relaxation at high doses (see review by Costall et al. 1990b). It remained an intriguing finding that doses one hundred to one thousand times or more the minimally effective dose were less effective in antagonising the aversive response.

Using the social interaction test in the rat (File, 1980), ondansetron was found to increase social interaction assessed in paired rats under high light unfamiliar conditions (Jones et al. 1988). Again, ondansetron and other 5-HT$_3$ receptor antagonists were much more potent than diazepam and failed to modify locomotor activity (see review by Costall et al. 1990b). The ability of 5-HT$_3$ receptor antagonists to release suppressed behaviour in the rat test was shown by the failure of ondansetron to increase social interaction under minimally aversive low light familiar conditions (Jones et al. 1988).

The anxiolytic profile of action of the 5-HT$_3$ receptor antagonists in the mouse and rat prompted assessments in the primate. In behavioural observations in cynomolgus monkey which were selected on the basis of a high basal level of emotionality, 10 out of 33 behavioural parameters, e.g. restlessness and agitation, were altered by diazepam and ondansetron in a direction which indicated reduced agitation (Jones et al. 1988). In another model developed by Costall et al. (1988a), marmosets were shown to respond to confrontation by a human observer by retreating from the cage front and the demonstration of characteristic postures, e.g. exposure of the genital region and anal scent marking. Diazepam and 5-HT$_3$ receptor antagonists caused a reduction in the number of postures and animals spend a greater proportion of time on the cage front (Jones et al. 1988; see review by Costall et al. 1990b).

The quantitative assessment of behaviour in the mouse and rat, and quantitative and semiquantitative measurements in primate, indicated that 5-HT$_3$ receptor antagonists had an anxiolytic profile. They were distinguished from benzodiazepines by a lack of sedative potential and, importantly, by the absence of withdrawal phenomena following withdrawal from chronic treatment in the mouse and rat (Costall et al. 1989). Subsequently in the mouse test, they were also shown to antagonise the anxiogenic profile caused by withdrawal from drugs of abuse, diazepam, nicotine, cocaine and alcohol (Costall et al. 1990a). Also in the mouse and rat, intracerebral injection studies have identified the amygdala and/or dorsal raphe nucleus as important loci of 5-HT$_3$ receptor antagonist action to antagonise aversive behaviour and to inhibit the consequences of withdrawal from drugs of abuse (Costall et al. 1988b; 1990; Higgins et al. 1989).

The ability of the 5-HT$_3$ receptor antagonists to reduce aversive behaviour has been confirmed in a number of laboratories in a number of models; rat social interaction (Johnston and File 1988; Piper et al. 1988) mouse black/white test (Kilfoil et al. 1989; Onaivi and Martin 1989; Young and Johnson 1988), gerbil social interaction (Cutler and Piper 1990), rat black/white test box (Morinan 1989), mouse ethological study-social investigation (Cutler and Dixon 1989) and rat passive avoidance test (Papp and Przegalinski 1989). Where inactivities or inconsistencies have been found (File and Johnston 1989; Piper et al. 1988) this may partly reflect differences between studies in the basal levels of anxiety responding.

The Role of 5-HT$_3$ Receptor Antagonists in Moderating Mesolimbic Dopamine Function

There is evidence that the 5-HT innervation to the nucleus accumbens can facilitate or inhibit locomotor activity (see review by Costall et al.

1990b), and a possible involvement of the 5-HT$_3$ receptor contributing to such actions has been investigated using three experimental paradigms. In the first procedure amphetamine was injected into the nucleus accumbens of the rat to increase locomotor activity; this was enhanced by the 5-HT$_3$ receptor agonist 2-methyl-5-HT and antagonised by ondansetron injected peripherally or into the nucleus accumbens (Costall et al. 1987). In a second experiment, locomotor activity was increased by the infusion of dopamine into the nucleus accumbens of rat and marmoset. The hyperactivity was antagonised by 5-HT$_3$ receptor antagonists and neuroleptic agents. But, whereas agents such as haloperiodol antagonised the hyperactivity to reduce levels of behaviour to below control values, the 5-HT$_3$ receptor antagonists caused a reduction in hyperactivity to return values to control levels. However, the 5-HT$_3$ receptor antagonists showed a reduced effectiveness in high doses (Costall et al. 1988b).

These two experimental approaches increased locomotor activity by the addition of exogenous drug or neurotransmitter into the forebrain. In the third approach, hyperactivity was induced by the stimulation of midbrain dopamine cells using the neurokinin agonist Di-Me-C7, causing a release of endogenous dopamine in the nucleus accumbens and amygdala (Hagan et al. 1987; 1990). Again, the hyperactivity (and biochemical) response was blocked by both neuroleptic drugs, ondansetron and other 5-HT$_3$ receptor antagonists. Data obtained in the three tests indicates that the 5-HT$_3$ receptor antagonists effectively inhibit a raised mesolimbic dopamine activity in the absence of sedation or potential to reduce normal levels of locomotor responding. More recently, Carboni et al. (1989a) have reported that ICS 205-930 selectively prevents the stimulation of dopamine release by drugs (alcohol, nicotine and morphine) known to stimulate the firing activity of dopamine neurones. The evidence obtained in the above behavioural models indicates that 5-HT$_3$ receptors may exert a permissive role to facilitate dopamine function in the limbic system.

The Role of 5-HT$_3$ Receptors in Reward

Pharmacological manipulations to enhance 5-HT synthesis and release or inhibit uptake have been shown to reduce voluntary alcohol intake (see Grupp et al. 1988). The 5-HT receptor mechanisms mediating these changes have not been elucidated although a 5-HT$_1$ receptor activation may appear to inhibit alcohol consumption (see Grupp et al. 1988). However, Oakley et al. (1988), using the marmoset, and Sellers et al. (1988) using the rat, found that ondansetron reduced alcohol intake. This may indicate a 5-HT$_3$ receptor involvement in the control of alcohol intake and possibly craving and reward, and may be related to the ability of 5-HT$_3$ receptor antagonists to inhibit a raised limbic dopamine function. It is important that such studies be extended to the use of other 5-HT$_3$ receptor antagonists to clarify the significance of 5-HT$_3$ receptors to the processes of craving and reward (see Carboni et al. 1989a; b).

The Role of 5-HT$_3$ Receptors in Cognition

The cortex and hippocampus play critical roles in cognition and there is evidence that 5-HT can interfere with the acquisition or retention of a conditioned or passive avoidance response (see Essman 1978). The use of compounds with selective actions on 5-HT receptor subtypes is allowing a clearer understanding of the role of 5-HT in cognition, and the presence of 5-HT$_3$ receptors in the cortex and hippocampus may provide at least one site of action.

In a habituation test in the mouse, ondansetron facilitated performance in young adult and aged animals and inhibited a scopolamine induced impairment (Barnes et al. 1990). The same group also showed that in a T-maze reinforced alternation test in rats, ondansetron antagonised a scopolamine impairment. The ability of ondansetron to enhance cognitive performance in the mouse and rat was also extended to the marmoset, ondansetron increasing performance in a reversal learning task using a Wisconsin General Test Apparatus. It was concluded that ondansetron is potent and effective in improving basal performance and inhibiting a scopolamine impairment. The ability of 5-HT$_3$ receptor agonist action 'in vitro' to reduce acetylcholine release in the cerebral cortex, an effect blocked by 5-HT$_3$ receptor antagonists (Barnes et al. 1989), would provide a mechanism of action to facilitate cholinergic transmission, whose function is considered essential for cognition (Bartus et al. 1982).

Conclusion

The 5-HT$_3$ receptor antagonists are revealed as exceedingly potent compounds in mouse, rat and marmoset models to inhibit anxiety, antagonise the behavioural and biochemical consequences of a raised mesolimbic dopamine function, and to increase performance in cognitive tests. The anxiolytic effects and ability to reduce mesolimbic dopamine function have been shown to involve drug action in the limbic system. Precisely how these effects are achieved remains the subject of many present investigations. For example, it may be relevant that the 5-HT$_3$ receptor antagonist, MDL 73147EF, can decrease the activity of dopamine neurones in the ventromedial tegmental area (Sorensen et al.

1989). It may also be important for the cognitive enhancing effects that 5-HT$_3$ receptor antagonists can antagonise the suppression of cell firing in the medial prefrontal cortex induced by 2-methyl-5-HT (Ashby et al. 1989). In any event, the data reveal an important and widespread involvement of 5-HT$_3$ receptors in the mediation of disturbed behaviour. The data obtained in animals indicates that in man 5-HT$_3$ receptor antagonists may be critical probes in an elucidation of the role of 5-HT$_3$ receptors in anxiety, schizophrenia and cognitive disorders.

References

Ashby, C. R., Edwards, F., Harkins, K., and Wang, R. Y. (1989). Characterisation of 5-hydroxytryptamine$_3$ receptors in the medial prefrontal cortex: a microiontopholetic study. Eur. J. Pharmacol. 173: 193–196.

Barnes, J. M., Barnes, N. M., Costall, B., Naylor, R. J., and Tyers, M. B. (1989). 5-HT$_3$ receptors mediate inhibition of acetylcholine release in cortical tissue. Nature 338: 762–763.

Barnes, J. M., Costall, B., Coughlan, J., Domeney, A. M., Gerrard, P. A., Kelly, M. E., Naylor, R. J., Onaivi, E. S., Tomkins, D. M., and Tyers, M. B. (1990). The effects of ondansetron, a 5-HT$_3$ receptor antagonist, on cognition in rodents and primates. Pharmacol. Biochem. Behav. 35: 955–962.

Barnes, N. M., Costall, B., Ironside, J. W., and Naylor, R. J. (1988b). Identification of 5-HT$_3$ recognition sites in human brain using ^{3}H.zacopride. J. Pharmac. Pharmacol. 40: 668.

Barnes, N. M., Costall, B., and Naylor, R. J. (1988a). ^{3}H.zacopride: a ligand for the identification of 5-HT$_3$ recognition sites. J. Pharm. Pharmacol. 40: 548–551.

Bartus, R. T., Dean, R. L., Beer, B., and Lippa, A. S. (1982). The cholinergic hypothesis of geriatric memory dysfunction. Science 217: 408–417.

Bradley, P. B., Engel, G., Feniuk, G., Fozard, J. R., Humphrey, P. P. A., Middlemiss, D. N., Mylecharane, E. J., Richardson, B. P., and Saxena, P. R. (1986). Proposals for the classification and nomenclature of functional receptors for 5-hydroxytryptamine. Neuropharmacol. 25: 563–576.

Carboni, E., Acquas, E., Frau, R., and DiChiara, G. (1989a). Differential inhibitory effects of a 5-HT$_3$ antagonist on drug-induced stimulation of dopamine release. Eur. J. Pharmacol. 164: 515–519.

Carboni, E., Acquas, E., Leone, P., and DiChiara, G. (1989b). 5-HT$_3$ receptor antagonists block morphine- and nicotine- but not amphetamine-induced reward. Psychopharmacology 97: 175–178.

Costall, B., Domeney, A. M., Jones, B. N. C., and Naylor, R. J. (1988a). Assessment of behavioural parameters associated with anxiogenesis in the common marmoset (Callithrix jacchus). Br. J. Pharmacol. 95: 670P.

Costall, B., Domeney, A. M., Naylor, R. J., and Tyers, M. B. (1987). Effects on the 5-HT$_3$ receptor antagonist GR 38032F on raised dopaminergic activity in the mesolimbic system of the rat and marmoset brain. Br. J. Pharmacol. 92: 881–894.

Costall, B., Jones, B. J., Kelly, M. E., Naylor, R. J., Oakley, N. R., Onaivi, E. S., and Tyers, M. B. (1989). The effects of ondanestron (GR 38032F) in rats and mice treated subchronically with diazepam. Pharmacol. Biochem. Behav. 34: 769–778.

Costall, B., Jones, B. J., Kelly, M. E., Naylor, R. J., Onaivi, E. S., and Tyers, M. B. (1990a). Sites of action of ondansetron to inhibit withdrawal from drugs of abuse. Pharmacol. Biochem. Behav. 36: 97–104.

Costall, B., Kelly, M. E., Naylor, R. J., Ónaivi, E. S., and Tyers, M. B. (1988b). Neuroanatomical sites of action of 5-HT$_3$ receptor antagonists to alter exploratory behaviour of the mouse. Br. J. Pharmacol. 96: 325–332.

Costall, B., and Naylor, R. J. (1990). 5-Hydroxytryptamine: new receptors and novel drugs for gastrointestinal motor disorders. Scand. J. Gastroenterol. 25: 769–787.

Costall, B., Naylor, R. J., and Tyers, M. B. (1990b). The psychopharmacology of 5-HT$_3$ receptors. Pharmac. Ther. 47: 181–202.

Cutler, M. G., and Dixon, A. K. (1989). Effects of the 5-HT₃ antagonist, ICS 205-930, on behaviour of mice during social encounters. Br. J. Pharmacol. 96: 12P.

Cutler, M. G., and Piper, D. C. (1990). Chronic administration of the 5-HT₃ receptor antagonist BRL 43694: effects on reflex epilepsy and social behaviour of the Mongolian gerbil. Psychopharmacology, in press.

Essman, W. B. (1978). Serotonin in learning and memory. In: Essman, W. B. (ed.) Serotonin in health and disease. Vol. III. The central nervous system. New York, Spectrum Publications, pp. 69–143.

File, S. E. (1980). The use of social interaction as a method for detecting anxiolytic activity of chlordiazepoxide-like drugs. J. Neurosci. Meth. 2: 219–238.

File, S. E., and Johnston, A. L. (1989). Lack of effects of 5-HT₃ receptor antagonists in the social interaction and elevated plus-maze tests of anxiety in the rat. Psychopharmacology 99: 248–251.

Fozard, J. R. (1984). MDL 72222: a potent and highly selective antagonist at neuronal 5-hydroxytryptamine receptors. Naunyn-Schmiedeberg's Arch. Pharmacol. 326: 36–44.

Grupp, L. A., Perlanski, E., and Stewart, R. B. (1988). Attenuation of alcohol intake by a serotonin uptake inhibitor: Evidence for mediation through the renin-angiotensin system. Pharmacol. Biochem. Behav. 30: 823–827.

Hagan, R. M., Butler, A., Hill, J. M., Jordan, C. C., Ireland, S. J., and Tyers, M. B. (1987). Effect of the 5-HT₃ receptor antagonist, GR 38032F, on responses to injection of a neurokinin agonist into the ventral tegmental area of the rat brain. Eur. J. Pharmacol. 138: 303–305.

Hagan, R. M., Jones, B. J., Jordan, C. C., and Tyers, M. B. (1990). Effect of 5-HT₃ receptor antagonists on responses to selective activation of mesolimbic dopaminergic pathways in the rat. Br. J. Pharmacol. 99: 227–232.

Higgins, G. A., Jones, B. J., Oakley, N. R., and Tyers, M. B. (1989). Evidence that the amygdala and not area postrema is involved in the putative anxiolytic effects of 5-HT₃ receptor antagonists. Br. J. Pharmacol. 98: 635P.

Ireland, S. J., and Tyers, M. B. (1987). Pharmacological characterisation of 5-hydroxytryptamine-induced depolarisation of the rat isolated vagus nerve. Br. J. Pharmacol. 90: 229–238.

Johnston, A. L., and File, S. E. (1988). Effects of 5-HT₃ antagonists in two animal tests of anxiety. Neurosci. Lett. 32: S44.

Jones, B. J., Costall, B., Domeney, A. M., Kelly, M. E., Naylor, R. J., Oakley, N. R., and Tyers, M. B. (1988). The potential anxiolytic activity of GR 38032F, a 5-HT₃ receptor antagonist. Br. J. Pharmacol. 93: 985–993.

Kilfoil, T., Michael, A., Montgomery, D., and Whiting, R. L. (1989). Effects of anxiolytic and anxiogenic drugs on exploratory activity in a simple model of anxiety in mice. Neuropharmacology 28: 901–905.

Kilpatrick, G. J., Jones, B. J., and Tyers, M. B. (1987). Identification and distribution of 5-HT₃ receptors in rat brain using radioligand binding. Nature 330: 746–748.

Morinan, A. (1989). Effects of the 5-HT₃ receptor antagonists, GR 38032F and BRL 24924 on anxiety in socially isolated rats. Br. J. Pharmacol. 97: 457P.

Oakley, N. R., Jones, B. J., Tyers, M. B., Costall, B., and Domeney, A. M. (1988). The effect of ondansetron on alcohol consumption in the marmoset. Br. J. Pharmacol. 95: 870P.

Onaivi, E. S., and Martin, B. R. (1989). Neuropharmacological and physiological validation of a computer-controlled two-compartment black and white test box for the assessment of anxiety. Prog. Neuropsychopharmacol. Biol. Psychiat. 13: 963–976.

Papp, M., and Przegalinski, E. (1989). The 5-HT₃ receptor antagonists ICS 205-930 and GR 38032F, putative anxiolytic drugs, differ from diazepam in their pharmacological profile. J. Psychopharmacol. 3: 14–20.

Piper, D., Upton, N., Thomas, D., and Nicholson, J. (1988). The effects of the 5-HT₃ receptor antagonists BRL 43694 and GR 38032F in animal behavioural models of anxiety. Br. J. Pharmacol. 94: 314P.

Richardson, B. P., Engel, S., Donatsch, P., and Stadler, P. A. (1985). Identification of serotonin M-receptor subtypes and their specific blockade by a new class of drugs. Nature 316: 125–131.

Sellers, E. M., Kaplan, H. L., Lawrin, M. O., Somer, G., Noranjo, C. A., and Frecker, R. C. (1988). The 5-HT$_3$ antagonist GR 38032F decreases alcohol consumption in rats. Soc. Neurosci. 18th An. Meet. Toronto Abstr. 21.4.

Sorensen, S. M., Humphreys, T. M., and Palfreyman, M. G. (1989). Effect of acute and chronic MDL 73, 147EF, a 5-HT$_3$ receptor antagonist, on A9 and A10 dopamine neurones. Eur. J. Pharmacol. 163: 115–118.

Vogel, J. R., Beer, B., and Clody, D. E. (1971). A simple and reliable conflict procedure for testing anti-anxiety agents. Psychopharmacology 21: 1–7.

Young, R., and Johnson, D. N. (1988). Comparative effects of zacopride, GR 38032F, buspirone and diazepam in the mouse light/dark exploratory model. Soc. Neurosci. Abst. 14: 207.

Serotonin: Molecular Biology, Receptors and Functional Effects
ed. by J. R. Fozard/P. R. Saxena
© 1991 Birkhäuser Verlag Basel/Switzerland

Utilization of Zacopride and its R- and S-Enantiomers in Studies of 5-HT$_3$ Receptor "Subtypes"

L. M. Pinkus* and J. C. Gordon†

Department of Molecular Biology, A. H. Robins Research Laboratories, Richmond, VA 23220, USA
Present address: National Institutes of Health, DRG, 5333 Westbard Avenue, Bethesda, MD 20892, USA;
†*Biology Department, Fisons Corporation, 755 Jefferson Road, Rochester, NY 14623, USA*

Summary. 5-HT$_3$ binding sites with high affinity (Kd = 0.3 − 1.0 nM) for [^{3}H]zacopride have been studied in brain, vagus nerve, enteric neurons, and PC12 cells. A novel 5-HT$_3$ binding site in guinea pig small bowel exhibits decreased affinity for zacopride and other 5-HT$_3$ antagonists but a similar rank order of potency to that of 5-HT$_3$ binding sites in rabbit bowel and vagus nerve. The (S)-enantiomer of zacopride exhibits a 10–40 fold greater affinity for 5-HT$_3$ binding sites than its (R)-enantiomer, a ratio which correlates with their antimetic potency. However (R)-zacopride is >1000 fold more potent that the (S)-enantiomer as an anxiolytic. Together with our ligand binding and radioautographic studies, this suggests that some of the therapeutic effects of the R isomer may not be due to binding at 5-HT$_3$ receptors. [^{3}H](S)-zacopride also labelled a 5-HT$_3$ binding site which is expressed in PC12 cells cultured with nerve growth factor. Finally, a 5-HT$_3$ binding site was solubilized from rabbit small bowel muscularis membranes, labelled with [^{3}H](S)-zacopride and found to retain the binding properties of the membrane-bound form.

Introduction

Zacopride [4-amino-N-(1-azabicyclo[2.2.2]oct-3yl)-5-chloro-2-methoxy-benzamide hydorchloride hydrate] is a racemic benzamide that exhibits high affinity (Kd < 1 nM) for 5-HT$_3$ binding sites. Autoradiographic (Barnes et al. 1990b; Waeber et al. 1990) and ligand binding studies (Pinkus et al. 1989; Gordon et al. 1989) have revealed high densities of 5-HT$_3$-receptor specific zacopride binding sites in the limbic system (amygdala, hippocampus), entorhinal cortex, nuclei of the solitary tract, area postrema and plasma membranes of enteric neurons and the vagus nerve. 5-HT$_3$ receptor binding accounts for 90%−99% of the total binding in partially purified membranes from intestinal muscularis and vagus nerve whereas values in the 50%−80% range have been found in crude brain membranes (Barnes et al. 1988). Zacopride has no significant affinity for a large number of receptors including dopamine, 5-HT$_1$

and 5-HT$_2$ (Pinkus et al. 1989). However, it is also a weak 5-HT$_4$ receptor agonist (Bockaert et al. 1990).

The studies described in the present paper were conducted to determine if different forms or states of 5-HT$_3$ binding sites could be identified by utilizing the two enantiomers (R+ and S−) of zacopride.

Materials and Methods

Zacopride and its (S)- and (R)-enantiomers were synthesized at A. H. Robins Chemical Research Department. The enantiomers were >99.9% pure as determined by chromatography over chiral columns (Demain and Gripshover 1989). Tissues were obtained from animals following an overnight fast. Studies were conducted with 10,000–100,000 × g membranes (except 500–10,000 × g from brain and 0–100,000 × g from PC12 cells) isolated by differential centrifugation as previously described (Gordon et al. 1989). Tritiated compounds were obtained from Dupont-NEN and reagent chemicals from Sigma. Ligand binding assays with [^{3}H]zacopride (49 Ci/mMol) and its [^{3}H](S)- or [^{3}H](R)-enantiomers (both 60 Ci/mmol) were performed in 50 mM Tris-HCl, pH 7.4 at 37°C for 30 min. For guinea pig membranes competition studies employed 2.5 nM [^{3}H](S)-zacopride at 4°C. 5-HT$_3$ specific binding was that displaced by 500 nM ICS 205-930 or 20 μM ICS 205-930 (guinea pig). Kd values were determined from Scatchard analysis and Ki values were calculated from the IC50 values (Cheng and Prusoff 1973). Protein was determined colorimetrically with bovine serum albumin as the standard (Wang and Smith 1975). Autoradiography of 10 μ sections of rat brain was performed as described (Waeber et al. 1990).

Results and Discussion

Antagonism of [^{3}H]zacopride Binding by its (S)- and (R)-enantiomers

Competition studies utilizing either 1 nM [^{3}H]zacopride or its [^{3}H](S)-enantiomer (as indicated) are summarized in Table 1. Regardless of the tissue examined, 5-HT$_3$ binding sites exhibited a marked preference for the (S)- over the (R)-enantiomer. The ratio of Ki's ((R)/(S)) was 20–40 in rabbit intestine, rabbit vagus, and human jejunum and 8–12 in rat intestine, rat brain, and guinea pig intestine. The Kd value for [^{3}H](S)-zacopride obtained from Scatchard analysis in rabbit (0.15 nM) was the same as the Ki value for the (S)-enantiomer in competition studies calculated from the Cheng-Prusoff equation. At low concentrations (0.5 nM in rabbit and 2.5 nM in guinea pig) [^{3}H](R)-zacopride exhibited negligible 5-HT$_3$ specific binding (<5% of [3H](S)-zacopride).

Table 1. Kd values (nM) for [^{3}H]zacopride or its [^{3}H](S)-enantiomer and Ki values (nM) for 5-HT$_3$ antagonists. In guinea pig intestine and NGF-PC12 cells, Ki values were determined against 2.5 nM and 0.5 nM of the [^{3}H](S)-enantiomer, respectively.

Tissue	5-HT$_3$ Antagonist							
	ZAC	(S)	ICS	GR	(R)	Q	BRL	MDL
	(Kd)				(Ki)			
Rabbit Intestine (8)	0.3	0.15	0.9	2.2	4.6	10.3	15	23
Rabbit Vagus (4)	0.4	0.16	1.3	2.6	7.1	5.4	13	17
Human Jejunum (3)	1.0	0.5	2.0	5.0	10.7	9.6	45	14
Rat Intestine (3)	0.6	0.7 (a)	1.5	3.4	5.4	1.7	11	12
Rat Cortex (3)	1.0	0.3 (a)	2.0	5.0	3.7	3.5	22	45
G. Pig Intestine (6)	16.0	7.5	15.7	253	65.4	542	583	3795
NGF-PC12 Cells (4)	– –	0.8	8.0	14	– –	5.8	33	36

ZAC = zacopride, ICS = ICS 205-930, GR = GR 38032F (ondansetron), Q = quipazine, BRL = BRL 24924 (renzapride), and MDL = MDL 72222. Standard deviations were ⩽10% of the values shown except for guinea pig ⩽15%. (a) Ki for (S) vs [^{3}H]zacopride. Number of individual values contributing to each mean value shown in parentheses.

Studies with guinea pig intestine were performed at 4°C to minimize a gradual loss of 5-HT$_3$ specific binding which occurred upon prolonged incubation at 37°C. At either 4 or 37°C, the Kd determined for [^{3}H](S)-zacopride binding (7.5 nM) was much greater than in other tissues. As the concentration of the (S)-enantiomer was increased from 1 to 20 nM the 5-HT$_3$ binding (displaced by 20 uM ICS 205-930) decreased from 80% to 55% of the total. At 2.5 nM (S)-zacopride it represented 71 + 5% (n = 4) of the total binding. This is a lower specific binding percentage than observed in rabbit muscularis membranes (83–93%). Radioautograms indicate that (S)-zacopride can be utilized to visualize those regions expected to contain 5-HT$_3$ receptors in rat (see below) and guinea pig (Waeber and Palacios, personal communication) brain.

Available data indicate that [^{3}H](S)-zacopride labels a 5-HT$_3$-like site in guinea pig intestine rather than a 5-HT$_4$ site. The potency ratio of [(R)/(S)] as antagonists is approximately 10; as 5-HT$_4$ receptor agonists their ratio is approximately 100 (Clarke and Craig 1990). Compared with zacopride, cisapride (Ki = 666 nM) is an extermely poor antagonist of [^{3}H](S)-zacopride binding (Table 2); however, it is equipotent (Clarke and Craig 1990) or more potent (Dumius et al. 1989) than zacopride as a 5-HT$_4$ receptor agonist. In contrast, GR 38032F, one of the more potent 5-HT$_3$ antagonists in guinea pig, is inactive at 5-HT$_4$ sites (Dumius et al. 1989). Finally, our Ki values appear to correlate with pA$_2$ values for 5-HT$_3$ antagonists which block serotonin-induced contraction of guinea pig ileum (Cohen et al. 1989).

By many criteria, including competition studies against 5-HT$_3$ antagonists and agonists, the 5-HT$_3$ binding sites in rabbit intestinal muscularis and rabbit vagus are indistinguishable (Gordon et al. 1989). A plot

Table 2. K_1 values for inhibitors of $[^3H](S)$-zacopride binding in guinea-pig intestine. Competition experiments were performed at 2.5 nM $[^3H](S)$-zacopride and six to eight concentrations of the compounds indicated. IC_{50} values were calculated as described under Methods. K_1 values were calculated from the corresponding IC_{50} values using a value of $K_D = 7.5$ nM and the formula: $K_1 = IC_{50}/(1 + (ligand/K_D))$ (Cheng and Prusoff 1973). K_i values are expressed as the mean $\pm$ S.E.M. of the indicated number of determinations

Compound	Ki (nM) $\pm$ S.E.M.	n
ICS 205-930	15.7 $\pm$ 3.2	4
BRL 43694	36.9 $\pm$ 11.5	5
R-zacopride	65.4 $\pm$ 21.7	4
GR 38032F	253 $\pm$ 49.9	4
Quipazine	542 $\pm$ 125	5
BRL 24924	583 $\pm$ 61.0	3
Cisapride	666 $\pm$ 113	4
Metoclopramide	1403 $\pm$ 75.7	2
MDL 72222	3795 $\pm$ 1559	6
2-Methyl-5-HT	8482 $\pm$ 2629	4
5-HT	12492 $\pm$ 1084	2
Mianserin	12873 $\pm$ 2853	3

of pKi values for these tissues yields a linear regression line with a correlation coefficient of 0.99 (Figure 1, bottom). A plot of pKi values for guinea pig vs rabbit intestinal muscularis yields a correlation of 0.91 (Figure 1, top). Therefore, by the criterion of rank order of potency there is no obvious difference between these 5-HT$_3$ binding sites. Hoyer et al. (1989) have compared different tissues but reported the same conclusion. Both the cause and significance of the apparently reduced affinity for 5-HT$_3$ compounds in guinea pig intestine remain to be established.

5-HT$_3$ Binding Sites in PC12 Cells Cultured with Nerve Growth Factor

Binding studies with $[^3H]$zacopride in PC12 cells cultured on a polylysine-coated plastic substrate found no detectable 5-HT$_3$ binding sites. However, culturing with nerve growth factor on collagen-coated dishes resulted in the time-dependent appearance (maximum in 8–10 days) of a high affinity 5-HT$_3$ binding site (Kd = 0.8 nM (S)-zacopride, Table 1) as the cells morphology changed to that of neurons. As antagonists of $[^3H](S)$-zacropride binding to NGF-PC12 cell membranes (Gordon and Rowland 1990) selected 5-HT$_3$ agonists (5-HT, 2-methyl 5-HT and phenylbiguanide) exhibited a significantly higher potency than was found in membranes from normal animal tissues. Comparing binding studies in PC12 cells with those in other cell lines is difficult because a different ligand ($[^3H]$ICS 205-930) and assay conditions were utilized. However, published data from N1E-115 cells (Hoyer and Neijt 1988) suggests that Ki values for 5-HT (160 nM), 2-methyl-5-HT (350 nM), and phenylbiguanide (110 nM) are 2.5–6 fold

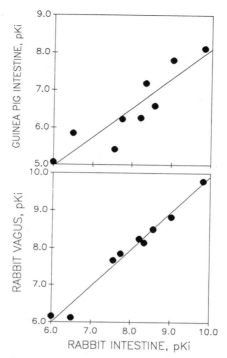

Figure 1. Linear regression analysis of a plot of pKi values for 5-HT$_3$ receptor ligands in rabbit vagus (bottom) and guinea pig intestinal muscularis (top) against pKi values in rabbit intestinal muscularis. pKi values were calculated from IC50 values (Cheng and Prusoff, 1973). The ligand was 1 nM [^{3}H]zacopride (rabbit tissues) and 2.5 nM [^{3}H] (S)-zacopride (guinea pig). Competition experiments for (S)-zacopride, ICS 205-930, GR 38032F, (R)-zacopride, quipazine, BRL 24924, MDL 72222, metoclopramide, and 2-methyl serotonin were performed as described under "Materials and Methods".

decreased in PC12 cells, whereas Ki values for BRL 24924 (33 nM) and MDL 7222 (36 nM) are 6–10 fold increased. Increases in Ki values of 6–9 fold for zacopride and 3–6 fold for serotonin, GR 38032F, BRL 24924, and quipazine are apparent in PC12 cells by comparison with NCB 20 cells (McKernan et al. 1990a). However the Ki for MDL 72222 was 2 fold decreased. These small differences do not allow any firm conclusion. Additional studies comparing the inducible 5-HT$_3$ binding site in NGF-treated PC12 cells with that constitutively present in other cultured cells are of interest.

Autoradiography of Rat Brain 5-HT$_3$ Receptors with [^{3}H]Enantiomers of Zacopride

Binding of [^{3}H](S)- and [^{3}H](R)-zacopride and [^{3}H]ICS 205-930 was compared in coronal sections of rat hippocampus and medulla (Waeber

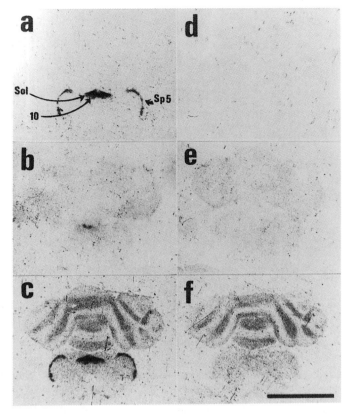

Figure 2. Coronal sections of rat lower medulla. Consecutive (10 μ) sections were incubated with 1.8 nM [³H]S-zacopride (a, d), or 4.5 nM [³H]R-zacopride (b, e), or 1.5 nM [³H]ICS 205-930 (c, f) and apposed to tritium-sensitive film. Total binding (a–c) and that obtained in the presence of 1 μM ICS 205-930 (d–f) are shown. Bar = 5 mm. Sol = nuclei of the solitary tract; 10 = vagus nerve; Sp5 = spinal tract of trigeminal nerve.

et al. 1990). The results for medulla are shown in Figure 2. In panel a the binding of 1.8 nM [³H](S)-zacopride is restricted to the solitary tract, vagal afferents, and Sp5 regions, the expected high density locations of 5-HT₃ receptors. ICS 205-930 (panel c) also binds in these regions. However, as shown in panel d, 5 μM ICS 205-930 completely blocks [³H](S)-zacopride binding, e.g. there is no significant binding to non-5-HT₃ sites, whereas, as shown in panel f, a large component of non-specific binding remains for [³H]ICS 205-930. Panel b shows binding of 4.5 nM [³H](R)-zacopride to 5-HT₃ sites. As would be expected for a ligand with 10-fold decreased affinity, the grain density is markedly reduced. There is also non-5-HT₃ receptor binding of [³H](R)-zacopride (panel e) but it appears to be widely distributed.

These autoradiographic results show the superiority of $[^3H](S)$-zacopride as a 5-HT$_3$ ligand and substantiate the conclusions obtained from ligand binding (Pinkus et al. 1990).

Evidence that zacopride can act centrally has been provided by anxiolytic responses at low doses in the mouse light dark transition model, rat social interaction test and marmoset threat test (Barnes et al. 1990a; c; Costall et al. 1988) and procognitive effects in monkeys (Barnes et al. 1990a). As an anxiolytic the (R)-enantiomer is reported to be at least 1000 fold more potent than the (S)-enantiomer (Young and Johnson 1989). This means that the activity of the (S)-enantiomer could be accounted for by contamination with as little as 0.1% of the (R)-enantiomer, raising the likelihood that the (S)-enantiomer is inactive as an anxiolytic. Furthermore, the anxiolytic activity of (R)- and (S)-zacopride does not correlate with their affinity for 5-HT$_3$ binding sites; ligand binding studies (Table 1) indicate a preference for (S) over (R) by 10–40 fold.

Several 5-HT$_3$ antagonists are active in anxiolytic paradigms, providing evidence for an involvement of 5-HT$_3$ receptors that cannot be easily dismissed (Barnes et al. 1990a). One may speculate that (S)-zacopride binds to additional sites pharmacologically masking its anxiolytic activity or that a subpopulation of receptors are linked to anxiolytic effects and that the (R)- but not the (S)-enantiomer interacts with these sites. The autoradiographic evidence (Figure 2) gives no indication of non-5-HT$_3$ binding sites for (S)-zacopride or of any discrete binding of the (R)-enantiomer to brain regions separate from those binding (S).

However, $[^3H](R)$-zacopride at 10 nM is bound to brain membranes at sites not inhibited by serotonin (unpublished observations) and "non-specific" binding of 5-HT$_3$ compounds is commonly reported (Pinkus et al. 1989; Kilpatrick et al. 1990). Also of interest are an unusual dose-response curve for (R)-zacopride in the mouse anxiolytic model (Young and Johnson 1989) and unexplained losses of gastrokinetic activity at higher doses of zacopride. The significance of these observations remains to be established.

Solubilization of a 5HT$_3$ Binding Site

Sodium cholate (0.5%) supplemented with 400 nM ammonium sulfate (or potassium sulfate) effectively solubilized 5-HT$_3$ binding activity at 4°C from rabbit small bowel muscularis membranes (Gordon et al. 1990). In competition studies against 0.5 nM $[^3H](S)$-zacopride the protein-detergent complex retained affinity for 5-HT$_3$ antagonists and agonists comparable to the membrane-bound form. It also retained binding activity after freezing and thawing. HPLC on a Waters SW300 column gave an estimated maximum Mr of 444,000–660,000. Two papers describing the solubilization of 5-HT₃ binding sites from rat entorhinal cortex and NCB 20 cells (McKernan et al. 1990a; b) appeared at about the same time as our initial report. They found a Mr of 245,000, suggesting that the protein-detergent complex that we isolated from rabbit bowel may have been a soluble aggregate. Available data indicate that 5-HT₃ binding sites are stable enough to be solubilized and purified from a variety of tissues. Comparisons between their component proteins should enhance our understanding of whether distinct 5-HT₃ receptors exist structurally, as well as functionally.

Importance of Zacopride

Originally developed to be a better antiemetic than metoclopramide (e.g. more potent and free of extrapyramidal side-effects), zacopride's efficacy in man remains to be established. Its potent and effective antiemetic activity in dogs, ferrets, and monkeys is firmly attributed to 5-HT₃ receptor antagonism, although the relative importance of central vs peripheral 5-HT₃ binding sites in preventing vomiting is still uncertain (see Robertson et al. 1990). The mechanism(s) governing its numerous central effects in animals remain to be elucidated. In the GI tract zacopride has promotility and gastrokinetic effects (Alphin et al. 1986), antagonizes an acid-mediated contractile response of the pylorus (Tougas et al. 1990), both contracts (Eglen et al. 1990) and antagonizes the contraction (Cohen et al. 1989) of isolated muscularis of guinea pig bowel, and contracts isolated gastric smooth muscle cells of guinea pig (Sancilio et al. 1991). Whether the gastrointestinal effects of zacopride can be attributed solely to 5-HT₃ receptor antagonism will remain problematic until the significance of the 5-HT₄ receptor and of serotonin (5-HT₁ₚ)-evoked enteric neuronal depolarization, mimicked by (S)-zacopride (Gershon et al. 1990) is clarified.

As a 5-HT₃-selective ligand [³H]zacopride is similar in potency to BRL 43694 (granisetron) and LY 278584 and markedly superior to GR 38032F (ondansetron) and BRL 24924 (renzapride). Unfortunately, [³H]zacopride has not been available commercially due to the reluctance of its patent holder, Delalande. However, it exhibits superior potency and a reduced level of non-specific binding compared with the now commercially-available ligands, GR 65630 and quipazine. The [³H](S)-enantiomer is twice as potent as its parent and appears to be highly selective for 5-HT₃ binding sites. Certain iodonated derivatives of zacopride retain its potency (Pinkus and Munson, unpublished observations) and should be useful in detecting very low densities of 5-HT₃ binding sites.

References

Alphin, R. S., Smith, W. L., Jackson, C. B., Droppleman, D. A., and Sancilio, L. F. (1986). Zacopride (AHR 11190B): A unique and potent gastrointestinal prokinetic and antiemetic agent in laboratory animals. Dig. Dis. Sci. 31: 4825.

Barnes, J. M., Barnes, N. M., Costall, B., Domeney, A. M., Kelly, M. E., and Naylor, R. J. (1990a). Influence of 5-HT$_3$ antagonists on limbic cortical circuitry. Second IUPHAR Satellite Meeting on Serotonin. abs. P102.

Barnes, J. M., Barnes, N. M., Costall, B., Naylor, I. L., Naylor, R. J., and Rudd, J. A. (1990b). Topographical distribution of 5-HT$_3$ receptor recognition sites in the ferret brain stem. Naunyn-Schmiedeberg's. Arch. Pharmacol. 342: 17–21.

Barnes, J. M., Barnes, N. M., Costall, B., Domeney, A. M., Kelly, M. E., and Naylor, R. J. (1990c). Profile of the stereoisomers of zacopride as 5-HT$_3$ receptor antagonists. Second IUPHAR Satellite Meeting on Serotonin. abs. P117.

Barnes, N. M., Costall, B., and Naylor, R. J. (1988). [^{3}H]Zacopride: A ligand for the identification of 5-HT$_3$ recognition sites. J. Pharm. Pharmacol. 40: 548–551.

Bockaert, I., Sebben, M., and Dumuis, A. (1990). Interrelationship between the interactions of 5-HT and benzamide derivatives on 5-HT$_4$ receptors in brain. Second IUPHAR Satellite Meeting on Serotonin. abs. P66.

Cheng, Y. C., and Prusoff, W. H. (1973). Relationship between the inhibitor constant (Ki) and the concentration of inhibitor which causes 50% inhibition (IC50) of an enzymatic reaction. Biochem. Pharmacol. 22: 3099–3108.

Clarke, D. E., and Craig, D. A. (1990). Pharmacological properties of the putative 5-HT$_4$ receptor in guinea pig ileum. IUPHAR Satellite Symposium on Serotonin. abs P67.

Cohen, M. L., Bloomquist, W., Gidda, J. S., and Lacefield, W. (1989). Comparison of the 5-HT$_3$ receptor antagonist properties of ICS 205-930, GR 38032F, and zacopride. J. Pharmacol. Exp. Ther. 248: 197–201.

Costall, B., Domeney, A. M., Gerrard, P. A., Kelly, M. E., and Naylor, R. J. (1988). Zacopride: anxiolytic profile in rodent and primate models of anxiety. J. Pharm. Pharmacol. 40: 302–305.

Demain, I., and Gripshover, D. F. (1989). Enantiomeric purity determination of 3-aminoquin-uclidine by diastereomeric derivatization and high performance liquid chromotographic separation. J. Chromatogr. 466: 415–420.

Dumuis, A., Sebben, M., and Bockaert, J. (1989). The gastrointestinal prokinetic benzamide derivatives are agonists at the non-classical 5-HT receptor (5-HT$_4$) positively coupled to adenylate cyclase in neurone. Naunyn-Schmiedeberg's. Arch. Pharmacol. 340: 403–409.

Eglen, R. M., Swank, S. R., Walsh, L. K. M., and Whiting, R. L. (1990). Characterization of 5-HT$_3$ and "atypical" 5-HT receptors mediating guinea-pig ileal contractions in vitro. Br. J. Pharmacol. 101: 513–520.

Gershon, M. D., Wade, P., Kirchgesser, A., Florica-Howells, E., and Tamir, H. (1990). 5-HT$_{1p}$ receptors in the bowel: G protein coupling, localization, and function. Second IUPHAR Satellite Meeting on Serotonin. abs. P40.

Gordon, J. C., Barefoot, D. S., Sarbin, N. S., and Pinkus, L. M. (1989). [^{3}H]zacopride binding to 5-hydroxytryptamine$_3$ sites on partially purified rabbit enteric neuronal membranes. J. Pharmacol. Exp. Ther. 251: 962–968.

Gordon, J. C., and Rowland, H. (1990). Nerve Growth Factor induces 5-HT$_3$ recognition sites in rat pheochromocytoma (PC12) cells. Life Sci. 46: 1435–42.

Gordon, J. C., Sarbin, N. S., Barefoot, D. S., and Pinkus, L. M. (1990). Solubilization of a 5-HT$_3$ receptor from rabbit small bowel muscularis membranes. Eur. J. Pharmacol. (Mol. Pharmacol. Sect.). 188: 313–319.

Hoyer, D., and Neijit, H. C. (1988). Identification of serotonin 5-HT$_3$ recognition sites in membranes of N1E-115 neuroblastoma cells by radioligand binding. Mol. Pharmacol. 33: 303–309.

Hoyer, D., Waeber, C., Karpf, A., Neijt, H., and Palacios, J. M. (1989). [^{3}H]ICS 205-930 labels 5-HT$_3$ recognition sites in membranes of cat and rabbit vagus nerve and superior cervical ganglion. Naunyn-Schmiedeberg's. Arch. Pharmacol. 340: 396–402.

Kilpatrick, G. J., Butler, A., Hagan, R. M., Jones, B. J., and Tyers, M. B. (1990). [^{3}H]GR 67330, a very high affinity ligant for 5-HT$_3$ receptors. Naunyn-Schmiedeberg's. Arch. Pharmacol. 342: 22–30.

448

McKernan, R. M., Biggs, C. S., Gillard, N., Quirk, K., and Ragan, C. I. (1990a). Molecular size of the 5-HT$_3$ receptor solubilized from NCB 20 cells. J. Neurochem. 269: 623–628.

McKernan, R. M., Quirk, K., Jackson, R. G., and Ragan, C. I. (1990b). Solubilization of the 5-hydroxytryptamine$_3$ receptor from pooled rat cortical and hippocampal membranes. J. Neurochem. 54: 924–930.

Pinkus, L. M., Sarbin, N. S., Barefoot, D. S., and Gordon, J. C. (1989). Association of [^{3}H]zacoprode binding to 5-HT$_3$ binding sites. Eur. J. Pharmacol. 168: 355–362.

Pinkus, L. M., Sarbin, N. S., Gordon, J. C., and Munson, H. R. (1990). Antagonism of [^{3}H]zacopride binding to 5-HT$_3$ recognition sites by its (R) and (S) enantiomers. Eur. J. Pharmacol. 179: 231–235.

Robertson, D. W., Cohen, M. L., Krushinski, J. H., Wong, D. T., Parli, C. J., and Gidda, J. S. (1990). LY 191617, A 5-HT$_3$ receptor antagonist which does not cross the blood brain barrier. Second IUPHAR Satellite Meeting on Serotonin. abs. P111.

Sancilio, L. F., Pinkus, L. M., Jackson, C. B., and Munson, H. R., Jr. (1990). Contraction of guinea pig gastric smooth muscle cells by 5-HT$_3$ antagonists and agonists. Eur. J. Pharmacol. submitted for publication.

Tougas, G., Woskowska, Z., Fox, J. E. T., and Daniel, E. E. (1990). Acid-induced pyroduodenal motor response: role of 5-HT$_3$ receptors. Second IUPHAR Satellite Meeting on Serotonin. abs. P76.

Wang, C., and Smith, R. L. (1975). Lowry determination of protein in the presence of Triton X-100. Anal. Biochem. 63: 414–420.

Waeber, C., Pinkus, L. M., and Palacios, J. M. (1990). The (S)-isomer of [^{3}H]zacopride labels 5-HT$_3$ receptors with high affinity in rat brain. Eur. J. Pharmacol. 181: 283–287.

Young, R., and Johnson, D. (1989). Effects of +, R(+), and S(−)-zacopride in the mouse light/dark exploratory model. Lorenzini Fdn. Symp.: 179, "Serotonin from Cell Biology to Pharmacology and Therapeutics". abs. P182.

Serotonin: Molecular Biology, Receptors and Functional Effects
ed. by J. R. Fozard/P. R. Saxena
© 1991 Birkhäuser Verlag Basel/Switzerland

CP-93,129:

A Potent and Selective Agonist for the Serotonin (5-HT$_{1B}$) Receptor and Rotationally Restricted Analog of RU-24,969

J. E. Macor, C. A. Burkhart, J. H. Heym, J. L. Ives, L. A. Lebel, M. E. Newman, J. A. Nielsen, K. Ryan, D. W. Schulz, L. K. Torgersen, and B. K. Koe

Central Research Division, Pfizer Inc. Groton, CT 06340, USA

Summary. The *in vitro* and *in vivo* characteristics of CP-93,129 [Structure 1 in Figure 1, 3-(1,2,5,6-tetrahydropyrid-4-yl)-pyrrolo[3,2-b]pyrid-5-one] are described. This rotationally restricted phenolic analog of RU-24,969 is a potent (15 nm) and selective (200 × vs. the 5-HT$_{1A}$ receptor, 150 × vs. the 5-HT$_{1D}$ receptor) functional agonist for the 5-HT$_{1B}$ receptor. Direct infusion of CP-93,129 into the paraventricular nucleus of the hypothalamus of rats significantly inhibits food intake, implicating the role of 5-HT$_{1B}$ receptors in regularing feeding behavior in rodents. CP-93,129 has also been shown to be biochemically discriminatory in its ability to selectively inhibit forskolin-stimulated adenylate cyclase activity only at the 5-HT$_{1B}$ receptor. The source of the selectivity of CP-93,129 appears to lie in the ability of a pyrrolo[3,2-b]pyrid-5-one to act as a rotationally restricted bioisosteric replacement for 5-hydroxyindole.

Introduction

The use of conformationally and/or rotationally restricted analogs of biologically important molecules represents an important tool in the arsenal of the bio-organic chemist for understanding the molecular recognition requirements of the receptor or enzyme to which the natural ligand binds. This approach is especially relevant when the natural ligand binds to a number of similar receptors of enzymes. Therefore, the use of conformationally and/or rotationally restricted analogs of serotonin can provide an understanding of the molecular recognition requirements of individual serotonin receptors since at this point in time, more than six subtypes of serotonin receptors have been identified (Peroutka, 1988).

The 5-HT$_{1B}$ receptor was first reported by Pedigo and colleagues in 1981 (Pedigo et al. 1981) and further described by Frazier in 1984 (Sills et al. 1984) and Palacios and Hoyer in 1985 (Pazos et al. 1985a; Hoyer et al. 1985). It is negatively linked to adenylate cyclase (Bouhelal et al. 1988), and relatively selective 5-HT$_{1B}$ receptor agonists have been claimed to produce anorexia in rodents (Kennett et al. 1987). Most reports to data suggest that the existence of 5-HT$_{1B}$ receptors appears to be limited to the brains of mice and rats (Heuring and Peroutka 1987), but the lack of a potent and selective ligand for that receptor has hampered further studies to test this hypothesis (Murphy and Bylund 1989). Serotonin, 5-carboxamidotryptamine, 2-cyanopindolol, and RU-24,969 [5-methoxy-3-(1,2,5,6-tetrahydropyridyl)indole, **2g**] bind to the 5-HT_{1B} receptor with high affinities (IC_{50} values < 10 nm), but not only RU-24,969 demonstrates any degree of selectivity for the 5-HT_{1B} receptor (5–10 times selective versus the 5-HT_{1A} receptor). In 1987, a report claimed that CGS-12066B [4-(4-methylpiperazinyl)-7-trifluoro-methylpyrrolo[1,2-a]quinoxaline, **3a**] was selective for the 5-HT_{1B} receptor (Neale et al. 1987); however, in our hands, **3a** appears to be slightly selective for the 5-HT_{1A} receptor (Table 1). This discrepancy is clearly a result of the methodological differences in the 5-HT_{1A} binding assays utilized in our studies and by Neale and co-workers (Neale et al. 1987). Specific 5-HT_{1A} binding measured using [^{3}H]-8-OH-DPAT has generally been accepted as a more accurate measure of 5-HT_{1A} binding than the older method employed by Neale and co-workers using [^{3}H]-5-HT. It should be noted that the desmethyl analog of CGS-12066B [4-piperazinyl-7-trifluoromethylpyrrolo[1,2-a]quinoxaline, **3b**] shows very slight selectivity for the 5-HT_{1B} receptor. The need for a more selective agent was apparent, and this led us to design and synthesize potent and selective 5-HT_{1B} ligands.

While RU-24,969 (**2g**, X = $-OCH_3$) has often been claimed to be the standard for selectivity for the 5-HT_{1B} receptor (Sills et al. 1984), in fact, other members of the 3-(1,2,5,6-tetrahydropyridyl)indole family are considerably more selective for the 5-HT_{1B} receptor versus the 5-HT_{1A} receptor (Table 1) (Macor et al. 1990b). The synthesis of these compounds has been covered extensively in the patent and primary literature (Guillaume et al. 1987; Taylor et al. 1988) and 3-(1,2,5,6-tetrahydropyrid-4-yl)indoles (**2**) arise from the basic condensation of the appropriate indole derivative with 4-piperidone (Freter 1975). Taylor and co-workers (Taylor et al. 1988) have proposed that steric fit of the C5-substituent of RU-24,969 analogs determined 5-HT_{1A} receptor potency. They hypothesized that the idealized volume of 24 cubic angstroms, or approximately the size occupied by the carboxamido substituent, ensured potent 5-HT_{1A} receptor binding. From our study of the SAR of the series of 5-substituted-3-tetrahydropyridylindoles (**2**) shown in Table 1, an important trend for 5-HT_{1B} receptor selectivity was seen. We noted that the smaller the volume occupied by the C5-substituent, the greater the 5-HT_{1B} versus 5-HT_{1A} receptor selectivity. Compounds **2a** (X = H, 5-HT_{1A}/5-HT_{1B} = 38) and **2b** (X = F, 5-HT_{1A}/5-HT_{1B} = 34) were the most 5-HT_{1B} receptor selective compounds when compared to RU-24,969 (**2g**, 5-HT_{1A}/5-HT_{1B} = 7). Compounds **2c** (X = Cl, 5-HT_{1A}/5-HT_{1B} = 8) and **2d** (X = Br, 5-HT_{1A}/5-HT_{1B} = 4) lost 5-HT_{1B} receptor selectivity with their increasing steric bulk. The

Table 1. Effects of serotonin receptor ligands at 5-HT receptor subtypes

Compound	5-HT_{1A}	5-HT_{1B}	Ratio (1A/1B)	5-HT_{1C}	5-HT_{1D}	5-HT_{2}
CP-93,129 (1)	3000 ± 400 [4]	15 ± 5 [5]	200	6400 ± 800 [5]	2200 ± 700 [3]	>10000 [3]
RU-24,969 (2g)	14 ± 7 [5]	2.0 ± 0.9 [5]	7	290 ± 60 [6]	39 ± 4 [3]	5400 ± 2000 [4]
2h (X = OH)	30 ± 11 [3]	1.2 ± 0.2 [3]	25	170 ± 40 [3]	22 ± 2 [4]	2200 ± 300 [3]
TFMPP	290 ± 60 [8]	27 ± 4 [10]	11	150 ± 50 [6]	610 ± 100 [4]	570 ± 60 [3]
CGS-12066B (3a)	19 ± 7 [6]	150 ± 70 [5]	0.1	>10000 [3]	35 ± 7 [4]	6800 ± 100 [3]
desmethyl-CGS-12066B (3b)	37 ± 11 [5]	21 ± 7 [5]	2	10000 ± 1000 [3]	190 ± 50 [3]	>10000 [3]
5-HT	5.2 ± 1.5 [15]	5.0 ± 1.7 [12]	1	81 ± 30 [3]	3.0 ± 0.3 [3]	4600 ± 2100 [3]
2a (X = H)	140 ± 20 [4]	3.7 ± 0.6 [4]	38	230 ± 30 [5]	190 ± 40 [4]	860 ± 30 [3]
2b (X = F)	65 ± 11 [4]	1.9 ± 0.4 [4]	34	35 ± 25 [6]	110 ± 13 [4]	420 ± 180 [4]
2c (X = Cl)	26 ± 2 [3]	3.3 ± 0.1 [3]	8	150 ± 70 [3]	—	—
2d (X = Br)	13 ± 2 [3]	3.2 ± 1.0 [3]	4	—	—	—
2e (X = CN)	32 ± 4 [4]	0.8 ± 0.1 [4]	40	1100 ± 500 [3]	24 ± 1 [3]	4900 ± 1200 [3]
2f (X = NO$_2$)	36 ± 5 [3]	3.5 ± 1.5 [4]	10	—	—	—

Presented are IC$_{50}$ values in nM (± SEM) of number of experiments shown in brackets.

452

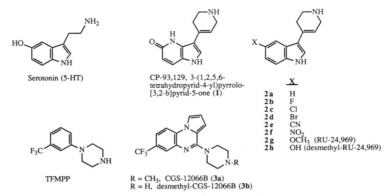

Figure 1. Structures of compounds referred to in manuscript.

comparison of **2g** (X = OCH$_3$, RU-24,969) and **2h** [X = OH, 5-HT$_{1A}$/5-HT$_{1B}$ = 25] further confirmed this trend. At first it appeared that **2e** (X = CN, 5-HT$_{1A}$/5-HT$_{1B}$ = 40) deviated from this trend, but since this substituent was linear, the actual volume occupied by it was still very small. Using this trend for 5-HT$_{1B}$ selective compounds (i.e. the smaller the C5 substituent, the greater the 5-HT$_{1B}$ selectivity), we hypothesized that either a hydrogen bonding or a steric interaction must be occurring in the 5-HT$_{1B}$ receptor with the C5-substituent of RU-24,969 analogs at a distance just past the first atom of the substituent. To test this hypothesis we desired to synthesize a 5-HT$_{1B}$ receptor selective RU-24,969 analog in which the C5-substituent: 1) occupied the least volume of space possible, while still mimicking the C5-oxygen substituent of serotonin, and 2) was rotationally restricted to optimize this apparently tight hydrogen bonding or steric interaction. Such a compound would be a rotationally restricted analog of 5-hydroxy-3-(1,2,5,6-tetrahydropy-ridyl)indole (**2h**). The pyrrolo[3,2-b]pyrid-5-one analog (i.e. CP-93,129) of RU-24,969 was seen as the target molecule since 2-hydroxypyridines exist almost exclusively in the amide tautomer (Tieckelmann, 1974). The pyridone oxygen, since it is reported to be almost entirely amide-like, cannot rotate, and represented for us the optimal structure for testing our hypothesis. This rationale led to the synthesis of CP-93,129.

Methods and Materials

CP-93,129 [3-(1,2,5,6-tetrahydropyrid-4-yl)pyrrolo[3,2-b]pyrid-5-one], RU-24,969, CGS-12066B, and derivatives of these compounds were obtained as previously described (Macor and Newman 1990a; Macor et al. 1990b; Taylor et al. 1988; Guillaume et al. 1987). Serotonin was obtained from Sigma Chemical Co., and TFMPP was obtained from Aldrich Chemical Co.

Binding Experiments: were conducted by methods previously reported in the literature: 5-HT$_{1A}$ using rat cortex and [^{3}H]8-OH-DPAT (Hoyer et al. 1985); 5-HT$_{1B}$ using rat cortex and [^{3}H]serotonin (Peroutka 1986); 5-HT$_{1C}$ using pig choroid plexus and [^{3}H]mesulergine (Pazos et al. 1985b); 5-HT$_{1D}$ using bovine caudate and [^{3}H]serotonin (Heuring and Peroutka 1987); 5-HT$_2$ using rat anterior cortex and [^{3}H]ketanserin (Leysen et al. 1982). The concentration of radioligand used in competition studies was approximately equal to the K$_D$ of the binding system.

Hypothalamic Infusion Experiments: Male CD rats (170–190 g, Charles River) were housed on a 12 h light/dark cycle. Food (PRO-LAB; rat, mouse, and hamster 3000, animal diet) and water were freely available unless otherwise stated. Rats were anesthetized with equithesin (3.25 mg/kg, ip) and a guide cannula (Plastic Products) was implanted approximately 1 mm above the paraventricular nucleus of the hypothalamus (PVN) using coordinates according to Paxinos and Watson (Paxinos, 1982) as follows: A, −1.1 mm (bregma); L, 0.3 mm; H, 7.5 mm below dura; incisor bar, −3.3 mm. The guide was fitted with a dummy cannula (Plastic Products) and the animals were transferred to individual cages (24 × 46 × 22 cm). Seven days after surgery, food pellets but not water were removed from their cages. Twenty-four hours later, the dummy cannula from each animal was removed and an infusion cannula (Plastic Products), projecting 1 mm past the tip of the guide cannula, was inserted into the PVN and connected via a length of PE-50 polyethylene tubing to a 5 μL Hamilton syrings. Rats were infused at 1 μL/min with either CP-93,129 (16 μg in 0.9% NaCl), serotonin (32 μg in 0.9% NaCl), or vehicle (0.9% NaCl). The rats were then placed individually in plastic test chambers (16 × 46 × 18 cm) with a weighed amount of food pellets on the floor of the chamber. Thirty minutes later the rats were returned to their home cages. The difference in food weight was the amount eaten. The results were analyzed by 2-tailed Student's t-tests. The verification of the cannula placement was performed as follows. Rats were killed by decapitation. Brains were removed and fixed in a 10% sucrose-buffered formalin solution for at least 2 days. Frozen sections (50 μm), taken in the coronal plane, were mounted and stained with neutral red. Sections were viewed relative to the stereotaxic atlas of Paxinos and Watson (Paxinos, 1982). The cannula track was noted to end in or alongside the PVN. Data from animals with cannula outside the PVN were not used.

Adenylate Cyclase Experiments: Washed membranes were prepared from freshly dissected guinea pig hippocampus, rat substantia nigra, or guinea pig substantia nigra. Tissue was incubated at 30°C in a reaction medium containing [α-^{32}P]-ATP, as described previously (Schulz, 1984).

Samples also contained buffer or 3 μm forskolin, along with varying concentrations of serotonin or CP-93,129. Following termination of the reaction by addition of 3% sodium dodecyl sulfate, [^{32}P]-cAMP was separated from [^{32}P]-ATP using a two column chromatographic procedure (Salomon, 1974). The amount of [^{32}P]-cAMP formed was determined by liquid scintillation counting, with results ultimately expressed in ρmol [^{32}P]-cAMP formed per mg protein per minute. EC$_{50}$ values (mean $\pm$ SEM) were determined by performing a minimum of three experiments for each compound at each receptor subtype.

Results

Table 1 summarizes the comparative receptor binding data obtained for CP-93,129 and the serotonergic literature standards: RU-24,969 (**2g**), CGS-12066B (**3a**), desmethyl-CGS-12066B (**3b**), desmethyl-RU-24,969 (**2h**), TFMPP (*m*-trifluoromethylphenylpiperazine), and serotonin (5-HT). CP-93,129 is considerably more selective for the 5-HT$_{1B}$ receptor than the 5-HT$_{1A}$ receptor (200 fold), 5-HT$_{1C}$ receptor (425 fold), 5-HT$_{1D}$ receptor (150 fold), and the 5-HT$_2$ receptor (over 700 fold). Additionally, CP-93,129 has not shown any substantial affinity for any other neurotransmitter receptor studied to date (i.e. dopamine, adrenergic, opiate).

The ability of serotonin and serotonergic agonists to inhibit forskolin-stimulated adenylate cyclase activity at 5-HT$_{1A}$, 5-HT$_{1B}$, and 5-HT$_{1D}$ receptors is well known (Schoeffter and Hoyer 1988a; Bouhelal et al. 1988; and Schoeffter et al. 1988b). Schoeffter and Hoyer reported that, while a number of serotonergic agents have been claimed to be selective for the 5-HT$_{1B}$ receptor based on receptor binding results, none of these compounds acted as functionally selective agonists at the 5-HT$_{1B}$ receptor (Schoeffter and Hoyer 1989). They concluded that there was no functionally selective agonist for the 5-HT$_{1B}$ receptor available in the literature. Therefore, the ability of CP-93,129 to inhibit forskolin-stimulated adenylate cyclase activity in 5-HT$_{1A}$ (guinea pig hippocampus) (De Vivo and Maayani 1986), 5-HT$_{1B}$ (rat substantia nigra) (Bouhelal et al. 1988), and 5-HT$_{1D}$ (guinea pig substantia nigra) (Hoyer and Schoeffter 1988) receptors was studied. As shown in Figures 2a (5-HT$_{1A}$), 2b (5-HT$_{1B}$), and 2c (5-HT$_{1C}$), CP-93,129 is selective in its ability to function as a serotonin agonist. CP-93,129 shows little or no ability to inhibit the forskolin-stimulated adenylate cyclase activity at the 5-HT$_{1A}$ and 5-HT$_{1D}$ receptors (EC$_{50}$ > 10 μm), while showing strong agonist activity at the 5-HT$_{1B}$ receptor (EC$_{50}$ = 56 $\pm$ 5 nm). These results clearly show that CP-93,129 is functionally discriminatory within the series of 5-HT$_1$ receptors, acting as an agonist only at the 5-HT$_{1B}$ receptor. CP-93,129 represents the first truly selective agonist for the 5-HT$_{1B}$ receptor.

Table 2. Effects of CP-93,129 given by infusion into the paraventricular nucleus on feeding in rats

	Food intake (grams/30 min)
Vehicle (saline)	2.76 ± 0.36 [10]
5-HT (32 μg)	1.15 ± 0.44* [5]
CP-93,129 (16 μg)	0.12 ± 0.01** [5]

Food intake ± SEM. Numbers of rats in parentheses. *, $p < 0.05$; **, $p < 0.01$ versus vehicle.

It has been reported that direct infusion of serotonin into the paraventricular medial hypothalamus (PVN) of rats leads to a dose dependent inhibition of feeding (Weiss et al. 1986; Hutson et al. 1988). Additionally, it has been suggested that this inhibitory role of 5-HT on feeding behavior results from activation of 5-HT$_{1B}$ receptors (Kennett et al. 1987). Therefore, the effect of direct injections of CP-93,129 and serotonin into the PVN of rats on feeding behavior was studied, and these results are summarized in Table 2. Injection of 32 μg of 5-HT directly into the PVN reduced food intake to 41% of control (saline injected) level, while injection of only 16 μg of CP-93,129 directly into the PVN reduced food intake to *less than* 5% of control level. This result is consistent with previous findings which ascribe satiety as a function of the rodent 5-HT$_{1B}$ receptor (Kennett, 1987). Further studies to fully investigate the behavioral and physiological effects of CP-93,129 are ongoing.

Discussion

CP-93,129 shows a selectivity for the 5-HT$_{1B}$ receptor that is unparalleled by any other known agent. Most significant is the comparison between CP-93,129 and desmethyl-RU-24,969 (**2h**). While desmethyl-RU-24,969 (**2h**) was potent at 5-HT$_1$ receptors, it demonstrated little selectivity for the 5-HT$_{1B}$ receptor. CP-93,129 is spatially an atom-for-atom analog of desmethyl-RU-24,969 (**2h**) with the only structural difference being the electronic configuration of the C5-oxygen substituent. The 5-hydroxy-4-azaindole portion (pyrrolo[3,2-b]pyrid-5-one) of CP-93,129 represents a rotationally restricted bioisosteric replacement of 5-hydroxyindole. The tautomerism of 2-pyridones (2-hydroxypyridines) has been well studied (Tieckelmann, 1974), and these molecules have been demonstrated to exist almost exclusively in the amide (i.e. pyridone) tautomer. Therefore, a 5-hydroxypyrrolo[3,2-b]-pyridine would exist as the pyrrolo[3,2-b]pyrid-5-one. High resolution X-ray analysis of CP-93,129 confirms that the pyrrolo[3,2-b]pyrid-5-one portion of the molecule exists as the amide tautomer. The amide C-O bond distance of 1.24 Å indicates that this bond is almost completely

456

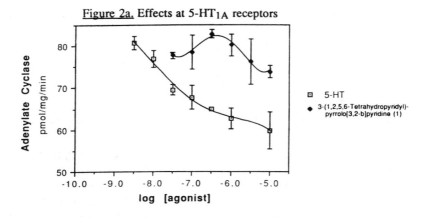

Figure 2a. Effects at 5-HT$_{1A}$ receptors

□ 5-HT
◆ 3-(1,2,5,6-Tetrahydropyridyl)-
   pyrrolo[3,2-b]pyridine (1)

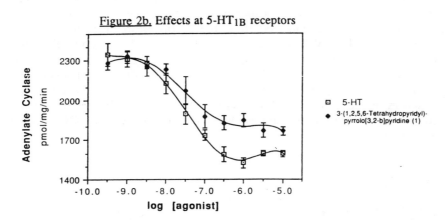

Figure 2b. Effects at 5-HT$_{1B}$ receptors

□ 5-HT
◆ 3-(1,2,5,6-Tetrahydropyridyl)-
   pyrrolo[3,2-b]pyridine (1)

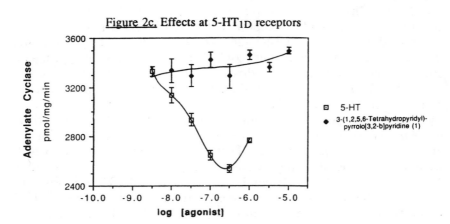

Figure 2c. Effects at 5-HT$_{1D}$ receptors

□ 5-HT
◆ 3-(1,2,5,6-Tetrahydropyridyl)-
   pyrrolo[3,2-b]pyridine (1)

Receptor	EC$_{50}$ @ 5-HT	EC$_{50}$ @ CP-93,129
5-HT$_{1A}$	30±13 [3]	>10000 [3]
5-HT$_{1B}$	26±4 [3]	56±5 [3]
5-HT$_{1D}$	16±7 [3]	>10000 [3]

Presented are EC$_{50}$ values (±SEM) of number of experiments shown in brackets.

Figure 2. Effects of CP-93,129 on forskolin-stimulated adenylate cyclase activity.

sp^2, and amide C-N bond distance of 1.39 Å indicates that this bond is primarily sp^3. Additionally, the location of the amide proton on the ring nitrogen further confirms that CP-93,129 exists primarily in the amide tautomer, at least in the crystalline state.

Herein lies the probable source of 5-HT$_{1B}$ receptor selectively of CP-93,129. The amide in CP-93,129 restricts hydrogen bond accepting interactions with a receptor to occur only in the plane of the aromatic ring (Macor et al. 1990b). This type of restricted interaction accommodates the 5-HT$_{1B}$ receptor only, almost to the exclusion of the other 5-HT$_1$ receptors. Desmethyl-RU-24,969 (2h) has free rotation about the C5-oxygen substituent, and this accommodates the hydrogen bonding interactions of all the 5-HT$_1$ receptors. Therefore, CP-93,129 represents an important breakthrough in the study of serotonin receptors as it helps to define important molecular recognition requirements for the individual 5-HT$_1$ receptors. In the 5-HT$_{1B}$ receptor, the directionality of the hydrogen bond accepting interaction between the receptor and the C5-oxygen functionality of the agonist seems to lie in the plane of the indole molecule. The same interaction in 5-HT$_{1A}$ and 5-HT$_{1D}$ receptors seems to require a different directionality, possibly above or below the plane of the indole ring. The rotationally restricted nature of the oxygen in the pyrrolo[3,2-b]pyridone of CP-93,129 appears to exclusively favor the directionality requirements of that area of the 5-HT$_{1B}$ receptor. These results indicate that a pyrrolo[3,2-b]pyrid-5-one, which is essentially locked in the amide tautomer, can represent a rotationally restricted bioisosteric replacement for 5-hydroxindole when its hydrogen bond accepting interactions occur exclusively in the plane of the aromatic ring.

In conclusion, CP-93,129 represents an important tool in defining the pharmacology and molecular recognition requirements of serotonin receptors, specifically the 5-HT$_{1B}$ receptor. Its potency at and virtual selectivity for the 5-HT$_{1B}$ receptor should prove useful in unequivocally defining the brain distribution and function of 5-HT$_{1B}$ receptors in rats and higher species. Since CP-93,129 inhibits forskolin-stimulated adenylate cyclase activity only at the 5-HT$_{1B}$ receptor, it represents the first functionally selective agonist for that receptor. Direct injections of low doses of CP-93,129 into the PVN of rats greatly reduces their food intake, strongly implicating a key physiological role for hypothalamic 5-HT$_{1B}$ receptors in the regulation of food intake in rodents. Future studies with CP-93,129 should further our understanding of the distribution and function of those receptors in mammals.

References

Bouhelal, R., Smounya, L., and Bockaert J. (1988). 5-HT$_{1B}$ receptors are negatively linked with adenylate cyclase in rat substantia nigra. Eur. J. Pharmacol. 151: 189–196.

De Vivo, M., and Maayani, S. (1986). Characterization of the 5-hydroxytryptamine$_{1A}$ receptor-mediated inhibition of forskolin-stimulated adenylate cyclase activity in guinea pig and rat hippocampal membranes. J. Pharmacol. Exp. Ther. 238: 248–253.

Freter, K. J. (1975). 3-Cycloalkenylindoles. J. Org. Chem. 40: 2525–2529.

Guillaume, J., Dumont, C., Laurent, J., and Nédélec, L. (1987). (Tétrahydro-1,2,3,6-pyridinyl-4)-3 1H indoles: synthèse, proriétés sérotoninergique et anti-dopaminergiques. Eur. J. Med. Chem. 22: 33–43.

Heuring, R. E., Schlegel, J. R., and Peroutka, S. J. (1986). Species variations in RU-24969 interactions with non-5-HT$_{1A}$ binding sites. Eur. J. Pharmacol. 122: 279–282.

Heuring, R. E., and Peroutka, S. J. (1987). Characterization of a novel ^{3}H-5-hydroxytryptamine binding site subtype in bovine brain membranes. J. Neurosci. 7: 894–903.

Hoyer, D., Engel, G., and Kalkman, H. O. (1985). Molecular pharmacology of 5-HT$_1$ and 5-HT$_2$ recognition sites in rat and pig brain membranes: radioligand binding studies with [^{3}H]5-HT, [^{3}H]8-OH-DPAT, (−)[^{125}I]iodocyanopindolol, [^{3}H]mesulergine, and [^{3}H]ketanserin. Eur. J. Pharm. 118: 13–23.

Hoyer, D., and Schoeffter, P. (1988). 5-HT$_{1D}$ receptor-mediated inhibition of forskolin-stimulated adenylate cyclase activity in calf substantia nigra. Eur. J. Pharmacol. 147: 145–146.

Hutson, P. H., Donohoe, T. P., and Curzon, G. (1988). Infusion of 5-hydroxytryptamine agonists RU-24969 and TFMPP into the paraventricular nucleus of the hypothalamus causes hypophagia. Psychopharm. 95: 550–552.

Kennett, G. A., Dourish, C. T., and Curzon G. (1987). 5-HT$_{1B}$ agonists induce anorexia at a postsynaptic site. Eur. J. Pharmacol. 141: 429–435.

Leysen, J. E., Niemegeers, C. J. E., Van Nueten, J. M., and Laduron, P. M. (1982). [^{3}H]Ketanserin (R 41 468), a selective ^{3}H-ligand for serotonin receptor binding sites: binding properties, brain distribution, and functional role. Mol. Pharmacol. 21: 301–314.

Macor, J. E., and Newman, M. E. (1990a). The synthesis of a rotationally restricted phenolic analog of 5-methoxy-3-(1,2,5,6-tetrahydropyrid-4-yl)indole (RU-24,969). Heterocycles. 31: 805–809.

Macor, J. E., Burkhart, C. A., Heym, J. H., Ives, J. L., Lebel, L. A., Newman, M. E., Nielsen, J. A., Ryan, K., Schulz, D. W., Torgersen, L. K., and Koe, B. K. (1990b). 3-(1,2,5,6-Tetrahydropyrid-4-yl)pyrrolo[3,2-b]pyrid-5-one: a potent and selective serotonin (5-HT$_{1B}$) agonist and rotationally restricted phenolic analogue of 5-methoxy-3-(1,2,5,6-tetrahydropyrid-4-yl)indole. J. Med. Chem. 33: 2087–2093.

Murphy, T. J., and Bylund, D. B. (1989). Characterization of serotonin-1B receptors negatively coupled to adenylate cyclase in OK cells, a renal epithelial cell line from the opossum. J. Pharmacol. Exp. Ther. 249: 535–543.

Neale, R. F., Fallon, S. L., Boyar, W. C., Wasley, J. W. F., Martin, L. L., Stone, G. A., Glaeser, B. S., Sinton, C. M., and Williams, M. (1987). Biochemical and pharmacological characterization of CGS-12066B, a selective serotonin-1B agonist. Eur. J. Pharmacol. 136: 1–9.

Paxinos, G., and Watson, C. (1982). The rat brain in stereotaxic coordinates. New York: Academic Press.

Pazos, A., Engel, G., and Palacios, J. M. (1985a). β-Adrenoceptor blocking agents recognize a subpopulation of serotonin receptors in brain. Brain Res. 343: 403–408.

Pazos, A., Hoyer, D., and Palacios, J. M. (1985b). The binding of serotonergic ligands to the porcine choroid plexus: characterization of a new type of serotonin recognition site. Eur. J. Pharmacol. 106: 539–546.

Pedigo, N. W., Yamamura, H. I., and Nelson, D. L. (1981). Discrimination of multiple [^{3}H]5-hydroxytryptamine binding sites by the neuroleptic spiperone in rat brain. J. Neurochem. 36: 220–226.

Peroutka, S. J. (1986). Pharmacological differentiation and characterization of 5-HT$_{1A}$, 5-HT$_{1B}$, and 5-HT$_{1C}$ binding sites in rat frontal cortex. J. Neurochem. 47: 529–540.

Peroutka, S. J. (1988). 5-Hydroxytryptamine receptor subtypes. Ann. Rev. Neurosci. 11: 45–60.

Salomon, Y., Londos, C., and Rodbell, M. (1974). A highly sensitive adenylate cyclase assay. Anal. Biochem. 58: 541–548.

Schoeffter, P., and Hoyer, D. (1988a). Centrally acting hypotensive agents with affinity for 5-HT$_{1A}$ binding sites inhibit forskolin-stimulated adenylate cyclase activity in calf hippocampus. Br. J. Pharmacol. 95: 975–985.

Schoeffter, P., Waeber, C., Palacios, J. M., and Hoyer, D. (1988b). The 5-hydroxytryptamine 5-HT$_{1D}$ receptor subtype is negatively coupled to adenylate cyclase in calf substantia nigra. Naunyn-Schmiedeberg's Arch. Pharmacol. 337: 602–608.

Schoeffter, P., and Hoyer, D. (1989). Interaction of arylpiperazines with 5-HT$_{1A}$, 5-HT$_{1B}$, 5-HT$_{1C}$ and 5-HT$_{1D}$ receptors: do discriminatory 5-HT$_{1B}$ receptor ligands exist? Naunyn-Schmiedeberg's Arch. Pharmacol. 339: 675–683.

Schulz, D. W. and Mailman, R. B. (1984). An improved, automated adenylate cyclase assay utilizing preparative HPLC: effects of phosphodiesterase inhibitors. J. Neurochem. 42: 764–774.

Sills, M. A., Wolfe, B. B., and Frazer, A. (1984). Determination of selective and nonselective compounds for the 5-HT$_{1A}$ and 5-HT$_{1B}$ receptor subtypes in rat frontal cortex. J. Pharmacol. Exp. Ther. 231: 480–487.

Taylor, E. W., Nikam, S. S., Lambert, G., Martin, A. R., and Nelson, D. L. (1988). Molecular determinants for recognition of RU-24969 analogs at central 5-hydroxytryptamine recognition sites: use of bilinear function and substituent volumes to describe steric fit. Molecular Pharmacology. 34: 42–53.

Tieckelmann, H. (1974). Pyridinols and Pyridones. In: Taylor, E. C., and Weissberger, E. (eds.), The Chemistry of Heterocyclic Compounds (vol. 14, part 3 supplemental). New York: John Wiley and Sons, pp. 733–734.

Weiss, G. F., Papadakos, P., and Leibowitz, S. F. (1986). Medial hypothalamic serotonin: effects on deprivation and norepinerphirine-induced eating. Pharm. Biochem. Behavior. 25: 1223–1230.

Serotonin: Molecular Biology, Receptors and Functional Effects
ed. by J. R. Fozard/P. R. Saxena
© 1991 Birkhäuser Verlag Basel/Switzerland

Differentiation of 8-OH-DPAT and Ipsapirone in Rat Models of 5-HT$_{1A}$ Receptor Function

T. Glaser, J. M. Greuel, E. Horvàth, R. Schreiber, and J. De Vry

Institute for Neurobiology, Troponwerke GmbH & Co., Köln, Germany

Summary. Because the anxiolytic/antidepressant profile of 5-HT$_{1A}$ receptor ligands may be related to their intrinsic activity at 5-HT$_{1A}$ receptors, we characterized the apparent intrinsic activities of 8-OH-DPAT and ipsapirone in several models reflecting pre- or postsynaptic 5-HT$_{1A}$ receptor activation. Presynaptically, as measured with electrophysiological techniques, 8-OH-DPAT and ipsapirone acted as full agonists; whereas in some behavioral models, which presumably reflect postsynaptic activity (i.e., 5-HT syndrome or circling behavior), as well as in the hippocampal adenylate cyclase assay, ipsapirone but not 8-OH-DPAT had mixed agonist/antagonist properties. In a drug discrimination and body temperature model, how-ever, 8-OH-DPAT and ipsapirone appeared to be full agonists both after systemic administra-tion and local application into the dorsal raphe nucleus and the hippocampus. Similar results have been obtained when 8-OH-DPAT and ipsapirone were applied after lesion of the brain serotonergic system by 5,7-DHT. These results suggest that ipsapirone may act as a full agonist at some post-synaptic 5-HT$_{1A}$ receptors. The experiments confirm earlier suggestions that the apparent intrinsic activity of a 5-HT$_{1A}$ ligand may differ at pre- vs postsynaptic 5-HT$_{1A}$ receptors. In addition, it is suggested that the compound's intrinsic activity also may differ at various postsynaptic 5-HT$_{1A}$ receptor populations. Whether this finding reflects the occurrence of regio-specific spare receptors and/or the existence of subtypes of 5-HT$_{1A}$ receptors is presently unclear.

Introduction

The identification of different serotonin (5-HT) receptor subtypes present in the central nervous system has evolved dramatically during the past several years. At least seven types – termed 5-HT$_{1A}$-, 5-HT$_{1B}$-, 5-HT$_{1C}$-, 5-HT$_{1D}$-, 5-HT$_2$-, 5-HT$_3$- and 5-HT$_4$ receptors – have been identified and pharmacologically characterized (for review see Glennon 1990). Due to the availability of relatively specific ligands, the 5-HT$_{1A}$ receptors have been the subject of intense research. By means of receptor autoradiographical and electrophysiological techniques it has been shown that they are located presynaptically on dendrites and cell bodies of serotonergic neurons in the raphe nuclei, as well as postsynap-tically in the projection areas, predominantly in structures of the limbic system (Verge et al. 1985; Weissmann-Nanopoulos et al. 1985; Glaser et al. 1985; Sprouse and Aghajanian 1987).

A substantial amount of preclinical and clinical evidence supports the idea that 5-HT$_{1A}$ receptor ligands provide a novel class of mixed anxiolytics/antidepressants with a unique profile of activity (Traber and

Glaser 1987; De Vry et al. 1991). In animal models, he found that the anxiolytic/antidepressant profile is not ide. the 5-HT$_{1A}$ ligands (De Vry et al. 1991). Moreover, from biochemical and electrophysiological studies it appeared that ti sic activity of these 5-HT$_{1A}$ ligands is not necessarily the same at postsynaptic 5-HT$_{1A}$ receptors and it was suggested that the int. activity at pre- and postsynaptic 5-HT$_{1A}$ receptors may determine anxiolytic/antidepressant profile of the compound (De Vry et al. 199). The present study further characterizes the prototypical 5-HT$_{1A}$ receptor ligands 8-hydroxy-2-(di-n-propylamino)tetralin (8-OH-DPAT) and ipsapirone (2-[4-(2-pyrimidinyl)-1-piperazinylbutyl]-1,2-benzoisothiazol-3-(2H)-one-1,1-dioxide hydrochloride) in some models reflecting activation of pre- and/or postsynaptic 5-HT$_{1A}$ receptors.

Materials and Methods

For all experiments male Wistar rats (Winkelmann, Borchen, F.R.G.) were used. 8-OH-DPAT and ipsapirone were synthesized by the Chemistry Department of Bayer AG, Wuppertal, F.R.G. All other chemicals were commercially purchased.

Electrophysiology: The brain stem of the rats was cut into 350 μm slices containing the dorsal raphe nucleus (DRN). Slices were transferred into an interface recording chamber that was oxygenated with carbogen. Temperature was held constant at 35°C. Potentials were recorded and processed by standard electrophysiological techniques. Putative serotonergic cells were identified by their characteristic, low frequency firing in the presence of 5 μmol/l noradrenaline and their responses to 5-HT$_{1A}$ ligands, applied by switching the perfusion solution (VanderMaelen and Aghajanian 1983).

For *in vivo* recordings, subjects were placed in a stereotactic apparatus after being anaesthetized with a mixture of xylazine-HCl 2 mg/kg, ketamine-HCl 100 mg/kg and chlorpromazine-HCl 10 mg/kg i.m. 1.3 mm lateral to the saggital suture the cortex was exposed and a microelectrode was advanced to the DRN (interaural line coordinates: F 1.2, H 3.5, L 0 at an angle of 15° from the vertical plane; Paxinos and Watson 1982). Putative serotonergic cells were identified by their spontaneous low frequency firing and by the inhibitory effect of 5-HT$_{1A}$ agonists. 8-OH-DPAT and ipsapirone were applied i.v. via a small cannula inserted into one of the main tail veins.

Circling Behavior: Rats received a unilateral injection of the serotonergic neurotoxin 5,7-dihydroxytryptamine (5,7-DHT) in the DRN (2 μl, 8 mg/kg in saline with 0.2% ascorbic acid) 45 min after pretreatment with desipramine (25 mg/kg, i.p.). Five to ten days later, rats were assessed for circling behavior in automated rotometers after administraion of 8-OH-DPAT (1 mg/kg, i.p.; Blackburn et al. 1984). In further experiments only rats which showed more than 100 contralateral and less than 20 ipsilateral turns over a 2 h period were used. Drugs were injected i.p. followed immediately by circling measurements during a 2 h period. In antagonism experiments, drugs were injected 10 min prior to 8-OH-DPAT (1 mg/kg).

Body Temperature: Body temperature was measured oesophagally, 5 min before and 15, 30, 60, 120, and 240 min after i.p. or s.c. application of the test compound. In the antagonism experiment, ipsapirone was administered 15 min prior to 8-OH-DPAT. Effects of local application of 8-OH-DPAT in the DRN were measured 5, 10, 20, 40 and 60 min after application (for cannulation details, see further). Effects of 5,7-DHT lesion on 8-OH-DPAT and ipsapirone induced hypothermia were assesed 10 days after lesion (for lesion details, see further). Drug induced temperature changes were expressed relative to vehicle control and compared to pre-application baseline values.

Drug Discrimination: Rats were trained to discriminate 0.1 mg/kg 8-OH-DPAT from vehicle (i.p., *t*-15 min) in a two lever fixed ratio 10 food reinforced operant procedure (Spencer et al. 1987). Drug generalization tests were performed 15 min and 5 min after i.p. and local application in the DRN or dorsal hippocampus, respectively. Results were expressed as drug lever selection percentages.

Cannulation: After reaching the drug discrimination criterion, rats received three cannula (23 gauge, 8.0 mm length), one aimed at the DRN (interaural line coordinates: F 1.2, H 7.1, L 1.7 at an angle of 25°, stereotactically implanted 4.0 mm above the centre of the DRN) and two aimed at the CA-4 region of the dorsal hippocampus (F 4.8, L +2.4/ −2.4, H 6.5, cannula tip 2.0 mm above the dorsal hippocampus). Cannula were shut by a stainless stylet (0.25 mm). A recovery period of one week was allowed before training restarted. Local injections were made through a 32 gauge Hamilton cannula; application volume was 0.5 μl, except in the case of the largest doses where it was increased to 2 μl (due to solubility problems). Animals used in the temperature experiment received a unilateral cannula to the DRN as described above.

Lesion: Rats got a unilateral (F 8.0, L 1.0, H 6.4) i.c.v. injection of 5,7-DHT (150 μg/15 μl/rat, 0.1% ascorbic acid) after pretreatment with desipramine (25 mg/kg, i.p.). Such treatment resulted in a reduction of 72% and 84% of 5-HT and 5-HIAA, respectively, in hippocampus (H. Sommermeyer, personal communication).

Results

8-OH-DPAT and ipsapirone completely and reversibly inhibited DRN firing in rat brain slices at concentrations of 50 and 250 nmol/l, respec-

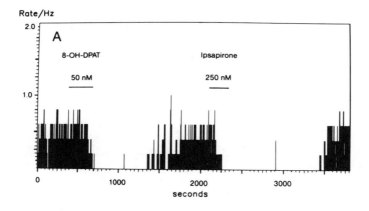

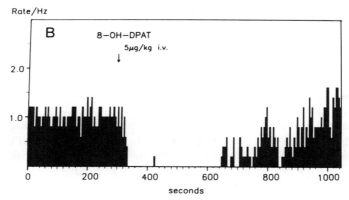

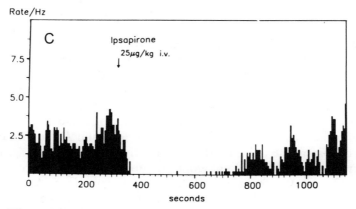

Figure 1. Effects of 8-OH-DPAT and ipsapirone on electrical activity of rat dorsal raphe neurons. Top panel (A) shows the effects obtained in brain slices. Bars indicate the times for which drugs were washed in the perfusion chamber (usually 5 min). Middle (B, 8-OH-DPAT) and lower panel (C, ipsapirone) show the effects when compounds were injected intravenously over 1 min. The times of injection are indicated by the arrows.

464

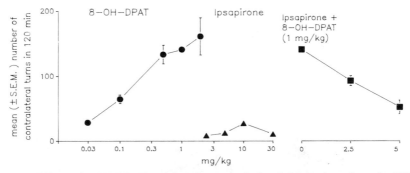

Figure 2. Effects of 8-OH-DPAT and ipsapirone on circling behavior in unilaterally DRN lesioned rats. Left panel shows the effects of the compounds alone, right panel shows the effects of the combination of 1 mg/kg 8-OH-DPAT with ipsapirone. Compounds were administered i.p.

tively (Figure 1A). When firing rates of serotonergic neurons were recorded *in vivo* from anaesthetized rats, similar results were obtained after intravenous administration of 8-OH-DPAT (Figure 1B) and ipsapirone (Figure 1C) at dosages of 5 and 25 μg/kg, respectively. These data suggest that both compounds are full agonists at the presynaptic 5-HT$_{1A}$ receptors.

However, in the unilaterally DRN lesioned rats – a behavioral model of postsynaptic 5-HT$_{1A}$ receptor function – 8-OH-DPAT induced dose-dependent contralateral circling (with no effect on ipsilateral circling), whereas ipsapirone showed only weak activity (Figure 2, left and middle panel). Moreover, ipsapirone was able to antagonize the effect of 1 mg/kg 8-OH-DPAT (Figure 2, right panel), suggesting that 8-OH-DPAT is a full agonist and ipsapirone is a partial agonist at the postsynaptic receptors involved in this model.

8-OH-DPAT and ipsapirone dose-dependently decreased body temperature after subcutaneous administration (Figures 3A, 3B). The effects were maximal between 30 and 60 min and disappeared within 4 h (8-OH-DPAT) and 2 h (ipsapirone) (data not shown). Under these conditions, 8-OH-DPAT appeared to be approximately 100 times more potent and slightly more efficient than ipsapirone. However, combination of a nearly maximal effective dose of 8-OH-DPAT with increasing doses of ipsapirone did not attenuate the hypothermic response (Figure 3B) indicating a lack of antagonistic properties of ipsapirone in this model. To study whether pre- and/or postsynaptic 5-HT$_{1A}$ sites are involved in the hypothermic effects, experiments with local application into the DRN and lesion studies were performed. Administration of 10 μg 8-OH-DPAT into the DRN caused a marked reduction of body temperature, suggesting that presynaptic 5-HT$_{1A}$ receptors are involved in the hypothermic effects of 5-HT$_{1A}$ ligands (Figure 3C). However,

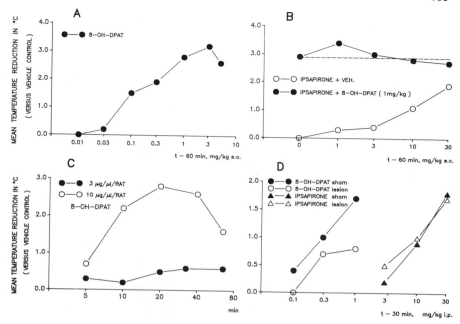

Figure 3. Effects of 8-OH-DPAT and ipsapirone on body temperature of rats. Panel (A) shows the dose response curve of 8-OH-DPAT. Panel (B) shows the dose response of ipsapirone alone (open circles) and in combination with 1 mg/kg 8-OH-DPAT (closed circles). Data represent effects obtained 60 min after subcutaneous administration. Panel (C) shows time course of hypothermia after injection of 3 μg (closed circles) or 10 μg (open circles) 8-OH-DPAT into the DRN. Panel (D) shows the dose response curves of 8-OH-DPAT ($\bullet$) and ipsapirone ($\blacktriangle$) in 5,7-DHT lesioned rats (open symbols) and sham-operated controls (closed symbols). Data represent effects obtained 30 min after intraperitoneal administration. Lesion did not affect basal body temperature.

lesion of the brain 5-HT system by administration of 5,7-DHT did not affect the hypothermia caused by ipsapirone and attenuated only to some extent that induced by 8-OH-DPAT, suggesting that postsynaptic 5-HT$_{1A}$ receptors are also involved in the hypothermic effects of 5-HT$_{1A}$ ligands (Figure 3D). Under these conditions, and especially after lesioning the 5-HT system, ipsapirone appeared to be at least as efficient as 8-OH-DPAT, suggesting that at the postsynaptic 5-HT$_{1A}$ receptors involved in 5-HT$_{1A}$ ligand induced hypothermia, ipsapirone may be a full agonist.

In rats trained to discriminate 8-OH-DPAT from saline, ipsapirone completely generalized to the 8-OH-DPAT cue; the ED$_{50}$ value being 0.44 mg/kg i.p. (equals about 150 μg/rat) and thus being 15-fold higher than that of 8-OH-DPAT itself (0.03 mg/kg $\simeq$ 10 μg/rat; Figure 4, upper panel). Complete generalization with 8-OH-DPAT and about 75% generalization with ipsapirone was obtained after administration of these compounds into either the DRN (Figure 4, middle panel) or into

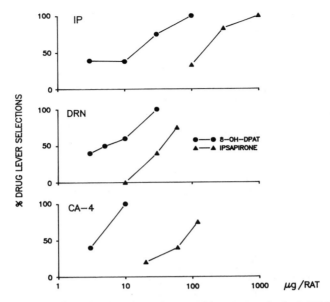

Figure 4. Involvement of the dorsal raphe nucleus and hippocampus in the 8-OH-DPAT cue. Upper panel shows the effects of 8-OH-DPAT (●) and ipsapirone (▲) after intraperitoneal administration. Middle panel shows the effects after infusion into the DRN (unilaterally) and lower panel shows the effects after infusion into the CA-4 region of the hippocampus (bilaterally).

the CA-4 region of the hippocampus (Figure 4, lower panel). Higher doses of ipsapirone could not be tested due to solubility problems. The potencies of both 8-OH-DPAT and ipsapirone were somewhat higher after direct infusion into the brain than after systemic injection. The ED_{50} values of 8-OH-DPAT in the DRN and hippocampus were 5 and 4 μg/rat, those of ipsapirone were 38 and 62 μg/rat, respectively.

Discussion

In the present study the intrinsic activities of 8-OH-DPAT and ipsapirone have been compared in several functional models reflecting activation of 5-HT_{1A} receptors. The electrophysiological data on DRN neurons confirm previous findings (Sprouse and Aghajanian 1987) and show that 8-OH-DPAT and ipsapirone completely block 5-HT neuron firing and thus act as full agonists at the presynaptic somatodendritic autoreceptors. Consequently, in models where the activation of presynaptic 5-HT_{1A} receptors is believed to play a major role, one would also expect agonistic properties for both compounds.

Postsynaptic 5-HT_{1A} receptors, e.g. in the hippocampus, are negatively coupled to the adenylate cyclase system (De Vivo and Maayani

1987; Bockaert et al. 1987). 8-OH-DPAT completely, but ipsapirone only partially, mimicks the inhibiting effects of 5-HT on the forskolin stimulated enzyme; whereas ipsapirone antagonizes the enzyme inhibition induced by 5-HT or 8-OH-DPAT (De Vry et al. 1991). This suggests that, in contrast to the situation at presynaptic receptors, ipsapirone is a partial agonist at postsynaptic receptors. It is therefore expected that, in models where the activation of postsynaptic 5-HT_{1A} receptors is believed to play a major role, 8-OH-DPAT is a full agonist and ipsapirone is partial agonist.

Such a characterization was obtained in two models which presumably reflect postsynaptic 5-HT_{1A} receptor activation. First, in the unilaterally DRN lesioned rats, 8-OH-DPAT behaves as a full agonist, whereas ipsapirone appears to be a partial agonist. The partial agonist profile of ipsapirone in this model is not confounded by the formation of its metabolite 1-(2-pyrimidinyl)piperazine (1-PP), as it was found that a) pretreatment with the metabolism inhibitor proadifen failed to affect the circling response of ipsapirone and b) ipsapirone was far more potent and efficient than 1-PP in antagonizing 8-OH-DPAT induced circling behavior (De Vry et al. 1991).

Second, differences in the ability of 8-OH-DPAT and ipsapirone to induce some behavioral elements which are believed to be mediated by activation of postsynaptic 5-HT_{1A} receptors (i.e., forepaw treading, Straub's tail, stimulation of locomotor activity, hindlimb abduction and flat body posture; Tricklebank et al. 1985) were observed (De Vry et al. 1990; 1991; unpublished). While 8-OH-DPAT dose-dependently induces these behaviors, ipsapirone is only very weakly, if at all, active. Ipsapirone, however, antagonizes 8-OH-DPAT induced forepaw treading, Straub's tail and locomotor activity, but not hindlimb abduction and flat body posture; suggesting a partial agonist profile of ipsapirone.

The finding, that ipsapirone completely generalizes to the 8-OH-DPAT cue after systemic application, is in accordance with previous findings (Cunningham et al. 1987; Glennon et al. 1989; De Vry et al. 1991) and suggests that presynaptic 5-HT_{1A} receptors are strongly involved in this behavioral model. This suggestion is supported by the fact that local administration of 8-OH-DPAT and ipsapirone into the DRN also leads to a considerable extent of generalization (Schreiber and De Vry 1989; this study). Direct infusion of 8-OH-DPAT and ipsapirone into the hippocampus, however, which contains postsynaptic 5-HT_{1A} receptors, also leads to a similar degree of generalization. These results suggest that a postsynaptic target may also contribute to the discriminative stimulus effects of 8-OH-DPAT and ipsapirone. Recent evidence from lesion and local application studies are in favour of this view (Gleeson and Barrett 1989; Kalkman 1990; De Vry and Schreiber 1990). Thus, the drug discrimination assay with 5-HT_{1A} ligands appears to be a mixed pre/postsynaptic model. An alternative explanation might be that the apparent presynaptic component may result from primary effects of these compounds on postsynaptic 5-HT$_{1A}$ receptors which in turn leads to a feedback inhibition of serotonergic cell activity (Blier and de Montigny 1987).

Rather similar conclusions can be drawn from the experiments on the hypothermic effects of 8-OH-DPAT and ipsapirone. There is some controversy whether the 5-HT$_{1A}$ receptor ligand induced hypothermia is presynaptically or postsynaptically mediated (Goodwin et al. 1987; Hutson et al. 1987). The decrease in body temperature after local infusion into the DRN, as well as after lesion of the brain 5-HT-system point to the notion, that both pre- and postsynaptic 5-HT$_{1A}$ receptors are involved. One argument against this conclusion could be that despite the 80% reduction of the 5-HT content after the 5,7-DHT lesion, there are enough – due to the suggested occurrence of a high number of spare receptors (Meller et al. 1990) – functional serotonergic neurons left in the DRN, to produce the effects. To resolve this issue experiments are currently underway with local infusion of drugs in 5,7-DHT lesioned animals.

In the hypothermia experiments – similar to the drug discrimination experiments – ipsapirone can be characterized as a full agonist with intrinsic activity close to 8-OH-DPAT. In view of the expected postsynaptic partial agonist activity of ipsapirone it was remarkable that, in the 5,7-DHT lesioned rats, ipsapirone retained its full agonist (hypothermic) activity. This finding probably cannot be ascribed to the development of (postsynaptic) receptor denervation supersensitivity, as a) the effect was identical in another group of rats lesioned 4 – instead of 10 – days before the experiment (data not shown) and b) the hypothermic effect of 8-OH-DPAT was not increased after lesion (in fact, it was even partially abolished). The findings that ipsapirone retained its effects after local application into the hippocampus (drug discrimination), or after lesioning of the 5-HT system (hypothermia), suggest that ipsapirone may be a full agonist at some postsynaptic receptor populations.

In conclusion, it appears that 8-OH-DPAT is a full agonist at both pre- and postsynaptic 5-HT$_{1A}$ receptors, whereas ipsapirone is a full agonist at pre-synaptic receptors and – depending on the model – a partial or full agonist at postsynaptic receptors. The fact that the different models used to characterize the intrinsic activity of a 5-HT$_{1A}$ ligand represent (partially) different populations of 5-HT$_{1A}$ receptors, suggests that the intrinsic activity of a 5-HT$_{1A}$ ligand may vary according to the particular 5-HT$_{1A}$ receptors involved in the model. This in turn, would be consistent with the suggested occurrence of spare receptors (Meller et al. 1990), presumably highly represented in the raphe nuclei and to a variable degree presented in the different postsynaptic receptor populations. An additional or alternative explanation which needs to be considered is the existence of subtypes of 5-HT$_{1A}$ receptors. Thus far, however, evidence for this hypothesis has not been convincing.

Acknowledgements

The authors wish to thank Mr. K.-H. Augstein, Ms. G. Eckel, Ms. W. Scheip, Ms. R. Schneider and Ms. S. Terhardt-Krabbe for excellent technical assistance. Ms. H. Otto is thanked for preparing the figures and Ms. H. Wodarz for typing the manuscript.

References

Blackburn, T. P., Kemp, J. D., Martin, D. A., and Cox, B. (1984). Evidence that 5-HT agonist-induced rotational behaviour in the rat is mediated via 5-HT$_1$ receptors. Psychopharmacology 83: 163–165.

Blier, P., and de Montigny, C. (1987). Modification of 5-HT neuron properties by sustained administration of the 5-HT$_{1A}$ agonist gepirone: electrophysiological studies in the rat brain. Synapse 1: 470–480.

Bockaert, J., Dumuis, A., Bouhelal, R., Sebben, M., and Cory, R. N. (1987). Piperazine derivatives including the putative anxiolytic drugs, buspirone and ipsapirone, are agonists at 5-HT$_{1A}$ receptors negatively coupled with adenylate cyclase in hippocampal neurons. Arch. Pharmacol. 335: 588–592.

Cunningham, K. A., Callaghan, P. M., and Appel, J. B. (1987). Discriminative stimulus properties of 8-hydroxy-2-(di-n-propylamino) tetralin (8-OH-DPAT): implications for understanding the actions of novel anxiolytics. Eur. J. Pharmacol. 138: 29–36.

De Vivo, M., and Maayani, S. (1985). Characterization of the 5-hydroxy-tryptamine$_{1A}$ receptor-mediated inhibition of forskolin-stimulated adenylate cyclase activity in guinea pig and rat hippocampal membranes. J. Pharmacol. Exp. Ther. 238: 248–253.

De Vry, J., and Schreiber, R. (1990). Neuro-anatomical correlate of the discriminative stimulus effects of the 5-HT$_{1A}$ receptor ligands 8-OH-DPAT and ipsapirone in the rat. Psychopharmacology 101: S67.

De Vry, J., Glaser, T., and Traber, J. (1990). 5-HT$_{1A}$ receptor partial agonists as anxiolytics. In: Paoletti, R., Vanhoutte, P. M., Brunello, N., and Maggi, F. M. (eds.). Serotonin: From Cell Biology to Pharmacology and Therapeutics. Dordrecht: Kluwer, pp. 517–522.

De Vry, J., Glaser, T., Schuurman, T., Schreiber, R., and Traber, J. 5-HT$_{1A}$ receptors in anxiety. In: Briley, M., and File, S. E. (eds.). New concepts in anxiety. London: Macmillan, in press.

Glaser, T., Rath, M., Traber, J., Zilles, K., and Schleicher, A. (1985). Autoradiographic identification and topographical analyses of high affinity serotonin receptor subtypes as a target for the novel putative anxiolytic TVX Q 7821. Brain Research 358: 129–136.

Glennon, R. (1990). Serotonin Receptors: Clinical implications. Neuroscience & Biobehavioral Reviews 14: 35–47.

Glennon, R. A., Young, R., and Pierson, M. E. (1989). Stimulus properties of arylpiperazine second generation anxiolytics. In: Bevan, P., Cools, A. R., and Archer, T. (eds.), Behavioural Pharmacology of 5-HT. Hillsdale, New Jersey: Lawrence Erlbaum Associates, pp. 445–448.

Gleeson, S., and Barrett, J. E. (1989). Discriminative stimulus effects of anxiolytics in the hippocampus. Soc. Neurosci. Abstr. 15: 633.

Goodwin, G. M., DeSouza, R. J., Green, A. R., and Heal, D. J. (1987). The pharmacology of the behavioural and hypothermic responses of the rats to 8-hydroxy-2-(di-n-propylamino)tetralin (8-OH-DPAT). Psychopharmacology 91: 506–511.

Hutson, P. H., Donohoe, T. P., and Curzon, G. (1987). Hypothermia induced by the putative 5-HT$_{1A}$ agonists LY165163 and 8-OH-DPAT is not prevented by 5-HT depletion. Eur. J. Pharmacol. 143: 221–228.

Kalkman, H. O. (1990). Discriminative stimulus properties of 8-OH-DPAT in rats are not altered by pretreatment with para-chlorophenylalanine. Psychopharmacology 101: 39–42.

Meller, E., Goldstein, M., and Bohmaker, K. (1990). Receptor reserve for 5-HT$_{1A}$-mediated inhibition of serotonin synthesis: possible relationship to anxiolytic properties of 5-HT$_{1A}$ agonists. Mol. Pharmacol. 37: 231–237.

Paxinos, G., and Watson, Ch. (1982). The rat brain in stereotaxic coordinates. Sydney: Academic Press.

Schreiber, R., and De Vry, J. (1989). Involvement of the dorsal raphe nucleus in the discriminative stimulus properties of the 5-HT$_{1A}$ receptor agonist 8-OH-DPAT. Soc. Neurosci. Abstr. 15: 222.

Spencer, D. G., Glaser, T., and Traber, J. (1987). Serotonin receptor subtype mediation of the interoceptive discriminative stimuli induced by 5-methoxy-N,N-dimethyltryptamine. Psychopharmacology 93: 158–166.

Sprouse, J. S., and Aghajanian, G. K. (1987). Electrophysiological responses of serotonergic dorsal raphe neurons to 5-HT$_{1A}$ and 5-HT$_{1B}$ agonists. Synapse 1: 3–9.

Traber, J., and Glaser, T. (1987). 5-HT$_{1A}$ receptor-related anxiolytics. Trends Pharmacol. Sci. 8: 432–437.

Tricklebank, M. D., Forler, C., and Fozard, J. R. (1985). The involvement of subtypes of the 5-HT$_1$ receptor and of catecholaminergic systems in the behavioral response to 8-hydroxy-2-(di-*n*-propylamino)tetralin in the rat. Eur. J. Pharmacol. 106: 271–282.

VanderMaelen, C. P., and Aghajanian, G. K. (1983). Electrophysiology and pharmacological characterization of serotonergic dorsal raphe neurons recorded extracellularly and intracellularly in rat brain slices. Brain Res. 289: 109–119.

Verge, D., Daval, G., Patey, A., Gozlan, H., El Mestikawy, S., and Hamon, M. (1985). Presynaptic 5-HT autoreceptors on serotonergic cell bodies and/or dendrites but not terminals are of the 5-HT$_{1A}$ subtype. Eur. J. Pharmacol. 113: 463–464.

Weissmann-Nanopoulos, D., Mach, E., Magre, J., Demassey, Y., and Pujol, J.-F. (1985). Evidence for the location of 5-HT$_{1A}$ binding sites on serotonin containing neurons in the raphe dorsalis and raphe centralis nuclei of the rat brain. Neurochem. Int. 7: 1061–1072.

Serotonin: Molecular Biology, Receptors and Functional Effects
ed. by J. R. Fozard/P. R. Saxena
© 1991 Birkhäuser Verlag Basel/Switzerland

Initial Studies in Man to Characterise MDL 73,005EF, a Novel 5-HT$_{1A}$ Receptor Ligand and Putative Anxiolytic

M. J. Boyce[1], C. Hinze[1], K. D. Haegele[1], D. Green[1], and P. J. Cowen[2]

[1]*Merrell Dow Research Institute, Clinical Research Dept., 67000 Strasbourg, France;*
[2]*University Dept. Psychiatry, Littlemore Hospital, Oxford, UK*

Summary. In healthy man, MDL 73,005EF reduced body temperature and antagonised buspirone-induced increases in plasma ACTH and growth hormone. The effect on temperature is consistent with agonist activity at 5-HT$_{1A}$ receptors, possibly located presynaptically on midbrain raphé neurones, whilst the effect on the hormone responses is consistent with antagonist activity at postsynaptic 5-HT$_{1A}$ receptors, possibly in the hypothalamus. The pharmacological profile of MDL 73,005EF in man differs from that of buspirone. MDL 73,005EF may have therapeutic advantages over buspirone and merits study in patients.

Introduction

Ligand binding and functional studies using the 5-HT$_{1A}$ receptor agonist, 8-hydroxy-2-(di-n-propylamino)tetralin (8-OH-DPAT) (Middlemiss & Fozard 1983), indicate that 5-HT$_{1A}$ receptors are located presynaptically on raphé neurones in the midbrain and postsynaptically in many other areas, such as the hippocampus (Gozlan et al. 1983; Vergé et al. 1985). The finding that buspirone, a non-benzodiazepine anxiolytic, is a ligand for 5-HT$_{1A}$ receptors (Traber and Glaser 1987) led to a search for more potent and selective compounds.

MDL 73,005EF (8-[2-(2,3-dihydro-1,4-benzodioxin-2-yl-methyl-amino)-ethyl]-8-azaspiro[4,5]decane-7,9-dione methyl sulphonate; Figure 1) evolved from a programme of work aimed at the rational design of 5-HT$_{1A}$ receptor ligands by computer modelling (Hibert et al. 1990; Hibert and Moser 1990). In binding studies, MDL 73,005EF showed high affinity (pIC$_{50}$ 8.6) and selectivity (> 100-fold compared to other monoamine and benzodiazepine receptor sites) for 5-HT$_{1A}$ receptor sites and was more potent and more selective than buspirone (Moser et al. 1990). Functional studies in the rat showed that MDL 73,005EF possesses anxiolytic activity and that the mechanism of action differs from that of buspirone, possibly reflecting greater selectivity of MDL 73,005EF for presynaptic 5-HT$_{1A}$ receptors on raphé neurones and/or a different balance of pre- and postsynaptic activities (Moser et al. 1990).

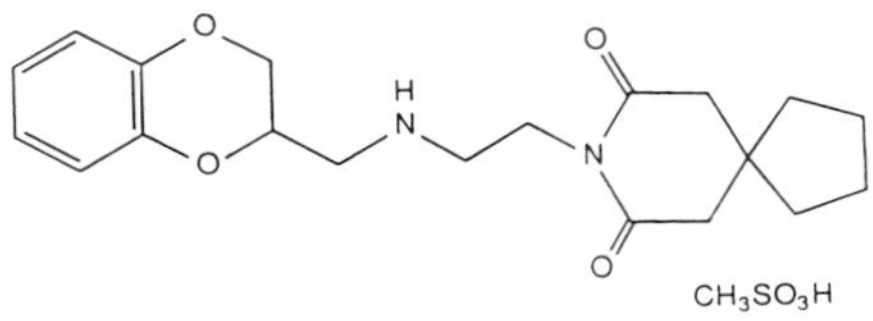

472

Figure 1. Chemical structure of MDL 73,005EF.

MDL 73,005EF lacks the aryl piperazine moiety of buspirone and two other 5-HT$_{1A}$ ligands in clinical development, gepirone and ipsapirone (Traber and Glaser 1987), and cannot be metabolised to 1-(2-pyrimidyl)-piperazine, an antagonist at α_2-adrenoceptors.

We describe the initial studies of MDL 73,005EF in man. The objectives were to assess the safety, pharmacology and pharmacokinetics of single doses in healthy man and to identify doses suitable for study in patients with anxiety disorders. Temperature and neuroendocrine variables were measured to assess central activity of MDL 73,005EF. In the rat, 8-OH-DPAT, buspirone, gepirone and ipsapirone all reduce temperature and increase plasma adreno-corticotrophic hormone (ACTH) and cortisol/corticosterone via 5-HT$_{1A}$ receptors (Green and Goodwin 1987; Gilbert et al. 1988). The aryl piperazine derivatives can induce similar responses in man (Anderson et al. 1990; Lesch et al. 1989, 1990a; b; Cowen et al. 1990). Buspirone (Meltzer et al. 1983), gepirone (Anderson et al. 1990) and ipsapirone (Lesch et al. 1989) can also increase plasma prolactin (PRL) and/or growth hormone (GH) in man but it is less clear whether these responses involve 5-HT$_{1A}$ receptors (Cowen et al. 1990).

Methods

Three separate studies of single oral doses of MDL 73,005EF were done in healthy male subjects at the Clinical Pharmacology Unit, Merrell Dow Research Institute, Kehl, Germany. Subjects were studied at rest in the supine position following an overnight fast. Protocols were reviewed by an ethics committee. Studies were done according to the principles of the Declaration of Helsinki.

Study 1

Thirty subjects, mean age 29 yr and mean weight 74 kg, participated in a double-blind, parallel-group study primarily to assess the safety of rising doses of MDL 73,005EF 2.5, 5, 10, 15 and 20 mg. Each dose was administered to five subjects. A sixth subject in each dose-group received placebo in a random manner. Active and placebo treatments were in identical capsules. Safety was assessed by monitoring cardiovascular variables (blood pressure, heart rate and electrocardiogram), routine tests of blood and urine and adverse events.

Study 2

Eight subjects, mean age 25 yr and mean weight 70 kg, participated in a double-blind, randomised, four-way crossover study of identical capsules of MDL 73,005EF 20, 40 and 80 mg and placebo at weekly intervals. The randomisation was such that no subject received the 80 mg dose as the first active treatment. Rectal temperature was measured by electronic thermometer (Novotherm) at 0, 0.5, 1, 1.5, 2, 3, 4, 5, 6 and 24 h after dosing. Venous blood was drawn at 0, 0.5, 1, 1.5, 2 and 3 h after dosing; serum or plasma was separated and stored at $-20°C$ for subsequent assay of ACTH, cortisol, GH and PRL by ITMC GmbH, Marburg, Germany using radioimmunoassay methods (RIA Kit). Plasma was also collected at 0, 0.5, 1, 1.5, 2, 4, 6, 8, 12, 24, 36 and 48 h after dosing and urine was collected throughout 0–24 h after dosing; samples were stored at $-20°C$ for subsequent measurement of MDL 73,005 by a combined gas chromatography – mass spectrometric method. Pharmacokinetic parameters were calculated. Safety was assessed as in the first study.

Study 3

This was a pilot study to assess the safety of co-administered MDL 73,005EF and buspirone and whether MDL 73,005EF can inhibit the known neuroendocrine responses to buspirone. Four subjects, mean age 25 yr and mean weight 73 kg, were each studied on four days separated by one-week intervals. Buspirone 60 mg (Buspar[R], 6×10 mg tablets) was administered alone on day 1 and one hour after a capsule of MDL 73,005EF 20, 40 and 10 mg on days 2, 3 and 4, respectively. Administration of buspirone was designated zero time. Venous blood was drawn at -30, -15, 0, 15, 30, 45, 60, 75, 90, 120, 150 and 180 min. Plasma was separated and stored at $-20°C$ for subsequent assay of ACTH, cortisol, GH and PRL by Littlemore Hospital, Oxford using radioimmunoassay methods, which together with inter- and intra-assay variations, have been published elsewhere (Anderson et al. 1990). At 30 min intervals, the subject marked a 10-point scale (0 = none; 9 = severe) for dizziness, headache and nervousness, the main adverse events associated with buspirone (Newton et al. 1986). Otherwise, safety was assessed as in the previous two studies.

474

Results

Study 1

Single doses of MDL 73,005EF up to 20 mg did not affect any of the safety variables and were well-tolerated. Adverse events were minor in nature, occurred after MDL 73,005EF or placebo and were not dose-dependent. Therefore, randomised doses ≥ 20 mg were administered in study 2.

Study 2

MDL 73,005EF significantly reduced temperature with respect to time and dose (Figure 2). Mean maximal changes for 20, 40 and 80 mg and placebo, respectively, were: -0.01, -0.14, -0.36 and $0.26°C$ (ANOVA; $p = 0.12$, 0.04 and 0.0002 for 20, 40 and 80 mg, respectively, vs placebo). MDL 73,005EF did not significantly affect ACTH (Figure 3a), cortisol (Figure 3b), growth hormone and PRL (Figure 3c) with respect to time and dose although after 80 mg, PRL increased in several subjects. Mean maximal changes in PRL for 20, 40 and 80 mg and placebo, respectively, were: 4.2, 14.0, 39.1 and -1.6 ng/ml (ANOVA; $p = 0.68$, 0.27 and 0.07 for 20, 40 and 80 mg, respectively, vs placebo).

One subject dropped out for social reasons after his first treatment (MDL 73,005EF 40 mg). No subject who received placebo or MDL 73,005EF 20 mg experienced an adverse event. Following MDL 73,005EF 40 mg, four of eight subjects developed one or more adverse

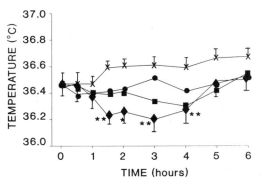

Figure 2. Mean ± sem body temperature before and after MDL 73,005EF 20 mg [●], 40 mg [■] and 80 mg [◆] and placebo [×]. Each treatment was administered by mouth immediately after the baseline measurement. $n = 8$ for 40 mg; $n = 7$ for other treatments. For clarity, sem are shown only for 80 mg and placebo. The overall dose × time interaction was highly significant (ANOVA; $F = 2.37$; df $= 24$, 14; $p = 0.0009$). There were significant differences between 80 mg and placebo at individual times ($*p < 0.05$, $**p < 0.01$).

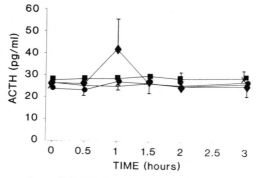

Figure 3a. Mean ± sem plasma ACTH before and after MDL 73,005EF. Symbols and numbers as for Figure 2. There was no significant dose × time interaction ($F = 0.62$; df = 12, 72; $p = 0.82$).

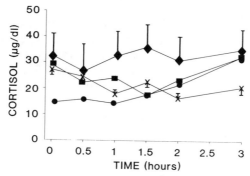

Figure 3b. Mean ± sem serum cortisol before and after MDL 73,005EF. Symbols and numbers as for Figure 2. There was no significant dose × time interaction ($F = 0.65$; df = 12, 72; $p = 0.79$).

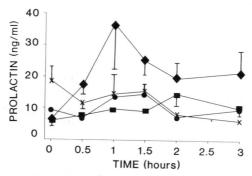

Figure 3c. Mean ± sem serum PRL before and after MDL 73,005EF. Symbols and numbers as for Figure 2. There was no significant dose × time interaction ($F = 0.85$; df = 12, 72; $p = 0.59$) but there was a significant dose effect ($F = 5.75$; df = 3, 18; $p = 0.006$).

Table 1. Pharmacokinetic parameters for MDL 73,005

Parameter	MDL 73,005EF dose		
	20 mg	40 mg	80 mg
C_{max} (nmol/ml)	1.6 ± 0.4	3.0 ± 0.5	8.1 ± 2.0
t_{max} (h)	0.8 ± 0.3	0.7 ± 0.4	0.7 ± 0.3
AUC (nmol.h.ml^{-1})	7.1 ± 1.4	12.2 ± 3.5	29.0 ± 11.2
$t_{1/2\beta}$ (h)	1.8 ± 0.5	1.5 ± 0.3	1.5 ± 0.5
$t_{1/2\gamma}$ (h)	5.3 ± 1.1	4.6 ± 0.6	4.0 ± 0.5

C_{max} = peak concentration; t_{max} = time to peak; $t_{1/2\beta}$ = initial phase half-life; $t_{1/2\gamma}$ = terminal elimination half-life.

events (dizziness, $n = 3$; headache, $n = 2$; loose stools, $n = 1$). All seven subjects who received MDL 73,005EF 80 mg developed one or more adverse events (dizziness, $n = 7$; headache, $n = 2$; vomiting, $n = 1$). MDL 73,005EF did not affect any of the safety variables.

Pharmacokinetic parameters for MDL 73,005 are summarised in Table 1. MDL 73,005EF was absorbed rapidly. Peak plasma concentrations and area under the plasma concentration-time curve (AUC) increased in a dose-linear fashion ($R = 0.92$ and 0.82, respectively; both $p < 0.001$). From peak, plasma concentrations decayed in a biphasic manner, with an initial, more rapid phase and a slower, terminal decay phase. Very little parent drug was recovered in the urine; mean excretion was $0.3 \pm 0.1\%$ of each administered dose.

Study 3

On control day, buspirone increased ACTH, GH and PRL in all four subjects and cortisol in three subjects. Figure 4 shows the mean concentration-time curves for control day and for pretreatment with MDL 73,005EF 10 mg. Time to maximum concentration varied more for ACTH than for other hormones; hence the biphasic shape of the control curve. Mean maximal responses to buspirone on control day and in the presence of MDL 73,005EF are shown in Figure 5.

Following buspirone in the presence of MDL 73,005EF 10 mg, there was no ACTH response in any subject and no GH response in two of the four subjects; cortisol and PRL responses were similar to those on the control day. ACTH, cortisol, GH and PRL concentrations in the one-hour period after MDL 73,005EF 10 mg and before buspirone were similar to those on the control day.

Following buspirone in the presence of MDL 73,005EF 20 mg, there was either no ACTH response or an attenuated response compared to control day; cortisol and PRL responses were similar to those on the control day. In the one-hour period before buspirone, ACTH, cortisol

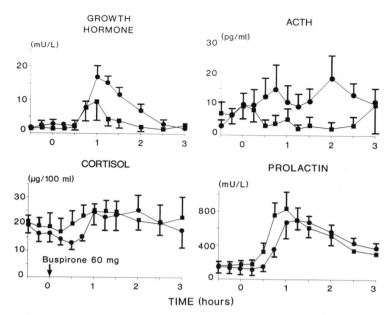

Figure 4. Mean ± sem ($n = 4$) plasma hormone responses to buspirone 60 mg in the absence [●] and presence of MDL 73,005EF 10 mg [■]. MDL 73,005EF was administered 1 h before buspirone which was administered at 0 h. MDL 73,005EF significantly reduced the area under the curve for ACTH (paired t test; $p = 0.02$) and GH ($p = 0.05$).

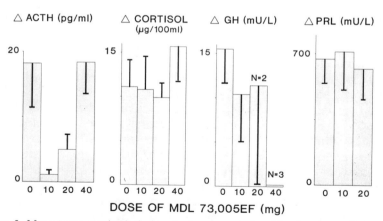

Figure 5. Mean ± sem maximal changes in hormones induced by buspirone 60 mg in the absence and presence of MDL 73,005EF. $N = 4$ unless stated otherwise; see text for explanation.

and PRL were similar to control day but GH was elevated in two subjects and their data were excluded (see Anderson et al. 1990). One of the other two subjects had no GH response following buspirone.

Following buspirone in the presence of MDL 73,005EF 40 mg, there was no GH response in three of four subjects; in the other subject, GH was elevated in the period before buspirone administration and his data were excluded. ACTH, cortisol and PRL responses to buspirone in the presence of MDL 73,005EF 40 mg were similar to those of control day. ACTH and cortisol in the one-hour period before buspirone were similar to controls but PRL increased, range 3–5 fold, in three subjects in that period and therefore PRL data for buspirone in the presence of MDL 73,005EF 40 mg were excluded.

All subjects developed one or more of the monitored symptoms (dizziness, $n = 4$; headache, $n = 2$; nervousness, $n = 2$) following buspirone on the control day. Dizziness was the main symptom; scores were maximum (range 3–5) at 60 min after buspirone and thereafter declined. One subject vomited after completion of the control day. Overall, the scores for monitored symptoms following buspirone in the presence of MDL 73,005EF were similar to those for control day. Dizziness developed in some subjects in the period after MDL 73,005EF and before buspirone and tended to be MDL 73,005EF dose-dependent (10 mg, $n = 0$; 20 mg, $n = 2$; 40 mg, $n = 3$); scores (range 1–3) were lower than those following buspirone on control day. The subject who vomited at the end of the control day also did so at the end of the other study days. Buspirone with or without MDL 73,005EF did not affect safety variables.

Discussion

MDL 73,005EF reduced body temperature in man and in that respect behaved like the aryl piperazine derivatives, buspirone (Cowen et al. 1990), gepirone (Anderson et al. 1990) and ipsapirone (Lesch et al. 1990a). MDL 73,005EF also caused hypothermia in the rat and mouse but to a greater extent in the rat, suggesting a species difference (Moser 1991). In rodents, the 5-HT$_{1A}$ receptor which mediates hypothermia induced by 5-HT$_{1A}$ ligands appears to be presynaptic in the mouse and postsynaptic in the rat (Green and Goodwin 1987). Thus, the reduction in temperature induced by MDL 73,005EF in man is consistent with stimulation of 5-HT$_{1A}$ receptors, location uncertain.

MDL 73,005EF 10–80 mg did not affect basal ACTH nor cortisol in man whereas buspirone 60 mg, twice the recommended total daily dose for patients with anxiety, increased both. The other aryl piperazine derivatives, gepirone (Anderson et al. 1990) and ipsapirone (Lesch et al. 1989; 1990b) can also increase ACTH and cortisol. In a previous study, buspirone 30 mg increased cortisol (Cowen et al. 1990) but did not clearly increase ACTH (Cowen, unpublished data). The ACTH response to buspirone 60 mg appears to have been antagonised completely by MDL 73,005EF 10 mg and to a lesser extent by 20 mg but not by 40 mg. In the rat, MDL 73,005EF did not affect basal ACTH but did antagonise the increase in ACTH stimulated by 8-OH-DPAT (Gartside et al. 1990). The 5-HT$_{1A}$ receptor which mediates ACTH release in the rat is postsynaptic (Gilbert et al. 1988); MDL 73,005EF appears to be an antagonist at that receptor site. The results for 10 and 20 mg but not 40 mg suggest that MDL 73,005EF can also act as an antagonist at that receptor site in man. Why such an effect should be lost at the higher dose of MDL 73,005EF is puzzling but there is a precedent for a bell-shaped dose-response curve to MDL 73,005EF in behavioural studies in the rat (Moser et al. 1990).

The lack of effect of MDL 73,005EF on the cortisol response to buspirone in man would appear to conflict with the concept that MDL 73,005EF is an antagonist at the postsynaptic 5-HT$_{1A}$ receptor which mediates ACTH release but is in accord with a study in the rat in which pindolol, a 5-HT$_{1A}$ antagonist, did not antagonise the corticosterone response to buspirone (Koenig et al. 1988). Pindolol, however, did antagonise the ACTH response to buspirone in the rat (Gilbert et al. 1988). The effect of MDL 73,005EF on the corticosterone response to 8-OH-DPAT or other 5-HT$_{1A}$ ligands in the rat is unknown.

MDL 73,005EF 10–80 mg did not affect basal GH in man, if the elevated values on three out of twelve occasions before administration of buspirone are excluded. They occurred in different subjects and were unrelated to dose. Basal GH > 10 mU/l is not uncommon in normals (see Anderson et al. 1990). Buspirone 60 mg clearly did increase basal GH, as did 30 mg in previous studies (Meltzer et al. 1983; Cowen et al. 1990). Gepirone (Anderson et al. 1990) and probably ipsapirone (Lesch et al. 1989) can also release GH in man. The GH response to buspirone 60 mg appears to have been antagonised by MDL 73,005EF, probably in a dose-dependent manner. Pindolol pretreatment also abolished the GH response to buspirone 30 mg in healthy subjects (Cowen et al. 1990). Thus, it would appear that a postsynaptic 5-HT$_{1A}$ receptor mediates GH release in man and MDL 73,005EF is an antagonist at that receptor. The rat does not appear to be a good model to assess the relevance of 5-HT$_{1A}$ receptors to GH secretion in man (Cowen et al. 1990).

Buspirone released PRL in man as expected (Meltzer et al. 1983; Cowen et al. 1990) but the response was not affected by MDL 73,005EF pretreatment. Indeed, the results for the high doses suggest that MDL 73,005EF can release PRL in man similar to buspirone and also to gepirone (Anderson et al. 1990). In the rat, MDL 73,005EF caused a dose-dependent increase in plasma PRL which was not affected by pindolol or metergoline, a non-selective 5-HT antagonist (Gartside et al. 1990), suggesting that 5-HT_{1A} receptors are not involved.

Studies in the rat *in vitro* and *in vivo* have shown MDL 73,005EF to be a partial agonist with intrinsic activity at 5-HT_{1A} receptors dependent on the model. Thus, MDL 73,005EF acts as an antagonist at the postsynaptic 5-HT_{1A} receptor mediating ACTH release (Gartside et al. 1990), as an antagonist at 5-HT_{1A} receptors mediating hyperpolarisation induced by 5-HT in hippocampal slices (Van den Hooff and Galvan 1991), as a weak partial agonist at the postsynaptic 5-HT_{1A} receptor mediating forepaw treading (Moser et al. 1990), as a strong partial agonist at 5-HT_{1A} receptors mediating adenylate cyclase inhibition in the hippocampus (Cornfield et al. 1989), as a full agonist at 5-HT_{1A} receptors mediating the 8-OH-DPAT discriminative stimulus (Moser et al. 1990), as a full agonist at 5-HT_{1A} receptors on raphé neurones mediating inhibition of 5-HT release in the microdialysed hippocampus (Gartside et al. 1990) and as an agonist at 5-HT_{1A} receptors on spontaneously firing dorsal raphé neurones (Sprouse et al. 1990). The simplest explanation for this spectrum of activity is that 5-HT_{1A} receptor reserve varies between the sites involved. A partial agonist like MDL 73,005EF would act as an agonist at a site where receptor reserve is large, such as raphé neurones (Meller et al. 1990), and as an antagonist at a site where there is little or no receptor reserve, such as hippocampus (Yocca et al. 1990), but where the compound would have high affinity for 5-HT_{1A} receptors (Sprouse et al. 1990). The same hypothesis can be used to explain the findings in man, a species in which MDL 73,005EF acts as an agonist at 5-HT_{1A} receptors mediating hypothermia and an antagonist at 5-HT_{1A} receptors mediating ACTH and GH release.

MDL 73,005EF has a favourable pharmacokinetic profile in man. The plasma concentrations of MDL 73,005EF are indicative of good bioavailability and the long half-life may permit twice daily dosing. The compound is extensively metabolised before urinary excretion. Mean $C_{\max}$ in man (1.6 nmol/ml) after 20 mg (0.3 mg/kg) exceeded that in the rat (1.0 nmol/ml; unpublished data) after 10 mg/kg, a 30-fold larger dose on a mg/kg basis. Doses of 0.03–1.0 mg/kg exhibited anxioltyic activity in the rat (Moser et al. 1990). Hence, doses much less than 20 mg may show anxiolytic activity in patients.

In general, MDL 73,005EF was well tolerated. Adverse events were dose-dependent and not troublesome; doses of ≤ 20 mg were essentially free of adverse events. Dizziness was the principal adverse event as it is for buspirone and the other aryl piperazine 5-HT_{1A} ligands (Newton et al. 1986; Jenkins et al. 1990).

Overall, the results of these early studies of MDL 73,005EF in man are encouraging. That MDL 73,005EF can antagonise buspirone-induced ACTH and GH responses requires confirmation in a randomised, controlled study with a larger number of subjects; doses of MDL

73,005EF < 10 mg should also be studied and the use of alternative 5-HT_{1A} probes should be explored. The pharmacological profile of MDL 73,005EF appears similar in man and rat but different from buspirone and the other aryl piperazine derivatives in both species. The combined pre- and postsynaptic actions of MDL 73,005EF may reduce 5-HT neurotransmission more effectively than buspirone and related compounds which could prove beneficial in terms of anxiolytic activity and adverse-event profile in patients.

References

Anderson, I. M., Cowen, P. J., and Grahame-Smith, D. G. (1990). The effects of gepirone on neuroendocrine function and temperature in humans. Psychopharmacology 100: 498–503.

Cornfield, L. J., Nelson, D. L., Taylor, E. W., and Martin, A. R. (1989). MDL 73,005EF: partial agonist at the 5-HT_{1A} receptor negatively linked to adenylate cyclase. Eur. J. Pharmacol. 173: 189–192.

Cowen, P. J., Anderson, I. M., and Grahame-Smith, D. G. (1990). Neuroendocrine effects of azapirones. J. Clin. Psychopharmacol. 10: 21S–25S.

Gartside, S. E., Cowen, P. J., and Hjorth, S. (1990). Effects of MDL 73005EF on central pre- and postsynaptic 5-HT_{1A} receptor function in the rat *in vivo*. Eur. J. Pharmacol. 191: 391–400.

Gilbert, F., Brazell, C., Tricklebank, M. D., and Stahl, S. M. (1988). Activation of the 5-HT_{1A} receptor subtype increases rat plasma ACTH concentration. Eur. J. Pharmacol. 147: 431–439.

Gozlan, H., El Mestikawy, S., Pichat, L., Glowinski, J., and Hamon, M. (1983). Identification of presynaptic serotonin autoreceptors using a new ligand: $^3\text{H-PAT}$. Nature 305: 140–142.

Green, A. R., and Goodwin, G. M. (1987). The pharmacology of the hypothermic response of rodents to 8-OH-DPAT administration and the effects of psychotropic drug administration on this response. In: Dourish, C. T., Ahlenius, S., and Hutson P. H. (eds.), Brain 5-HT_{1A} receptors. Chichester: Ellis Horwood, pp. 161–176.

Hibert, M. F., Mir, A. K., and Fozard, J. R. (1990). Serotonin (5-HT) receptors. In: Emmett J. C. (ed.), Comprehensive medicinal chemistry, The rational design, mechanistic study and therapeutic application of chemical compounds. Vol. 3, Membranes and receptors. Oxford: Pergamon Press, pp. 567–600.

Hibert, M., and Moser, P. (1990). MDL 72832 and MDL 73005EF: novel, potent and selective 5-HT_{1A} receptor ligands with different pharmacological properties. Drugs of the future. 15: 159–170.

Jenkins, S. W., Donald, S., Robinson, M. D., Fabre, L. F., Andary, J. J., Messina, M. E., and Reich, L. A. (1990). Gepirone in the treatment of major depression. J. Clin. Psychopharmacol. 10: 77S–85S.

Koenig, J. I., Meltzer, H. Y., and Gudelsky, G. A. (1988). 5-Hydroxytryptamine$_{1A}$ receptor-mediated effects of buspirone, gepirone and ipsapirone. Pharmacol. Biochem. Behav. 29: 711–715.

Lesch, K. P., Rupprecht, R., Poten, B., Müller, U., Söhnle, K., Fritze, J., and Schulte, H. M. (1989). Endocrine responses to 5-hydroxytryptamine$_{1A}$ receptor activation by ipsapirone in humans. Biol. Psychiatry 26: 203–205.

Lesch, K. P., Poten, B., Söhnle, K., and Schulte, H. M. (1990a). Pharmacology of the response to 5-HT_{1A} receptor activation in humans. Eur. J. Clin. Pharmacol. 39: 17–19.

Lesch, K. P., Söhnle, K., Poten, B., Schoellnhammer, G., Rupprecht, R., and Schulte, H. M. (1990b). Corticotropin and cortisol secretion after central 5-hydroxytryptamine$_{1A}$ receptor activation: effects of 5-HT receptor and β-adrenoceptor antagonists. J. Clin. Endocrinol. Metab. 70: 670–674.

Meller, E., Goldstein, M., and Bohmaker, K. (1990). Receptor reserve for 5-hydroxytryptamine$_{1A}$-mediated inhibition of serotonin synthesis: possible relationship to anxiolytic properties of 5-hydroxytryptamine$_{1A}$ agonists. Mol. Pharmacol. 37: 231–237.

482

Meltzer, H. Y., Flemming, R., and Robertson, A. (1983). The effect of buspirone on prolactin and growth hormone secretion in man. Arch. Gen. Psychiatry 40: 1099–1102.

Middlemiss, D. N., and Fozard, J. R. (1983). 8-hydroxy-2-(di-n-propylamino)tetralin discriminates between subtypes of the 5-HT$_1$ recognition site. Eur. J. Pharmacol. 90: 151–153.

Moser, P. C. (1991). The effects of MDL 73005EF, ipsapirone, buspirone and (+) and (−)-MDL 72832A on 8-OH-DPAT-induced hypothermia in the rat and mouse (submitted for publication).

Moser, P. C., Tricklebank, M. D., Middlemiss, D. N., Mir, A. K., Hibert, M. F., and Fozard, J. R. (1990). Characterization of MDL 73005EF as a 5-HT$_{1A}$ selective ligand and its effects in animal models of anxiety: comparison with buspirone, 8-OH-DPAT and diazepam. Br. J. Pharmacol. 99: 343–349.

Newton, R. E., Marunycz, J. D., and Alderdice, M. T. (1986). Review of the side-effect profile of Buspirone. Amer. J. Med. 80: 17–21.

Sprouse, J. S., McCarty, D. R., and Dudley, M. W. (1990). Inhibition of dorsal raphé cell firing by MDL 73005EF: relationship to 5-HT$_{1A}$ receptor binding. Soc. Neurosci. Abstr. 16: 1034.

Traber, J., and Glaser, T. (1987). 5-HT$_{1A}$ receptor-related anxiolytics. Trends Pharmacol. Sci. 8: 432–437.

Van den Hooff, P., and Galvan, M. (1991). Electrophysiology of the 5-HT$_{1A}$ ligand MDL 73005EF in the rat hippocampal slice. Eur. J. Pharmacol. (in press).

Vergé, D., Duval, G., Patey, A., Gozlan, H., El Mestikawy, S., and Hamon, M. (1985). Presynaptic 5-HT autoreceptors on serotonergic cell bodies and/or dendrites but not terminals are of the 5-HT$_{1A}$ subtype. Eur. J. Pharmacol. 113: 463–464.

Yocca, F. D., Iben, L., and Meller, E. (1990). Lack of receptor reserve at postsynaptic 5-HT$_{1A}$ receptors linked to adenylyl cyclase in rat hippocampus. Soc. Neurosci. Abstr. 16: 1034.

Serotonin: Molecular Biology, Receptors and Functional Effects
ed. by J. R. Fozard/P. R. Saxena
© 1991 Birkhäuser Verlag Basel/Switzerland

8-OH-DPAT-Induced Hypothermia in Rodents. A Specific Model of 5-HT$_{1A}$ Autoreceptor Function?

K. F. Martin and D. J. Heal

Boots Pharmaceuticals Research Department, Nottingham NG2 3AA, UK

Summary. 8-OH-DPAT induces hypothermia in rodents. In mice, this response is selectively attenuated by various antagonists of 5-HT$_1$, but not 5-HT$_2$, receptors. Studies with 5,7-DHT or PCPA demonstrate that the 5-HT receptors involved are located on central 5-HT neurones. Although the hypothermia is probably mediated by activation of 5-HT$_{1A}$ somatodendritic autoreceptors in the raphe nucleus, the response is also influenced by antagonism of dopaminergic and noradrenergic receptors. Repeated antidepressant administration attenuates 8-OH-DPAT-induced hypothermia in mice, albeit at high dose in some instances. In addition, down-regulation is also observed with certain other classes of psychotropic agent. In rats, although there is reasonable evidence to show that 8-OH-DPAT decreases body temperature by activating 5-HT$_{1A}$ receptors, their synaptic location is uncertain. There have been few studies which have investigated the adaptation of these receptors in rats after prolonged administration of psychotropic drugs and those effects reported have been inconsistent. Overall, the response in mice, but not rats, provides a valuable, functional index of somato-dendritic 5-HT$_{1A}$ autoreceptor function and, in addition, is a useful predictive measure of potential antidepressant activity.

Introduction

It has been known for many years that changes in core body temperature can be elicited in a number of species by drugs which act upon central 5-hydroxytryptamine (5-HT) receptors (e.g. Jacob and Girault 1979). In 1979, Cox and Lee reported that injection of 5-HT into rat hypothalamus elicited a dose-dependent hypothermia which was prevented by 5-HT receptor antagonists. These workers subsequently proposed that this hypothermic response could be used as a quantitative model for investigating central 5-HT receptors (Cox and Lee 1981). At this time, work in the area was hampered by the absence of selective agonists and antagonists for the then recently described 5-HT$_1$ and 5-HT$_2$ receptor subtypes (Peroutka and Snyder 1979). Subsequently, 5-HT$_1$ receptors have been divided into numerous sub-types (see Fozard 1987 for review), a process facilitated by the synthesis of selective ligands such as 8-hydroxy-2-(di-n-propylamino) tetralin (8-OH-DPAT), a specific agonist at central 5-HT$_{1A}$ receptors (Hjorth et al. 1982). This compound is active when given peripherally, making it an ideal tool for the investigation of 5-HT$_{1A}$ receptors in the central nervous system.

Hypothermia in mice and rats following administration of 8-OH-DPAT was first reported in 1985 (Hjorth 1985; Goodwin and Green 1985; Goodwin et al. 1985a) and this finding has subsequently been confirmed by many authors (Gudelsky et al. 1986; Wozniak et al. 1988; Higgins et al. 1988; Martin et al. 1989). Pharmacological evaluation of the hypothermia following 8-OH-DPAT has led to general acceptance of the view that this response is specifically mediated by 5-HT$_{1A}$ receptors located on 5-HT neuronal cell bodies in the mid-brain raphe complex. However, a careful examination of the literature suggests that matters may not be so clear-cut.

Effects of 5-HT$_{1A}$ Receptor Agonists on Temperature

All studies in mice have shown the log dose-response relationship for 8-OH-DPAT-induced hypothermia to be sigmoidal (e.g. Goodwin and Green 1985). Similarly, the 5-HT$_{1A}$ partial agonists, ipsapirone, buspirone and gepirone, also dose-dependently induce hypothermia in this species (Goodwin et al. 1986; Figure 1). In the rat, however, the temperature response to 8-OH-DPAT is bell-shaped with the decrease in hypothermia coinciding with the onset of the serotonin syndrome (Goodwin et al. 1986). Ipsapirone and LY165163 (1-[2-(4-amino-phenyl)ethyl]-4-(3-trifluoromethylphenyl)piperazine, PAPP) are the only other 5-HT$_{1A}$ agonists which have been reported to elicit hypothermia in this species (Goodwin et al. 1986; Hutson et al. 1987).

Anatomical Location of 5-HT$_{1A}$ Receptors Mediating Hypothermia

a) Mice

Destruction of 5-HT neurones with the neurotoxin, 5,7-dihydroxy-tryptamine (5,7-DHT), abolishes the hypothermia induced by 8-OH-DPAT (Goodwin et al. 1985a). Similarly, inhibition of 5-HT synthesis by chronic administration of p-chlorophenylalanine (PCPA) also prevents the decrease in rectal temperature following 8-OH-DPAT administration (Goodwin et al. 1985a). These data, therefore, demonstrate that an intact 5-HT neuronal system is essential for the expression of 8-OH-DPAT-induced hypothermia and by implication strongly argue that the 5-HT$_{1A}$ receptors involved are located on 5-HT neurones.

b) Rats

Whilst the anatomical location of the 5-HT$_{1A}$ receptors responsible for hypothermia in mice is quite clear, the findings in the rat are much less

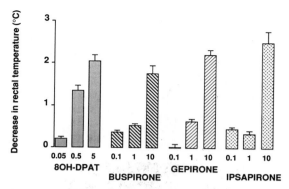

Figure 1. The effect of 8-OH-DPAT, buspirone, gepirone and ipsapirone on the rectal temperature of C57/B16/Ola mice. Measurements were made immediately before and 20 minutes after s.c. drug injection. The doses used, in mg/kg, are shown under each column. Histobars represent mean temperature decrease + s.e. mean for $n = 10$ in each group.

consistent. For example, Goodwin et al. (1987) have reported that 5,7-DHT lesions resulting in a 38% decrease in hypothalamic 5-HT concentrations have no effect on 8-OH-DPAT-induced hypothermia, although a second 5,7-DHT injection protocol, which produced a 52% depletion of 5-HT, did result in a partial reversal of this response. Complementary studies using the 5-HT synthesis inhibitor, PCPA, have not helped to clarify the issue. Thus, Hjorth (1985) and Hutson et al. (1987) have claimed that three days of PCPA treatment potentiates the hypothermia induced by 8-OH-DPAT or LY165163, whereas Goodwin et al. (1987) observed no effect after four days of administration and inhibition after 14 days of treatment. Further evidence against the hypothesis that hypothermia is mediated by 5-HT_{1A} autoreceptors in the raphe complex has been provided by Higgins et al. (1988) who showed that 8-OH-DPAT infusion into the dorsal raphe is only effective at high doses and that buspirone is inactive. Viewed overall, therefore, there is insufficient evidence to support the view that 8-OH-DPAT elicits hypothermia in the rat via activation of 5-HT_{1A} autoreceptors on 5-HT neurones.

Pharmacological Characterisation of 8-OH-DPAT-Induced Hypothermia

a) Mice

In their early work, Goodwin et al. (1985a, 1986), reported that only ipsapirone, a partial agonist at 5-HT_{1A} receptors (Martin and Mason 1986) and low doses of quipazine inhibit the temperature response to 8-OH-DPAT. Recently we have confirmed these two findings and have

486

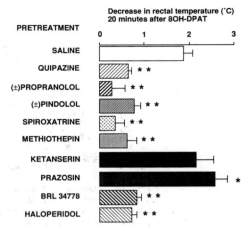

Figure 2. The effects of 5-HT and catecholamine receptor antagonists on 8-OH-DPAT-induced hypothermia in mice. Mice received i.p. injections of one of the following: saline (0.25 ml), quipazine (3 mg/kg), ($\pm$)propranolol (20 mg/kg), ($\pm$)pindolol (10 mg/kg), spiroxatrine (0.5 mg/kg), methiothepin (0.05 mg/kg), ketanserin (0.2 mg/kg), BRL 34778 (0.5 mg/kg), haloperidol (1 mg/kg) or prazosin (1 mg/kg). They were injected 30 minutes later with 8-OH-DPAT (0.5 mg/kg s.c.). The temperature change 20 minutes after 8-OH-DPAT injection was determined. Each histobar represents the mean + s.e. mean, $n = 6-12$ animals per group. $*P < 0.05$, $**P < 0.01$ versus saline pretreatment group (Student's t-test).

extended them by showing that 8-OH-DPAT-induced hypothermia is reversed by other 5-HT$_1$ receptor antagonists, including propranolol, pindolol, and low dose methiothepin (Figure 2) which Goodwin et al. (1985a) found to be ineffective. In addition, ketanserin and ritanserin, 5-HT$_2$/5-HT$_{1C}$ antagonists, have no effect (Goodwin et al. 1985a; Figure 2), providing good evidence to support the view that 8-OH-DPAT-induced hypothermia in mice is mediated by 5-HT$_{1A}$ receptors.

Pharmacological characterisation studies have also shown that hypothermia is not expressed solely through 5-HT-containing neurones and that other neurotransmitter systems are involved in this response. The dopamine D2 receptor antagonists, BRL 34778 (exo-n-{9-[(4-fluorophenyl)methyl]-9-azabicyclo (3.3.1)-non-3-yl}-4-amino-5-chloro-2-methoxybenzamide, Brown et al. 1988) and haloperidol both strongly attenuate the effects of 8-OH-DPAT in mice (Figure 2). Similarly Goodwin et al. (1985a) have also reported that haloperidol is a potent antagonist of the response, but ignored this observation in their discussion of the pharmacology.

Central adrenoceptors are also implicated in the expression of 5-HT$_{1A}$-mediated hypothermia, because prazosin, an α_1-adrenoceptor antagonist, has been consistently shown to potentiate the effects of 8-OH-DPAT (Goodwin et al. 1985a; Figure 2). Similarly, the β-adrenoceptor agonist, clenbuterol, has also been shown to increase the efficacy of 8-OH-DPAT in this model by a mechanism sensitive to ICI 118,551(erythro-DL-1-(7-methylindan-4-yloxy)-3-(isopropylamino)butan-2-ol), a β_2-adrenoceptor antagonist, indicating that the adrenoceptors involved are probably of the β_2 sub-type (Goodwin et al. 1985a).

b) Rats

Only limited investigations of neurotransmitter receptor antagonists have been carried out in rats. Experiments have shown potent antagonism by pindolol (Gudelsky et al. 1986; Wozniak et al. 1988), propranolol (Goodwin et al. 1987) and quipazine (Goodwin and Green, 1985) and weak antagonism by ipsapirone (Goodwin et al. 1986) indicating 5-HT$_1$ receptor mediation of hypothermia in this species. This hypothesis is supported by the findings that 5-HT$_2$ receptor antagonists are without effect (Wozniak et al. 1988).

The involvement of dopaminergic and noradrenergic neurons in the expression of 5-HT$_{1A}$-mediated hypothermia cannot be discounted because haloperidol and clenbuterol respectively attenuate and potentiate the response to 8-OH-DPAT (Wozniak et al. 1988; Goodwin et al. 1987).

Effects of Repeated Psychotropic Drug Administration on the Hypothermic Responses to 8-OH-DPAT

a) Mice

Goodwin et al. (1985b) claimed that 8-OH-DPAT-induced hypothermia is selectively attenuated after repeated administration of various types of antidepressant, including electroconvulsive shock. However, the numbers and types of antidepressant tested were strictly limited and the authors failed to investigate the effects of any other class of psychotropic agent. Hence their conclusion that this effect was an important adaptive response to antidepressant drug treatment was rather presumptious.

We have now re-evaluated this hypothesis by investigating a wide range of antidepressant, anxiolytic and other psychotropic drugs. Although our findings generally agree with those of Goodwin et al. (1985b), the results (Table 1) reveal two important caveats. First, although all of the antidepressants evaluated do attenuate presynaptic 5-HT$_{1A}$ receptor function, this effect is often only apparent after the administration of very high doses. Second, this response is also down-regulated after prolonged administration of a broad selection of other psychotropic drugs.

488

Table 1. The effect of repeated (14 days) administration of various psychotropic drugs on the hypothermia induced by 8-OH-DPAT in mice and rats

Treatment	Dose (mg/kg/day)	Effect on 8-OH-DPAT response	
		Mice	Rats
Antidepressants			
Desipramine	10	0	—
Desipramine	20	↓↓[1,6]	↓↓[2]
Maprotiline	10	0[6]	—
Maprotiline	20	↓↓[6]	—
Clomipramine	5	—	0[3]
Zimeldine	10	0[4]	—
Zimeldine	20	↓↓[1,6]	↓↓[2]
Citalopram	10	0[6]	—
Citalopram	20	↓↓[6]	—
Imipramine	5	—	0[3]
Amitriptyline	10	↓↓[4]	—
Amitriptyline	20	↓↓[1]	—
Dothiepin	30	↓↓[4]	—
Sibutramine	3	0[4]	—
Sibutramine	10	↓[6]	—
Clorgyline	1	—	0[3]
Tranylcypromine	6[1]/5[4]	↓↓[1,4]	—
Tranylcypromine	20	—	↓↓[2]
Mianserin	2	↓[1]	—
Mianserin	5	↓↓[4]	—
Mianserin	10	↓↓[1]	—
Idazoxan	0.1	↓↓[6]	—
Idazoxan	1	↓↓[6]	—
Anxiolytics			
Diazepam	1	↓[6]	—
Diazepam	2.5	↓[4]	—
Chlordiazepoxide	1	0[6]	—
Chlordiazepoxide	3	0[6]	—
Flurazepam	10	—	0[2]
Buspirone	10	↓↓[6]	—
Gepirone	10	↓↓[6]	—
Others			
Haloperidol	0.5	0[4]	—
Haloperidol	1	↓[6]	—
Ketanserin	0.5	↓[4]	—
Pindolol	10	↓↓[6]	—
8-OH-DPAT	3[5]/1[7]	↓↓[5]	↓↓[7]

0 = no effect, ↓ = weak attenuation, ↓↓ = potent attenuation, — = not tested
1 = Goodwin et al. (1985); 2 = Goodwin et al. (1987); 3 = Wozniak et al. (1988), 4 = Martin et al. (1989); 5 = Luscombe et al. (1989); 6 = Martin et al. (unpublished observations); 7 = Larsson et al. (1990).

While it is possible to postulate that mechanisms apart from down-regulation of central 5-HT$_{1A}$ receptors, such as anticholinergic activity, are responsible for the observed effects, the lack of any acute changes (Goodwin et al. 1985b; Martin et al. 1989) argues against this hypothesis.

b) Rats

Little attention has been paid to the possible effects of repeated antidepressant administration on 8-OH-DPAT-induced hypothermia in rats. Those studies which have been conducted (Table 1), show no consistent action on the response. One possible explanation for the discrepancies is that the dose used by Wozniak et al. (1988) were insufficient to induce down-regulation. Alternatively, the inconsistent effects of the antidepressant may reflect the probability that 8-OH-DPAT-induced hypothermia in rats is not mediated by 5-HT$_{1A}$ autoreceptors in the raphe nucleus.

Conclusions

When the data are viewed overall, it has to be accepted that there is little conclusive evidence to demonstrate that 8-OH-DPAT-induced hypothermia in rats is initiated by activation of 5-HT$_{1A}$ autoreceptors in the raphe nucleus. However, the response in mice shows pharmacological characteristics which are entirely compatible with the hypothesis that this effect reflects activation of 5-HT$_{1A}$ autoreceptors. While the most recent evidence demonstrates that 8-OH-DPAT-induced hypothermia in mice is not exclusively attenuated by prolonged administration of antidepressants, the data nevertheless suggest that this paradigm provides a valuable predictor of potential antidepressant activity.

Acknowledgements

We would like to thank Ian Phillips and Mitchell Hearson for technical assistance with the experiments carried out in our laboratory. We are grateful to Miss Sharon Hyde for help in preparation of this manuscript.

References

Brown, F., Campbell, W., Clark, M. S. G., Graves, D. S., Hadley, M. S., Hatcher, J. Mitchell, P., Needham, P., Riley, G. and Semple, J. (1988). The selective dopamine antagonist properties of BRL 34778: a novel substituted benzamide. Psychopharmacol. 94: 350–358.

Cox, B. and Lee, T. F. (1979). Possible involvement of 5-hydroxytryptamine in dopamine receptor-mediated hypothermia in the rat. J. Pharm. Pharmacol. 31: 352–354.

Cox, B. and Lee, T. F. (1981). 5-Hydroxytryptamine-induced hypothermia in rats as an *in vivo* model for the quantitative study of 5-hydroxytryptamine receptors. J. Pharmacol. Meth. 5: 43–51.

Fozard, J. R. (1987). 5-HT: the enigma variations. Trends Pharmacol. Sci. 8: 501–506.

Goodwin, G. M., De Souza R. J. and Green, A. R. (1985a). The pharmacology of the hypothermic response in mice to 8-hydroxy-2-(di-n-propylamino)tetralin (8-OH-DPAT): a model of presynaptic 5-HT$_1$ function. Neuropharmacology 24: 1187–1194.

490

Goodwin, G. M., De Souza, R. J. and Green, A. R. (1985b). Presynaptic serotonin receptor-mediated response in mice attenuated by antidepressant drugs and electro-convulsive shock. Nature 317: 531–533.

Goodwin, G. M., De Souza, R. J. and Green, A. R. (1986). The effects of the 5-HT₁ receptor ligand ipsapirone (TVX Q 7821) on 5-HT synthesis and the behavioural effects of 5-HT agonists in mice and rats. Psychopharmacol. 89: 382–387.

Goodwin, G. M., De Souza, R. J., Green, A. R. and Heal, D. J. (1987). The pharmacology of the behavioural and hypothermic responses of rats to 8-hydroxy-2-(di-n-propy-lamino)tetralin (8-OH-DPAT). Psychopharmacol. 91: 506–511.

Goodwin, G. M. and Green, A. R. (1985). A behavioural and biochemical study in mice and rats of putative selective agonists and antagonists for 5-HT₁ and 5-HT₂ receptors. Brit. J. Pharmacol. 84: 743–753.

Gudelsky, G. A., Koenig, J. J. and Meltzer, H. Y. (1986). Thermoregulatory responses to serotonin (5-HT) receptor stimulation in the rat. Evidence for opposing roles of 5-HT₂ and 5-HT₁A receptors. Neuropharmacology 25: 1307–1313.

Higgins, G. A., Bradbury, A. J., Jones, B. J. and Oakley, M. R. (1988). Behavioural and biochemical consequences following activation of 5-HT₁-like and GABA receptors in the dorsal raphe nucleus of the rat. Neuropharmacology 27: 993–1001.

Hjorth, S. (1985). Hypothermia in the rat induced by the potent serotonergic agent 8-OH-DPAT. J. Neural Transm. 61: 131–135.

Hjorth, S., Carlsson, A., Lindberg, P., Sanchez, D., Wikstrom, H., Arvidsson, L. E., Hacksell, U. and Nilsson, J. L. G. (1982). 8-Hydroxy-2-(di-n-propylamino)tetralin, 8-OH-DPAT, a potent and selective simplified ergot congener with central 5-HT receptor stimulating activity. J. Neural Transm. 55: 169–188.

Hutson, P. H., Donohoe, T. P. and Curzon, G. (1987). Hypothermia induced by the putative 5-HT₁A agonists LY-165163 and 8-OH-DPAT is not prevented by 5-HT depletion. Eur. J. Pharmacol. 143: 221–228.

Jacob, J. J. and Girault, J. M. (1979). Serotonin In: Body Temperature Regulation, Drug Effects and Therapeutic Implications, P. Lomax and E. Schonbaum, (eds.), pp. 183–220. Marcel Dekker, New York.

Larsson, L.-G., Renyi, L., Ross, S. B., Svensson, T. and Angeby-Moller, K. (1990). Different effects on the responses of functional pre- and post-synaptic 5-HT₁A receptors by repeated treatment of rat with the 5-HT₁A receptor agonist 8-OH-DPAT. Neuropharmacology 29: 85–91.

Luscombe, G. P., Martin, K. F., Hutchins, L. J. and Buckett, W. R. (1989a). Attenuation of the hypothermia but not the antidepressant-like effect of 8-OH-DPAT following its repeated administration to mice. Brit. J. Pharmacol. 96: 307P.

Martin, K. F. and Mason, R. (1986). Ipsapirone is a partial agonist at 5-hydroxytryptamine₁A (5-HT₁A) receptors in the rat hippocampus: electrophysiological evidence. Eur. J. Pharma-col. 141: 479–483.

Martin, K. F., Phillips, I., Heal, D. J. and Buckett, W. R. (1989). Is down-regulation of 5-HT₁A receptor function a common feature of the action of antidepressant drugs? Brit. J. Pharmacol. 97: 407P.

Peroutka, S. J. and Snyder, S. H. (1979). Multiple serotonin receptors: differential binding of [³H]5-hydroxytryptamine, [³H]lysergic acid diethylamide and [³H]spiroperidol. Mol. Phar-macol. 16: 687–699.

Wozniak, K. M., Aulakh, C. S., Hill, J. L. and Murphy, D. L. (1988). The effect of 8-OH-DPAT on temperature in the rat and its modification by chronic antidepressant treatments. Pharmacol. Biochem. Behav. 30: 451–456.

Serotonin: Molecular Biology, Receptors and Functional Effects
ed. by J. R. Fozard/P. R. Saxena
© 1991 Birkhäuser Verlag Basel/Switzerland

Serotonin Release is Responsible for the Locomotor Hyperactivity in Rats Induced by Derivatives of Amphetamine Related to MDMA

C. W. Callaway[1], D. E. Nichols[2], M. P. Paulus[1], and M. A. Geyer[1]

[1]Department of Psychiatry, UCSD School of Medicine, La Jolla, CA 92093-0804, USA;
[2]Department of Medicinal Chemistry and Pharmacognosy, School of Pharmacy and
Pharmacal Sciences, Purdue University, West Lafayette, IN 47907, USA

Summary. A structural derivative of amphetamine, 3,4-methylenedioxymethamphetamine
(MDMA), is a potent serotonin (5-HT) releasing drug. When administered to humans,
MDMA produces psychological effects that are distinct from the effects of classical psychostimulants and hallucinogens. In rats, MDMA produces locomotor hyperactivity, but the
spatial pattern of locomotion and the suppression of investigatory behaviors in MDMA-
treated rats differs qualitatively from the pattern of exploration produced by other psychostimulants. Previous experiments have indicated that the motor activating effects of MDMA
in rats depend upon 5-HT release. Antagonism of MDMA-induced norepinephrine release by
pretreatment with nisoxetine was less effective than serotonergic manipulations at reducing the
behavioral effects of MDMA, suggesting a minor contribution of norepinephrine release to the
behavioral effects of MDMA. A derivative of MDMA with greater selectivity for serotonergic
systems over catecholaminergic systems, 5,6-(methylenedioxy)-2-aminoindan (MDAI), produced a biphasic effect in rats consisting of an early suppression of exploratory activity and
a later motor activation. Only the motor activation by MDAI is antagonized by treatments
designed to prevent drug-induced 5-HT release, consistent with the importance of 5-HT
release for the activating effects of MDMA. Finally, the locomotor activation, suppression of
investigatory responding and detailed spatial patterning of activity produced by MDMA was
not diminished in a familiar environment. Similar effects produced by hallucinogens are
attenuated in a familiar testing environment, suggesting that these effects of MDMA do not
reflect hallucinogenic properties of this drug. These results demonstrate the potential utility of
5-HT releasing drugs in future studies of the influence of endogenous 5-HT on unconditioned
exploratory activity.

Introduction

Many behavioral effects have been attributed to stimulation of central
serotonin (5-HT) receptors (Glennon and Lucki 1988). This diversity of
responses is not surprising considering the large number of central 5-HT
receptor subtypes and the anatomical heterogeneity of central 5-HT
systems. However, one conceptual difficulty in the interpretation of the
behavioral effects of selective 5-HT ligands is extrapolating from the
actions of these drugs to the physiological actions of the endogeneous
neurotransmitter. One approach to unraveling this difficulty is to assess

the behavioral actions of drugs that are 5-HT releasing agents. Presumably, the behavioral effects of 5-HT releasing drugs will result from stimulation of only those 5-HT receptors that are innervated by serotonergic terminals and from coordinated activation of various 5-HT receptor subtypes in the same proportions that they are normally stimulated by the endogenous transmitter.

Structural derivatives of amphetamine have varying potencies as 5-HT releasing agents. One such drug, 3,4-methylenedioxymethamphetamine (MDMA), releases 5-HT from synaptosomes and striatal slices (Nichols et al. 1982; Schmidt et al. 1987). *In vivo*, MDMA produces an acute decrease in tissue 5-HT levels that is consistent with depletion of 5-HT stores by release (Schmidt et al. 1987; Johnson and Nichols 1989). Both the 5-HT release *in vitro* and the decrease in 5-HT levels *in vivo* are blocked by 5-HT uptake inhibitors (Schmidt et al. 1987; Hekmatpanah and Peroutka 1990). These data suggest that the MDMA-induced release of 5-HT is dependent upon the functional integrity of the 5-HT uptake carrier, and that MDMA releases 5-HT via a mechanism similar to that whereby amphetamine releases dopamine (Fischer and Cho 1979; Raiteri et al. 1979).

Both biochemical and electrophysiological studies reveal that, like amphetamine, MDMA affects dopamine (DA) neurotransmission. For example, MDMA inhibits the uptake of DA into synaptosomes (Steele et al. 1987). Similarly, MDMA releases DA from striatal slices with the *S*-isomer being more potent (Johnson et al. 1986). *In vivo*, MDMA increases tissue levels of DA and decreases levels of the primary DA metabolite DOPAC (Matthews et al. 1989). This pattern of biochemical changes was interpreted as reflecting decreased DA utilization ("turnover"), but is also consistent with inhibition of DA reuptake. *In vivo* voltammetry with a stearate electrode reveals MDMA-induced increases in DA levels in both the nucleus accumbens and caudate of conscious rats (Yamamoto and Spanos 1988). In contrast, a study using graphite electrodes in chloral hydrate anesthetized rats revealed decreased DA levels after MDMA (Gazzara et al. 1989). The different results obtained in these two studies may result from the use of anesthetic, because a separate study also found that MDMA increased caudate DA levels using *in vivo* dialysis in conscious rats (Hiramatsu and Cho 1990). The differential effect of anesthesia should also be considered in reports that MDMA decreases DA neuronal firing in the substantia nigra of anesthetized rats (Kelland et al. 1989; Matthews et al. 1989). Most interestingly, the effects of MDMA on indices of DA neurotransmission are attenuated by prior depletion of central 5-HT, suggesting that drug-induced release of 5-HT indirectly influences dopaminergic neurotransmission (Gazzara et al. 1989; Kelland et al. 1989).

In humans, MDMA produces novel mood-altering effects that have promoted its recreational use by some individuals. Originally the structural similarity of MDMA to the hallucinogenic amphetamine derivatives 3,4-methylenedioxyamphetamine (MDA) and 2,5 dimethoxy-4-methyl-amphetamine (DOM), prompted the hypothesis that the unique subjective effects of MDMA resulted from a combination of hallucinogenic and stimulant-like stimulus cues (Anderson et al. 1978). However, the subjective effects of MDMA that are associated with its recreational use consist of alterations in the perception of emotions and in the ability to communicate with other people that are qualitatively different from the euphoria produced by amphetamine and other psychostimulants, while perceptual distortions associated with hallucinogenic amphetamine derivatives are reported infrequently (Peroutka et al. 1988). Animals also discriminate MDMA from both psychostimulants and hallucinogens. For example, MDMA is not recognized by rats trained to discriminate either psychostimulants or hallucinogens from saline (Glennon et al. 1982; Oberlender and Nichols 1988; Broadbent et al. 1989). Thus, subjective reports and drug-discrimination data reveal that the acute effects of MDMA differ from psychostimulants and hallucinogens, prompting some to consider MDMA as a representative of a new drug class, separate from both psychostimulants and hallucinogens (Nichols et al. 1986).

MDMA resembles classical psychostimulants by virtue of producing behavioral hyperactivity during tests of unconditioned exploratory behavior in rats (Spanos and Yamamoto 1989). However, multivariate behavioral studies reveal that the detailed pattern of MDMA-induced behavioral activation is markedly different from that produced by psychostimulants (Gold et al. 1988). One hypothesis to explain these observations and the drug-discrimination data is that the release of DA by MDMA produces amphetamine-like behavioral stimulation. The concurrent release of 5-HT modifies particular components of this behavioral activation and accounts for the unique stimulus cues of MDMA. An approach to testing this hypothesis would be to selectively block DA or 5-HT release by MDMA, thereby unmasking the effects of 5-HT or DA release respectively. The following experiments were designed, therefore, to identify the respective contributions of 5-HT and DA activation to the effects of MDMA on exploratory behavior.

Materials and Methods

In the present studies, the neurochemical basis for the behavioral effects of MDMA in rats was investigated using a Behavioral Pattern Monitor (BPM) that records both locomotor activity and investigatory behaviors (Flicker and Geyer 1982). Briefly, the BPM consists of a 30.5 × 61.0 cm Plexiglas box. Three 2.5 cm holes are placed in each long wall and in the long axis of the floor, and a single hole is placed in one short wall.

Infrared photobeams detect the X-Y position of the animal with a resolution of 3.5 cm and separate photobeams detect nosepokes into the holes. A metal touchplate 15.2 cm above the floor detects rearings. A computer continuously monitors and records the status of all photobeams and the touchplate. Transitions between eight equal 15.25 cm × 15.25 cm regions in the BPM are calculated as an index of motor activity (crossings). The geometrical pattern of the locomotor path followed by an individual rat can be reconstructed from the raw data for subsequent analysis. Moreover, the number and duration of entries into the center of the BPM are calculated, along with the number and durations of all holepokes and rearings.

In addition to recording the time spent in various regions of the BPM, the raw data are used to calculate measures that describe the spatial character of the rat locomotor path. One such measure, the coefficient of variation (CV), is influenced by the variety or predictability of the locomotor path (Geyer et al. 1986). Briefly, the number of each of 40 possible transitions between any of 9 regions in the BPM are tabulated over the entire 60 min test session for an individual animal. The CV for that animal is then calculated as the standard deviation of the 40 transition frequencies divided by the mean of these frequencies. Values for the CV increase when certain transitions are preferentially repeated and decrease when the locomotor path becomes more evenly distributed throughout the BPM. A second measure to describe the spatial character of the locomotor path is the "spatial d" or "d" (Paulus and Geyer 1991). This measure is based conceptually on fractal geometry and describes the relative smoothness or roughness of the locomotor path. Briefly, the length of the entire locomotor path taken by an individual rat through the BPM is calculated from the raw data using several different spatial resolutions. Using scaling arguments, the rate with which the calculated path length decreases as a function of decreasing spatial resolution is fitted to an exponential curve, and the fitted coefficient for the exponent of this function is taken as d. Values for d increase when the locomotor path is rougher (many direction changes per path length) and decrease when the path is smoother (fewer direction changes).

Results

Acute Behavioral Effects of MDMA

In the BPM, MDMA produces behavioral stimulation consisting primarily of increased forward locomotion (Gold et al. 1988). The behaviorally activating effects of MDMA have been confirmed in other studies (Spanos and Yamamoto 1989), yet several measures distinguish

MDMA from amphetamine. (1) MDMA-treated rats frequently circle the periphery and avoid the center of the testing chamber (Gold et al. 1988), whereas amphetamine-treated rats exhibit more varied spatial patterns of locomotion and explore the entire chamber (Geyer et al. 1987; Paulus and Geyer 1991). (2) Investigatory behaviors (holepokes and rearings) are suppressed by the same doses of MDMA that increase locomotion (Gold et al. 1988). In contrast, amphetamine increases the absolute number of holepokes and rearings (Geyer et al. 1987). (3) As the dose of MDMA is increased, focussed stereotypy is not observed (Gold et al. 1988; Spanos and Yamamoto 1989), although focussed stereotyped behaviors become the primary activity at high doses of amphetamine. Thus, MDMA and amphetamine produce qualitatively different behavioral effects in rats across a range of doses, supporting the hypothesis that different neurochemical mechanisms mediate the effects of MDMA and amphetamine.

The spatial character of MDMA-induced locomotor hyperactivity has been characterized more closely using both CV and d measures. Both racemic MDMA and its S-isomer produce dose-dependent increases in CV (Gold et al. 1988; Callaway et al. 1990). Because changes in CV occurred at the same doses that produced the repetitive circling of the periphery of the BPM, it was hypothesized that the high predictability of this rotational locomotor pattern accounted for the changes in CV. The d statistic, however, is markedly reduced by MDMA and its isomers, even at doses that are too low to evoke rotational locomotor paths or changes in CV (Paulus et al. unpublished). Thus, MDMA appears to stimulate straighter locomotor paths without necessarily altering the spatial distribution of these paths.

Neurochemical Basis for Behavioral Effects of MDMA

We attempted to identify the contribution of 5-HT release to the acute behavioral effects of MDMA by examining the response to MDMA after pretreatment with selective 5-HT uptake inhibitors (Callaway et al. 1990). Biochemical studies *in vitro* indicated that these pretreatments will selectively prevent release of 5-HT by MDMA without affecting the direct actions of MDMA on catecholamine release (Schmidt et al. 1987; Hekmatpanah and Peroutka 1990). Surprisingly, pretreatment with selective 5-HT uptake inhibitors prevents the locomotor hyperactivity induced by S-MDMA, suggesting that 5-HT release rather than DA release mediates the behavioral activating effects of MDMA. Providing converging evidence for the role of 5-HT release, depletion of central 5-HT with p-chlorophenylalanine also attenuates the response to MDMA, while depletion of catecholamines with alpha-methyl-p-tyrosine is ineffective (Callaway et al. 1990). The effects of MDMA on the spatial pattern of locomotor hyperactivity is also sensitive to fluoxetine. The conclusion that MDMA produces its stimulant-like behavioral effects via release of 5-HT rather than DA is consistent with the previous observations that MDMA-induced hyperactivity is qualitatively different from the stimulation produced by amphetamine.

While data indicate that 5-HT release mediates the behavioral stimulation by MDMA, this evidence does not preclude a contribution of catecholamine release to the MDMA-induced behavioral syndrome. Preliminary studies have been undertaken to identify the contribution of DA release to the behavioral effects of MDMA using pretreatment with selective DA uptake inhibitors. This pretreatment was designed to prevent the drug-induced release of DA (Fischer and Cho 1979; Raiteri et al. 1979), just as 5-HT uptake inhibitors were employed to prevent the release of 5-HT (Callaway et al. 1990). However, DA uptake inhibitors including nomifensine, methylphenidate and GBR12909 produce significant behavioral stimulation by themselves (unpublished data), thus obscuring any interaction between these drugs and MDMA. Biochemical data suggest that structural relatives of MDMA also can alter the release of norepinephrine (NE) (Marquardt et al. 1978). Therefore, the following experiment was designed to assess the contribution of NE release to the behavioral effects of MDMA, using pretreatment with the norepinephrine uptake inhibitor nisoxetine.

Nisoxetine (10 mg/kg, i.p.) or saline was administered to rats 50 min prior to the administration of S-MDMA (3 mg/kg, s.c.) or saline. At 10 min after the second injection, rats were placed into the BPM and behavior was monitored for 60 min. Nisoxetine produced no change in

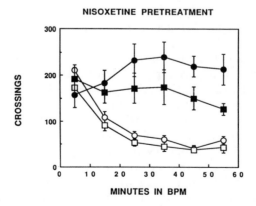

Figure 1. Effect of nisoxetine pretreatment on MDMA-induced locomotor hyperactivity. Rats were pretreated with saline (circles) or 10 mg/kg nisoxetine (squares) i.p. 50 min prior to receiving saline (open symbols) or 3 mg/kg of S-MDMA s.c. (closed symbols). After 10 min, animals were placed into the BPM. Crossings per 10 min (mean ± SEM) are indicated at times after being placed into the BPM ($n = 8$ per group).

the number of crossings recorded for control animals (Figure 1). Furthermore, no significant interaction between pretreatment and treatment was observed. A significant interaction between pretreatment, treatment and time ($F = 2.86$; df $= 5, 140$; $p < 0.05$) reflected a decrease in the mean number of crossings observed for the nisoxetine + MDMA group relative to the saline + MDMA group during the last 20 min of the session. Nisoxetine did not affect holepokes or rearings, and did not alter the reductions of these measures by S-MDMA (data not shown). Furthermore, the decrease in the d statistic produced by MDMA (saline: 1.630 ± 0.011; MDMA: 1.467 ± 0.020) was not attenuated by nisoxetine (nisoxetine: 1.649 ± 0.011; nisoxetine $\pm$ MDMA: 1.524 ± 0.020). The antagonism of MDMA by nisoxetine is much smaller in magnitude than the decrease observed after fluoxetine pretreatment (Callaway et al. 1990). Thus, these data may indicate that NE release contributes to MDMA-induced hyperactivity, but does not account for as large a component of the behavioral stimulation as does 5-HT release. Release of NE probably does not contribute to the primary effects of MDMA on investigatory behavior or on the spatial pattern of locomotor activity.

Behavioral Effects of More Selective 5-HT Releasing Drugs

The behavioral and biochemical profiles of MDMA congeners provide an alternative approach to determine the relative contributions of 5-HT and DA release to the acute effects of MDMA. Particular structural features of phenethylamines can increase or decrease the selectivity of these drugs for DA versus 5-HT systems. For example, derivatives of MDMA with elongated alpha-side chains are less potent at releasing DA from synaptosomes, but have undiminished potency as 5-HT releasers (Johnson et al. 1986). Thus, a spectrum of phenethylamines with varying selectivities for 5-HT and DA release are available as tools for identifying the relative contributions of DA and 5-HT release to behavior. In particular, the indan derivative of MDMA, 5,6-(methylenedioxy)-2-aminoindan (MDAI) retains the 5-HT releasing properties of MDMA, but produces no direct effects on DA release *in vitro* (Nichols et al. 1990). The following experiments characterized the acute behavioral effects of MDAI in the BPM to determine what components of the MDMA behavioral syndrome were also produced by this selective 5-HT releasing agent.

Rats were injected with saline or MDAI (2.5, 5 or 10 mg/kg, s.c.). After 10 min, each rat was placed into the BPM and behavior monitored for 60 min. Figure 2A depicts the resulting data. At every dose, MDAI reduced the amount of locomotor activity during the first 20 min in the testing apparatus. During the second half of the testing session,

498

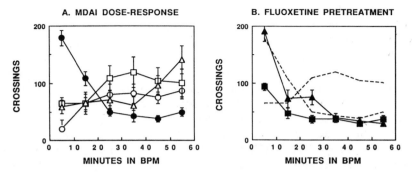

Figure 2. Effect of MDAI on exploratory behavior. Crossings per 10 min are indicated at times after being placed into the BPM. (A) Rats were treated with saline, 2.5, 5 or 10 mg/kg MDAI (closed circles, open circles, squares or triangles, respectively) and were placed into the BPM 10 min later. (B) Rats were treated with 10 mg/kg fluoxetine 50 min prior to receiving saline (triangles) or 5 mg/kg MDAI (squares), and were placed into the BPM 10 min later. Saline and 5 mg/kg MDAI groups from (A) are shown as dotted lines in (B). ($n = 7-10$ per group).

however, MDAI-treated animals maintained or slightly increased their modest initial levels of locomotor activity and were thus more active than the saline-treated animals that had habituated to the BPM (time × drug interaction for crossings: $F = 9.33$; df = 15, 165; $p < 0.0001$). Holepokes were reduced relative to saline controls (114.9 ± 26.2 holepokes per 60 min) after 2.5, 5.0 and 10.0 mg/kg MDAI (56.5 ± 7.0; 42.4 ± 9.2 and 55.0 ± 9.6 respectively) ($F = 11.39$; df = 3, 33; $p < 0.0001$). Similarly, normal levels of rearing (48.4 ± 17.9 rearings per 60 min) were reduced after administration of these doses of MDAI (24.1 ± 8.2; 2.9 ± 1.5 and 0.0 ± 0.0 respectively) ($F = 7.33$; df = 3, 33; $p < 0.001$). Thus, MDAI produced weaker locomotor activation than MDMA but resembled MDMA by decreasing investigatory responding.

In order to test the hypothesis that these behavioral effects of MDAI were mediated via carrier-dependent release of 5-HT, the behavioral response to MDAI was examined in animals pretreated with fluoxetine (10 mg/kg, i.p., 60 min). As shown in Figure 2B, fluoxetine pretreatment attenuated the MDAI-induced locomotor hyperactivity observed during the second half of the test session but did not reverse the initial suppression of locomotor activity by MDAI (time × pretreatment × treatment interaction for crossings: $F = 6.71$; df = 5, 150; $p < 0.0001$). As was the case for MDMA, fluoxetine pretreatment did not alter the MDAI-induced suppression of holepokes and rearings. Thus, it appears that the response to MDAI is multiphasic, consisting of intervals of both decreased and increased locomotor activity, but only the period of increased activity can be attributed to fluoxetine-sensitive 5-HT release.

The behavioral effects of other 5-HT releasing drugs also resemble MDMA in the BPM. Within a particular dose range, MDA and *p*-chloroamphetamine, as well as the alpha-ethyl derivative of MDMA, N-methyl-1-(1,3-benzodioxol-5-yl)-2-butanamine, produce locomotor hyperactivity with decreased investigatory behaviors (Callaway et al., 1991). Furthermore, the spatial pattern of locomotor hyperactivity consisted of straight paths with fewer direction changes than exhibited by control rats, resulting in a decrease in the d statistics. In some cases, the straight paths were restricted to the periphery of the chamber ("rotation" about the periphery). As with MDMA and MDAI, the behavioral hyperactivity induced by these drugs was attenuated by fluoxetine pretreatment, suggesting that these agents also affect behavior via serotonergic mechanisms. These data confirm the importance of 5-HT release in generating the locomotor hyperactivity produced by MDMA and its congeners.

Comparison of MDMA Effects with Hallucinogens

The unique behavioral properties of MDMA were originally hypothesized to reflect combined hallucinogen-like and stimulant-like actions. The data so far indicate that the behavioral stimulant effects of MDMA and its congeners result from 5-HT release rather than DA release and therefore differ from classical psychostimulants like amphetamine. However, it is possible that the release of endogeneous 5-HT also produces effects that resemble the hallucinogenic effects of many direct 5-HT agonists. In particular, avoidance of the center of the chamber and the decrease in investigatory behaviors which distinguish MDMA from amphetamine are also produced by lysergic acid diethylamide and other hallucinogens that are more selective 5-HT$_2$ receptor agonists (Adams and Geyer 1985; Wing et al. 1990). Notably, the behavioral effects of hallucinogenic drugs are affected by the novelty of the testing environment, and both the decrements in investigatory responding and the avoidance of the center are less pronounced when the animal has been familiarized with the testing chamber previously (Wing et al. 1990). In order to assess whether the behavioral effects of MDMA resemble those of these direct 5-HT receptor agonists, the following experiments determined the influence of prior exposure to the BPM on the behavioral response to MDMA.

Rats were assigned to one of three treatment groups. Each rat was injected with saline or MDMA (5 mg/kg, s.c.) on three occasions at 48 h intervals. The first treatment group received saline for all three injections (SSS). The second group received saline for the first two injections, and MDMA for the final injection (SSM). The third group received MDMA on all three occasions (MMM). Each rat was placed

Table 1. Effect of 5.0 mg/kg MDMA on locomotor activity and investigatory behaviors in novel and familiar environments ($n = 13-14$ per group)

	Day 1	Day 2	Day 3
Crossings			
SSS*	452.2 ± 16.8	375.8 ± 28.9	311.8 ± 26.4
SSM*	517.0 ± 25.6	397.1 ± 24.5	$881.9 \pm 83.2^{a,c}$
MMM*	1221.2 ± 71.6^a	1024.9 ± 53.6^a	911.2 ± 68.4^a
Holepokes			
SSS	156.5 ± 12.0	111.4 ± 12.2	93.9 ± 8.9
SSM	143.5 ± 10.9	111.3 ± 12.6	42.5 ± 4.3^b
MMM	75.3 ± 15.8^a	49.1 ± 12.6^a	48.5 ± 8.7
Rearings			
SSS	137.4 ± 12.4	104.8 ± 11.1	82.2 ± 9.1
SSM	155.2 ± 9.5	119.2 ± 6.8	43.3 ± 9.4
MMM	20.8 ± 6.4^a	39.4 ± 10.0^a	55.5 ± 17.3

[a] Significantly different from SSS, $p < 0.01$
[b] Significantly different from SSS, $p < 0.05$
[c] Significantly different from MMM, day 1, $p < 0.05$
*See text for explanation of abbreviations

into the BPM 10 min after being injected with saline or MDMA, and behavior was monitored for 60 min.

Results from this study are depicted in Table 1. First, the locomotor activating effects of MDMA persisted in the familiar environment, as revealed by the significant elevation of crossings in the SSM group on the third day. However, the level of activity in rats first exposed to MDMA in a familiar environment (SSM, day three) was significantly less than the level of activity in rats first exposed to MDMA in a novel environment (MMM, day one). This decrease was proportional to the similar decrease observed in the activity of control animals (group SSS). Thus, the influence of between-sessions habituation to the testing environment on this component of the behavioral response to MDMA cannot be discounted.

The MDMA-induced suppression of investigatory responding was not diminished in a familiar environment. Holepokes were significantly reduced in the SSM group relative to the SSS group on day three, and were also decreased in rats receiving MDMA on all three testing days (group MMM). Thus, the decrease in this component of investigatory responding does not exhibit tolerance. Rearings, however, were reduced significantly only after acute administration of MDMA in a novel environment (group MMM, day one). The failure to observe a significant reduction in rearings by MDMA on day three is consistent with the development of tolerance to this effect of MDMA, although the habituation of control animals across testing days may have obscured any drug effect. In contrast, the suppression of investigatory responding by 5-HT$_2$ receptor agonists does not persist in a familiar environment and exhibits rapid tolerance (Adams and Geyer 1985; Wing et al. 1990). Interestingly, rearings and holepokes also are regulated differentially by hallucinogens, with tolerance developing to the suppression of rearings but not to the suppression of holepokes. Conversely, 5-HT_{1A} receptor agonists diminish investigatory responding in both novel and familiar environments (Mittman and Geyer 1989). These data indicate that the effects of MDMA on investigatory behavior more closely resemble the actions of 5-HT_{1A} receptor agonists.

The different measures employed to describe the spatial pattern of MDMA-induced locomotor hyperactivity were differentially affected by the familiarity of the testing environment and the repeated drug administration (Table 2). Confirming the tendency of MDMA to decrease the number of entries into the center of the BPM relative to the amount of locomotor activity, the ratio of center entries to crossings was decreased by MDMA in both novel (group MMM, day one) and familiar (group SSM, day three) environments. However, the decrease in this ratio was not evident upon the third administration of MDMA (group MMM, day three), consistent with the development of tolerance to this action of MDMA. Thigmotaxis (maintaining contact with the walls) is part of the behavioral response to hallucinogenic drugs that also results in a relative decrease of center entries in the BPM (Adams and Geyer 1985). As with the MDMA-induced thigmotaxis observed here, hallucinogen-induced thigmotaxis persists in a familiar environment and rapidly develops tolerance. Thus, the thigmotaxic component of the MDMA response may actually resemble behavioral actions similar to those of classical hallucinogens.

Table 2. Effect of 5.0 mg/kg MDMA on spatial patterning of locomotor activity in novel and familiar environments ($n = 13-14$ per group)

	Day 1	Day 2	Day 3
Center Entries/Crossings			
SSS*	0.32 ± 0.03	0.36 ± 0.02	0.39 ± 0.02
SSM*	0.32 ± 0.02	0.32 ± 0.02	0.25 ± 0.02^a
MMM*	0.17 ± 0.03^a	0.19 ± 0.03^a	0.30 ± 0.04
Coefficient of Variation ("CV")			
SSS	1.05 ± 0.03	1.02 ± 0.02	1.10 ± 0.04
SSM	0.96 ± 0.02	1.05 ± 0.03	1.06 ± 0.03
MMM	1.15 ± 0.07	1.18 ± 0.07	1.25 ± 0.05^b
Spatial d ("d")			
SSS	1.54 ± 0.02	1.58 ± 0.01	1.59 ± 0.01
SSM	1.54 ± 0.01	1.55 ± 0.01	1.43 ± 0.02^a
MMM	1.40 ± 0.02^a	1.41 ± 0.02^a	1.47 ± 0.02^a

[a]Significantly different from SSS, $p < 0.01$
[b]Significantly different from SSS, $p < 0.05$
*See text for explanation of abbreviations

In contrast to the center entry ratio, the increased predictability (increased CV) of the locomotor path exhibited by MDMA-treated rats was only marginally evident following this dose of MDMA. The CV was not increased significantly by MDMA in a familiar environment (group SSM, day three) or in a novel environment (group MMM, day one). Moreover, the significant increase of CV on the third administration of MDMA (group MMM, day three) certainly indicates the absence of tolerance to the effects of MDMA on this measure. Hallucinogens produce almost the opposite effect: a decrease in CV for which tolerance rapidly develops (Adams and Geyer 1985). This dissociation between CV and center entry ratio suggests that the MDMA-induced increases in the predictability of the rat locomotor paths are not merely a function the rotational (thigmotaxic) locomotor paths produced by some animals.

The most robust change in the spatial pattern of locomotion induced by MDMA was captured by the d statistic. MDMA-treated animals exhibited significantly lower values for d reflecting a tendency to make straighter locomotor paths (fewer direction changes). This decrease in d was evident in both novel (group MMM, day one) and familiar (group SSM, day three) environments, and did not exhibit any tolerance (group MMM, day three). One interpretation of the decrease in d is that the rotational pattern of MDMA-induced locomotion produces a relatively straighter locomotor path because it is constrained by the walls of the BPM. The fact that the decreased center entries/crossings ratio was attenuated after multiple administrations of MDMA, when the drug still effectively decreased d, indicates a dissociation between the d statistic and the rotational pattern of locomotion. Thus, each of the measures examined in this study capture a different aspect of the spatial pattern of MDMA-induced locomotor activity, and these various patterns are differentially regulated. Future studies may reveal the neurochemical mechanisms for these effects.

Other Mechanisms of MDMA Action: Future Directions

The present data do not identify the 5-HT receptor subtype(s) that mediate the various behavioral effects of 5-HT releasing agents. In the same paradigm, direct $5-HT_{1A}$ and $5-HT_2$ receptor agonists decrease activity (Mittman and Geyer 1989; Wing et al. 1990), indicating that activation of these two receptors probably does not account for the behavioral activation observed following MDMA. Increased locomotor activity has been reported following administration of a direct agonist with affinity for $5-HT_{1B}$ receptors (Oberlander et al. 1986). Future experiments with selective antagonists at the various 5-HT receptor subtypes may identify the receptor subtypes that mediate the behavioral activation by MDMA.

Although 5-HT release is necessary for these behavioral effects of MDMA, other observations point to a role for the mesolimbic DA system in the locomotor stimulant effects of MDMA. Lesions of the mesolimbic DA system with 6-hydroxydopamine, for example, reduce the locomotor simulation produced by MDMA (Gold et al. 1989). In recent biochemical studies, infusion of 5-HT into the region of the midbrain DA nuclei increased release of DA in the nucleus accumbens (Guan and McBride 1989). Taken together, these data prompt the hypothesis that the behavioral activating effects of MDMA-induced 5-HT release are mediated indirectly via serotonergic stimulation of DA release rather than via direct release of DA by MDMA, as originally postulated. Alternatively, DA neurotransmission may facilitate the expression of hyperactivity even if the effects of MDMA on motor activity are mediated through nondopaminergic systems. Future studies with specific neurochemical lesions and selective receptor antagonists aim to explore the respective contributions of 5-HT and DA systems to the acute behavioral effects of MDMA and its more selective congeners.

References

Adams, L. M., and Geyer, M. A. (1985). A proposed animal model for hallucinogens based on LSD's effects on patterns of exploration in rats. Behav. Neurosci. 99: 881–900.

Anderson, G. M., Braun, G., Braun, U., Nichols, D. E., and Shulgin, A. T. (1978). Absolute configuration and psychotomimetic activity. NIDA Research Monograph 22: 8–15.

Broadbent, J., Michael, E. K., and Appel, J. B. (1989). Generalization of cocaine to isomers of 3,4-methylenedioxyamphetamine and 3,4-methylenedioxymethamphetamine: effects of training dose. Drug Devel. Res. 16: 443–450.

Callaway, C. W., Johnson, M. P., Gold, L. H., Nichols, D. E., and Geyer, M. A. (1991). Amphetamine derivatives produce locomotor hyperactivity by acting as indirect serotonin agonists. Psychopharm., In Press.

Callaway, C. W., Wing, L. L., and Geyer, M. A. (1990). Serotonin release contributes to the stimulant effects of 3,4-methylenedioxymethamphetamine in rats. J. Pharm. Exp. Ther. 254: 456–464.

Fischer, J. F., and Cho, A. K. (1979). Chemical release of dopamine from striatal homogenates: evidence for an exchange diffusion model: J. Pharmacol. Exp. Ther. 208: 203–209.

Flicker, C., and Geyer, M. A. (1982). Behavior during hippocampal microinfusions: I. Norepinephrine and diversive exploration. Brain. Res. Rev. 4: 79–103.

Gazzara, R. A., Takeda, H., Cho, A. K., and Howard, S. G. P. (1989). Inhibition of dopamine release by methylenedioxymethamphetamine is mediated by serotonin. Eur. J. Pharm. 168: 209–217.

Geyer, M. A., Russo, P., and Masten, V. L. (1986). Multivariate assessment of locomotor behavior: pharmacological and behavioral analyses. Pharm. Biochem. Behav. 28: 393–399.

Geyer, M. A., Russo, P. V., Segal, D. S., and Kuczenski, R. (1987). Effects of apomorphine and amphetamine on patterns of locomotor and investigatory behavior in rats. Pharm. Biochem. Behav. 28: 393–399.

Glennon, R. A., and Lucki, I. (1988). Behavioral models of serotonin receptor activation. In: Sanders-Bush, E. (ed.), The Serotonin Receptors. Clinton: Humana, pp. 253–293.

Glennon, R. A., Young, R., Rosecrans, J. A., and Anderson, G. M. (1982). Discriminative stimulus properties of MDA and related agents. Biol. Psychiat. 17: 807–814.

Gold, L. H., Hubner, C. B., and Koob, G. F. (1989). A role for the mesolimbic dopamine system in the psychostimulant actions of MDMA. Psychopharm. 99: 40–47.

Gold, L. H., Koob, G. F., and Geyer, M. A. (1988). Stimulant and hallucinogenic behavioral profiles of 3,4-methylenedioxymethamphetamine and N-ethyl-3,4-methylenedioxyamphetamine in rats. J. Pharm. Exp. Ther. 247: 547–555.

Guan, X. M., and McBride, W. J. (1989). Serotonin microinfusion into the ventral tegmental area increases accumbens dopamine release. Brain. Res. Bull. 23: 541–547.

Hekmatpanah, C. R., and Peroutka, S. J. (1990). 5-Hydroxytryptamine uptake blockers attenuate the 5-hydroxytryptamine-releasing effect of 3,4-methylenedioxymethamphetamine and related agents. Eur. J. Pharm. 177: 95–98.

Hiramatsu, M., and Cho, A. K. (1990). Enantiomeric differences in the effects of 3,4-methylenedioxymethamphetamine on extracellular monoamines and metabolites in the striatum of freely-moving rats: an in vivo microdialysis study. Neuropharm. 29: 269–275.

Johnson, M. P., Hoffman, A. J., and Nichols, D. E. (1986). Effects of enantiomers of MDA, MDMA and related analogues on [^{3}H]serotonin and [^{3}H]dopamine release from superfused rat brain slices. Eur. J. Pharm. 132: 269–276.

Johnson, M. P., and Nichols, D. E. (1989). Neurotoxic effects of the alpha-ethyl homologue of MDMA following subacute administration. Pharm. Biochem. Behav. 33: 105–108.

Kelland, M. D., Freeman, A. S., and Chiodo, L. A. (1989). (±)3,4-Methylenedioxymethamphetamine-induced changes in basal activity and pharmacological responsiveness of nigrostriatal dopamine neurons. Eur. J. Pharm. 169: 11–21.

Marquardt, G. M., DiStefano, V., and Ling, L. L. (1978). Pharmacological effects of (±)-, (S)- and (R)-MDA. In: Stillman, R. C., and Willett, R. E. (eds.), The Psychopharmacology of Hallucinogens. New York: Pergamon Press, pp. 84–104.

Matthews, R. T., Champney, T. H., and Frye, G. D. (1989). Effects of (±)3,4-methylenedioxymethamphetamine (MDMA) on brain dopaminergic activity in rats. Pharmacol. Biochem. Behav. 33: 741–747.

Mittman, S. M., and Geyer, M. A., (1989). Effects of 5-HT$_{1A}$ agonists on locomotor and investigatory behaviors in rats differ from those of hallucinogens. Psychopharm. 98: 321–329.

Nichols, D. E., Brewster, W. K., Johnson, M. P., Oberlander, R., and Riggs, R. M. (1990). Nonneurotoxic tetralin and indan analogues of 3,4-(methylenedioxy)amphetamine (MDA). J. Med. Chem. 33: 703–710.

Nichols, D. E., Hoffman, A. J., Oberlender, R. A., Jacob, P., and Shulgin, A. T. (1986). Derivatives of 1-(1,3-benzodioxol-5-yl)-2-butanamine: representatives of a novel therapeutic class. J. Med. Chem. 29: 2009–2015.

Nichols, D. E., Lloyd, D. H., Hoffman, A. J., Nichols, M. B., and Yim, G. K. W. (1982). Effects of certain hallucinogenic amphetamine analogues on the release of [^{3}H]serotonin from rat brain synaptosomes. J. Med. Chem. 25: 530–535.

Oberlander, C., Blaquiere, B., and Pujol, J.-F. (1986). Distinct functions for dopamine and serotonin in locomotor behaviour: evidence using the 5-HT$_1$ agonist RU 24969 in globus pallidus-lesioned animals. Neurosci. Lett. 67: 113–118.

Oberlander, R., and Nichols, D. E. (1988). Drug discrimination studies with MDMA and amphetamine. Psychopharmacol. 95: 71–76.

Paulus, M., and Geyer, M. A. (1991). A temporal and spatial scaling hypothesis for the behavioral effects of psychostimulants. Psychopharm. (in press).

Peroutka, S. J., Newman, H., and Harris, H. (1988). Subjective effects of 3,4-methylenedioxymethamphetamine in recreational users. Neuropsychopharm. 1: 273–277.

Raiteri, M., Cerrito, A. M., Cervoni, A. M., and Levi, G. (1979). Dopamine can be released by two mechanisms differentially affected by the dopamine transport inhibitor nomifensine. J. Pharm. Exp. Ther. 208: 195–202.

Schmidt, C. J., Levin, J. A., and Lovenberg, W. (1987). In vitro and in vivo neurochemical effects of methylenedioxymethamphetamine on striatal monoaminergic systems in the rat brain. Biochem. Pharm. 36: 747–755.

Spanos, L. J., and Yamamoto, B. K. (1988). Acute and subchronic effects of methylenedioxymethamphetamine (±MDMA) on locomotion and serotonin syndrome behavoir in the rat. Pharm. Biochem. Behav. 32: 835–840.

Steele, T. D., Nichols, D. E., and Yim, G. K. W. (1987). Stereochemical effects of 3,4-methylenedioxymethamphetamine (MDMA) and related amphetamine derivatives on inhibition of uptake of [^{3}H]-monoamines into synaptosomes from different regions of rat brain. Biochem. Pharm. 36: 2297–2303.

Wing, L. L., Tapson, G. S., and Geyer, M. A. (1990). 5-HT$_2$ mediation of the acute behavioral effects of hallucinogens in rats. Psychopharm. 100: 417–425.

Yamamoto, B. K., and Spanos, L. J. (1988). The acute effects of methylenedioxymethamphetamine on dopamine release in the awake behaving rat. Eur. J. Pharm. 148: 195–203.

Index

acetylcholine 435
ACTH 45, 330, 471
adenylate cyclase 23, 26, 96, 97, 99, 100, 103
adrenoceptors 486, 487
aging 43, 276, 395
(−)-alprenolol 346
aminotetralin 187
ammonium sulfate 445
amphetamine 491
analgesia 347
antibodies 95, 96, 102, 103, 104, 105
    anti-idiotypic antibodies 133
antidepressant 488, 489
antinociception 347
anxiety 431
arteriovenous anastomoses 192, 193, 197, 198
arteriosclerosis 398, 409
autoradiography 6, 102, 107, 443
    autoradiographic labelling 98
autoreceptors 104
azabicycloalkylbenzamine derivatives 220, 255
basal ganglia 7
basilar artery, guinea pig 399
[125]BH-8-MeO-N-PAT 95, 102
blood brain barrier 422
blood flow 197
blood pressure 300
BMY 7378 346
body temperature 462, 471
brainstem 6, 311
BRL 20627 246
bufotenine 189
buspirone 334, 471
calcium 72, 404
calcium channel blocker 398
cAMP 133, 220
cGMP 74
5-carboxamidotriptamine (5-CT) 107, 108, 109, 112, 243, 275, 282, 312, 317, 355, 423
caudal ventrolaterial medulla 291
cells, natural killer 390
    astroglial 44
    cultured vascular smooth 398
    transfected 23
    medial prefrontal cortical (mPFc) 174
central sympathetic outflow 289
cerebral blood vessels 265
channels 204

m-chlorophenylpiperazine (mCPP) 319, 335
cholera toxin 133
cinanserin 318
circling behavior 461
cisapride 237, 246
cocaine 84
cognition 435
collateral sprouting 50
colliculi neurons 220
colon 382
conductance 310
    leak potassium 310
    voltage sensitive potassium 310
    inwardly rectifying K+ 310
constipation 385
coronary artery 275, 406
corticosterone 330
CP-93,129 449
current, inward rectifier 311
current-voltage plots 312
cyproheptadine 318
desensitization 174, 234
development 43
5,7-Dihydroxytryptamine 50
dipropyl-5-CT 312, 313, 317
[125I]DOI 107, 109, 114
dog, conscious 374
dopamine 46, 355, 434
drug discrimination 462
electrophysiology 461
    techniques 203
endothelin-1 411
endothelium 287, 406
endothelium-derived relaxing factor (EDRF) 75, 407
enteramine 3
enteric neurones 319
entorhinal cortex 439
extravasation 428
facial motoneurones 310
fecal pellets 382
feeding behaviour 449
D-fenfluramine 389
Fischer 344 rats 389
flesinoxan 300
flumezapine 335
forskolin 97, 100, 103
gastric acid 374
gene expression 38
glioma 70
granisetron 382
growth factors 398

growth hormone 471
guanine nucleotides 28, 29, 95
guanylate cyclase 74
G proteins 95, 96, 102, 104
    G-protein coupled receptors 22, 27, 29
GTPP-γ-S 133
guinea pig 220, 232, 399, 441
hallucinogens 491
heart 193, 194, 254, 258, 300
hippocampus 7, 50, 319, 462
    pyramidal neurones 310
homotryptamine 188
human coronary artery 370
human cerebrovascular bed 270, 271
hybridization, in situ 403
hypothalamus 300, 323
hypothermia 462, 483, 484, 485, 487, 489
homotryptamine 188
5-HT 3, 194, 195, 197, 264, 265, 267, 268, 269, 270, 271, 300, 374, 389
    neurons 50
    neurobiology 203
    release 46, 491
5-HT$_1$ receptor 161, 192, 193, 195, 196, 197, 198, 217, 275, 282, 403, 427, 370
5-HT$_{1A}$ receptor 44, 95, 96, 97, 98, 99, 100, 101, 102, 103, 104, 105, 133, 209, 300, 339, 346, 483, 486, 487, 488
    5-HT$_{1A}$ receptor, postsynaptic 464
    5-HT$_{1A}$ receptor, presynaptic 464
    5-HT$_{1B}$ receptor 398, 449
    5-HT$_{1C}$ receptor 144, 146, 326
    5-HT$_{1D}$ receptor 154, 264, 269, 270, 271, 355
    5-HT$_{1P}$ receptor 133
5-HT$_2$ receptor 161, 214, 276, 326, 339, 398, 419, 370
5-HT$_3$ receptor 46, 133, 174, 205, 322, 382, 431
    5-HT$_3$ receptor antagonist 374
5-HT$_4$ receptor 133, 220, 232, 260, 367
    5-HTQ 189
    5-HT, N,N,N-trimethyl 190
5-HTP-DP 133
ICS 205–930 234, 246, 257, 312, 317, 319, 325, 382, 442
ileum, guinea pig 220, 232, 441
immuno-autoradiography 94, 98, 100, 101, 103
immunoprecipitation 95, 96, 97, 98, 102, 103, 104
indole derivatives 256

inositolphosphates 74
in situ hybridization 6
intestinal motility 243
investigatory behaviors 491
ipsapirone 460
ketanserin 312, 315, 318, 319
low density lipoproteins 410
[^{3}H]LSD 108, 109, 114, 186
LY-53857 312, 314, 315, 317
lymphocyte, large granular 390
B lymphocyte 390
T lymphocyte 390
medulla oblongata 289
MDL 73,005EF 471
meningeal circulation 422, 426
metergoline 318, 332
methiothepin 312, 317, 427
5-methoxy-l-n-propyl-α-methyl-tryptamine 187
5-methoxytryptamine 317
5-methoxy-N,N-dimethyltryptamine 318
α-methyl-5-HT 322
2-methyl-5-HT 174, 186, 312, 314, 317, 322
3,4-methylenedioxyamethamphetamine (MDMA) 491
methysergide 312, 314, 317
metoclopramide 246
mianserin 38, 319
microiontophoresis 174
migraine 265, 427
mitogenesis 398
MK 212 336
morphine 346
motoneurones 210, 310, 311, 312
mRNA 36, 38, 146
NAN-190 346
nerve growth factor 442
neurite extension 50
neuroblastoma cells 70, 84, 115
neurones 310
    cortical 310
    nodose ganglion 84
    nucleus accumbens 310
    sympathetic preganglionic 310
neurotrophic factor 51
nitric oxide 80, 409
noradrenaline 310, 312, 322, 491
nucleus 310
    accumbens neurones 310
    dorsal lateral geniculate 311
    dorsal raphe 461
    nuclei raphé 6
    prepositus hypoglossi 311
    subretrofacial 289

8-OH-DPAT   289, 300, 312, 313, 314, 317, 330, 346, 483, 484, 485, 486, 487, 489
   [³H]8-OH-DPAT   95, 97, 98, 99, 100, 101, 102, 103, 104
oesophagus rat   232
ondansetron   382, 446, 431
Org GC 94   153
Page, Irvine H.   3
parapyramidal region   291
PC12 cells   440
PCR analysis   146
pepsin secretion   374
peristalsis   235, 243
phenylbiguanide   174
plasticity   43
potassium channels   311, 317, 319
prokinetic activity   235
psychostimulants   491
psychotropic drug   487
pulmonary embolism   414, 419
quaternary amines   188
quipazine   330
QDOB   188
receptor reserve   104
recombinant interleukin-2   389
renal vasoconstrictor responses   282
renzapride   133, 237, 246, 319, 374, 446
ritanserin   318, 324
rostral ventrolateral medulla   290
rotation   355
RU 24,969   449

saphenous vein   192, 193, 195, 196, 198
schizophrenia   430, 436
second messenger   117
SNTF   50
spinal cord   210
spiperone   312, 315, 318, 319, 331
spleen   390
stomach fundus rat   144, 153
striatum   109, 115
striosomes   115
substantia nigra   7, 108, 113, 114, 355
sumatriptan   243, 264, 265, 282, 284, 355
superior colliculus   108
sympathoadrenal system   339
tachycardia   192, 193, 194, 195, 198, 253
tail-flicks, spontaneous   346
temperature   161
thrombin   398
thymidine   398
trans-acting factors   41
Trendelenburg rechnique   243
trigeminal nerve   426
tubocurarine   75, 84
turnover   174
uptake   50
vascular responses   414, 419
vasodilator effects   282
vasospasm   411
vasotonin   3
voltage-dependent K+ channel   220
zacopride   174, 237, 246, 317, 325

BIRKHÄUSER

LIFE SCIENCES

Agents and Actions Supplements

H. Timmermann and H. van der Goot
(Eds)

New Perspectives in Histamine Research

1991. 434 pages. Hardcover
ISBN 3-7643-2507-0
AAS 33

Histamine and its antagonists are intriguing compounds. The classical H_1 antagonists were developed in the 1950s and 1960s. The new, nonsedating H_1 antagonists of the 1980s stemmed from insights into structural features causing sedation. In the 1970s, the H_2 receptors became important targets for developing new anti-ulcer agents. New properties of histamine and histaminergic systems were discovered by pharmacologists and medicinial chemists in the 1980s. The role of histamine as a neurotransmitter, the existence of a presynaptic receptor (H_3) and the particular characteristics of this receptor became important for designing novel therapeutic agents. This book addresses, and seeks to answer, recent questions regarding the histaminergic systems. For instance, what is the role of the H_3 receptor, and can we expect new therapeutics based on interactions with this system? Is a distinct histamine receptor involved in immunological features of histamine? Does histamine play a part in the prolactin axis? Are histamine receptors intracellularly present in certain cells, and what is their possible role?

M.J. Parnham / M.A. Bray and
W.B. van den Berg (Eds)

Drugs in Inflammation

1991. 247 pages. Hardcover
ISBN 3-7643-2504-6
AAS 32

The search for more efficacious and better tolerated anti-inflammatory drugs has received a new impetus in recent years with advances in our understanding of new mediators (e.g. cytokines and tachykines) and previously poorly understood processes such as synovial cartilage breakdown. This book presents invited review lectures and research communications given at the inaugural symposium of the International Association of Inflammation Societies in June, 1990. The topics include the latest developments in basic research and approaches to anti-inflammatory drug therapy, concentrating on minimization of side effects, analgesics, treatment of synovial damage and modulation of cytokine release.

Please order through your bookseller or

Birkhäuser Verlag
P.O. Box 133, CH-4010 Basel/Switzerland
or, for originating from the USA and Canada, through
Birkhäuser Boston, Inc.., c/o Springer Verlag New
York, Inc.., 44 Hartz Way, Secaucus, NJ O7096-2491/
USA

Birkhäuser

Birkhäuser Verlag AG
Basel · Boston · Berlin